Fields Institute Communications

Volume 84

The Communications series features conference proceedings, surveys, and lecture notes generated from the activities at the Fields Institute for Research in the Mathematical Sciences. The publications evolve from each year's main program and conferences. Many volumes are interdisciplinary in nature, covering applications of mathematics in science, engineering, medicine, industry, and finance.

More information about this series at http://www.springer.com/series/10503

Spiro Karigiannis · Naichung Conan Leung ·
Jason D. Lotay
Editors

Lectures and Surveys on G_2-Manifolds and Related Topics

Editors
Spiro Karigiannis
Department of Pure Mathematics
University of Waterloo
Waterloo, ON, Canada

Naichung Conan Leung
Institute of Mathematical Sciences
Chinese University of Hong Kong
Shatin, New Territories, Hong Kong

Jason D. Lotay
Mathematical Institute
University of Oxford
Oxford, UK

ISSN 1069-5265 ISSN 2194-1564 (electronic)
Fields Institute Communications
ISBN 978-1-0716-0579-0 ISBN 978-1-0716-0577-6 (eBook)
https://doi.org/10.1007/978-1-0716-0577-6

Drawing of J.C. Fields by Keith Yeomans

This Springer imprint is published by the registered company Springer Science+Business Media, LLC part of Springer Nature.
The registered company address is: 1 New York Plaza, New York, NY 10004, U.S.A.

Preface

A central theme in differential geometry is the existence and classification of 'optimal geometric' structures. An important class of optimal metrics in Riemannian geometry is the Einstein metrics, especially metrics with special holonomy. These include, in particular, the Calabi–Yau and more generally the Kähler–Einstein metrics, the study of which is by now a classical area of research at the intersection of geometric analysis and algebraic geometry. More recently, there has been a steadily increasing interest in G_2 manifolds and other Riemannian manifolds with special geometric structures, such as Spin(7) manifolds or nearly Kähler manifolds, whose study are *not* amenable to classical tools of algebraic geometry. This rise in interest is partly due to the fact that all of these manifolds play important roles as ingredients for compactifications in string theory and M-theory. The area began to really develop in the 1980s and 1990s with the pioneering work of Bryant, Salamon, Joyce, Kovalev, Hitchin, and others. It is now one of the most exciting frontiers in modern geometry, especially in geometric analysis. Some of the key objectives include constructing new complete and compact examples (both by elliptic glueing methods and by parabolic flow techniques); finding topological invariants and obstructions to existence; and understanding the local and global structure of the moduli spaces of such structures.

Very recently, there has been a veritable explosion of research activity on G_2 manifolds, guided in part by a general programme initiated by Sir Simon Donaldson intending to establish analogues in the G_2 and Spin(7) settings of certain gauge-theoretic and enumerative invariants from classical low-dimensional geometry and topology. A small sample of some of these recent results include the spectacular generalization of the Kovalev glueing construction by Corti–Haskins–Nördstrom–Pacini to increase its applicability by orders of magnitude; new topological and analytic invariants of G_2 structures that allow one to detect diffeomorphic but non-isotopic G_2 structures by Crowley–Nordström and Crowley–Goette–Nordström; significant progress by Haydys, Oliveira, Sá Earp, and Walpuski in understanding the analytic aspects of G_2 instantons as defined by Donaldson–Thomas; an analysis of the obstructedness and virtual dimension of the moduli space of G_2 conifolds by Karigiannis–Lotay; and results about short-time existence and stability for various natural flows in G_2 geometry by Witt–Weiss, Bryant–Xu, Grigorian, Lotay–Wei, and others.

Because of this incredible surge in research activity, it was decided to host a 'Minischool' and Workshop on G_2 *manifolds and related topics* at the Fields Institute in August 2017, as part of the Major Thematic Program on Geometric Analysis. The event began on a weekend with a series of introductory lectures (which are almost all collected in this volume, in expanded form) by recognized experts in the field, and was followed by five days worth of workshop talks on many different analytic aspects of current research. Most of the speakers very graciously agreed to submit survey articles on the topics of their presentations, and these are also collected here. Despite the focus of the Fields Institute Workshop on *analytic* aspects of the theory, we were also very fortunate to solicit a contribution for the present volume from Crowley–Goette–Nordström on new *topological* results.

We express our sincerest gratitude to all the contributors to this volume, which would not exist without their hard work. A special thanks is also due to Shubham Dwivedi and Ragini Singhal for their myriad efforts at assisting us with proof-reading and quality control. We very much hope that this collection will be helpful to many readers, both to beginners looking for an accessible entry point into the field and to experts looking for a useful reference.

Waterloo, Canada — Spiro Karigiannis
Shatin, Hong Kong — Naichung Conan Leung
Oxford, UK — Jason D. Lotay

About the Conferences

Miniscool Participants: August 19–20, 2017[1]

Afiny Akdemir, University of Toronto
S. Ali Aleyasin, CIRGET
Yassine Amri
Leonardo Bagaglini, Università degli Studi di Firenze
Ahmad Barari, University of Ontario Institute of Technology
Ki Fung Chan, Chinese University of Hong Kong
Hanci Chi, McMaster University
Xianzhe Dai, University of California, Santa Barbara
Xinle Dai, University of Waterloo
Joe Driscoll, University of Leeds
Shubham Dwivedi, University of Waterloo
Lorenzo Foscolo, Stony Brook University
Udhav Fowdar, University College London
Hossein Gohari Bahabadi, University of Ontario Institute of Technology
Giulia Gugiatti, University College London (LSGNT)
Max Hallgren, Cornell University
Jiyuan Han, University of Wisconsin Madison
Andriy Haydys, Universität Bielefeld
Shaosai Huang, Stony Brook University
ShuTing Huang, National Taiwan University
Spiro Karigiannis*, University of Waterloo
Kotaro Kawai, Gakushuin University
Matt Kazakov, University of Guelph
Ilyas Khan, University of Wisconsin-Madison
Alexander Kolpakov
Eva Kopfer, Universität Bonn

[1](* indicates minischool invited speaker.)

Alexei Kovalev*, University of Cambridge
Adela Latorre, Centro Universitario de la Defensa
Jorge Lauret, Universidad Nacional de Córdoba
Naichung Conan Leung*, Chinese University of Hong Kong
Christopher Lin, University of South Alabama
Jason Lotay*, University College London
Siyuan Lu, McGill University
Jesse Madnick, Stanford University
Eric Massoud, University of Toronto
Kim Moore, University of Cambridge
Amir Mohammad Mostaed, Amirkabir University of Technology
Ákos Nagy, University of Waterloo
Goncalo Oliveira, Duke University
Alberto Raffero, Università degli Studi di Firenze
Sumayyah Saadi, University of Karachi
Anthonath Roslin Sagaya Mary, Universitat Rovira I Virgili
Ragini Singhal, University of Waterloo
Chi Cheuk Tsang, Chinese University of Hong Kong
Leo Tzou, University of Sydney
Raquel Villacampa, Centro Universitario de la Defensa
Thomas Walpuski*, Michigan State University
Guofang Wei, UC Santa Barbara
Yong Wei, Australia National University
Semin Yoo, University of Rochester

Workshop Participants: August 21–25, 2017[2]

S. Ali Aleyasin, CIRGET
Leonardo Bagaglini, Università degli Studi di Firenze
Ahmad Barari, University of Ontario Institute of Technology
Hanci Chi, McMaster University
Andrew Clarke*, Federal University of Rio de Janeiro
Xianzhe Dai, University of California, Santa Barbara
Joe Driscoll, University of Leeds
Shubham Dwivedi, University of Waterloo
Marisa Fernàndez, Universidad del Pais Vasco
Anna Fino*, University of Torino
Lorenzo Foscolo*, Stony Brook University
Udhav Fowdar, University College London
Hossein Gohari Bahabadi, University of Ontario Institute of Technology
Sergey Grigorian*, University of Texas Rio Grande Valley

[2](* indicates workshop invited speaker.)

Giulia Gugiatti, University College London (LSGNT)
Jiyuan Han, University of Wisconsin Madison
Andriy Haydys*, Universität Bielefeld
ShuTing Huang, National Taiwan University
Spiro Karigiannis*, University of Waterloo
Kotaro Kawai*, Gakushuin University
Matt Kazakov, University of Guelph
Ilyas Khan, University of Wisconsin-Madison
Alexei Kovalev*, University of Cambridge
Adela Latorre, Centro Universitario de la Defensa
Jorge Lauret*, Universidad Nacional de Córdoba
Naichung Conan Leung, Chinese University of Hong Kong
Christopher Lin, University of South Alabama
Jason Lotay*, University College London
Siyuan Lu, McGill University
Jesse Madnick, Stanford University
Thomas Bruun Madsen, Aarhus University
Eric Massoud, University of Toronto
Kim Moore*, University of Cambridge
Ákos Nagy, University of Waterloo
Makoto Narita, National Institute of Technology, Okinawa College
Goncalo Oliveira*, Duke University
Tommaso Pacini*, University of Torino
Paolo Piccinni, Sapienza Università di Roma
Alberto Raffero, Università degli Studi di Firenze
Henrique Sá Earp*, Unicamp—University of Campinas
Tim Talbot, University of Cambridge
Raquel Villacampa, Centro Universitario de la Defensa
Thomas Walpuski*, Michigan State University
Changliang Wang, McMaster University
McKenzie Wang, McMaster University
Guofang Wei, UC Santa Barbara
Yong Wei*, Australia National University
Chengjian YAO, Universite Libre de Bruxelles
Semin Yoo, University of Rochester

Minischool and Workshop on G_2-Manifolds

Fields Institute, Toronto, Canada—August 19–25, 2017

Schedule

The Minischool is Sat–Sun (Aug 19–20) and the Workshop is Mon–Fri (Aug 21–25).

Saturday, 2017-Aug-19 (Minischool Day 1)

09:00–10:20	**Spiro Karigiannis** (University of Waterloo) "*Introduction to* G_2*-geometry, Part I*"
10:20–10:40	BREAK
10:40–12:00	**Spiro Karigiannis** (University of Waterloo) "*Introduction to* G_2*-geometry, Part II*"
12:00–14:00	LUNCH
14:00–15:20	**Alexei Kovalev** (University of Cambridge) "*Constructions of* G_2*-manifolds, Part I*"
15:20–15:40	BREAK
15:40–17:00	**Alexei Kovalev** (University of Cambridge) "*Constructions of* G_2*-manifolds, Part II*"

Sunday, 2017-Aug-20 (Minischool Day 2)

09:00–10:20	**Thomas Walpuski** (Michigan State University) "*Introduction to* G_2*-Gauge Theory, Part I*"
10:20–10:40	BREAK
10:40–12:00	**Thomas Walpuski** (Michigan State University) "*Introduction to* G_2*-Gauge Theory, Part II*"
12:00–14:00	LUNCH
14:00–15:20	**Naichung Conan Leung** (Chinese University of Hong Kong) "*Calibrated Submanifolds in* G_2*-geometry*"

15:20–15:40 BREAK
15:40–17:00 **Jason Lotay** (University College London)
"Geometric Flows of G_2*-structures"*

Monday, 2017-Aug-21 (Workshop Day 1)

09:30–10:30 **Foscolo, Lorenzo (Stony Brook)**
"Non-compact G_2*-manifolds from asymptotically conical Calabi-Yau 3-folds"*
10:30–11:00 COFFEE BREAK
11:00–12:00 **Clarke, Andrew (Federal do Rio de Janeiro)**
"Infinitesimal moduli of the G_2*-Strominger system"*
12:00–14:00 LUNCH
14:00–15:00 **DISCUSSION PERIOD: TBA**
15:00–15:30 COFFEE BREAK
15:30–16:30 **Sá Earp, Henrique (Unicamp)**
"Gauge theory and G_2*-geometry on Calabi-Yau links"*
16:30–17:00 BREAK
17:00–18:00 RECEPTION
18:00–19:00 **Jason Lotay** (University College London)
PUBLIC LECTURE: *"Adventures in the 7th Dimension"*

Tuesday, 2017-Aug-23 (Workshop Day 2)

09:30–10:30 **Fino, Anna (Torino)**
"G_2*-structures and Ricci solitons"*
10:30–11:00 COFFEE BREAK
11:00–12:00 **Kovalev, Alexei (Cambridge)**
"Compact holonomy Spin(7) *manifolds as generalised connected sums"*
12:00–14:00 LUNCH
14:00–15:00 **Oliveira, Gonçalo (Duke)**
"G_2*-Instantons on noncompact* G_2*-manifolds"*
15:00–15:30 COFFEE BREAK
15:30–16:30 **Wei, Yong (ANU)**
"Laplacian flow for closed G_2*-structures"*

Wednesday, 2017-Aug-23 (Workshop Day 3)

09:30–10:30 **Kawai, Kotaro (Gakushuin)**
"Frölicher–Nijenhuis cohomology on G_2*- and* Spin(7)*-manifolds"*
10:30–11:00 COFFEE BREAK

11:00–12:00 **Haydys, Andriy (Bielefeld)**
"Degenerations of the Seiberg–Witten monopoles with multiple spinors and G_2*-instantons"*
12:00–14:00 LUNCH
14:00–15:00 **Grigorian, Sergey (Texas Rio Grande Valley)**
"G_2*-structures and octonion bundles"*
15:00–15:30 COFFEE BREAK
15:30–16:30 **DISCUSSION PERIOD: Open Problems**
16:30–18:00 BREAK
18:00–19:00 **WORKSHOP DINNER** (For those who registered in advance)

Thursday, 2017-Aug-24 (Workshop Day 4)

09:30–10:30 **Walpuski, Thomas (Michigan State)**
"The $(1, k)$*-ADHM Seiberg–Witten equation and* k*-fold covers of associatives"*
10:30–11:00 COFFEE BREAK
11:00–12:00 **Lauret, Jorge (Córdoba)**
"Laplacian flow and Laplacian solitons among closed G_2*-structures on solvable Lie groups"*
12:00–14:00 LUNCH
14:00–15:00 **Pacini, Tommaso (Torino)**
"New facts and tools concerning minimal Lagrangian submanifolds"
15:00–15:30 COFFEE BREAK
15:30–16:30 **Moore, Kim (Cambridge)**
"Deformation theory of Cayley submanifolds"

Friday, 2017-Aug-25 (Workshop Day 5)

09:30–10:30 **Lotay, Jason (University College London)**
"Laplacian flow and 4D geometry"
10:30–11:00 COFFEE BREAK
11:00–12:00 **Karigiannis, Spiro (Waterloo)**
"A new construction of compact G_2*-manifolds by gluing families of Eguchi–Hanson spaces"*
12:00–14:00 LUNCH
14:00–16:30 Informal discussions/Departures

Minischool and Workshop on G_2-Manifolds

Fields Institute, Toronto, Canada—August 19–25, 2017

Minischool Abstracts

Speaker: **Karigiannis, Spiro (Waterloo)**
Title: **Introduction to G_2-Geometry**

Abstract: We begin with a very brief review of Berger's list of Riemannian holonomy groups and of the more well-known $\mathrm{U}(m)$-structures. Then we will introduce the octonions, cross products, and the exceptional calibrations on $\mathbb{R}^7$, which will allow us to define G_2-structures. Next, we will study the concrete representation theory of G_2, which will allow us to define the torsion forms and define various classes of G_2-structures. Finally, we will end part one by expressing the Ricci tensor in terms of the torsion, and give a concrete computational proof of the theorem of Fernández–Gray relating parallel and harmonic calibration forms. In part two, we will briefly mention Joyce's perturbative existence theorem of torsion-free G_2-structures given appropriate initial data used for compact constructions of smooth compact G_2-manifolds. This topic will be treated in great detail in the later lectures of Kovalev. Finally, we will establish the smoothness of the moduli space of compact G_2-manifolds and discuss some special geometric structures on this moduli space.

Speaker: **Kovalev, Alexei (Cambridge)**
Title: **Constructions of Compact G_2-Manifolds**

Abstract: The exceptional Lie group G_2-occurs as the holonomy group of Riemannian metrics on 7-dimensional manifolds. In these lectures I shall explain the constructions for two geometrically different classes of examples of *compact* Riemannian 7-manifolds with holonomy G_2. One method uses resolutions of singularities of appropriately chosen 7-dimensional orbifolds with the help of ALE spaces. Another method uses the gluing of two asymptotically cylindrical pieces and requires a certain matching condition for their 'boundaries at infinity'.

Speaker: **Leung, Naichung Conan (Chinese University of Hong Kong)**
Title: **Calibrated Submanifolds in G_2-Geometry**

Abstract: Calibrated submanifolds in G_2-manifolds are associative submanifolds and coassociative submanifolds. This lecture will give an introduction to this important class of submanifolds.

Speaker: **Lotay, Jason (University College London)**
Title: **Geometric Flows of G_2-Structures**

Abstract: Geometric flows have proved to be a powerful geometric analysis tool, perhaps most notably in the study of 3-manifold topology, the differentiable sphere theorem and Kähler metrics. In the context of G_2-geometry, there are several geometric flows which arise. Each flow provides a potential means to study the geometry and topology associated with a given class of G_2-structures. I will introduce these flows and describe some of the key known results and open problems in the field.

Speaker: **Walpuski, Thomas (Michigan State University)**
Title: **Introduction to G_2-Gauge Theory**

Abstract: A central object of interest in the study of gauge theory are Yang–Mills connections. Special classes of solutions exist on many classes of manifolds: flat connections, ASD instantons on 4-manifolds, Hermitian Yang–Mills connections on Kähler manifolds, etc. The purpose of this minischool is to familiarize the participants with G_2-instantons (which are certain Yang–Mills connections on G_2-manifolds). I will discuss the basic properties of G_2-instantons, discuss their deformation theory, and explain one construction method (based on joint work with Henrique Sá Earp). Time permitting, I will discuss some points of the known compactness results regarding G_2-instantons (due to Uhlenbeck, Price, Nakajima, Tian).

Workshop Abstracts

Speaker: **Clarke, Andrew (Federal do Rio de Janeiro)**
Title: **Infinitesimal Moduli of the G_2-Strominger System**

Abstract: We consider G_2-structures with torsion coupled with G_2-instantons, on a compact 7-dimensional manifold. The coupling is via an equation for 4-forms which appears in supergravity and generalized geometry, known as the Bianchi identity. The resulting system of partial differential equations can be regarded as an analogue of the Strominger system in 7-dimensions. We initiate the study of the moduli space of solutions and show that it is finite dimensional using elliptic operator theory.

Speaker: **Fino, Anna (Torino)**
Title: **G_2-Structures and Ricci Solitons**

Abstract: In this talk we present some general results about G_2-structures whose underlying Riemannian metric is Einstein, as well results on the existence of left invariant closed G_2-forms determining a Ricci soliton metric on nilpotent Lie groups. For closed G_2-structures, we will also show some results related to the Laplacian flow.

Speaker: **Foscolo, Lorenzo (Stony Brook)**
Title: **Non-compact G_2-Manifolds from Asymptotically Conical Calabi-Yau 3-Folds**

Abstract: Only four examples of complete non-compact G_2-manifolds are currently known. In joint work with Mark Haskins and Johannes Nordström we construct infinitely many families of new complete non-compact G_2-holonomy manifolds.

The underlying smooth 7-manifolds are all circle bundles over asymptotically conical (AC) Calabi–Yau manifolds of complex dimension 3. The metrics are circle-invariant and their geometry at infinity is that of a circle bundle over a Calabi–Yau cone with fibres of fixed finite length. The G_2-manifolds we construct are therefore 7-dimensional analogues of 4-dimensional ALF hyperKähler metrics. The dimensional reduction of the equations for G_2-holonomy in the presence of a Killing field was considered by Apostolov–Salamon and by several groups of physicists. We reinterpret the dimensionally reduced equations in terms of a pair consisting of an $SU(3)$ structure on the 6-dimensional orbit space coupled to an abelian Calabi–Yau monopole on this 6-manifold. We solve this coupled system of non-linear PDEs by considering the adiabatic limit in which the circle fibres of the associated circle-invariant G_2-holonomy metrics collapse. The G_2-holonomy metrics we construct should be thought of as arising from abelian Hermitian–Yang–Mills connections on AC Calabi–Yau 3-folds, especially AC Calabi–Yau metrics on crepant resolutions of Calabi–Yau cones.

Speaker: **Grigorian, Sergey (Texas Rio Grande Valley)**
Title: **G_2-Structures and Octonion Bundles**

Abstract: We use a G_2-structure on a 7-dimensional Riemannian manifold to define an octonion bundle with a fiberwise non-associative product. We then define a metric-compatible octonionic covariant derivative on this bundle that is also compatible with the octonion product. The torsion of the G_2-structure is then shown to be an octonionic connection for this covariant derivative with curvature given by the component of the Riemann curvature that lies in the 7-dimensional representation of G_2. The choice of a particular G_2-structure within the same metric class is then interpreted as a choice of gauge and we show that under a change of this gauge, the torsion transforms as an octonion-valued connection 1-form. We then also define an energy functional for octonion sections, the critical points of which are shown to correspond to an octonionic analogue of the Coulomb gauge. The gradient flow for this functional is an octonionic harmonic map heat flow that minimizes the torsion within the same metric class.

Speaker: **Haydys, Andriy (Bielefeld)**
Title: **Degenerations of the Seiberg–Witten Monopoles with Multiple Spinors and G_2-Instantons**

Abstract: Both G_2-instantons and the Seiberg–Witten monopoles with multiple spinors can degenerate to Fueter sections, which are $\mathbb{Z}/2$ harmonic spinors in the simplest case. I will show that there are obstructions for $\mathbb{Z}/2$ harmonic spinors to be realizable as degenerations of the Seiberg–Witten monopoles.

Speaker: **Karigiannis, Spiro (Waterloo)**
Title: **A New Construction of Compact G_2-Manifolds by Gleuing Families of Eguchi–Hanson Spaces**

Abstract: I will give an overview of the proof of a new construction of compact G_2-manifolds (joint work with Dominic Joyce). We resolve $(X^6 \times S^1)/\mathbb{Z}_2$ by glueing in a family of Eguchi–Hanson spaces parametrized by the singular set, two copies of a special Lagrangian submanifold L^3 in X^6. There are two key differences from the previous glueing constructions of Joyce and Kovalev/CHNP. First, there are three pieces being glued together rather than two, and second, two of the three pieces do *not* admit torsion-free G_2-structures to start with, so we need to work harder to construct a closed G_2-structure with sufficiently small torsion on the resolved space in order to apply Joyce's fundamental existence theorem. I plan to explain all of the main ideas and to give a few of the details of each of the principal steps in the proof.

Speaker: **Kawai, Kotaro (Gakushuin)**
Title: **Frölicher–Nijenhuis Cohomology on G_2- and $Spin(7)$-Manifolds**

Abstract: We show that a parallel differential form Ψ of even degree on a Riemannian manifold allows to define a natural differential both on $\Omega(M)$ and $\Omega^*(M, TM)$, defined via the Frölicher–Nijenhuis bracket. For instance, on a Kähler manifold, these operators are the complex differential and the Dolbeault differential, respectively. We investigate this construction when taking the differential with respect to the canonical parallel 4-form on a G_2- and $Spin(7)$-manifold, respectively. We calculate the cohomology groups of $\Omega^*(M)$ and give a partial description of the cohomology of $\Omega^*(M, TM)$. This is joint work with Hông Vân Lê and Lorenz Schwachhöfer.

Speaker: **Kovalev, Alexei (Cambridge)**
Title: **Compact Holonomy Spin(7) Manifolds as Generalised Connected Sums**

Abstract: I will explain a construction of compact 8-manifolds with holonomy $\mathrm{Spin}(7)$ from pairs of asymptotically cylindrical $\mathrm{Spin}(7)$ manifolds with compatible cross-sections 'at infinity'. The cross-sections are G_2-manifolds which may, but need not, in general, have full holonomy G_2. I will discuss examples of the construction, including topologically new examples, and a relation to compact $\mathrm{Spin}(7)$ manifolds previously constructed by Dominic Joyce by a different method.

Speaker: **Lauret, Jorge (Córdoba)**
Title: **Laplacian Flow and Laplacian Solitons Among Closed G_2-Structures on Solvable Lie Groups**

Abstract: We will present some results on the Laplacian flow of G_2-structures and its solitons in the homogeneous case, including the following:

- Long time existence for any closed Laplacian flow solution in the context of solvable Lie groups with a codimension-one abelian normal subgroup.

- Many examples of closed expanding Laplacian solitons which are not eigenfunctions.
- First examples of closed Laplacian solitons which are shrinking and, in particular, produce closed Laplacian flow solutions with a finite-time singularity.
- Extremally Ricci pinched G_2-structures (introduced by Bryant) which are steady Laplacian solitons.

Speaker: **Lotay, Jason (University College London)**
Title: **Laplacian Flow and *4D* Geometry**

Abstract: Using $4D$ geometry, we can reduce the Laplacian flow in G_2-geometry to a flow in 3 or 4 dimensions. In this expository talk, I will describe Fine–Yao's recent work on a flow in 4 dimensions, which is related to a conjecture of Donaldson on hyperKähler 4-manifolds. Time permitting, I will also describe a flow in 3 dimensions, which is connected to space-like submanifolds in Euclidean spaces with mixed signature.

Speaker: **Moore, Kim (Cambridge)**
Title: **Deformation Theory of Cayley Submanifolds**

Abstract: Calibrated submanifolds are volume minimizing submanifolds that occur naturally in manifolds with special holonomy. This talk will focus on the deformation theory of Cayley submanifolds, which live inside Spin(7)-manifolds, with an emphasis on the relationship between Cayley submanifolds and complex surfaces inside four-dimensional Calabi–Yau manifolds.

Speaker: **Oliveira, Gonçalo (Duke)**
Title: **G_2-Instantons on Noncompact G_2-Manifolds**

Abstract: I shall start by reporting what is known about G_2-instantons on noncompact G_2-manifolds. Then, I will focus on joint work with Jason Lotay concerning existence and classification results for these instantons. That work investigates the particular case of $\mathbb{R}^4 \times S^3$, with its two explicitly known distinct G_2-holonomy metrics, exhibiting the different existence/behaviour of G_2-instantons. We also give an explicit example of sequences of G_2-instantons where 'bubbling' and 'removable singularity' phenomena occur in the limit. If time permits, I will state some quite accessible (I hope) open problems. (This is joint work with Jason Lotay.)

Speaker: **Pacini, Tommaso (Torino)**
Title: **New Facts and Tools Concerning Minimal Lagrangian Submanifolds**

Abstract: I will present recent work, joint with Jason Lotay (UCL), on the subject of minimal Lagrangian submanifolds in Kähler geometry. Analogous results may hold in other geometric settings, including G_2. I will try to explain this. Some parts of the talk will be based on the preprint arXiv:1704.08226.

Speaker: **Sá Earp, Henrique (Unicamp)**
Title: **Gauge Theory and G_2-Geometry on Calabi-Yau Links**

Abstract: The 7-dimensional link K of a weighted homogeneous hypersurface on the round 9-sphere in $\mathbb{C}^5$ has a nontrivial null Sasakian structure which is contact Calabi–Yau, in many cases. It admits a canonical co-closed G_2-structure φ induced by the Calabi–Yau 3-orbifold basic geometry. We distinguish these pairs (K, φ) by the Crowley–Nordström $\mathbb{Z}_{48}$-valued ν invariant, for which we prove odd parity and provide an algorithmic formula.

We describe moreover a natural Yang–Mills theory on such spaces, with many important features of the torsion-free case, such as a Chern–Simons formalism and topological energy bounds. In fact compatible G_2-instantons on holomorphic Sasakian bundles over K are exactly the transversely Hermitian Yang Mills connections. As a proof of principle, we obtain G_2-instantons over the Fermat quintic link from stable bundles over the smooth projective Fermat quintic, thus relating in a concrete example the Donaldson–Thomas theory of the quintic threefold with a conjectural G_2-instanton count.

This is joint work with Omegar Calvo-Andrade and Lazaro Rodriguez.

Speaker: **Walpuski, Thomas (Michigan State)**
Title: **The $(1,k)$-ADHM Seiberg–Witten Equation and k-Fold Covers of Associatives**

Abstract: The $(1, k)$-ADHM Seiberg–Witten equations are a class of generalized Seiberg–Witten equations associated with the hyperKähler quotient appearing in the Atiyah, Drinfeld, Hitchin, and Manin's construction of the framed moduli space of ASD instantons on $\mathbb{R}^4$. Formally, degenerating solutions of this equation are related to Fueter sections of bundles of symmetric products of k copies of $\mathbb{R}^4$. In this talk I will explain this relation in more detail and discuss why we believe these equations to be relevant to issues of multiply covered associatives. This is joint work in progress with Aleksander Doan.

Speaker: **Wei, Yong (ANU)**
Title: **Laplacian Flow for Closed G_2-Structures**

Abstract: We will discuss the Laplacian flow for closed G_2-structures. This flow was introduced by R. Bryant in 1992 to study the geometry of G_2-structures, inspired by Hamilton's Ricci flow in studying the generic Riemannian structures and the Kähler Ricci flow in studying Kähler structures. The primary goal is to understand the conditions under which the Laplacian flow can converge to a torsion-free G_2-structure, and thus to a Ricci flat metric with holonomy G_2. I will start with the background of G_2-structures and the motivation of introducing the Laplacian flow, and then describe my recent progress on this flow (Joint work with Jason D. Lotay).

Contents

Lectures

Surveys

Lectures

Introduction to G_2 Geometry

Spiro Karigiannis

Abstract These notes give an informal and leisurely introduction to G_2 geometry for beginners. A special emphasis is placed on understanding the special linear algebraic structure in 7 dimensions that is the pointwise model for G_2 geometry, using the octonions. The basics of G_2-structures are introduced, from a Riemannian geometric point of view, including a discussion of the torsion and its relation to curvature for a general G_2-structure, as well as the connection to Riemannian holonomy. The history and properties of torsion-free G_2 manifolds are considered, and we stress the similarities and differences with Kähler and Calabi–Yau manifolds. The notes end with a brief survey of three important theorems about compact torsion-free G_2 manifolds.

1 Aim and Scope

The purpose of these lecture notes is to give the reader a gentle introduction to the basic concepts of G_2 geometry, including a brief history of the important early developments of the subject.

At present, there is no "textbook" on G_2 geometry. (This is on the author's to-do list for the future.) The only references are the classic monograph by Salamon [39] which emphasizes the representation theoretic aspects of Riemannian holonomy, and the book by Joyce [23] which serves as both a text on Kähler and Calabi–Yau geometry as well as a monograph detailing Joyce's original construction [22] of compact manifolds with G_2 Riemannian holonomy. Both books are excellent resources, but are not easily accessible to beginners. The book by Harvey [17] is at a more appropriate level for new initiates, but is much broader in scope, so it is

S. Karigiannis (✉)
Department of Pure Mathematics, University of Waterloo, Waterloo, Canada
e-mail: karigiannis@uwaterloo.ca

S. Karigiannis et al. (eds.), *Lectures and Surveys on G_2-Manifolds and Related Topics*, Fields Institute Communications 84, https://doi.org/10.1007/978-1-0716-0577-6_1

less focused on G_2 geometry. Moreover, both [17, 39] predate the important analytic developments that started with Joyce's seminal contributions.

The aim of the present notes is to attempt to at least psychologically prepare the reader to access the recent literature in the field, which has undergone a veritable explosion in the past few years. The proofs of most of the deeper results are only sketched, with references given to where the reader can find the details, whereas most of the simple algebraic results are proved in detail. Some important aspects of G_2 geometry are unfortunately only briefly mentioned in passing, including the relations to Spin(7)-structures and the intimate connection with spinors and Clifford algebras. Good references for the connection with spinors are Harvey [17], Lawson–Michelsohn [31, Chap. IV. 10], and the more recent paper by Agricola–Chiossi–Friedrich–Höll [2].

These notes are written in a somewhat informal style. In particular, they are meant to be read leisurely. The punchline is sometime spoiled for the benefit of motivation. In addition, results are sometimes explained in more than one way for clarity, and results are not always stated in the correct logical order but rather in an order that (in the humble opinion of the author) is more effective pedagogically. Finally, there is certainly a definite bias towards the personal viewpoint of the author on the subject. In fact, a distinct emphasis is placed on the explicit details of the linear algebraic aspects of G_2 geometry that are consequences of the *nonassociativity of the octonions* $\mathbb{O}$, as the author believes that this gives good intuition for the striking differences between G_2-structures and $U(m)$-structures in general and $SU(m)$-structures in particular.

The reader is expected to be familiar with graduate level smooth manifold theory and basic Riemannian geometry. Some background in complex and Kähler geometry is helpful, especially to fully appreciate the distinction with G_2 geometry, but is not absolutely essential.

1.1 History of These Notes

These lecture notes have been gestating for many years. In their current form, they are a synthesis of lecture notes for several different introductions to G_2 and Spin(7) geometry that have been given by the author at various institutions or workshops over the past decade. Specifically, these are the following, in chronological order:

- October 2006: Seminar; Mathematical Sciences Research Institute; Berkeley.
- November 2008: Seminar series; University of Oxford.
- January/February 2009: Seminar series; University of Waterloo.
- August 2014: 'G_2-*manifolds*'; Simons Center for Geometry and Physics; Stony Brook.
- September 2014: '*Special Geometric Structures in Mathematics and Physics*'; Universität Hamburg.
- August 2017: Minischool on 'G_2-*manifolds and related topics*'; Fields Institute; Toronto.

The current version of these notes is the first part of the "minischool lectures" on G_2-geometry collected in the book *Lectures and Surveys on* G_2*-geometry and related topics*, published in the *Fields Institute Communications* series by Springer. The other parts of the minischool lectures are

- "Constructions of compact G_2-holonomy manifolds" by Alexei Kovalev [30]
- "Geometric flows of G_2 structures" by Jason Lotay [36]
- "Calibrated Submanifolds in G_2 geometry" by Ki Fung Chan and Naichung Conan Leung [34]
- "Calibrated Submanifolds" by Jason Lotay [35]

1.2 Notation

Let (M, g) be a smooth oriented Riemannian n-manifold. We use both vol and μ to denote the Riemannian volume form induced by g and the given orientation. We use the Einstein summation convention throughout. We use $S^2(T^*M)$ to denote the second symmetric power of T^*M.

Given a vector bundle E over M, we use $\Gamma(E)$ to denote the space of smooth sections of E. These spaces are denoted in other ways in the following cases:

- $\Omega^k = \Gamma(\Lambda^k(T^*M))$ is the space of smooth k-forms on M;
- $\mathcal{S} = \Gamma(S^2(T^*M))$ is the space of smooth symmetric 2-tensors on M.

With respect to the metric g on M, we use $\mathcal{S}_0$ to denote those sections h of $\mathcal{S}$ that are traceless. That is, $\mathcal{S}_0$ consists of those sections of $\mathcal{S}$ such that $\operatorname{Tr} h = g^{ij}h_{ij} = 0$ in local coordinates. Then $\mathcal{S} \cong \Omega^0 \oplus \mathcal{S}_0$, where $h \in \mathcal{S}$ is decomposed as $h = \frac{1}{n}(\operatorname{Tr} h)g + h_0$. Then we have $\Gamma(T^*M \otimes T^*M) = \Omega^0 \oplus \mathcal{S}_0 \oplus \Omega^2$, where the splitting is pointwise orthogonal with respect to the metric on $T^*M \otimes T^*M$ induced by g.

2 Motivation

Let (M^n, g) be an n-dimensional smooth Riemannian manifold. For all $p \in M$, we have an n-dimensional real vector space T_pM equipped with a positive-definite inner product g_p, and these "vary smoothly" with $p \in M$. A natural question is the following:

What other "natural structures" can we put on Riemannian manifolds?

What we would like to do is to attach such a "natural structure" to each tangent space T_pM, for all $p \in M$, in a "smoothly varying" way. That is, such a structure corresponds to a smooth section of some tensor bundle of M, satisfying some natural algebraic condition at each $p \in M$. Let us consider two examples. Let V be an n-dimensional real vector space equipped with a positive-definite inner product g. Note that if we fix an isomorphism $(V, g) \cong (\mathbb{R}^n, \bar{g})$, where $\bar{g}$ is the standard Euclidean inner product on $\mathbb{R}^n$, then the subgroup of $\mathrm{GL}(n, \mathbb{R})$ preserving this structure is $\mathrm{O}(n)$.

Example 2.1 An *orientation* on V is a nonzero element μ of $\Lambda^n V^*$. Let $\beta = \{e_1, \ldots, e_n\}$ be an ordered basis of V. Then $e^1 \wedge \cdots \wedge e^n = \lambda\mu$ for some $\lambda \neq 0$. We say that β is positively (respectively, negatively) oriented with respect to μ if $\lambda > 0$ (respectively, $\lambda < 0$). To demand some kind of compatibility with g, we can rescale μ so that $g(\mu, \mu) = 1$. Thus (V, g) admits precisely two orientations. Note that if we fix an isomorphism $(V, g) \cong (\mathbb{R}^n, \bar{g})$, then the subgroup of $\mathrm{O}(n)$ preserving this structures is $\mathrm{SO}(n)$.

On a smooth manifold, an orientation is thus a nowhere-vanishing smooth section μ of $\Lambda^n(T^*M)$. That is, it is a nowhere-vanishing top form. Such a structure *does not always exist*. Specifically, it exists if and only if the real line bundle $\Lambda^n(T^*M)$ is smoothly trivial. In terms of characteristic classes, this condition is equivalent to the vanishing of the first Stiefel-Whitney class $w_1(TM)$ of the tangent bundle. (See [38], for example.) To demand compatibility with g, we can rescale μ by a positive function so that $g_p(\mu_p, \mu_p) = 1$ for all $p \in M$. This normalized μ is the *Riemannian volume form* associated to the metric g and the chosen orientation on M. It is given locally in terms of a positively oriented orthonormal frame $\{e_1, \ldots, e_n\}$ of TM by $\mu = e^1 \wedge \cdots \wedge e^n$.

An orientation compatible with the metric is called a $\mathrm{SO}(n)$-structure on M. It is equivalent to a reduction of the structure group of the frame bundle of TM from $\mathrm{GL}(n, \mathbb{R})$ to $\mathrm{SO}(n)$. ▲

Example 2.2 A *Hermitian structure* on (V, g) is an orthogonal complex structure J. That is, J is a linear endomorphism of V such that $J^2 = -I$ and $g(Jv, Jw) = g(v, w)$ for all $v, w \in V$. It is well-known and easy to check that such a structure exists on V if and only if $n = 2m$ is even. Such a structure allows us to identify the $2m$-dimensional real vector space V with a m-dimensional complex vector space, where the linear endomorphism J corresponds to multiplication by $\sqrt{-1}$. Note that if we fix an isomorphism $(V, g) \cong (\mathbb{R}^{2m}, \bar{g})$, then the subgroup of $\mathrm{O}(n)$ preserving this structures is $\mathrm{U}(m) = \mathrm{SO}(2m) \cap \mathrm{GL}(m, \mathbb{C})$.

On a Riemannian manifold (M, g), a Hermitian structure is a smooth section J of the tensor bundle $TM \otimes T^*M = \mathrm{End}(TM)$ such that $J^2 = -I$ (which is called an *almost complex structure*) and such that $g_p(J_p X_p, J_p Y_p) = g_p(X_p, Y_p)$ for all $X_p, Y_p \in T_p M$ (which makes it *orthogonal*). As in Example 2.1, such a structure does not always exist, even if $n = \dim M = 2m$ is even. There are *topological obstructions* to the existence of an almost complex structure, which is equivalent to a reduction of the structure group of the frame bundle of TM from $\mathrm{GL}(2m, \mathbb{R})$ to $\mathrm{GL}(m, \mathbb{C})$. See Massey [37] for discussion on this question.

Further demanding that J be compatible with the metric g (that is, orthogonal) is a reduction of the structure group of the frame bundle of TM from $\mathrm{GL}(2m, \mathbb{R})$ to $\mathrm{U}(m)$. For this reason a Hermitian structure on M^{2m} is sometimes also called a $\mathrm{U}(m)$*-structure*. Readers can consult [12] for a comprehensive treatment of the geometry of general $\mathrm{U}(m)$-structures. ▲

Again, let V be an n-dimensional real vector space equipped with a positive-definite inner product g. A G_2-*structure* is a special algebraic structure we can consider on (V, g) only when $n = 7$. In this case, if we fix an isomorphism $(V, g) \cong (\mathbb{R}^7, \bar{g})$, then G_2 is the subgroup of SO(7) preserving this special algebraic structure. In order to describe this structure at the level of linear algebra, we first need to discuss the *octonions*, which we do in Sect. 3. Then G_2-structures are defined and studied in Sect. 4. For the purposes of this motivational section, all the reader needs to know is that a "G_2-structure" corresponds to a special kind of 3-form φ on M^7.

Suppose we have a "natural structure" on a Riemannian manifold (M^n, g), such as that of Examples 2.1 or 2.2 or the mysterious G_2-structure that is the subject of the present notes. Since we have a Riemannian metric g, we have a Levi-Civita connection ∇ and we can further ask for the "natural structure" to be *parallel* or *covariantly constant* with respect to ∇. For example:

- If μ is an orientation (Riemannian volume form) on (M^n, g), then it is *always* parallel.
- If J is an orthogonal almost complex structure on (M^{2m}, g), then if we have $\nabla J = 0$, we say that (M, g, J) is a *Kähler* manifold. Such manifolds have been classically well-studied.
- If φ is a G_2-structure on (M^7, g), then if we have $\nabla \varphi = 0$, we say that (M, g, φ) is a *torsion-free* G_2 *manifold*. Such manifolds are discussed in Sect. 6 of the present notes.

3 Algebraic Structures from the Octonions

In this section we give an introduction to the algebra of the *octonions* $\mathbb{O}$, an 8-dimensional real normed division algebra, and to the induced algebraic structure on Im $\mathbb{O}$, the 7-dimensional space of *imaginary octonions*. We do this by discussing both normed division algebras and spaces equipped with a cross product, and then relating the two concepts. This is not strictly necessary if the intent is to simply consider G_2-structures, but it has the pedagogical benefit of putting both G_2 and Spin(7) geometry into the proper wider context of geometries associated to real normed division algebras. (See Leung [32] for more on this perspective.)

We do not discuss all of the details here, but we do prove many of the important simple results. More details on the algebraic structure of the octonions can be found in Harvey [17], Harvey–Lawson [18], and Salamon–Walpuski [40], for example.

3.1 Normed Division Algebras

Let $\mathbb{A} = \mathbb{R}^n$ be equipped with the standard Euclidean inner product $\langle \cdot, \cdot \rangle$.

Definition 3.1 We say that $\mathbb{A}$ is a *normed division algebra* if $\mathbb{A}$ has the structure of a (*not necessarily associative!*) algebra over $\mathbb{R}$ with multiplicative identity $1 \neq 0$ such that

$$\|ab\| = \|a\| \, \|b\| \qquad \text{for all } a, b \in \mathbb{A}, \tag{3.1}$$

where $\|a\|^2 = \langle a, a\rangle$ is the usual Euclidean norm on $\mathbb{R}^n$ induced from $\langle \cdot, \cdot \rangle$. Equation (3.1) imposes a *compatibility condition* between the inner product and the algebra structure on $\mathbb{A}$. ▲

Remark 3.2 This is not the most general definition possible, but it suffices for our purposes. See [18, Appendices IV.A and IV.B] for more details. ▲

We discuss examples of normed division algebras later in this section, although it is clear that the standard algebraic structures on $\mathbb{R}$ and $\mathbb{C} \cong \mathbb{R}^2$ give examples. We now define some additional structures and investigate some properties of normed division algebras. It is truly remarkable how many far reaching consequences arise solely from the fundamental identity (3.1).

Definition 3.3 Let $\mathbb{A}$ be a normed division algebra. Define the *real part* of $\mathbb{A}$, denoted $\operatorname{Re}\mathbb{A}$, to be the span over $\mathbb{R}$ of the multiplicative identity $1 \in \mathbb{A}$. That is, $\operatorname{Re}\mathbb{A} = \{t1 : t \in \mathbb{R}\}$. Define the *imaginary part* of $\mathbb{A}$, denoted $\operatorname{Im}\mathbb{A}$, to be the orthogonal complement of $\operatorname{Re}\mathbb{A}$ with respect to the Euclidean inner product on $\mathbb{A} = \mathbb{R}^n$. That is, $\operatorname{Im}\mathbb{A} = (\operatorname{Re}\mathbb{A})^{\perp} \cong \mathbb{R}^{n-1}$. Given $a \in \mathbb{A}$, there exists a unique decomposition

$$a = \operatorname{Re} a + \operatorname{Im} a, \qquad \text{where } \operatorname{Re} a \in \operatorname{Re}\mathbb{A} \text{ and } \operatorname{Im} a \in \operatorname{Im}\mathbb{A}.$$

We define the *conjugate* $\overline{a}$ of a to be the element

$$\overline{a} = \operatorname{Re} a - \operatorname{Im} a.$$

▲

Note that the map $a \mapsto \overline{a}$ is a linear involution of $\mathbb{A}$, and is precisely the isometry that is minus the reflection across the hyperplane $\operatorname{Im}\mathbb{A}$ of $\mathbb{A}$. It is clear that

$$\operatorname{Re} a = \tfrac{1}{2}(a + \overline{a}) \qquad \text{and} \qquad \operatorname{Im} a = \tfrac{1}{2}(a - \overline{a}). \tag{3.2}$$

As a result, we deduce that

$$\overline{a} = -a \quad \text{if and only if} \quad a \in \operatorname{Im}\mathbb{A}. \tag{3.3}$$

We now derive a slew of important identities that are all consequences of the defining property (3.1).

Lemma 3.4 *Let $a, b, c \in \mathbb{A}$. Then we have*

$$\langle ac, bc\rangle = \langle ca, cb\rangle = \langle a, b\rangle \|c\|^2, \tag{3.4}$$

and

$$\langle a, bc\rangle = \langle a\bar{c}, b\rangle, \qquad \langle a, cb\rangle = \langle \bar{c}a, b\rangle. \tag{3.5}$$

Moreover, we also have

$$\overline{ab} = \bar{b}\bar{a}. \tag{3.6}$$

Proof First observe that

$$\begin{aligned}\|(a+b)c\|^2 &= \|ac+bc\|^2 = \|ac\|^2 + 2\langle ac, bc\rangle + \|bc\|^2,\\ \|a+b\|^2\,\|c\|^2 &= (\|a\|^2 + 2\langle a, b\rangle + \|b\|^2)\|c\|^2.\end{aligned}$$

Equating the left hand sides above using the fundamental identity (3.1), and again using (3.1) to cancel the corresponding first and third terms on the right hand sides, we obtain

$$\langle ac, bc\rangle = \langle a, b\rangle\|c\|^2. \tag{3.7}$$

Similarly we can show

$$\langle ca, cb\rangle = \langle a, b\rangle\|c\|^2. \tag{3.8}$$

Thus we have established (3.4). Consider the first equation in (3.5). It is clearly satisfied when c is real, since then $\bar{c} = c$ and the inner product $\langle\cdot,\cdot\rangle$ is bilinear over $\mathbb{R}$. Because both sides of the equation are linear in c, it is enough to consider the situation when c is purely imaginary, in which case $\bar{c} = -c$. Then c is orthogonal to 1, so $\|1+c\|^2 = 1 + \|c\|^2$. Applying (3.1) and (3.4), we find

$$\begin{aligned}\langle a, b\rangle(1+\|c\|^2) &= \langle a, b\rangle\|1+c\|^2 = \langle a(1+c), b(1+c)\rangle\\ &= \langle a+ac, b+bc\rangle = \langle a, b\rangle + \langle ac, bc\rangle + \langle a, bc\rangle + \langle ac, b\rangle\\ &= \langle a, b\rangle + \langle a, b\rangle\|c\|^2 + \langle a, bc\rangle + \langle ac, b\rangle.\end{aligned}$$

Thus we have $\langle a, bc\rangle = -\langle ac, b\rangle = \langle a\bar{c}, b\rangle$. This establishes the first equation in (3.5). The other is proved similarly. Using (3.5) and the fact that conjugation is an isometry, we have

$$\langle \overline{ab}, c\rangle = \langle ab, \bar{c}\rangle = \langle b, \bar{a}\bar{c}\rangle = \langle bc, \bar{a}\rangle = \langle c, \bar{b}\bar{a}\rangle.$$

Since this holds for all $c \in \mathbb{A}$, we deduce that $\overline{ab} = \bar{b}\bar{a}$. □

Lemma 3.4 has several important corollaries.

Corollary 3.5 *Let $a, b, c \in \mathbb{A}$. Then we have*

$$a(\bar{b}c) + b(\bar{a}c) = 2\langle a, b\rangle c, \tag{3.9}$$

$$(a\bar{b})c + (a\bar{c})b = 2\langle b, c\rangle a, \tag{3.10}$$

$$a\bar{b} + b\bar{a} = 2\langle a, b\rangle 1. \tag{3.11}$$

Proof Polarizing (3.4), we have

$$\begin{aligned}\langle a, b\rangle \|c+d\|^2 &= \langle a(c+d), b(c+d)\rangle \\ \langle a, b\rangle(\|c\|^2 + 2\langle c, d\rangle + \|d\|^2) &= \langle ac, bc\rangle + \langle ad, bc\rangle + \langle ac, bd\rangle + \langle ad, bd\rangle,\end{aligned}$$

and hence upon using (3.4) to cancel the corresponding first and last terms on each side, we get

$$\langle ad, bc\rangle + \langle ac, bd\rangle = 2\langle a, b\rangle\langle c, d\rangle. \tag{3.12}$$

Using (3.5), we can write the above as

$$\langle d, \overline{a}(bc)\rangle + \langle \overline{b}(ac), d\rangle = 2\langle a, b\rangle\langle c, d\rangle.$$

Since the above holds for any $d \in \mathbb{A}$, we deduce that

$$\overline{a}(bc) + \overline{b}(ac) = 2\langle a, b\rangle c.$$

Replacing $a \mapsto \overline{a}$ and $b \mapsto \overline{b}$ and using the fact that conjugation is an isometry, we obtain (3.9). Equation (3.10) is obtained similarly. Alternatively, one can take the conjugate of (3.9) and use the relation (3.6). Finally, (3.11) is the special case of (3.9) when $c = 1$. □

Corollary 3.6 *Let $a, b, c \in \operatorname{Im}\mathbb{A}$. Then we have*

$$a(bc) + b(ac) = -2\langle a, b\rangle c, \tag{3.13}$$

$$(ab)c + (ac)b = -2\langle b, c\rangle a, \tag{3.14}$$

$$ab + ba = -2\langle a, b\rangle 1. \tag{3.15}$$

Proof These are immediate from Corollary 3.5 and Eq. (3.3). □

Corollary 3.7 *Let $a, b \in \mathbb{A}$. Then we have*

$$\langle a, b\rangle = \operatorname{Re}(a\overline{b}) = \operatorname{Re}(b\overline{a}) = \operatorname{Re}(\overline{b}a) = \operatorname{Re}(\overline{a}b) \tag{3.16}$$

and

$$\|a\|^2 = a\overline{a} = \overline{a}a. \tag{3.17}$$

Proof Using (3.5), we have $\langle a, b\rangle = \langle a, b1\rangle = \langle a\overline{b}, 1\rangle = \operatorname{Re}(a\overline{b})$. The remaining equalities in (3.16) follow from the symmetry of $\langle\cdot,\cdot\rangle$ and the fact that conjugation is an isometry. From (3.6), we find $\overline{\overline{a}a} = \overline{a}\,\overline{\overline{a}} = \overline{a}a$, so $\overline{a}a$ is real. Equation (3.17) thus follows from (3.16). □

Corollary 3.8 *Let $a \in \mathbb{A}$. Then $a^2 = aa$ is real if and only if a is either real or imaginary.*

Proof Write $a = \operatorname{Re} a + \operatorname{Im} a$. Since $\overline{\operatorname{Im} a} = -\operatorname{Im} a$, from (3.17) we have $(\operatorname{Im} a)^2 = -(\operatorname{Im} a)(\overline{\operatorname{Im} a}) = -\|\operatorname{Im} a\|^2$. Thus we have

$$a^2 = (\operatorname{Re} a + \operatorname{Im} a)(\operatorname{Re} a + \operatorname{Im} a) = \big((\operatorname{Re} a)^2 - \|\operatorname{Im} a\|^2\big)1 + 2(\operatorname{Re} a)(\operatorname{Im} a).$$

Since the first term on the right hand side above is always real and the second term is always imaginary, we conclude that a^2 is real if and only if $(\operatorname{Re} a)(\operatorname{Im} a) = 0$, which means that either $\operatorname{Re} a = 0$ or $\operatorname{Im} a = 0$. □

Corollary 3.9 *Let $a, c \in \mathbb{A}$. Then we have*

$$\begin{aligned} (ac)\bar{c} = a(c\bar{c}) = \|c\|^2 a = a(\bar{c}c) = (a\bar{c})c, \\ a(\bar{a}c) = (a\bar{a})c = \|a\|^2 c = (\bar{a}a)c = \bar{a}(ac). \end{aligned} \tag{3.18}$$

Proof Observe from (3.5) and (3.4) that

$$\langle (ac)\bar{c}, b\rangle = \langle ac, bc\rangle = \langle a, b\rangle \|c\|^2 = \langle a\|c\|^2, b\rangle = \langle a(c\bar{c}), b\rangle.$$

Since this holds for all $b \in \mathbb{A}$, we deduce that

$$(ac)\bar{c} = a(c\bar{c}).$$

The rest of the first identity in (3.18) follows by interchanging c and $\bar{c}$. The second identity in (3.18) is proved similarly. □

We now introduce two fundamental $\mathbb{A}$-valued multilinear maps on $\mathbb{A}$.

Definition 3.10 Let $\mathbb{A}$ be a normed division algebra. Define a bilinear map $[\cdot,\cdot] : \mathbb{A}^2 \to \mathbb{A}$ by

$$[a, b] = ab - ba \qquad \text{for all } a, b \in \mathbb{A}. \tag{3.19}$$

The map $[\cdot,\cdot]$ is called the *commutator* of $\mathbb{A}$.

Define a trilinear map $[\cdot,\cdot,\cdot] : \mathbb{A}^3 \to \mathbb{A}$ by

$$[a, b, c] = (ab)c - a(bc) \qquad \text{for all } a, b, c \in \mathbb{A}. \tag{3.20}$$

The map $[\cdot,\cdot,\cdot]$ is called the *associator* of $\mathbb{A}$. ▲

It is clear that the commutator vanishes identically on $\mathbb{A}$ if and only if $\mathbb{A}$ is commutative, and similarly the associator vanishes identically on $\mathbb{A}$ if and only if $\mathbb{A}$ is associative.

Proposition 3.11 *The commutator and associator are both* alternating. *That is, they are totally skew-symmetric in their arguments.*

Proof The commutator is clearly alternating. Because $\mathbb{A}$ is an algebra over $\mathbb{R}$, the associator clearly vanishes if any of the arguments are purely real. Thus, because the associator is trilinear it suffices to show that $\mathbb{A}$ is alternating on $(\operatorname{Im} \mathbb{A})^3$. If $a, b \in \operatorname{Im} \mathbb{A}$, then $\overline{a} = -a$ and $\overline{b} = -b$. Thus by (3.18) we find that

$$-[a, a, b] = [a, \overline{a}, b] = (a\overline{a})b - a(\overline{a}b) = 0.$$

Similarly we have $-[a, b, b] = [a, \overline{b}, b] = (a\overline{b})b - a(\overline{b}b) = 0$. Thus $[\cdot, \cdot, \cdot]$ is alternating in its first two arguments and in its last two arguments. Thus $[a, b, a] = -[a, a, b] = 0$ as well. □

The next result says that both the commutator and the associator restrict to vector-valued alternating multilinear forms on $\operatorname{Im} \mathbb{A}$.

Lemma 3.12 *If $a, b, c \in \operatorname{Im} \mathbb{A}$, then $[a, b] \in \operatorname{Im} \mathbb{A}$ and $[a, b, c] \in \operatorname{Im} \mathbb{A}$.*

Proof We need to show that $[a, b]$ and $[a, b, c]$ are orthogonal to 1. Using the fact that $\overline{a} = -a$ for any $a \in \operatorname{Im} \mathbb{A}$, and the identities (3.5) and (3.30), we compute

$$\begin{aligned} \langle [a, b], 1 \rangle &= \langle ab - ba, 1 \rangle = \langle b, \overline{a} \rangle - \langle a, \overline{b} \rangle \\ &= -\langle b, a \rangle + \langle a, b \rangle = 0. \end{aligned}$$

Similarly, noting that $\overline{bc} = \overline{c}\,\overline{b} = (-c)(-b) = cb$, we have

$$\begin{aligned} \langle [a, b, c], 1 \rangle &= \langle (ab)c - a(bc), 1 \rangle = \langle ab, \overline{c} \rangle - \langle bc, \overline{a} \rangle \\ &= -\langle ab, c \rangle + \langle bc, a \rangle = -\langle a, c\overline{b} \rangle + \langle bc, a \rangle \\ &= \langle a, cb + bc \rangle = \langle a, bc + \overline{bc} \rangle = 2\langle a, \operatorname{Re}(bc) \rangle = 0, \end{aligned}$$

as claimed. □

Proposition 3.13 *Let $a, b, c, d \in \mathbb{A}$. The multilinear expressions $\langle a, [b, c] \rangle$ and $\langle a, [b, c, d] \rangle$ are both totally skew-symmetric in their arguments.*

Proof The commutator and the associator are alternating by Proposition 3.11. Thus we need only show that $\langle a, [a, b] \rangle = 0$ and $\langle a, [a, b, c] \rangle = 0$. Using the identity (3.4) we compute

$$\langle a, [a, b] \rangle = \langle a, ab - ba \rangle = \|a\|^2 \langle 1, b \rangle - \|a\|^2 \langle 1, b \rangle = 0,$$

and similarly using (3.4) and (3.5) we have

$$\begin{aligned} \langle a, [a, b, c] \rangle &= \langle a, (ab)c - a(bc) \rangle = \langle a\overline{c}, ab \rangle - \|a\|^2 \langle 1, bc \rangle \\ &= \|a\|^2 \langle \overline{c}, b \rangle - \|a\|^2 \langle \overline{c}, b \rangle = 0 \end{aligned}$$

as claimed. □

3.2 *Induced Algebraic Structure on* Im $\mathbb{A}$

Let $\mathbb{A} \cong \mathbb{R}^n$ be a normed division algebra with imaginary part $\operatorname{Im}\mathbb{A} \cong \mathbb{R}^{n-1}$. We define several objects on $\operatorname{Im}\mathbb{A}$ induced from the algebra structure on $\mathbb{A}$. Motivated by Lemma 3.12 and Proposition 3.13 the following definition is natural. The factor of $\frac{1}{2}$ is for convenience, to avoid factors of 2 in Eqs. (3.29) and (3.39).

Definition 3.14 Recall the statement of Proposition 3.13. Define a 3-form φ and a 4-form ψ on $\operatorname{Im}\mathbb{A}$ as follows:

$$\varphi(a, b, c) = \tfrac{1}{2}\langle a, [b, c]\rangle = \tfrac{1}{2}\langle [a, b], c\rangle \qquad \text{for } a, b, c \in \operatorname{Im}\mathbb{A}, \tag{3.21}$$

$$\psi(a, b, c, d) = \tfrac{1}{2}\langle a, [b, c, d]\rangle = -\tfrac{1}{2}\langle [a, b, c], d\rangle \quad \text{for } a, b, c, d \in \operatorname{Im}\mathbb{A}. \tag{3.22}$$

The form $\varphi \in \Lambda^3(\operatorname{Im}\mathbb{A})^*$ is called the *associative* 3-form, and the form $\psi \in \Lambda^4(\operatorname{Im}\mathbb{A})^*$ is called the *coassociative* 4-form for reasons that become clear in the context of calibrated geometry [34, 35].

The 3-form φ is intimately related to another algebraic operation on $\operatorname{Im}\mathbb{A}$ that is fundamental in G_2-geometry, given by the following definition. ▲

Definition 3.15 Define a bilinear map $\times : \mathbb{A}^2 \to \mathbb{A}$ by

$$a \times b = \operatorname{Im}(ab) \qquad \text{for all } a, b \in \operatorname{Im}\mathbb{A}. \tag{3.23}$$

Essentially, since the product in $\mathbb{A}$ of two imaginary elements need not be imaginary, we project to the imaginary part to define $\times$. The bilinear map $\times$ is called the *vector cross product* on $\operatorname{Im}\mathbb{A}$ induced by the algebraic structure on $\mathbb{A}$. ▲

The vector cross product $\times$ has several interesting properties.

Lemma 3.16 *Let $a, b \in \operatorname{Im}\mathbb{A}$. The we have*

$$a \times b = -b \times a, \tag{3.24}$$

$$\langle a \times b, a\rangle = 0, \qquad \text{so } (a \times b) \perp a \text{ and } (a \times b) \perp b, \tag{3.25}$$

$$\operatorname{Re}(ab) = -\langle a, b\rangle 1. \tag{3.26}$$

Proof Recall that $\bar{a} = -a$ and $\bar{b} = -b$. Thus from (3.2) and (3.6), we have

$$2a \times b = ab - \overline{ab} = ab - ba = [a, b]. \tag{3.27}$$

Thus $a \times b = -b \times a$.

Since $a \in \operatorname{Im}\mathbb{A}$, Eq. (3.23) show that $\langle a \times b, a\rangle = \langle \operatorname{Im}(ab), a\rangle = \langle ab, a\rangle$. Thus, using (3.4) we get $\langle a \times b, a\rangle = \langle ab, a\rangle = \langle ab, a1\rangle = \|a\|^2\langle b, 1\rangle = 0$ because $b \in \operatorname{Im}\mathbb{A}$ is orthogonal to $1 \in \operatorname{Re}\mathbb{A}$.

Since $\bar{b} = -b$, Eq. (3.16) gives $\langle a, b\rangle = \operatorname{Re}(a\bar{b}) = -\operatorname{Re}(ab)$, which is (3.26). □

Combining Eqs. (3.26) and (3.23) gives us that the decomposition of $ab \in \mathbb{A}$ into real and imaginary parts is given by

$$ab = -\langle a, b \rangle 1 + a \times b. \tag{3.28}$$

It then follows from (3.27) and (3.21) that

$$\varphi(a, b, c) = \langle a \times b, c \rangle \qquad \text{for } a, b, c \in \operatorname{Im} \mathbb{A}. \tag{3.29}$$

Note that since $a \times b - ab$ is real by (3.28), we can equivalently write (3.29) as

$$\varphi(a, b, c) = \langle ab, c \rangle \qquad \text{for } a, b, c \in \operatorname{Im} \mathbb{A}. \tag{3.30}$$

Lemma 3.17 *Let $a, b, c \in \operatorname{Im} \mathbb{A}$. Then we have*

$$a(bc) = -\tfrac{1}{2}[a, b, c] - \varphi(a, b, c)1 - \langle b, c \rangle a + \langle a, c \rangle b - \langle a, b \rangle c. \tag{3.31}$$

Proof Applying the three identities in Corollary 3.6 repeatedly, we compute

$$\begin{aligned} a(bc) &= -b(ac) - 2\langle a, b \rangle c \\ &= -b\big(- ca - 2\langle a, c \rangle 1\big) - 2\langle a, b \rangle c \\ &= b(ca) + 2\langle a, c \rangle b - 2\langle a, b \rangle c \\ &= -c(ba) - 2\langle b, c \rangle a + 2\langle a, c \rangle b - 2\langle a, b \rangle c. \end{aligned}$$

Also, putting $c \mapsto c$ and $b \mapsto ab$ in (3.11) and using (3.30) gives

$$c(ba) - (ab)c = c(\overline{ab}) + (ab)\overline{c} = 2\langle ab, c \rangle 1 = 2\varphi(a, b, c)1.$$

Combining the above two expressions gives

$$a(bc) = -(ab)c - 2\varphi(a, b, c)1 - 2\langle b, c \rangle a + 2\langle a, c \rangle b - 2\langle a, b \rangle c.$$

Using $[a, b, c] = (ab)c - a(bc)$ to eliminate $(ab)c$ above and rearranging gives (3.31). □

Equation (3.31) is used to establish the following two corollaries.

Corollary 3.18 *Let $a, b, c \in \operatorname{Im} \mathbb{A}$. The vector cross product $\times$ on $\operatorname{Im} \mathbb{A}$ satisfies*

$$\|a \times b\|^2 = \|a\|^2 \, \|b\|^2 - \langle a, b \rangle^2 = \|a \wedge b\|^2, \tag{3.32}$$

and

$$a \times (b \times c) = -\langle a, b \rangle c + \langle a, c \rangle b - \tfrac{1}{2}[a, b, c]. \tag{3.33}$$

Proof Let $a, b \in \operatorname{Im} \mathbb{A}$. Using (3.28) we have $ab = -\langle a, b \rangle 1 + a \times b$ and $ba = -\langle b, a \rangle 1 + b \times a = -\langle a, b \rangle - a \times b$. Thus we have

$$\langle ab, ba\rangle = \langle -\langle a, b\rangle 1 + a \times b, -\langle a, b\rangle 1 - a \times b\rangle = \langle a, b\rangle^2 - \|a \times b\|^2.$$

Using the above expression and Eq. (3.27) and (3.1), we compute

$$\begin{aligned} 4\|a \times b\|^2 &= \langle ab - ba, ab - ba\rangle = \|ab\|^2 + \|ba\|^2 - 2\langle ab, ba\rangle \\ &= \|a\|^2\|b\|^2 + \|b\|^2\|a\|^2 - 2(\langle a, b\rangle^2 - \|a \times b\|^2), \end{aligned}$$

which simplifies to (3.32). Again using (3.28) twice and (3.29) we compute

$$\begin{aligned} a \times (b \times c) &= \langle a, b \times c\rangle 1 + a(b \times c) \\ &= \varphi(a, b, c)1 + a(\langle b, c\rangle + bc) \\ &= a(bc) + \varphi(a, b, c)1 + \langle b, c\rangle a\rangle. \end{aligned}$$

Substituting (3.31) for $a(bc)$ above gives (3.33). □

Corollary 3.19 *Let $a, b, c, d \in \operatorname{Im} \mathbb{A}$. Then the following identity holds:*

$$\|\tfrac{1}{2}[a, b, c]\|^2 + \big(\varphi(a, b, c)\big)^2 = \|a \wedge b \wedge c\|^2. \tag{3.34}$$

Proof Recall from Lemma 3.12 and Proposition 3.13 that $[a, b, c]$ is purely imaginary and is orthogonal to a, b, c. Thus, taking the norm squared of (3.31) and using the fundamental relation (3.1), we have

$$\begin{aligned} \|a\|^2\|b\|^2\|c\|^2 &= \|a\|^2\|bc\|^2 = \|a(bc)\|^2 \\ &= \|\tfrac{1}{2}[a, b, c]\|^2 + \big(\varphi(a, b, c)\big)^2 + \|a\|^2\langle b, c\rangle^2 + \|b\|^2\langle a, c\rangle^2 + \|c\|^2\langle a, b\rangle^2 \\ &\quad - 2\langle b, c\rangle\langle a, c\rangle\langle a, b\rangle + 2\langle b, c\rangle\langle a, b\rangle\langle a, c\rangle - 2\langle a, c\rangle\langle a, b\rangle\langle b, c\rangle. \end{aligned}$$

This can be rearranged to yield

$$\begin{aligned} \|\tfrac{1}{2}[a, b, c]\|^2 + \big(\varphi(a, b, c)\big)^2 &= \|a\|^2\|b\|^2\|c\|^2 - \|a\|^2\langle b, c\rangle^2 - \|b\|^2\langle a, c\rangle^2 \\ &\quad - \|c\|^2\langle a, b\rangle^2 + 2\langle a, b\rangle\langle a, c\rangle\langle b, c\rangle, \end{aligned}$$

which is precisely (3.34). □

Remark 3.20 Comparing Eqs. (3.21) and (3.29), one is tempted from (3.22) to think of the expression $\frac{1}{2}[a, b, c]$ as some kind of 3-fold vector cross product $P(a, b, c)$, as it is a trilinear vector valued alternating form on $\operatorname{Im} \mathbb{A}$. However, Eq. (3.34) says that $\|a \wedge b \wedge c\|^2 - \|P(a, b, c)\|^2$ is nonzero in general, whereas (3.32) says that $\|a \wedge b\|^2 - \|a \times b\|^2 = 0$ always. There *is* a notion of *3-fold vector cross product* (see Remark 3.23 below) but $[\cdot, \cdot, \cdot]$ on $\operatorname{Im} \mathbb{A}$ is not one of them. In fact, Eq. (3.34) is the *calibration inequality* for φ. It says that $|\varphi(a, b, c)| \leq 1$ whenever a, b, c are orthonormal, with equality if and only if $[a, b, c] = 0$. See [34, 35] in the present volume for more about the aspects of G_2 geometry related to *calibrations*. ▲

Equations (3.25) and (3.32) for the vector cross product $\times$ induced from the algebraic structure on $\mathbb{A}$ motivate the following general definition.

Definition 3.21 Let $\mathbb{V} = \mathbb{R}^m$, equipped with the usual Euclidean inner product. We say that V has a *vector cross product*, which we usually simply call a *cross product*, if there exists a *skew-symmetric* bilinear map $\times : \mathbb{V}^2 \to \mathbb{V}$ such that, for all $a, b, c \in \mathbb{V}$, we have

$$\langle a \times b, a \rangle = 0, \qquad \text{so } (a \times b) \perp a \text{ and } (a \times b) \perp b, \tag{3.35}$$

$$\|a \times b\|^2 = \|a\|^2 \, \|b\|^2 - \langle a, b \rangle^2 = \|a \wedge b\|^2. \tag{3.36}$$

Note that (3.35) and (3.36) are precisely (3.25) and (3.32), respectively. ▲

Remark 3.22 The fact that $\times$ is skew-symmetric and bilinear is equivalent to saying that $\times$ is a *linear map*

$$\times : \Lambda^2(\mathbb{V}) \to \mathbb{V}.$$

Then (3.36) says that $\times$ is *length preserving on decomposable elements of* $\Lambda^2(\mathbb{V})$. ▲

Remark 3.23 In Definition 3.21 we have really defined a special class of vector cross product, called a 2*-fold vector cross product*. A more general notion of *k-fold vector cross product* [3] exists. When $k = 1$ such a structure is an an orthogonal complex structure. When $k = 3$ such a structure is related to Spin(7)-geometry. See also Lee–Leung [33] for more details. Another extensive recent reference for general vector cross products is Cheng–Karigiannis–Madnick [7, Sect. 2]. ▲

We have seen that any normed division algebra $\mathbb{A}$ gives a vector cross product on $\mathbb{V} = \operatorname{Im} \mathbb{A}$. In the next section we show that we can also go the other way.

3.3 One-to-One Correspondence and Classification

We claim that the normed division algebras are in *one-to-one correspondence* with the spaces admitting cross products. The correspondence is seen explicitly as follows. Let $\mathbb{A}$ be a normed division algebra. In Sect. 3.2 we showed that $\mathbb{V} = \operatorname{Im} \mathbb{A}$ has a cross product. Conversely, suppose $\mathbb{V} = \mathbb{R}^m$ has a cross product $\times$. Define $\mathbb{A} = \mathbb{R} \oplus \mathbb{V} = \mathbb{R}^{m+1}$, with the Euclidean inner product. That is,

$$\langle (s, v), (t, w) \rangle = st + \langle v, w \rangle \qquad \text{for } s, t \in \mathbb{R} \text{ and } v, w \in \mathbb{V}.$$

Define a multiplication on $\mathbb{A}$ by

$$(s, v)(t, w) = (st - \langle v, w \rangle, sw + tv + v \times w). \tag{3.37}$$

The multiplication defined in (3.37) is clearly bilinear over $\mathbb{R}$, so it gives $\mathbb{A}$ the structure of a (not necessarily associative) algebra over $\mathbb{R}$. It is also clear from (3.37) that $(1, 0)$ is a multiplicative identity on $\mathbb{A}$. We need to check that (3.1) is satisfied. We compute:

$$\begin{aligned}\|(s, v)(t, w)\|^2 &= (st - \langle v, w\rangle)^2 + \|sw + tv + v \times w\|^2 \\ &= s^2t^2 - 2st\langle v, w\rangle + \langle v, w\rangle^2 + s^2\|w\|^2 + t^2\|v\|^2 + \|v \times w\|^2 \\ &\quad + 2st\langle v, w\rangle + 2s\langle w, v \times w\rangle + 2t\langle v, v \times w\rangle.\end{aligned}$$

Using (3.33) and (3.35), the above expression simplifies to

$$\begin{aligned}\|(s, v)(t, w)\|^2 &= s^2t^2 + s^2\|w\|^2 + t^2\|v\|^2 + \|v\|^2\|w\|^2 \\ &= (s^2 + \|v\|^2)(t^2 + \|w\|^2) = \|(s, v)\|\,\|(t, w)\|,\end{aligned}$$

verifying (3.1).

Normed division algebras were classified by Hurwitz in 1898. A proof using the *Cayley–Dickson doubling construction* can be found in [17, Chap. 6] or [18, Appendix IV.A]. There are exactly four possibilities, up to isomorphism.

The four normed division algebras are given by the following table:

$n = \dim \mathbb{A}$	1	2	4	8
Symbol	$\mathbb{R}$	$\mathbb{C} \cong \mathbb{R}^2$	$\mathbb{H} \cong \mathbb{R}^4$	$\mathbb{O} \cong \mathbb{R}^8$
Name	Real numbers	Complex numbers	Quaternions or Hamilton numbers	Octonions or Cayley numbers

Each algebra in the above table is a *subalgebra* of the next one. In particular, the multiplicative unit in all cases is the usual multiplicative identity $1 \in \mathbb{R}$. Moreover, as we enlarge the algebras $\mathbb{R} \to \mathbb{C} \to \mathbb{H} \to \mathbb{O}$, we lose some algebraic property at each step. From $\mathbb{R}$ to $\mathbb{C}$ we lose the *natural ordering*. From $\mathbb{C}$ to $\mathbb{H}$ we lose *commutativity*. And from $\mathbb{H}$ to $\mathbb{O}$ we lose *associativity*.

The octonions $\mathbb{O}$ are also called the *exceptional* division algebra and the geometries associated to $\mathbb{O}$ are known as *exceptional geometries*.

By the one-to-one correspondence between normed division algebras and spaces admitting cross products, we deduce that there exist precisely four spaces with cross product, given by the following table:

$m = \dim \mathbb{V}$	0	1	3	7
Symbol	$\{0\} \simeq \operatorname{Im}\mathbb{R}$	$\mathbb{R} \cong \operatorname{Im}\mathbb{C}$	$\mathbb{R}^3 \simeq \operatorname{Im}\mathbb{H}$	$\mathbb{R}^7 \cong \operatorname{Im}\mathbb{O}$
Cross product $\times$	trivial	trivial	standard (Hodge star)	exceptional
the 3-form $\varphi \in \Lambda^3(\mathbb{V}^*)$	0	0	$\varphi = \mu$ is the standard volume form	φ is the associative 3-form
the $(m-3)$-form $\star\varphi \in \Lambda^{m-3}(\mathbb{V}^*)$	0	0	$\star\varphi = 1 \in \Lambda^0(\mathbb{V}^*) \cong \mathbb{R}$ is the multiplicative unit	$\star\varphi = \psi \in \Lambda^4(\mathbb{V}^*)$ is the coassociative 4-form

Remark 3.24 Here are some remarks concerning the above table:

(i) When $m = 0, 1$, the cross product $\times$ is trivial because $\Lambda^2(\mathbb{V}) = \{0\}$ in these cases.
(ii) When $m = 3$ we recover the standard cross product on $\mathbb{R}^3$. It is well-known that the standard cross product can be obtained from quaternionic multiplication by (3.23), and that $\langle u \times v, w\rangle = \mu(u, v, w)$ is the volume form μ evaluated on the 3-plane $u \wedge v \wedge w$. Equivalently, the cross product is given by the Hodge star on $\mathbb{R}^3$. That is, $u \times v = \star(u \wedge v)$. In this case $\star\varphi = \star\mu = 1$.
(iii) The cross product on $\mathbb{R}^7$ is induced in the same way from octonionic multiplication, and is called the *exceptional cross product*. In this case φ is a nontrivial 3-form on $\mathbb{R}^7$, and $\star\varphi = \psi$ is a nontrivial 4-form on $\mathbb{R}^7$. We discuss these in more detail in Sect. 4.1. ▲

3.4 Further Properties of the Cross Product in $\mathbb{R}^3$ and $\mathbb{R}^7$

Let us investigate some further properties of the cross product. First, note that for $\mathbb{V} = \mathbb{R}^3 \cong \operatorname{Im}\mathbb{H}$, Eq. (3.33) reduces to the familiar $a \times (b \times c) = -\langle a, b\rangle c + \langle a, c\rangle b$, because $\mathbb{H}$ is associative. However, for $\mathbb{V} = \mathbb{R}^7 \cong \operatorname{Im}\mathbb{O}$, we have

$$a \times (b \times c) = -\langle a, b\rangle c + \langle a, c\rangle b - \tfrac{1}{2}[a, b, c] \tag{3.38}$$

where the last term *does not vanish in general*. In fact using (3.22) we can write (3.38) as

$$a \times (b \times c) = -\langle a, b\rangle c + \langle a, c\rangle b + \big(\psi(a, b, c, \cdot)\big)^\sharp \tag{3.39}$$

where $\alpha^\sharp$ is the vector in $\mathbb{V}$ that is metric-dual to the 1-form $\alpha \in \mathbb{V}^*$ via the inner product. Explicitly, $\langle \alpha^\sharp, b\rangle = \alpha(b)$ for all $b \in \mathbb{V}$.

Remark 3.25 The nontriviality of the last term in (3.38) or (3.39) is equivalent to the nonassociativity of $\mathbb{O}$ and is the source of the inherent *nonlinearity* in geometries defined using the octonions. See also Remark 4.15 below. ▲

We obtain a number of important consequences from the fundamental identity (3.38). The remaining results in this section hold for both the cases $\mathbb{V} = \mathbb{R}^3 \cong \operatorname{Im}\mathbb{H}$ and $\mathbb{V} = \mathbb{R}^7 \cong \operatorname{Im}\mathbb{O}$, with the understanding that the associator term vanishes in the $\mathbb{R}^3$ case.

Corollary 3.26 *Let $a, c \in \mathbb{V}$. Then we have*

$$a \times (a \times c) = -\|a\|^2 c + \langle a, c\rangle a. \tag{3.40}$$

Proof Let $a = b$ in (3.38). The associator term vanishes by Proposition 3.11. □

Remark 3.27 From Corollary 3.26 we deduce the following observation. Let $a \in \mathbb{V}$ satisfy $\|a\| = 1$. Consider the codimension one subspace $\mathbb{U} = (\mathrm{span}\{a\})^{\perp}$ orthogonal to a. Since $a \times c$ is orthogonal to c for all c, the linear map $J_a : \mathbb{V} \to \mathbb{V}$ given by $J_a(c) = a \times c$ leaves $\mathbb{U}$ invariant, and by (3.40) we have $(J_a)^2 = -I$ on $\mathbb{U}$, so J_a is a complex structure on $\mathbb{U}$. ▲

Corollary 3.28 *Let $a, b, c \in \mathbb{V}$ be orthonormal, with $a \times b = c$. Then $b \times c = a$ and $c \times a = b$.*

Proof Take the cross product of $a \times b = c$ on both sides with a or b and use (3.40). □

Corollary 3.29 *Let $a, b, c, d \in \mathbb{V}$. Recall that*

$$\langle a \wedge b, c \wedge d \rangle = \det \begin{pmatrix} \langle a, c \rangle & \langle a, d \rangle \\ \langle b, c \rangle & \langle b, d \rangle \end{pmatrix} = \langle a, c \rangle \langle b, d \rangle - \langle a, d \rangle \langle b, c \rangle.$$

Then we have

$$\langle a \times b, c \times d \rangle = \langle a \wedge b, c \wedge d \rangle - \tfrac{1}{2} \langle a, [b, c, d] \rangle, \tag{3.41}$$

$$\langle a \times b, a \times d \rangle = \langle a \wedge b, a \wedge d \rangle = \|a\|^2 \langle b, d \rangle - \langle a, b \rangle \langle a, d \rangle. \tag{3.42}$$

Proof Equation (3.42) follows from (3.41) by setting $c = a$ and using Proposition 3.13. To establish (3.41), we compute using (3.29) and the skew-symmetry of φ as follows:

$$\langle a \times b, c \times d \rangle = \varphi(a, b, c \times d) = -\varphi(a, c \times d, b) = -\langle a \times (c \times d), b \rangle.$$

Using (3.38), the above expression becomes

$$\begin{aligned} \langle a \times b, c \times d \rangle &= -\langle -\langle a, c \rangle d + \langle a, d \rangle c - \tfrac{1}{2}[a, c, d], b \rangle \\ &= \langle a, c \rangle \langle b, d \rangle - \langle a, d \rangle \langle b, c \rangle + \tfrac{1}{2} \langle b, [a, c, d] \rangle. \end{aligned}$$

Using Proposition 3.13, the above expression equals (3.41). □

Remark 3.30 Using (3.22), when $n = 7$ we can also write (3.41) as

$$\langle a \times b, c \times d \rangle = \langle a \wedge b, c \wedge d \rangle - \psi(a, b, c, d). \tag{3.43}$$

Recall from Remark 3.25 that the nontriviality of ψ is equivalent to the nonassociativity of $\mathbb{O}$. Thus the above equation says that the nonassociativity of $\mathbb{O}$ is also equivalent to the fact that

$$\langle a \times b, c \times d \rangle \neq \langle a \wedge b, c \wedge d \rangle$$

in general.

By contrast, when $n = 3$ the associator vanishes, and we do have $\langle a \times b, c \times d \rangle = \langle a \wedge b, c \wedge d \rangle$ in this case. This corresponds, by (ii) of Remark 3.24, to the fact that $a \times b = \star(a \wedge b)$ and $\star$ is an isometry. ▲

4 The Geometry of G_2-Structures

In this section we discuss G_2-structures, first on $\mathbb{R}^7$ and then on smooth 7-manifolds, including a discussion of the decomposition of the space of differential forms and of the torsion of a G_2-structure.

4.1 The Canonical G_2-Structure on $\mathbb{R}^7$

In this section we describe in more detail the canonical G_2-structure on $\mathbb{R}^7 \cong \operatorname{Im} \mathbb{O}$. This standard "$G_2$-package" on $\mathbb{R}^7$ consists of the standard Euclidean metric g_o, for which the standard basis $e_1, \ldots, e_7$ is orthonormal, the standard volume form $\mu_o = e^1 \wedge \cdots \wedge e^7$ associated to g_o and the standard orientation, the "associative" 3-form φ_o, the "coassociative" 4-form ψ_o, and finally the "cross product" $\times_o$ operation. We use the "o" subscript for the standard G_2-package $(g_o, \mu_o, \varphi_o, \psi_o, \times_o)$ on $\mathbb{R}^7$ to distinguish it from a general G_2-structure on a smooth 7-manifold which is defined in Sect. 4.2. We also use $\| \cdot \|_o$ to denote both the norm on $\mathbb{R}^7$ induced from the inner product g_o and also the induced norm on $\Lambda^\bullet(\mathbb{R}^7)^*$.

We identify $\mathbb{R}^7 \cong \operatorname{Im} \mathbb{O}$. Recall from Definition 3.14 that the associative 3-form φ_o and the coassociative 4-form ψ_o are given by

$$\begin{aligned} \varphi_o(a, b, c) &= \tfrac{1}{2}\langle [a, b], c \rangle \qquad \text{for } a, b, c \in \mathbb{R}^7, \\ \psi_o(a, b, c, d) &= -\tfrac{1}{2}\langle [a, b, c], d \rangle \quad \text{for } a, b, c, d \in \mathbb{R}^7. \end{aligned}$$

Using the octonion multiplication table, one can show that with respect to the standard dual basis $e^1, \ldots, e^7$ on $(\mathbb{R}^7)^*$, and writing $e^{ijk} = e^i \wedge e^j \wedge e^k$ and similarly for decomposable 4-forms, we have

$$\begin{aligned} \varphi_o &= e^{123} - e^{167} - e^{527} - e^{563} - e^{415} - e^{426} - e^{437}, \\ \psi_o &= e^{4567} - e^{4523} - e^{4163} - e^{4127} - e^{2637} - e^{1537} - e^{1526}. \end{aligned} \tag{4.1}$$

It is immediate that

$$\psi_o = *_o \varphi_o,$$

where $*_o$ is the Hodge star operator induced from (g_o, μ_o). The explicit expressions for φ_o and $\psi_o = *_o \varphi_o$ in (4.1) are not enlightening and need not be memorized by the reader. There is a particular method to the seeming madness in which we have written

φ_o and ψ_o, which is explained in [27] in relation to the standard SU(3)-structure on $\mathbb{R}^7 = \mathbb{C}^3 \oplus \mathbb{R}$, where $z^1 = x^1 + ix^5$, $z^2 = x^2 + ix^6$, $z^3 = x^3 + ix^7$ are the complex coordinates on $\mathbb{C}^3$ and x^4 is the coordinate on $\mathbb{R}$.

One piece of information to retain from (4.1) is that

$$\|\varphi_o\|_o^2 = \|\psi_o\|_o^2 = 7, \tag{4.2}$$

which is equivalent to the identity $\varphi_o \wedge \psi_o = 7\mu_o$. (These facts are analogous to the identities $\|\omega_o\|_o^2 = 2m$ and $\frac{1}{m!}\omega_o^m = \mu_o$ for the standard Kähler form ω_o on $\mathbb{C}^m$ with respect to the Euclidean metric.)

We now use this standard "G_2-package" on $\mathbb{R}^7$ to give a definition of the group G_2.

Definition 4.1 The group G_2 is defined to be the subgroup of $GL(7, \mathbb{R})$ that preserves the standard G_2-package on $\mathbb{R}^7$. That is,

$$G_2 = \{A \in GL(7, \mathbb{R}) : A^* g_o = g_o, A^*\mu_o = \mu_o, A^*\varphi_o = \varphi_o\}.$$

Note that because g_o and μ_o determine the Hodge star operator $*_o$, which in turn from φ_o determines ψ_o, and because g_o and φ_o together determine $\times_o$, it follows that any $A \in G_2$ also preserves ψ_o and $\times_o$. (But see Theorem 4.2 below.) Moreover, since by definition $A \in G_2$ preserves the standard Euclidean metric and orientation on $\mathbb{R}^7$, we see that G_2 as defined above is a *subgroup* of $SO(7, \mathbb{R})$. ▲

Theorem 4.2 (Bryant [4]) *Define $K = \{A \in GL(7, \mathbb{R}) : A^*\varphi_o = \varphi_o\}$. Then in fact $K = G_2$. That is, if $A \in GL(7, \mathbb{R})$ preserves φ_o, then it also automatically preserves g_o and μ_o as well.*

Proof One can show using the explicit form (4.1) for φ_o in terms of the standard basis $e^1, \ldots, e^7$ of $(\mathbb{R}^7)^*$ that

$$(a \lrcorner \varphi_o) \wedge (b \lrcorner \varphi_o) \wedge \varphi_o = -6 g_o(a, b)\mu_o. \tag{4.3}$$

It follows from (4.3) that if $A^*\varphi_o = \varphi_o$, then

$$(A^* g_o)(a, b) A^*\mu_o = g_o(Aa, Ab)(\det A)\mu_o = g_o(a, b)\mu_o. \tag{4.4}$$

Thus we have $(\det A) g_o(Aa, Ab) = g_o(a, b)$, or equivalently in terms of matrices, $g_o = (\det A) A^T g_o A$. Taking determinants of both sides, and observing that these are all 7×7 matrices, gives $\det g_o = (\det A)^9 \det g_o$, so $\det A = 1$ and $A^*\mu_o = \mu_o$. But then (4.4) says that $A^* g_o = g_o$ as claimed. □

Remark 4.3 In Bryant [4] the Eq. (4.3) has a $+6$ on the right hand side rather than our -6, because of a different orientation convention. See also Remark 4.17 below. ▲

Remark 4.4 Theorem 4.2 appears in [4]. Robert Bryant claims that it is a much older result, due to Élie Cartan. While this is almost certainly true, most mathematicians know this result as "Bryant's Theorem" as [4] is the earliest accessible reference for this result that is widely known. See Agricola [1] for more about this history of the group G_2. ▲

Corollary 4.5 *The group* G_2 *can equivalently be defined as the* automorphism group $\mathrm{Aut}(\mathbb{O})$ *of the normed division algebra* $\mathbb{O}$ *of octonions.*

Proof Let $A \in \mathrm{Aut}(\mathbb{O})$. Since A is an algebra automorphism we have $A(1) = 1$ and thus $A(t1) = t$ for all $t \in \mathbb{R}$. Now suppose $p \in \mathrm{Im}\,\mathbb{O}$. Then $p^2 = -p\overline{p} = -\|p\|_o^2$ is real. Thus we have

$$(A(p))^2 = A(p)A(p) = A(p^2) = A(-\|p\|_o^2) = -\|p\|_o^2$$

is real. By Corollary 3.8 we deduce that $A(p)$ must be real or imaginary. Suppose it is real. Then $A(p) = t1$ for some $t \in \mathbb{R}$. But then $A(p) = A(t1)$ and $p \neq t1$ since p is imaginary. This contradicts the invertibility of A. Thus $A(p)$ must be imaginary. This means $\overline{A(p)} = -A(p)$ whenever p is imaginary.

Now let $p = (\mathrm{Re}\,p)1 + (\mathrm{Im}\,p)$. Since A is linear over $\mathbb{R}$ and $A(1) = 1$, we get $A(p) = (\mathrm{Re}\,p)1 + A(\mathrm{Im}\,p)$. But then $\overline{A(p)} = (\mathrm{Re}\,p)1 - A(\mathrm{Im}\,p) = A(\overline{p})$. It follows that

$$\|A(p)\|_o^2 = A(p)\overline{A(p)} = A(p)A(\overline{p}) = A(p\overline{p}) = A(\|p\|_o^2) = \|p\|_o^2.$$

Thus $\|A(p)\|_o = \|p\|_o$, and from $A(1) = 1$ and $A(\mathrm{Im}\,\mathbb{O}) \subseteq (\mathrm{Im}\,\mathbb{O})$ we conclude that $A \in \mathrm{O}(7)$. Finally, from (3.30), if $a, b, c \in \mathrm{Im}\,\mathbb{O}$ we get

$$\begin{aligned}(A^*\varphi_o)(a, b, c) &= \varphi_o(Aa, Ab, Ac) = \langle (Aa)(Ab), Ac\rangle \\ &= \langle A(ab), Ac\rangle = \langle ab, c\rangle = \varphi_o(a, b, c).\end{aligned}$$

Thus $A^*\varphi_o = \varphi_o$, so by Theorem 4.2 we deduce that $A \in G_2$.

Conversely, if $A \in G_2$, then A preserves the cross product and the inner product, so if we extend A linearly from $\mathbb{R}^7 \cong \mathrm{Im}\,\mathbb{O}$ to $\mathbb{O} = \mathbb{R} \oplus \mathbb{R}^7$ by setting $A(1) = 1$, then it follows immediately from (3.37) that $A(ab) = A(a)A(b)$ for all $a, b \in \mathbb{O}$, so $A \in \mathrm{Aut}(\mathbb{O})$. □

Remark 4.6 Theorem 4.2 is an *absolutely crucial* ingredient of G_2-geometry. It says that the 3-form φ_o *determines* the orientation μ_o and the metric g_o in a *highly nonlinear way*. This is in stark contrast to the situation of the standard $\mathrm{U}(m)$-structure on $\mathbb{C}^m$, which consists of the Euclidean metric g_o, the standard complex structure J_o on $\mathbb{C}^m$, and the associated Kähler form ω_o, which are all related by

$$\omega_o(a, b) = g_o(J_o a, b). \tag{4.5}$$

Moreover, the standard volume form is $\mu_o = \frac{1}{m!}\omega_o^m$. Equation (4.5) should be compared to (3.29). The almost complex structure J_o is the analogue of the cross product $\times_o$, and the 2-form ω_o is the analogue of the 3-form φ_o. However, in the case of the standard $U(m)$-structure, the 2-form ω_o *does not* determine the metric. (Although it *does* determine the orientation.) The correct way to think about (4.5) is that knowledge of any two of g_o, J_o, ω_o uniquely determines the third. This is encoded by the following Lie group relation:

$$O(2m, \mathbb{R}) \cap GL(m, \mathbb{C}) = O(2m, \mathbb{R}) \cap Sp(m, \mathbb{R}) = GL(m, \mathbb{C}) \cap Sp(m, \mathbb{R}) = U(m).$$

Colloquially, we say that the intersection of any two of Riemannian, complex, and symplectic geometry is Kähler geometry. By constrast, G_2 geometry *does not "decouple"* in any such way. It is *not* the intersection of Riemannian geometry with any other "independent" geometry. The 3-form φ_o determines *everything else.* ▲

Let us consider how we should think about the group G_2, which by Theorem 4.2 is described as a particular *subgroup* of $SO(7, \mathbb{R})$. Before we can do that, we need a preliminary result.

Lemma 4.7 *Let f_1, f_2, f_4 be a triple of orthonormal vectors in $\mathbb{R}^7$ such that $\varphi_o(f_1, f_2, f_4) = 0$. Define*

$$f_3 = f_1 \times_o f_2, \qquad f_5 = f_1 \times_o f_4, \qquad f_6 = f_2 \times_o f_4, \qquad f_7 = f_3 \times_o f_4 = (f_1 \times_o f_2) \times_o f_4. \tag{4.6}$$

Then the ordered set $\{f_1, \ldots, f_7\}$ is an oriented orthonormal basis *of* $\mathbb{R}^7$.

Proof One can check using Eqs. (3.25), (3.29), and (3.42), together with the hypotheses that $\{f_1, f_2, f_4\}$ are orthonormal and $\varphi_o(f_1, f_2, f_4) = 0$, that $\langle f_i, f_j \rangle = \delta_{ij}$ for all i, j so the set is orthonormal. Most of these are immediate. We demonstrate one of the less trivial cases. Using Corollary 3.28, we deduce that $f_3 \times_o f_1 = f_2$. Thus we have

$$\begin{aligned} g_o(f_1, f_7) &= g_o(f_1, f_3 \times_o f_4) = \varphi_o(f_1, f_3, f_4) \\ &= -\varphi_o(f_3, f_1, f_4) = -g_o(f_3 \times_o f_1, f_4) = -g_o(f_2, f_4) = 0. \end{aligned}$$

It remains to show $\{f_1, \ldots, f_7\}$ induces the same orientation as $\{e_1, \ldots, e_7\}$. When $f_k = e_k$ for $k = 1, 2, 4$, then it follows from the octonion multiplication table and (4.6) that $f_k = e_k$ for all $k = 1, \ldots, 7$. It is then not hard to see that the matrix in $A \in O(7)$ given by

$$A = \begin{pmatrix} f_1 \mid f_2 \mid f_3 \mid f_4 \mid f_5 \mid f_6 \mid f_7 \end{pmatrix}$$

can be obtained from the identity matrix by a product of three elements of $SO(7)$. Thus $A \in SO(7)$ and hence $\{f_1, \ldots, f_7\}$ is oriented. □

Corollary 4.8 *The group G_2 can be viewed explicitly as the subgroup of* $SO(7)$ *consisting of those elements $A \in SO(7)$ of the form*

$$A = \big(f_1 \mid f_2 \mid f_1 \times_o f_3 \mid f_4 \mid f_1 \times_o f_4 \mid f_2 \times_o f_4 \mid (f_1 \times_o f_2) \times_o f_4\big) \tag{4.7}$$

where $\{f_1, f_2, f_4\}$ *is an orthonormal triple satisfying* $\varphi_o(f_1, f_2, f_4) = 0$. *(This means that the cross product of any two of* $\{f_1, f_2, f_4\}$ *is orthogonal to the third.)*

Proof By Lemma 4.7, every matrix of the form (4.7) does lie in SO(7). By Theorem 4.2, a matrix $A \in \mathrm{SO}(7)$ is in G_2 if and only if A preserves the vector cross product $\times_o$. Since A takes e_k to f_k, it follows from the fact that

$$e_3 = e_1 \times_o e_2, \qquad e_5 = e_1 \times_o e_4, \qquad e_6 = e_2 \times_o e_4, \qquad e_7 = e_3 \times_o e_4 = (e_1 \times_o e_2) \times_o e_4,$$

that such the elements of G_2 are precisely the matrices of the form (4.7). □

Remark 4.9 We can argue from Corollary 4.8 that $\dim \mathrm{G}_2 = 14$, as follows. We know G_2 corresponds to the set of orthonormal triples $\{f_1, f_2, f_4\}$ such that f_4 is orthogonal to f_1, f_2, and $f_1 \times_o f_2$. Thus f_1 is any unit vector in $\mathbb{R}^7$, so it lies on S^6. Then f_2 must be orthogonal to f_1, so it lies on the unit sphere S^5 of the $\mathbb{R}^6$ that is orthogonal to f_1. Finally, f_4 must be orthogonal to f_1, f_2, and $f_1 \times_o f_2$, so it lies on the unit sphere S^3 of the $\mathbb{R}^4$ that is orthogonal to these three vectors. Thus we have $6 + 5 + 3 = 14$ degrees of freedom, so $\dim \mathrm{G}_2 = 14$.

(In fact, G_2 is a connected, simply-connected, compact Lie subgroup of SO(7).) ▲

4.2 G₂-Structures on Smooth 7-Manifolds

In this section, as discussed in Sect. 2, we equip a smooth 7-manifold with the "G_2 package" at each tangent space, in a smoothly varying way.

Definition 4.10 Let M^7 be a smooth 7-manifold. A G_2-structure on M is a smooth 3-form φ on M such that, at every $p \in M$, there exists a linear isomorphism $T_p M \cong \mathbb{R}^7$ with respect to which $\varphi_p \in \Lambda^3(T_p^* M)$ corresponds to $\varphi_o \in \Lambda^3(\mathbb{R}^7)^*$. Therefore, because φ_o induces g_o and μ_o, a G_2-structure φ on M induces a Riemannian metric g_φ and associated Riemannian volume form μ_φ. These in turn induce a Hodge star operator $\star_\varphi$ and dual 4-form $\psi = \star_\varphi \varphi$. ▲

Thus if φ is a G_2-structure on M, then at every point $p \in M$, there exists a basis $\{e_1, \ldots, e_7\}$ of $T_p M$ with respect to which $\varphi_p = \varphi_o$ from (4.1). Note that in general we *cannot* choose a *local frame* on an open set U in M with respect to which φ takes the standard form in (4.1), *we can only do this at a single point.* This is analogous to how, in a manifold (M^{2m}, g, J, ω) with $\mathrm{U}(m)$-structure, we can always find a basis of $T_p M$ for any $p \in M$ in which the "$\mathrm{U}(m)$ package" assumes the standard form on $\mathbb{C}^m$, but we cannot in general do this on an open set. (See [12] for a comprehensive treatment of $\mathrm{U}(m)$-structures.)

Not every smooth 7-manifold admits G_2-structures. A G_2-structure is equivalent to a reduction of the structure group of the frame bundle of M from $GL(7, \mathbb{R})$ to $G_2 \subset SO(7)$. As such, the existence of a G_2-structure is entirely a topological question.

Proposition 4.11 *A smooth 7-manifold M admits a G_2-structure if and only if M is both* orientable *and* spinnable. *This is equivalent to the vanishing of the first two Stiefel-Whitney classes $w_1(TM)$ and $w_2(TM)$.*

Proof See Lawson–Michelsohn [31, Chap. IV, Theorem 10.6] for a proof. □

Therefore, while not all smooth 7-manifolds admit G_2-structures, there are many that do and they are completely characterized by Proposition 4.11.

There is a much more concrete way to understand when a 3-form φ on M is a G_2-structure. It can be considered as a "working differential geometer's definition of G_2-structure", and is described as follows. Let $\varphi \in \Omega^3(M^7)$. Let $x^1, \ldots, x^7$ be local coordinates on an open set U in M. For $i, j \in \{1, \ldots, 7\}$, define a smooth function B_{ij} on U by

$$-6B_{ij}\, dx^1 \wedge \cdots \wedge dx^7 = \left(\frac{\partial}{\partial x^i} \lrcorner \varphi\right) \wedge \left(\frac{\partial}{\partial x^j} \lrcorner \varphi\right) \wedge \varphi. \tag{4.8}$$

Since 2-forms commute, we have $B_{ij} = B_{ji}$. In fact, comparison with (4.3) shows that if φ is a G_2-structure, we must have $B_{ij} = g_{ij}\sqrt{\det g}$, since $\mu = \sqrt{\det g}\, dx^1 \wedge \cdots \wedge dx^7$ is the Riemannian volume form in local coordinates. Hence $\det B = (\sqrt{\det g})^7 \det g = (\det g)^{\frac{9}{2}}$ and thus $\sqrt{\det g} = (\det B)^{\frac{1}{9}}$. Solving for g_{ij} gives

$$g_{ij} = \frac{1}{(\det B)^{\frac{1}{9}}} B_{ij}. \tag{4.9}$$

We say that $\varphi \in \Omega^3(M^7)$ is a G_2-structure if this recipe *actually works* to construct a Riemannian metric. Thus we must have both:

(i) $\det B$ must be nonzero everywhere on U,
(ii) g_{ij} as defined in (4.9) must be positive definite everywhere on U.

Of course, these two conditions must hold in any local coordinates $x^1, \ldots, x^7$ on M. But the advantage of this way of thinking about G_2-structures (besides it being very concrete) is that it allows us to easily see that the condition of φ being a G_2-structure is an *open condition*. That is, if φ is a G_2-structure, and $\tilde{\varphi}$ is another smooth 3-form on M sufficiently close to φ (in the C^0-norm with respect to any Riemannian metric on M) then $\tilde{\varphi}$ will also be a G_2-structure. This is because both conditions (i) and (ii) above are open conditions at each point p of M.

We conclude that, if the space of G_2-structures on M is *nonempty*, then it can be identified with a space Ω^3_+ of smooth sections of a fibre bundle $\Lambda^3_+(T^*M)$ whose fibres are open subsets of the corresponding fibres of the bundle $\Lambda^3(T^*M)$. The space Ω^3_+ is also called the space of *nondegenerate* or *positive* or *stable* 3-forms on M.

Remark 4.12 There is another way of seeing that the condition of being a G_2-structure is open. At any point $p \in M$, the space of all G_2-structures $\Lambda^3_+(T^*_p M)$ can be identified with the orbit of φ_o in $\Lambda^3(\mathbb{R}^7)^*$ by the action of $\mathrm{GL}(7, \mathbb{R})$ quotiented by the stabilizer subgroup of φ_o, which is G_2 by Theorem 4.2. Since $\dim \mathrm{GL}(7, \mathbb{R}) = 49$, and $\dim G_2 = 14$, we have $\dim \Lambda^3_+(T^*_p M) = 49 - 14 = 35 = \dim \Lambda^3(T^*_p M)$, and thus $\Lambda^3_+(T^*_p M)$ is an open set of $\Lambda^3(T^*_p M)$. See Hitchin [20] for a general discussion of *stable forms*. ▲

Remark 4.13 The nonlinear map $\varphi \to g$ is not one-to-one. In fact, given a metric g on M induced from a G_2-structure φ, at each point $p \in M$, the space of G_2-structures at p inducing g_p is diffeomorphic to $\mathbb{RP}^7$. Thus the G_2-structures inducing the same metric g correspond to sections of an $\mathbb{RP}^7$-bundle over M. See [6, p. 10, Remark 4] for more details on *isometric G_2-structures*. ▲

Let (M, φ) be a manifold with G_2-structure, and let g be the induced metric. Let $\psi = \star_\varphi \varphi$ denote the dual 4-form. The vital relation (3.39), which is equivalent to (3.43) leads to fundamental local coordinate identities relating φ, ψ, and g.

Theorem 4.14 *In local coordinates on M, the tensors φ, ψ, and g satisfy the following relations:*

$$\varphi_{ijk}\varphi_{abc}g^{kc} = g_{ia}g_{jb} - g_{ib}g_{ja} - \psi_{ijab}, \tag{4.10}$$

$$\varphi_{ijk}\varphi_{abc}g^{jb}g^{kc} = 6g_{ia}, \tag{4.11}$$

$$\varphi_{ijk}\psi_{abcd}g^{kd} = g_{ia}\varphi_{jbc} + g_{ib}\varphi_{ajc} + g_{ic}\varphi_{abj} - g_{aj}\varphi_{ibc} - g_{bj}\varphi_{aic} - g_{cj}\varphi_{abi}, \tag{4.12}$$

$$\varphi_{ijk}\psi_{abcd}g^{jc}g^{kd} = -4\varphi_{iab}, \tag{4.13}$$

$$\psi_{ijkl}\psi_{abcd}g^{kc}g^{ld} = 4g_{ia}g_{jb} - 4g_{ib}g_{ja} - 2\psi_{ijab}, \tag{4.14}$$

$$\psi_{ijkl}\psi_{abcd}g^{jb}g^{kc}g^{ld} = 24g_{ia}. \tag{4.15}$$

Proof These are derived from the relation (3.39) or equivalently (3.43). Indeed, the first identity (4.10) is precisely (3.43) expressed in local coordinates. The explicit details can be found in [26, Sect. A.3]. □

Of course, there are many other possible contractions of φ, ψ, and g. In Theorem 4.14 we only list those that show up most frequently in practice.

Remark 4.15 The identities for G_2-structures in Theorem 4.14 should be contrasted with the analogue for $\mathrm{U}(m)$-structures. First, we have only a single form ω, as opposed to the two forms φ and ψ. Moreover, from $\omega_{ab} = J_a^c g_{cb}$, which comes from (4.5), and the fact that $J^2 = -I$, we find that $\omega_{ia}\omega_{jb}g^{ab} = g_{ij}$. This is much simpler than (4.10) as the right hand side only involves the metric g. This again illustrates the "increased nonlinearity" of G_2 geometry, as mentioned in Remark 3.25 above. ▲

4.3 Decomposition of $\Omega^\bullet$ into Irreducible G_2 Representations

Let (M, φ) be a manifold with G_2-structure. The bundle $\Lambda^\bullet(T^*M) = \oplus_{k=1}^7 \Lambda^k(T^*M)$ decomposes into irreducible representations of G_2. This in turn induces a decomposition of the space $\Omega^k = \Gamma(\Lambda^k(T^*M))$ of smooth k-forms on M. This is entirely analogous to how, on a manifold with almost complex structure, the space $\Omega^\bullet_{\mathbb{C}} = \Gamma(\Lambda^\bullet(T^*M) \otimes \mathbb{C})$ of complex-valued forms decomposes into "forms of type (p, q)".

By Theorem 4.2, all the tensors determined by φ will be invariant under G_2 and hence any subspaces of Ω^k defined using φ, ψ, g, and $\star$ will be G_2 representations. The space Ω^k is *irreducible* if $k = 0, 1, 6, 7$. However, for $k = 2, 3, 4, 5$ we have a nontrivial decomposition. Since $\Omega^k = \star\Omega^{7-k}$, the decompositions of Ω^5 and Ω^4 are obtained by taking $\star$ of the decompositions of Ω^2 and Ω^3, respectively.

In fact we have

$$\begin{aligned}\Omega^2 &= \Omega^2_7 \oplus \Omega^2_{14},\\ \Omega^3 &= \Omega^3_1 \oplus \Omega^3_7 \oplus \Omega^3_{27},\end{aligned}$$

where Ω^k_l has (pointwise) dimension l and these decompositions are *orthogonal* with respect to g. These spaces are described invariantly as follows:

$$\begin{aligned}\Omega^2_7 &= \{X \lrcorner \varphi \mid X \in \Gamma(TM)\} = \{\beta \in \Omega^2 \mid \star(\varphi \wedge \beta) = -2\beta\},\\ \Omega^2_{14} &= \{\beta \in \Omega^2 \mid \beta \wedge \psi = 0\} = \{\beta \in \Omega^2 \mid \star(\varphi \wedge \beta) = \beta\},\end{aligned} \tag{4.16}$$

and

$$\begin{aligned}\Omega^3_1 &= \{f\varphi \mid f \in \Omega^0\}, \qquad \Omega^3_7 = \{X \lrcorner \psi \mid X \in \Gamma(TM)\},\\ \Omega^3_{27} &= \{\gamma \in \Omega^3 \mid \gamma \wedge \varphi = 0, \gamma \wedge \psi = 0\}.\end{aligned} \tag{4.17}$$

It is sometimes necessary to get our hands dirty, so we need to describe these subspaces in terms of local coordinates. Consider first the G_2-invariant linear map $P : \Omega^2 \to \Omega^2$ given by $P\beta = \star(\varphi \wedge \beta)$. If we write $\beta = \frac{1}{2}\beta_{ij} dx^i \wedge dx^j$ and $P\beta = \frac{1}{2}(P\beta)_{ab} dx^a dx^b$, then one can show [26, Sect. 2.2] that

$$(P\beta)_{ab} = \tfrac{1}{2}\psi_{abcd}\, g^{ci} g^{dj} \beta_{ij}. \tag{4.18}$$

That is, up to the factor of $\frac{1}{2}$, the map P is given by contracting the 2-form with the 4-form ψ on two indices. It is easy to check that P is self-adjoint and thus orthogonally diagonalizable with real eigenvalues. Using the fundamental identity (4.14) for the contraction of ψ with itself on two indices, we find

$$\begin{aligned}(P^2\beta)_{ab} &= \tfrac{1}{2}\psi_{abcd}\, g^{ci} g^{dj} (P\beta)_{ij} = \tfrac{1}{4}\psi_{abcd}\, g^{cl} g^{dj} \psi_{ijst}\, g^{sp} g^{tq} \beta_{pq}\\ &= \tfrac{1}{4}(4 g_{as} g_{bt} - 4 g_{at} g_{bs} - 2\psi_{abst}) g^{sp} g^{tq} \beta_{pq}\\ &= \beta_{ab} - \beta_{ba} - \tfrac{1}{2}\psi_{abst}\, g^{sp} g^{tq} \beta_{pq} = 2\beta_{ab} - (P\beta)_{ab}.\end{aligned}$$

Thus we deduce that $P^2 = 2I - P$, so $(P + 2I)(P - I) = 0$. Thus the eigenvalues of P are -2 and $+1$, in agreement with (4.16). To verify that $\lambda = -2$ corresponds to Ω^2_7 as given in (4.16), we let $\beta_{ij} = (X \lrcorner \varphi)_{ij} = X^m \varphi_{mij}$. Then using (4.13) we have

$$(P\beta)_{ab} = \tfrac{1}{2}\psi_{abcd} g^{ci} g^{dj} X^m \varphi_{mij} = -2X^m \varphi_{mab} = -2\beta_{ab},$$

as claimed. Also, the condition that $\Omega^2_{14} = (\Omega^2_7)^\perp$ gives that $\beta \in \Omega^2_{14}$ must satisfy $X^m \varphi_{mij} \beta_{ab} g^{ia} g^{jb} = 0$ for all X^m. This is equivalent to $\varphi_{mij} \beta_{ab} g^{ia} g^{jb} = 0$. Thus, we can describe the decomposition (4.16) of Ω^2 in local coordinates as

$$\begin{aligned} \beta_{ij} \in \Omega^2_7 &\iff \beta_{ij} = X^m \varphi_{mij} &&\iff \tfrac{1}{2}\psi_{abcd} g^{ci} g^{dj} \beta_{ij} = -2\beta_{ab}, \\ \beta_{ij} \in \Omega^2_{14} &\iff \beta_{ij} g^{ia} g^{jb} \varphi_{abc} = 0 &&\iff \tfrac{1}{2}\psi_{abcd} g^{ci} g^{dj} \beta_{ij} = \beta_{ab}. \end{aligned} \tag{4.19}$$

Moreover, it is easy to check using (4.11) that for $\beta \in \Omega^2_7$ we have

$$\beta_{ij} = X^m \varphi_{mij} \iff X^m = \tfrac{1}{6}\beta_{ab} g^{ai} g^{bj} \varphi_{ijk} g^{km}. \tag{4.20}$$

Remark 4.16 The description of the orthogonal splitting $\Omega^2 = \Omega^2_7 \oplus \Omega^2_{14}$ in terms of the $-2, +1$ eigenspaces of the operator $\beta \mapsto \star(\varphi \wedge \beta)$ is analogous to the orthogonal splitting $\Omega^2 = \Omega^2_+ \oplus \Omega^2_-$ into self-dual and anti-self-dual 2-forms on an oriented Riemannian 4-manifold with respect to the operator $\beta \mapsto \star\beta$. This analogy is important in G_2 gauge theory. ▲

Remark 4.17 Many authors prefer to use the opposite orientation than we do for the orientation induced by φ. (See [27] for more details.) This changes the sign of $\star$. The upshot is that the eigenvalues $(-2, +1)$ in (4.16) and (4.19) are replaced by $(+2, -1)$. Readers should take care to be aware of any particular paper's sign conventions. ▲

The local coordinate description of the decomposition (4.17) of Ω^3 can be understood by considering the infinitesimal action of the $(1, 1)$ tensors $\Gamma(T^*M \otimes TM)$ on φ. Let $A = A^i_l \in \Gamma(T^*M \otimes TM)$. At each point $p \in M$, we have $e^{At} \in \mathrm{GL}(T_pM)$, and thus

$$e^{At} \cdot \varphi = \tfrac{1}{6}\varphi_{ijk} (e^{At} dx^i) \wedge (e^{At} dx^j) \wedge (e^{At} dx^k). \tag{4.21}$$

Define $A \diamond \varphi \in \Omega^3$ by

$$(A \diamond \varphi) = \left.\frac{d}{dt}\right|_{t=0} (e^{At} \cdot \varphi). \tag{4.22}$$

From (4.21) we compute

$$(A \diamond \varphi) = \tfrac{1}{6}(A^l_i \varphi_{ljk} + A^l_j \varphi_{ilk} + A^l_k \varphi_{ijl}) dx^i \wedge dx^j \wedge dx^k,$$

and hence

$$(A \diamond \varphi)_{ijk} = A^l_i \varphi_{ljk} + A^l_j \varphi_{ilk} + A^l_k \varphi_{ijl}. \tag{4.23}$$

Use the metric g to identify $A \in \Gamma(T^*M \otimes TM)$ with a bilinear form $A \in \Gamma(T^*M \otimes T^*M)$ by $A_{ij} = A_i^l g_{lj}$. Recall from Sect. 1.2 that there is an orthogonal splitting

$$\Gamma(T^*M \otimes T^*M) \cong \Omega^0 \oplus \mathcal{S}_0 \oplus \Omega^2.$$

By the orthogonal decomposition (4.16) on Ω^2 discussed above, we can further decompose this as

$$\Gamma(T^*M \otimes T^*M) \cong \Omega^0 \oplus \mathcal{S}_0 \oplus \Omega^2_7 \oplus \Omega^2_{14}.$$

With respect to this splitting, we can write $A = \frac{1}{7}(\operatorname{Tr} A)g + A_0 + A_7 + A_{14}$, where A_0 is a traceless symmetric tensor.

By (4.23), we have a *linear map* $A \mapsto A \diamond \varphi$ from $\Omega^0 \oplus \mathcal{S}_0 \oplus \Omega^2_7 \oplus \Omega^2_{14}$ to the space Ω^3.

Proposition 4.18 *The kernel of $A \mapsto A \diamond \varphi$ is Ω^2_{14}, and the remaining three summands Ω^0, $\mathcal{S}_0$, Ω^2_7, of $\Gamma(T^*M \otimes T^*M)$ are mapped isomorphically onto Ω^3_1, Ω^3_{27}, Ω^3_7, respectively. Explicitly, if $A = \frac{1}{7}(\operatorname{Tr} A)g + A_0 + A_7 + A_{14}$, then*

$$A \diamond \varphi = \underbrace{\tfrac{3}{7}(\operatorname{Tr} A)\varphi}_{\Omega^3_1} + \underbrace{A_0 \diamond \varphi}_{\Omega^3_{27}} + \underbrace{X \lrcorner \psi}_{\Omega^3_7},$$

where

$$X^m = -\tfrac{1}{2} A_{ij} g^{ia} g^{jb} \varphi_{abc} g^{cm}.$$

Proof This can be established using the various contraction identities of Theorem 4.14. The explicit details can be found in [26, Sect. 2.2]. □

Remark 4.19 The fact that Ω^2_{14} is the kernel of $A \mapsto A \diamond \varphi$ is a consequence of the fact that G_2 is the Lie group that preserves φ. Thus the infinitesimal action, which is the action of the Lie algebra $\mathfrak{g}_2$, annihilates φ. This is consistent with the fact that $G_2 \subset SO(7)$, so $\mathfrak{g}_2 \subset \mathfrak{so}(7) \cong \Lambda^2(\mathbb{R}^7)^*$. Thus, at every point $p \in M$, the space $\Lambda^2_{14}(T^*_p M)$ is isomorphic to $\mathfrak{g}_2$. ▲

4.4 The Torsion of a G2-Structure

Let (M, φ) be a manifold with G_2-structure. Since φ determines a Riemannian metric φ, we get a Levi-Civita covariant derivative ∇. Thus it makes sense to consider the tensor $\nabla\varphi \in \Gamma(T^*M \otimes \Lambda^3 T^*M)$.

Definition 4.20 The G_2-structure φ is called *torsion-free* if $\nabla\varphi = 0$. Although this appears to be a linear equation, recall that ∇ is induced from g which itself depends nonlinearly on φ. Thus the equation $\nabla\varphi = 0$ is in fact a fully nonlinear first order partial differential equation for φ. We say (M, φ) is a *torsion-free* G_2 *manifold* if φ is a torsion-free G_2-structure on M. For brevity, we sometimes use the term "G_2 manifold" when we mean "torsion-free G_2 manifold". ▲

The fundamental observation about the torsion of any G_2-structure is the following.

Theorem 4.21 *Let X be a vector field on M. Then the 3-form $\nabla_X \varphi$ lies in the subspace Ω^3_7 of Ω^3. Thus the covariant derivative $\nabla \varphi$ is a smooth section of $T^*M \otimes \Lambda^3_7(T^*M)$.*

Proof By Proposition 4.18, any 3-form γ can be written as $\gamma = A \diamond \varphi$ for a unique $A = \frac{1}{7}(\operatorname{Tr} A)g + A_0 + A_7$. We take the inner product of $A \diamond \varphi$ with $\nabla_X \varphi$. Using (4.23), this is

$$\begin{aligned}
\langle A \diamond \varphi, \nabla_X \varphi \rangle &= \tfrac{1}{6}(A \diamond \varphi)_{ijk} (\nabla_X \varphi)_{abc} g^{ia} g^{jb} g^{kc} \\
&= \tfrac{1}{6}(A_i^l \varphi_{ljk} + A_j^l \varphi_{ilk} + A_k^l \varphi_{ijl}) X^m \nabla_m \varphi_{abc} g^{ia} g^{jb} g^{kc} \\
&= \tfrac{1}{2} A_i^l \varphi_{ljk} X^m \nabla_m \varphi_{abc} g^{ia} g^{jb} g^{kc} \\
&= \tfrac{1}{2} A_{ip} X^m \varphi_{qjk} \nabla_m \varphi_{abc} g^{pq} g^{ia} g^{jb} g^{kc}.
\end{aligned}$$

Taking the covariant derivative of (4.11) and using that g is parallel, we get

$$\nabla_m \varphi_{qjk} \varphi_{abc} g^{jb} g^{kc} = -\varphi_{qjk} \nabla_m \varphi_{abc} g^{jb} g^{kc}.$$

This says that $\nabla_m \varphi_{qjk} \varphi_{abc} g^{jb} g^{kc}$ is skew in q, a. Thus the symmetric part of A_{ip} does not contribute to $\langle A \diamond \varphi, \nabla_X \varphi \rangle$ above. That is, $\nabla_X \varphi$ is orthogonal to any element of $\Omega^3_1 \oplus \Omega^3_{27}$, as claimed. □

Theorem 4.21 motivates the following definition.

Definition 4.22 Because $\nabla_X \varphi \in \Omega^3_7$, by (4.17) we can write

$$\nabla_X \varphi = T(X) \lrcorner \psi$$

for some vector field $T(X)$ on M. That is, there exists a tensor $T \in \Gamma(T^*M \otimes T^*M)$ such that

$$\nabla_m \varphi_{ijk} = T_{mp} g^{pq} \psi_{qijk}. \tag{4.24}$$

We call T the *full torsion tensor* of φ. ▲

By contracting (4.24) with ψ_{nijk} on i, j, k and using (4.15), we obtains

$$T_{mn} = \tfrac{1}{24} \nabla_m \varphi_{ijk} \psi_{nabc} g^{ia} g^{jb} g^{kc}. \tag{4.25}$$

Moreover, taking the covariant derivative of (4.10) and using (4.24) and (4.12), one can compute that

$$\nabla_p \psi_{ijkl} = -T_{pi}\varphi_{jkl} + T_{pj}\varphi_{ikl} - T_{pk}\varphi_{ijl} + T_{pl}\varphi_{ijk}. \tag{4.26}$$

Observe that Eqs. (4.24) and (4.25) show that $\nabla\varphi = 0$ if and only if $T = 0$. (In this case (4.26) shows that $\nabla\psi = 0$ as well, which is also clear because $\psi = \star\varphi$ and ∇ commutes with $\star$.)

Hence φ is torsion-free if and only if $T = 0$. The tensor T is a more convenient measure of the failure of φ to be parallel, because we can easily decomposes it into four independent pieces in $\Gamma(T^*M \otimes T^*M) \cong \Omega^0 \oplus \mathcal{S}_0 \oplus \Omega^2_7 \oplus \Omega^2_{14}$, as

$$T = T_1 + T_{27} + T_7 + T_{14}, \tag{4.27}$$

where $T_1 = \frac{1}{7}(\mathrm{Tr}\, T)g$ and $T_{27} = T_0$ is the traceless symmetric part of T.

Corollary 4.23 *Let φ be a* G_2*-structure on M. Then φ is torsion-free if and only if both* $\mathrm{d}\varphi = 0$ *and* $\mathrm{d}\psi = 0$.

Proof Note that $\mathrm{d}\psi = \mathrm{d} \star \varphi = - \star \mathrm{d}^\star\varphi$, so $\mathrm{d}\psi = 0$ if and only if $\mathrm{d}^\star\varphi = 0$. Because both the exterior derivative d and its adjoint $\mathrm{d}^\star$ can be written in terms of ∇, any parallel form is always closed and coclosed. It is the converse that is nontrivial here. In fact, $\mathrm{d}\varphi$ and $\mathrm{d}^\star\varphi$ are *linear* in $\nabla\varphi$ and hence linear in T. Since $\mathrm{d}\varphi \in \Omega^4 = \Omega^4_1 \oplus \Omega^4_7 \oplus \Omega^4_{27}$ and $\mathrm{d}^\star\varphi \in \Omega^2 = \Omega^2_7 \oplus \Omega^2_{14}$, it follows by Schur's Lemma that the independent components of $\mathrm{d}\varphi$ and $\mathrm{d}^\star\varphi$ must correspond to the 1, 7, 14, 27 components of T as in (4.27), up to constant factors. Thus if $\mathrm{d}\varphi = 0$ and $\mathrm{d}^\star\varphi = 0$, we must have $T = 0$. □

Corollary 4.23 is a classical theorem of Fernàndez–Gray [14]. The present proof is an extremely abridged version of the argument in [26, Sect. 2.3].

Remark 4.24 Recall that a differential form γ on (M, g) is *harmonic* if $\Delta_{\mathrm{d}}\gamma = (\mathrm{d}\mathrm{d}^\star + \mathrm{d}^\star\mathrm{d})\gamma = 0$. On a compact manifold, by integration by parts harmonicity is equivalent to $\mathrm{d}\gamma = 0$ and $\mathrm{d}^\star\gamma = 0$. Thus Corollary 4.23 says that in the compact case, a G_2-structure φ is torsion-free if and only if it is harmonic *with respect to its induced metric.* ▲

Since the torsion T of φ decomposes into four independent components as in (4.27), each component can be zero or nonzero. This gives $2^4 = 16$ distinct classes of G_2-structures. Some of the more interesting classes of G_2-structures are given in the following table.

T_1	T_{27}	T_7	T_{14}	G_2-*structure*
0	0	0	0	$\nabla\varphi = 0$ (torsion-free)
0	0	0	$*$	$d\varphi = 0$ (closed)
$*$	$*$	0	0	$d^\star\varphi = 0$ (coclosed)
$*$	0	0	0	$d\varphi = \lambda\psi$ ($\lambda \neq 0$)

The last class in the table above is called *nearly parallel*, and one can show that λ is constant and that the induced metric is *positive Einstein*, with $R_{ij} = \frac{3}{8}\lambda^2 g_{ij}$. (For example, see [26, after Remark 4.19].)

More details on the 16 classes of G_2-structures can be found in [8, 14, 25, 26]. In particular it is worth remarking [26, Theorem 2.32] that with respect to conformal changes of G_2-structure, the component T_7 plays a very different role than the other three components T_1, T_{27}, T_{14}.

Aside. There is an equivalent approach to studying G_2-structures using *spin geometry*. Let (M^7, g) be a Riemannian 7-manifold equipped with a spin structure and associated *spinor bundle* $\$(M)$. This is a real rank 8 vector bundle over M. Since $8 > 7$, by algebraic topology, this bundle always admits nowhere vanishing sections. Such a section s determines a 3-form φ on M by $\varphi(a, b, c) = \langle a \cdot b \cdot c \cdot s, s\rangle$, where $\cdot$ denotes the Clifford multiplication of tangent vectors to M on spinors. Using the fact that s is nowhere zero, one can show that the 3-form φ is always a G_2-structure. Moreover, φ is torsion-free if and only if s is a *parallel spinor*, with respect to the spin connection on $\$(M)$ induced from the Levi-Civita connection of g. (The existence of a parallel spinor for torsion-free G_2 manifolds is precisely why they are of interest in theoretical physics, as this is related to *supersymmetry*.) Similarly, φ is nearly parallel in the sense defined above if and only if s is a *Killing spinor*. The reader is directed to Harvey [17], Lawson–Michelsohn [31, Chap. IV. 10], and the more recent paper by Agricola–Chiossi–Friedrich–Höll [2] for more on this point of view. This approach is also very important in the construction of *invariants* of G_2-structures, as discussed by Crowley–Goette–Nordström [11] in the present volume.

4.5 Relation Between Curvature and Torsion for a G_2-Structure

Let (M, φ) be a manifold with G_2-structure. Since φ determines a Riemannian metric g_φ, we have a Riemann curvature tensor R. There is an important relation between the tensors R and ∇T, called the "G_2 *Bianchi identity*" that originally appeared in [26, Theorem 4.2].

Theorem 4.25 *The* G_2-Bianchi identity *is the following:*

$$\nabla_i T_{jk} - \nabla_j T_{ik} = (T_{ip} T_{jq} + \tfrac{1}{2} R_{ijpq}) g^{pa} g^{qb} \varphi_{abk}. \tag{4.28}$$

Proof Equation (4.28) can be derived by combining the covariant derivative of (4.24) with (4.26) to get an expression for $\nabla_m \nabla_p \varphi_{ijk}$ in terms of φ, ψ, and T, and ∇T. Then applying the Ricci identity to the difference

$$\nabla_m \nabla_p \varphi_{ijk} - \nabla_p \nabla_m \varphi_{ijk}$$

introduces Riemann curvature terms. Simplifying further using the identities of Theorem 4.14 eventually results in (4.28). □

An important consequence of Theorem 4.25 is the following.

Corollary 4.26 *The Ricci curvature R_{jk} of the metric g induced by a G_2-structure φ can be expressed in terms of the torsion T and its covariant derivative ∇T as follows:*

$$\begin{aligned} R_{jk} = {} & (\nabla_i T_{jp} - \nabla_j T_{ip}) \varphi_{lqk} g^{pq} g^{il} - T_{jp} g^{pq} T_{qk} + (\operatorname{Tr} T) T_{jk} \\ & - T_{jl} T_{ab} g^{ap} g^{bq} \psi_{pqmk} g^{lm}. \end{aligned} \tag{4.29}$$

Proof Equation (4.29) can be obtained from (4.28) by combining the first Bianchi identity of Riemannian geometry together with the identities of Theorem 4.14. The details can be found in [26, Sect. 4.2]. □

Remark 4.27 Equation (4.29) shows that the metric of a torsion-free G_2-structure is always *Ricci-flat*. (See also item (vi) of Remark 5.6 below.) ▲

On a general Riemannian manifold (M^n, g), the Riemann curvature tensor R decomposes into the scalar curvature, the traceless Ricci curvature, and the conformally invariant *Weyl curvature*. When g is induced from a G_2-structure φ, the Weyl tensor W decomposes further intro three independent components W_{27}, W_{64}, and W_{77} as irreducible G_2-representations. A detailed discussion of the curvature decomposition of G_2-structures can be found in Cleyton–Ivanov [9] and in the forthcoming [13].

5 Exceptional Riemannian Holonomy

In this section we briefly review the notion of the *holonomy* of a Riemannian manifold (M, g), and place the geometry of torsion-free G_2-structures in this context, as one of the geometries with *exceptional Riemannian holonomy*.

5.1 Parallel Transport and Riemannian Holonomy

Let (M^n, g) be a Riemannian manifold, and let ∇ be the Levi-Civita connection of the metric g. We review without proof the well-known basic properties of Riemannian holonomy. See, for example, [23, Chaps. 2 & 3] for a more detailed discussion.

Definition 5.1 Fix $p \in M$. Let γ be loop based at p. This means that $\gamma : [0, 1] \to M$ is a continuous path, and piecewise smooth, such that $\gamma(0) = \gamma(1) = p$. Then, with respect to ∇, the *parallel transport* $\Pi_\gamma : T_{\gamma(0)}M \to T_{\gamma(1)}M$ around the loop γ is a linear isomorphism of T_pM with itself, which depends on γ. We define the *holonomy* of the metric g at the point p, denoted $\mathsf{Hol}_p(g)$, to be the set of all such isomorphisms. That is,

$$\mathsf{Hol}_p(g) = \{\Pi_\gamma : T_pM \cong T_pM : \gamma \text{ is a loop based at } p\}.$$

It follows from the existence and uniqueness of parallel transport (which itself is a consequence of existence and uniqueness for systems of first order linear ordinary differential equations) that $\Pi_{\gamma\cdot\beta} = \Pi_\gamma \circ \Pi_\beta$, where $\gamma \cdot \beta$ is the concatenation of paths, β followed by γ. Consequently, it is easy to see that $\mathsf{Hol}_p(g)$ is closed under multiplication and inversion. That is, $\mathsf{Hol}_p(g)$ is a *subgroup* of $\mathrm{GL}(T_pM)$.

If we instead consider the restricted class of *contractible* loops at p, which is closed under concatenation of paths, we obtain the *restricted holonomy* of g at p, denoted $\mathsf{Hol}^0_p(g)$. The group $\mathsf{Hol}^0_p(g)$ is a normal subgroup of $\mathsf{Hol}_p(g)$, and is the connected component of the identity. If M is simply-connected, then $\mathsf{Hol}^0_p(g) = \mathsf{Hol}_p(g)$ for all $p \in M$.

Because ∇ is the Levi-Civita connection, we have $\nabla g = 0$. Thus parallel transport with respect to ∇ preserves the inner product, and we conclude that in fact $\mathsf{Hol}_p(g)$ is a subgroup of $\mathrm{O}(T_pM, g_p)$, the group of isometries of the inner product space (T_pM, g_p). Similarly $\mathsf{Hol}^0_p(g)$ is a subgroup of $\mathrm{SO}(T_pM, g_p)$, the group of orientation-preserving isometries of (T_pM, g_p). ▲

The following proposition is straightforward to prove using the definitions.

Proposition 5.2 *The holonomy group $\mathsf{Hol}_p(g)$ satisfies the following properties.*

- *Let $p, q \in M$ lie in the same connected component of M. Then $\mathsf{Hol}_p(g) \cong \mathsf{Hol}_q(g)$. In fact, if γ is a piecewise smooth continuous path from p to q, and $P = \Pi_\gamma : T_pM \cong T_qM$ is the parallel transport isomorphism from T_pM to T_qM, then $\mathsf{Hol}_q(g) = P \cdot \mathsf{Hol}_p(g) \cdot P^{-1}$.*
- *Fix $p \in M$, and fix an isomorphism $T_pM \cong \mathbb{R}^n$. Then $\mathrm{GL}(T_pM) \cong \mathrm{GL}(n)$ and $\mathrm{O}(T_pM, g_p) \cong \mathrm{O}(n)$. With respect to this identification, $\mathsf{Hol}_p(g)$ corresponds to a subgroup $H \subseteq \mathrm{O}(n)$. If we choose any other isomorphism $T_pM \cong \mathbb{R}^n$, then the resulting subgroup $\tilde{H}$ of $\mathrm{O}(n)$ is in the same conjugacy class as H.*
- *Suppose M is connected. Then $\mathsf{Hol}_p(g) \cong \mathsf{Hol}_q(g)$ for all $p, q \in M$. Moreover, there exists a subgroup H of $\mathrm{O}(n)$ such that $\mathsf{Hol}_p(g) \cong H$ for all $p \in M$, and this subgroup H is unique up to conjugation.*

Analogous statements hold for the restricted holonomy group $\mathsf{Hol}^0_p(g)$, determining (when M is connected) a subgroup H^0 of $\mathrm{SO}(n)$, unique up to conjugation.

Consequently, if M is connected, we abuse notation and call H *the holonomy group* and H^0 *the restricted holonomy group* of (M, g). Observe that H and H^0 are not just abstract groups, but that they come naturally equipped with isomorphism classes of *representations on* T_pM for all $p \in M$.

Recall that a tensor S on M is called *parallel* if $\nabla S = 0$. There is a fundamental relationship between the holonomy group of g and the *parallel tensors* on M, given by the following.

Proposition 5.3 *Let (M, g) be a Riemannian manifold. Fix $p \in M$. Let $H \subseteq \mathrm{GL}(T_p M)$ be the subgroup that leaves invariant $S|_p$ for all parallel tensors S on M.*

- *We always have $\mathsf{Hol}_p(g) \subseteq H$. Moreover, these two subgroups are* usually equal. *For example, this is the case if $\mathsf{Hol}_p(g)$ is a closed subgroup of $\mathrm{GL}(T_p M)$.*
- *If the group H fixes an element S_0 in some tensor space of $T_p M$, then there exists a* parallel tensor *S on M such that $S|_p = S_0$.*

The way to think about Proposition 5.3 is as follows. The Riemannian holonomy H of a Riemannian manifold (M, g) is strictly smaller than $\mathrm{O}(n)$ if and only if there exist nontrivial parallel tensors on M other than the metric g.

Remark 5.4 If M is simply-connected, then $H = H^0$ and consequently $H \subseteq \mathrm{SO}(n)$. This means there exists a (necessarily parallel) Riemannian volume form $\mu \in \Omega^n(M)$ on M. This is consistent with the well-known fact from topology that any simply-connected manifold is orientable. ▲

5.2 *The Berger Classification of Riemannian Holonomy*

In 1955, Marcel Berger classified the possible Lie subgroups of $\mathrm{O}(n)$ that could occur as Riemannian holonomy groups of a metric g, subject to the following technical hypotheses.

- We restrict to *simply-connected* manifolds. In general if (M, g) is not simply-connected then the holonomy H of (M, g) is a finite cover of the reduced holonomy H^0. That is, the quotient H/H^0 is a discrete group.
- We must exclude the case when (M, g) is *locally reducible*. A locally reducible Riemannian manifold is *locally* a Riemannian product $(M_1, g_1) \times (M_2, g_2)$. In this case the Riemanian holonomy of (M, g) is a product of the holonomies of (M_1, g_1) and (M_2, g_2).
- We must exclude the case when (M, g) is *locally symmetric*. A locally symmetric Riemannian manifold is *locally* isometric to a symmetric space $(G/H, g)$ where G is a group of isometries acting transitively on G/H with isotropy group H at any point. In this case the Riemannian holonomy of (M, g) is H.

Theorem 5.5 (Berger classification) *Let (M, g) be a* simply-connected *smooth Riemannian manifold of dimension n that is* not locally reducible *and* not locally symmetric. *Then the Riemannian holonomy $H \subseteq \mathrm{SO}(n)$ can only be one of the following seven possibilities:*

$n = \dim M$	H	Parallel tensors	Name	Curvature
n	$\mathrm{SO}(n)$	g, μ	*orientable*	
$2m$ $(m \geq 2)$	$\mathrm{U}(m)$	g, ω	*Kähler*	
$2m$ $(m \geq 2)$	$\mathrm{SU}(m)$	g, ω, Ω	*Calabi–Yau*	*Ricci-flat*
$4m$ $(m \geq 2)$	$\mathrm{Sp}(m)$	$g, \omega_1, \omega_2, \omega_3, J_1, J_2, J_3$	*hyper-Kähler*	*Ricci-flat*
$4m$ $(m \geq 2)$	$(\mathrm{Sp}(m) \times \mathrm{Sp}(1))/\mathbb{Z}_2$	g, Υ	*quaternionic-Kähler*	*Einstein*
7	G_2	g, φ, ψ	G_2	*Ricci-flat*
8	$\mathrm{Spin}(7)$	g, Φ	$\mathrm{Spin}(7)$	*Ricci-flat*

Sketch of proof. Berger arrived at this classification by studying the *holonomy algebra* $\mathfrak{h}$ of the holonomy group H. There is an intimate relation between $\mathfrak{h}$ and the Riemann curvature operator $R \in \mathcal{S}^2(\mathfrak{so}(n))$ of g. First, because the Riemann curvature operator can be viewed as "infinitesimal holonomy", it must be that $R \in \mathcal{S}^2(\mathfrak{h})$. Since it also satisfies the first Bianchi identity, this says that $\mathfrak{h}$ cannot be too big. Second, by the Ambrose–Singer holonomy theorem, the span of the image of R at any point in M must generate $\mathfrak{h}$ as a vector space, so $\mathfrak{h}$ cannot be too small. Finally, for certain possible $\mathfrak{h}$, the fact that R must also satisfy the second Bianchi identity forces $\nabla R = 0$, in which case (M, g) is locally symmetric. Much more detailed discussion of this argument can be found in Joyce [23, Sect. 3.4]. □

Remark 5.6 We make some remarks concerning the above table.

(i) The four restrictions $m \geq 2$ in the first column are mostly to eliminate redundancy, as we have the isomorphisms $\mathrm{U}(1) \cong \mathrm{SO}(2)$, $\mathrm{Sp}(1) \cong \mathrm{SU}(2)$, and $(\mathrm{Sp}(1) \times \mathrm{Sp}(1))/\mathbb{Z}_2 \cong \mathrm{SO}(4)$. The case $\mathrm{SU}(1)$ does not occur because $\mathrm{SU}(1) \cong \{1\}$ is trivial and such a space is flat and thus symmetric.

(ii) Because $\mathrm{Sp}(k) \subseteq \mathrm{SU}(2k) \subseteq \mathrm{U}(2k)$, all hyper-Kähler manifolds are Calabi–Yau, and all Calabi–Yau manifolds are Kähler.

(iii) Note that quaternionic-Kähler manifolds are in fact *not* Kähler. This ill-advised nomenclature has unfortunately stuck and is here to stay.

(iv) Usually, the term *special holonomy* refers to any of the holonomy groups above other than the first two, perhaps because Kähler manifolds exist in sufficient abundance to not be that special.

(v) The last two groups above, namely G_2 and $\mathrm{Spin}(7)$, are called the *exceptional holonomy groups*. These Lie groups are both intimately related to the octonions $\mathbb{O}$. The connection between G_2 and $\mathbb{O}$ is explained in Sect. 4.1 above. The connection between $\mathrm{Spin}(7)$ and $\mathbb{O}$ can be found, for example, in Harvey [17, Lemma 14.61] or Harvey–Lawson [18, Sect. IV.1.C.].

(vi) The fact that metrics with special holonomy are all Einstein (including Ricci-flat) follows from consideration of the constraints on the Riemann curvature due to its relation with the holonomy algebra $\mathfrak{h}$, as explained in the sketch proof above. (See also Remark 4.27 above for the G_2 case.) ▲

It is interesting to note that Berger did not actually prove that all these groups *can actually occur* as Riemannian holonomy groups. He only excluded all other possibilities. It was widely suspected that the exceptional holonomies could not actually occur, only they could not be excluded using Berger's method. We now know, of course, that *all* of the possibilities in the above table do occur, both in compact and in complete noncompact examples. See Sect. 6.2 for a brief survey of this history in the case of G_2.

6 Torsion-Free G_2 Manifolds

In this section we discuss torsion-free G_2 manifolds, including a brief history of the search for irreducible examples, the known topological obstructions to existence in the compact case, and a comparison with Kähler and Calabi–Yau manifolds.

6.1 Irreducible and Reducible Torsion-Free G_2 Manifolds

Let (M, φ) be a torsion-free G_2 manifold. That is, φ is a torsion-free G_2-structure as in Definition 4.20, and thus by Proposition 5.3 the holonomy $\mathsf{Hol}(g_\varphi)$ of the induced Riemannian metric g_φ lies in G_2.

Definition 6.1 We say (M, φ) is an *irreducible* torsion-free G_2 manifold if $\mathsf{Hol}(g_\varphi) = G_2$. ▲

A torsion-free G_2 manifold could have *reduced holonomy*. That is, we could have $\mathsf{Hol}(g_\varphi) \subsetneq G_2$. In fact there are some simple constructions that yield such reducible examples:

- If g_φ is flat, then $\mathsf{Hol}(g_\varphi) = \{1\}$. In this case M is locally isomorphic to Euclidean $\mathbb{R}^7$ with the standard G_2-structure φ_o.
- Let L^4 be a manifold with holonomy $SU(2) \cong Sp(1)$. This is a hyper-Kähler 4-manifold with hyper-Kähler triple $\omega_1, \omega_2, \omega_3$. Let X^3 be a flat Riemannian 3-manifold with global orthonormal parallel coframe e^1, e^2, e^3. Let $M^7 = X^3 \times L^4$, and define a smooth 3-form φ on M by

$$\varphi = e^1 \wedge \omega_1 + e^2 \wedge \omega_2 + e^3 \wedge \omega_3 - e^1 \wedge e^2 \wedge e^3.$$

  Then φ is a torsion-free G_2-structure with $\mathsf{Hol}(g_\varphi) = SU(2) \subsetneq G_2$. In this case we have

$$\psi = e^2 \wedge e^3 \wedge \omega_1 + e^3 \wedge e^1 \wedge \omega_2 + e^1 \wedge e^2 \wedge \omega_3 - \mathsf{vol}_L$$

  where $\mathsf{vol}_L = \frac{1}{2}\omega_1^2 = \frac{1}{2}\omega_2^2 = \frac{1}{2}\omega_3^2$ is the volume form of L.
- Let L^6 be a manifold with holonomy $SU(3)$. This is a Calabi–Yau complex 3-fold with Kähler form ω and holomorphic volume form Ω. Let X^1 be a Riemannian 1-manifold with global unit parallel 1-form e^1. Let $M^7 = X^1 \times L^6$, and define a smooth 3-form φ on M by

$$\varphi = e^1 \wedge \omega - \operatorname{Re} \Omega.$$

Then φ is a torsion-free G_2-structure with $\mathsf{Hol}(g_\varphi) = \mathrm{SU}(3) \subsetneq \mathrm{G}_2$. In this case we have

$$\psi = \tfrac{1}{2}\omega^2 + e^1 \wedge \operatorname{Im} \Omega.$$

Remark 6.2 If (M, φ) is a torsion-free G_2 manifold, then some criteria are known to determine if (M, φ) is irreducible. Here are two examples:

(i) If M is *compact* with $\mathsf{Hol}(g_\varphi) \subseteq \mathrm{G}_2$, then $\mathsf{Hol}(g_\varphi) = \mathrm{G}_2$ if and only if the fundamental group $\pi_1(M)$ is *finite*. (See Joyce [23, Proposition 10.2.2].)
(ii) If M is connected and *simply-connected*, with $\mathsf{Hol}(g_\varphi) \subseteq \mathrm{G}_2$, then $\mathsf{Hol}(g_\varphi) = \mathrm{G}_2$ if and only if there are *no nonzero parallel* 1*-forms*. (See Bryant–Salamon [5, Theorem 2].) ▲

6.2 A Brief History of Irreducible Torsion-Free G_2 Manifolds

The search for examples of *irreducible* torsion-free G_2 manifolds (that is, Riemannian metrics with holonomy exactly G_2) has a long history. As explained in Sect. 5.2, it was originally believed such metrics could not exist. In this section we give a very brief and far from exhaustive survey of some of this history.

The first local (that is, incomplete) examples were found by Bryant [4] in 1987, using methods of exterior differential systems and Cartan-Kähler theory.

Then in 1989, Bryant–Salamon [5] found the first *complete noncompact* examples of G_2 holonomy metrics. These were metrics on the *total spaces of vector bundles*. Explicitly, these metrics were found on the bundles $\Lambda^2_-(S^4)$ and $\Lambda^2_-(\mathbb{CP}^2)$, which are rank 3 bundles over 4-dimensional bases, and on the bundle $\$(S^3)$, the spinor bundle of S^3, which is a rank 4 bundle over a 3-dimensional base. These Riemannian manifold are all *asymptotically conical*. That is, the metrics approach Riemannian cone metrics at some particular rate at infinity. These torsion-free G_2-structures are *cohomogeneity one*. That is, there is a Lie group of symmetries acting on (M, φ) with generic orbits of codimension one. Such symmetry reduces the partial differential equation $\nabla\varphi = 0$ to a (fully nonlinear) system of *ordinary differential equations*, which can be explicitly solved. The fact that the metrics have holonomy exactly G_2 was verified by using the criterion in item (ii) of Remark 6.2.

Remark 6.3 Since then, several explicit examples and a great many nonexplicit examples of complete noncompact holonomy G_2 metrics have been discovered, with various prescribed asymptotic geometry at infinity, such as asymptotically conical (AC), asymptotically locally conical (ALC), and others. In fact, very recent work of Foscolo–Haskins–Nordström [15, 16] has produced a spectacular new plethora of such examples. ▲

The first construction of *compact* irreducible torsion-free G_2 manifolds was given by Joyce [22] in 1994, and pushed further in the monograph [23]. The idea is the following. Start with the flat 7-torus T^7, and take the quotient by a discrete group of isometries preserving the G_2-structure φ_o. The quotient is a singular *orbifold* with torsion-free G_2-structure. Joyce then resolved the singularities by gluing in (quasi)-asymptotically locally Euclidean spaces with SU(2) or SU(3) holonomy, to produce a smooth compact 7-manifold M with *closed* G_2-structure and "small" torsion. He then used analysis (see Sect. 7.1 below) to prove that M admits a torsion-free G_2-structure. Finally, he showed the metrics had holonomy exactly G_2 by using the criterion (i) of Remark 6.2. This first construction is explained in more detail by Kovalev [30] in the present volume.

The second construction of *compact* irreducible torsion-free G_2 manifolds was introduced by Kovalev [29] in 2001 and pushed significantly further by Corti–Haskins–Nordström–Pacini [10] in 2015. It is called the "twisted connect sum construction". The ideas is the following. Start with two noncompact *asymptotically cylindrical* Calabi–Yau complex 3-folds L_1 and L_2, which are both asymptotic to $X^4 \times T^2$ where X^4 is a K3 complex surface. Take $L_1 \times S^1$ and $L_2 \times S^1$ and glue them together with a "twist" by identifying different factors of S^1 in order to obtain a smooth compact 7-manifold. The goal is then to construct a closed G_2-structure on M with "small" torsion that can be perturbed using analysis to a torsion-free G_2-structure (see Sect. 7.1 below). Being able to do this is a very delicate problem in algebraic geometry involving "matching data". This second construction is also explained in more detail by Kovalev [30] in the present volume.

More recently, a third construction of *compact* irreducible torsion-free G_2 manifolds appeared in Joyce–Karigiannis [24], involving glueing 3-dimensional families of Eguchi-Hanson spaces. This construction differs from the previous two because some of the noncompact "pieces" that are being glued together this time *do not* come equipped with torsion-free G_2-structures. This is dealt with by solving a linear elliptic PDE on the noncompact Eguchi-Hanson space using weighted Sobolev spaces.

All three of the currently known constructions of compact irreducible torsion-free G_2 manifolds are similar in that they all use *glueing techniques* to construct a closed G_2-structure φ with "small" torsion, and then invoke a general existence theorem of Joyce to prove that it can be *perturbed* to a nearby torsion-free G_2-structure $\widetilde{\varphi}$. This existence theorem is the subject of Sect. 7.1 below.

Thus, we know that Riemannian metrics with holonomy exactly G_2 *do exist* on compact manifolds, but they are not explicit. This is analogous to the case of Riemannian metrics with holonomy exactly $SU(m)$ (also called Calabi–Yau metrics) on compact manifolds. By Yau's proof of the Calabi conjecture, we know that many such metrics exist, but we cannot describe them explicitly. In fact, special holonomy metrics on compact manifolds should in some sense be thought of as "transcendental" objects.

So far we have only found G_2-holonomy metrics that are "close to the edge of the moduli space". That is, these metrics are close to either developing singularities or tearing apart into two disjoint noncompact pieces. That is, the three known constructions of compact irreducible torsion-free G_2 manifolds are very likely producing only a very small part of the "landscape" of holonomy G_2 metrics.

6.3 Cohomological Obstructions to Existence in the Compact Case

There are several known *cohomological obstructions* to the existence of torsion-free G_2-structures on a compact manifold. We describe some of these in this section. Let (M, φ) be a *compact* manifold with a *torsion-free* G_2-structure φ. Let g_φ be the Riemannian metric induced by φ. Thus $\mathsf{Hol}(g_\varphi) \subseteq \mathrm{G}_2$. Since (M, g_φ) is a compact oriented Riemannian manifold, the Hodge Theorem applies. That is, any deRham cohomology class has a unique harmonic representative.

Since φ is torsion-free, by Corollary 4.23, the form φ is closed and coclosed and thus harmonic. Because $\varphi \neq 0$, we deduce from the Hodge Theorem that $[\varphi]$ is a nontrivial class in $H^3(M, \mathbb{R})$. Hence we find our first cohomological obstruction:

$$b^3 \geq 1 \qquad \text{if } M \text{ admits a torsion-free } \mathrm{G}_2\text{-structure.}$$

where $b^k = \dim H^k(M, \mathbb{R})$ is the kth Betti number of M. The same argument applies to ψ, so $b^4 \geq 1$, but $b^4 = b^3$ by Poincaré duality, so this is not new information.

Suppose $\mathsf{Hol}(g_\varphi) = \mathrm{G}_2$. Then by item (i) of Remark 6.2 we must have $\pi_1(M)$ is finite. It follows from algebraic topology that $H^1(M, \mathbb{R}) = \{0\}$. Hence we find our second cohomological obstruction:

$$b^1 = 0 \qquad \text{if } M \text{ admits an \textit{irreducible} torsion-free } \mathrm{G}_2\text{-structure.} \tag{6.1}$$

Before we can discuss the two other cohomological obstructions, we need to explain the interaction of the representation-theoretic decompositions of Sect. 4.3 with the Hodge Theorem.

Because φ is torsion-free, one can show that the Hodge Laplacian Δ_{d} *commutes* with the orthogonal projection operators onto the irreducible summands of the decomposition of $\Omega^\bullet$ described in Sect. 4.3. (See Joyce [23, Theorem 3.5.3] for details.) Combining this fact with the Hodge Theorem, we conclude that the decompositions of Sect. 4.3 *descend to deRham cohomology*. That is, if we define

$$\mathcal{H}^k_l = \{\gamma \in \Omega^k_l \mid \Delta_{\mathrm{d}} \gamma = 0\}$$

to be the space of *harmonic* Ω^k_l*-forms*, and $\mathcal{H}^k$ to be the space of harmonic k-forms, then we have

$$\begin{aligned} \mathcal{H}^2 &= \mathcal{H}^2_7 \oplus \mathcal{H}^2_{14}, \\ \mathcal{H}^3 &= \mathcal{H}^3_1 \oplus \mathcal{H}^3_7 \oplus \mathcal{H}^3_{27}. \end{aligned} \tag{6.2}$$

Moreover, it follows from the explicit descriptions of Ω^k_l in Sect. 4.3 and the fact that Δ_{d} commutes with the projections and with the Hodge star $\star$ that

$$\Delta_{\mathrm{d}}(f\varphi) = (\Delta_{\mathrm{d}} f)\varphi, \qquad \Delta_{\mathrm{d}}(\alpha \wedge \varphi) = (\Delta_{\mathrm{d}} \alpha) \wedge \varphi,$$

for all $f \in \Omega^0$ and all $\alpha \in \Omega^1$. These identities imply that

$$\begin{aligned} &\mathcal{H}_1^0 \cong \mathcal{H}_1^3 \cong \mathcal{H}_1^4 \cong \mathcal{H}_1^7, \qquad && \mathcal{H}_7^1 \cong \mathcal{H}_7^2 \cong \mathcal{H}_7^3 \cong \mathcal{H}_7^4 \cong \mathcal{H}_7^5 \cong \mathcal{H}_7^6, \\ &\mathcal{H}_{14}^2 \cong \mathcal{H}_{14}^5, && \mathcal{H}_{27}^3 \cong \mathcal{H}_{27}^4. \end{aligned}$$

Define $b_l^k = \dim \mathcal{H}_l^k$ to be the "refined Betti numbers" of (M, φ). Then we have shown that

$$b_l^k = b_{l'}^{k'} \quad \text{if } l = l'.$$

In particular $b_7^k = b_7^1 = b^1$ for $k = 2, \ldots, 6$, and $b_1^k = b_1^0 = b^0$ for $k = 3, 4, 7$. We deduce that

$$\begin{aligned} b^2 &= b_7^2 + b_{14}^2 = b^1 + b_{14}^2, \\ b^3 &= b_1^3 + b_7^3 + b_{27}^3 = b^0 + b^1 + b_{27}^3. \end{aligned}$$

(Note that if M is connected then $b^0 = 1$, and if in addition φ is irreducible then by (6.1) we get $b^2 = b_{14}^2$ and $b^3 = 1 + b_{27}^3$.)

There exists a real quadratic form Q on $H^2(M, \mathbb{R})$ defined as follows. Let $[\beta] \in H^2(M, \mathbb{R})$. Define

$$Q([\beta]) = \int_M \beta \wedge \beta \wedge \varphi. \tag{6.3}$$

(In fact, it is easy to see using Stokes's Theorem that Q is well-defined as long as $d\varphi = 0$, and that Q depends only on the *cohomology class* $[\varphi] \in H^3(M, \mathbb{R})$. We do not need torsion-free to define Q.)

But now suppose that φ is not only torsion-free, but also irreducible. Then by (6.1) and the discussion above, we have $\mathcal{H}^2 = \mathcal{H}_{14}^2$. Given a cohomology class $[\beta] \in H^2(M, \mathbb{R})$, the Hodge theorem gives us a unique harmonic representative β_H, which must lie in Ω_{14}^2. By (4.16), we have $\beta_H \wedge \varphi = *\beta_H$, and hence

$$Q([\beta]) = \int_M \beta_H \wedge \beta_H \wedge \varphi = \int_M \beta_H \wedge *\beta_H = \int_M \|\beta_H\|^2 \mathsf{vol} \geq 0$$

with equality if and only if $\beta_H = 0$, which is equivalent to $[\beta] = 0$. Hence we find our third cohomological obstruction:

- Let φ be a *closed* G_2-structure on a compact manifold M with $\pi_1(M)$ finite. (So that any torsion-free G_2-structure on M must necessarily be irreducible.) If there exists a torsion-free G_2-structure in the cohomology class $[\varphi] \in H^3(M, \mathbb{R})$, then the quadratic form Q defined in (6.3) must be *positive definite*.

Remark 6.4 Recall Remark 4.17. If the other orientation is used, then Q must be *negative definite*. So we can unambiguously state the third cohomological obstruction as saying that Q must be *definite*. Moreover, if φ is merely torsion-free but not irreducible, it is easy to see from (4.16) that (with our convention for the orientation), the *signature* of Q is $(b^2 - b^1, b^1)$. ▲

Finally, recall from Chern–Weil theory that a compact 7-manifold M has a *real first Pontryagin class* $p_1(TM) \in H^4(M, \mathbb{R})$, defined as the cohomology class represented by the closed 4-form $\frac{1}{8\pi^2} \operatorname{Tr}(R \wedge R)$ where $R \in \Gamma(\operatorname{End}(TM) \otimes \Lambda^2 T^*M)$ is the curvature form of *any* connection on TM. If φ is torsion-free, then g_φ has holonomy contained in G_2, and hence, because Riemann curvature is "infinitesimal holonomy" we have that in fact $R \in \Gamma(\operatorname{End}(TM) \otimes \Lambda^2_{14} T^*M)$. That is, the 2-form part of R lies in Ω^2_{14}. But then by (4.16) we have

$$\operatorname{Tr}(R \wedge R) \wedge \varphi = \operatorname{Tr}(R \wedge *R) = |R|^2 \mathsf{vol},$$

and thus

$$(p_1(TM) \cup [\varphi]) \cdot [M] = \frac{1}{8\pi^2} \int_M \operatorname{Tr}(R \wedge R) \wedge \varphi = \frac{1}{8\pi^2} \int_M |R|^2 \mathsf{vol},$$

where $[M] \in H_7(M)$ is the fundamental class of M and $\cdot$ denotes the canonical pairing between $H^7(M, \mathbb{R})$ and $H_7(M)$. This is clearly positive unless R is identically zero. Hence we find our fourth cohomological obstruction:

$$p_1(TM) \neq 0 \qquad \text{if } M \text{ admits a } \textit{nonflat} \text{ torsion-free } \mathrm{G}_2\text{-structure.}$$

6.4 Comparison with Kähler and Calabi–Yau Manifolds

In this section we make some remarks about the similarities and the differences between torsion-free G_2 manifolds and Kähler manifolds in general and Calabi–Yau manifolds in particular. A good reference for Kähler and Calabi–Yau geometry is Huybrechts [21].

Manifolds with $\mathrm{U}(m)$-structure are in some ways analogous to manifolds with G_2-structure, as detailed in the following table.

	$\mathrm{U}(m)$-structure on (M^{2m}, g)	G_2-structure on (M^7, g)
Nondegenerate form	$\omega \in \Omega^2$	$\varphi \in \Omega^3$
Vector cross product	$J \in \Gamma(TM \otimes T^*M)$	$\times \in \Gamma(TM \otimes \Lambda^2 T^*M)$
Fundamental relation	$\omega(u, v) = g(Ju, v)$	$\varphi(u, v, w) = g(u \times v, w)$

One very important difference between $\mathrm{U}(m)$-structures and G_2-structures was already mentioned in Remark 4.6, but it is so crucial that it is worth repeating here. For a $\mathrm{U}(m)$-structure, the metric g and the nondegenerate 2-form ω are essentially independent, subject only to mild compatibility conditions, and together they determine J. In contrast, for a G_2-structure the nondegenerate 3-form φ *determines* the metric g and consequently the cross product $\times$ as well.

Now consider the *torsion-free* cases of such structures. A $\mathrm{U}(m)$-structure is torison-free if $\nabla\omega = 0$. Such manifolds are called *Kähler* and have Riemannian holonomy contained in the Lie subgroup $\mathrm{U}(m)$ of $\mathrm{SO}(2m)$. A G_2-structure is torsion-free if $\nabla\varphi = 0$. Such manifolds have Riemannian holonomy contained in the Lie subgroup G_2 of $\mathrm{SO}(7)$. In the torsion-free cases, both ω and φ are *calibrations*. (See [34, 35] in the present volume for more about calibrations.) Both Kähler manifolds and torsion-free G_2 manifolds also admit special *connections* on vector bundles, namely the Hermitian–Yang–Mills connections and the G_2-instantons, respectively.

Here is where we see another very important difference. As we saw in Remark 4.27, the metric g_φ of a torsion-free G_2-structure is always Ricci-flat. But the metric g of a Kähler manifold need *not* be Ricci-flat. In fact, the Calabi–Yau Theorem, gives a topological characterization (in the compact case) of exactly which Kähler metrics are Ricci flat. They are precisely those metrics with holonomy contained in the Lie subgroup $\mathrm{SU}(m)$ of $\mathrm{U}(m)$. The precise statement of the Calabi–Yau theorem is as follows.

Theorem 6.5 *Let M be a* compact *Kähler manifold, with Kähler form ω. Then there exists a* Ricci-flat *Kähler metric $\widetilde{\omega}$ in the same cohomology class as ω if and only if $c_1(TM) = 0$, where $c_1(TM)$ is the* first Chern class *of TM. Moreover, when it exists the Ricci-flat Kähler metric is unique.*

We are very far from having an analogous theorem in G_2 geometry. In fact, we do not even have any idea of what the correct conjecture might be. The main tool that allowed Yau to reformulate the Calabi conjecture into a statement about existence and uniqueness of solutions to a complex Monge–Ampère equation is the $\partial\bar{\partial}$-lemma in Kähler geometry. There is no close analogue of this result for torsion-free G_2 manifolds.

Heuristically, the Calabi–Yau Theorem allows us to go from $\mathrm{U}(m)$ holonomy to $\mathrm{SU}(m)$ holonomy, which is a reduction in the dimension of the holonomy group from m^2 to $m^2 - 1$, a difference of 1 dimension, and it corresponds to an (albeit fully nonlinear) *scalar* partial differential equation. In contrast, to obtain a Riemannian metric with holonomy G_2, we must start with $\mathrm{SO}(7)$ holonomy. Thus we need to reduce the dimension of the holonomy group from 21 to 14, so we expect a system of 7 *equations*, or equivalently a single partial differential equation for an unknown *1-form* rather than for an unknown function as in the Calabi–Yau Theorem. Precisely how such a heuristic discussion can be made into a precise mathematical conjecture remains a mystery at present.

In fact, a better analogy is the following. Let M^{2m} be a compact manifold that admits $\mathrm{U}(m)$-structures. What are *necessary and sufficient* topological conditions that ensure that M^{2m} admits a Kähler structure? We know many necessary conditions. (See Huybrechts [21], for example.) But we are very far from knowing sufficient conditions.

7 Three Theorems About Compact Torsion-Free G_2-Manifolds

In this final section we briefly discuss three important theorems about compact torsion-free G_2 manifolds: an existence theorem of Joyce, the smoothness of the moduli space (also due to Joyce), and a variational characterization of compact torsion-free G_2 manifolds due to Hitchin. Only the main ideas of the proofs are sketched. We refer the reader to the original sources for the details.

7.1 *An Existence Theorem for Compact Torsion-Free* G_2 *Manifolds*

In Sect. 6.2 we discussed known constructions of compact irreducible torsion-free G_2 manifolds. These constructions invoke the only analytic existence theorem that is know for torsion-free G_2-structures, which is a result of Joyce that originally appeared in [22] but which can also be found in [23, Sect. 11.6]. As mentioned in Sect. 6.2, the idea is that if one has a *closed* G_2-structure φ on M whose torsion is sufficiently small, the theorem guarantees the existence of a "nearby" torsion-free G_2-structure $\widetilde{\varphi}$ that is in the same cohomology class as φ. The statement of the theorem that we give here is a slightly modified version given in [24, Theorem 2.7].

Theorem 7.1 (Existence Theorem of Joyce) *Let α, K_1, K_2, and K_3 be any positive constants. Then there exist $\varepsilon \in (0, 1]$ and $K_4 > 0$, such that whenever $0 < t \leq \varepsilon$, the following holds.*

Let (M, φ) be a compact *7-manifold with G_2-structure φ satisfying $\mathrm{d}\varphi = 0$. Suppose there exists a* closed *4-form η such that*

(i) $\| \star_{\varphi}\varphi - \eta\|_{C^0} \leq K_1\, t^{\alpha}$,
(ii) $\| \star_{\varphi}\varphi - \eta\|_{L^2} \leq K_1\, t^{\frac{7}{2}+\alpha}$,
(iii) $\|\mathrm{d}(\star_{\varphi}\varphi - \eta)\|_{L^{14}} \leq K_1\, t^{-\frac{1}{2}+\alpha}$,
(iv) the injectivity radius inj *of g_{φ} satisfies* $\mathrm{inj} \geq K_2\, t$,
(v) the Riemann curvature *tensor* Rm *of g_{ph} satisfies* $\|\mathrm{Rm}\|_{C^0} \leq K_3\, t^{-2}$.

Then there exists a smooth torsion-free G_2-*structure $\widetilde{\varphi}$ on M such that $\|\widetilde{\varphi} - \varphi\|_{C^0} \leq K_4\, t^{\alpha}$ and $[\widetilde{\varphi}] = [\varphi]$ in $H^3(M, \mathbb{R})$. Here all norms are computed using the original metric g_{φ}.*

We make some remarks about the conditions *(i)–(iii)* of the theorem. Since φ is closed, it would be torsion-free if and only if $\star_{\varphi}\varphi$ were also closed. The hypotheses *(i)–(iii)* above say that $\star_{\varphi}\varphi$ is *almost closed*, in that there exists a *closed* 4-form η that is close to $\star_{\varphi}\varphi$ in various norms, namely the C^0, L^2, and (essentially) the $W^{14,1}$ norms.

The idea of the proof of Theorem 7.1 is as follows. Since we want $\widetilde{\varphi}$ is to be in the same cohomology class as φ, we must have $\widetilde{\varphi} = \varphi + \mathrm{d}\sigma$ for some $\sigma \in \Omega^2$, and by Hodge theory we can assume that $\mathrm{d}^*\sigma = 0$. Joyce shows that the torsion-free condition

$$\mathrm{d}\big(\star_{\varphi+\mathrm{d}\sigma}(\varphi + \mathrm{d}\sigma)\big) = 0$$

can be rewritten as

$$\Delta_{\mathrm{d}}\sigma = \mathcal{Q}(\sigma, \mathrm{d}\sigma) \tag{7.1}$$

where $\mathcal{Q}$ is some nonlinear expression that is at least order two in $\mathrm{d}\sigma$. Joyce shows that the above equation can be solved by iteration. Explicitly, taking $\sigma_0 = 0$, then for each $k \geq 1$, Joyce solves the series of *linear* equations

$$\Delta_{\mathrm{d}}\sigma_k = \mathcal{Q}(\sigma_{k-1}, \mathrm{d}\sigma_{k-1}).$$

Using the *a priori estimates (i)–(iii)*, Joyce then shows that $\lim_{k\to\infty}\sigma_k$ exists as a smooth 2-form satisfying (7.1). This is essentially a "fixed-point theorem" type of argument. The complete details can be found in [23, Sect. 11.6].

7.2 *The Moduli Space of Compact Torsion-Free G2-Structures*

Whenever one studies a certain type of structure in mathematics, it is natural to consider the "set of all possible such structures", modulo a reasonable notion of equivalence. Usually this "moduli space" of structures has its own special structure, and an understanding of the special structure on the moduli space sometimes yields information about the original object on which such structures are defined.

In our setting, consider a *compact torsion-free* G_2 *manifold* (M, φ). We want to consider *the set of all possible torsion-free* G_2*-structures* on the same underlying smooth 7-manifold M, modulo a reasonable notion of equivalence. The usual notion of equivalence in differential geometry is *diffeomorphism*. Indeed, if φ is a torsion-free G_2-structure on M and $F : M \to M$ is a diffeormorphism, then it is easy to see that $F^*\varphi$ is also a torsion-free G_2-structure on M, with metric $g_{F^*\varphi} = F^*g_\varphi$.

In fact, it is more convenient to consider only those diffeomorphisms of M that are *isotopic to the identity*. That is, those diffeomorphisms that are connected to the identity map on M by a continuous path in the space Diff of diffeomorphisms of M. This is the *connected component of the identity* in Diff, and we denote it by Diff_0. The reason we prefer the space Diff_0 is because *it acts trivially on cohomology*. That is, suppose $[\alpha] \in H^k(M, \mathbb{R})$ and let $F \in \mathrm{Diff}_0$. Then we claim that $[F^*\alpha] = [\alpha]$. To see this, let F_t be a continuous path in Diff with $F_0 = \mathrm{Id}_M$ and $F_1 = F$, given by the flow of the vector field X_t on M. Since α is a closed form, we have

$$\begin{aligned} F^*\alpha - \alpha &= \int_0^1 \frac{d}{dt}(F_t^*\alpha) = \int_0^1 \mathcal{L}_{X_t}\alpha \\ &= \int_0^1 (\mathrm{d}X_t \lrcorner \alpha + X_t \lrcorner \mathrm{d}\alpha) = \int_0^1 \mathrm{d}X_t \lrcorner \alpha \\ &= \mathrm{d}\Big(\int_0^1 X_t \lrcorner \alpha\Big), \end{aligned}$$

and thus $F^*\alpha - \alpha$ is exact.

Definition 7.2 Let (M, φ_0) be a compact torsion-free G_2 manifold. Let $\mathcal{T}$ be the set of all torsion-free G_2-structures on M. That is,

$$\mathcal{T} = \{\varphi \in \Omega^3_+ \mid \mathrm{d}\varphi = 0, \mathrm{d} \star_\varphi \varphi = 0\}.$$

The *moduli space* $\mathcal{M}$ of torsion-free G_2-structures on M is defined to be the quotient topological space

$$\mathcal{M} = \mathcal{T}/\mathrm{Diff}_0$$

of $\mathcal{T}$ by the action of Diff_0. ▲

Remark 7.3 The space $\mathcal{M}$ in Definition 7.2 should probably more properly be called the *Teichmüller space*, and then the "moduli space" would be the quotient $\mathcal{T}/\mathrm{Diff}$ by the full diffeomorphism group, in analogy with the usage of terminology for Riemann surfaces. However, the nomenclature we have given in Definition 7.2 is standard in the field of G_2 geometry. ▲

The first important result that was established about the moduli space was the following theorem of Joyce, that originally appeared in [22] but which can also be found in [23, Sect. 10.4].

Theorem 7.4 (Moduli Space Theorem of Joyce) *Let M be a compact 7-manifold with torsion-free G_2-structure φ_0. The moduli space $\mathcal{M}$ of torsion-free G_2-structures on M is a* smooth manifold *of dimension $b^3 = \dim H^3(M, \mathbb{R})$. In fact, the "period map" $\mathcal{P} : \mathcal{M} \to H^3(M, \mathbb{R})$ that takes an equivalence class $[\varphi]_{\mathcal{M}}$ in the quotient space $\mathcal{M} = \mathcal{T}/\mathrm{Diff}_0$ to the deRham cohomology class $[\varphi]$ is a* local diffeomorphism.

The idea of the proof of Theorem 7.4 is as follows. Joyce constructs a "slice" $\mathcal{S}_\varphi$ for the action of Diff_0 on $\mathcal{T}$ in a neighbourhood of any $\varphi \in \mathcal{T}$. A slice $\mathcal{S}_\varphi$ is a submanifold of $\mathcal{T}$ containing φ that is locally transverse to the orbits of Diff_0 near φ. This means that all nearby orbits of Diff_0 each intersect $\mathcal{T}$ at only one point. Then $\mathcal{M} = \mathcal{T}/\mathrm{Diff}_0$ is locally homeomorphic in a neighbourhood of $[\varphi]_{\mathcal{M}} \in \mathcal{M}$ to $\mathcal{S}_\varphi$. Since $\varphi \in \mathcal{T}$ is arbitrary, we deduce that $\mathcal{M}$ is a smooth manifold of dimension $\dim \mathcal{S}$.

In fact a slice $\mathcal{S}_\varphi$ is given by

$$\mathcal{S}_\varphi = \{\widetilde{\varphi} \in \Omega^3_+ \mid \mathrm{d}\widetilde{\varphi} = 0, \mathrm{d} \star_{\widetilde{\varphi}} \widetilde{\varphi} = 0, \pi_7(\mathrm{d}^* \widetilde{\varphi}) = 0\}, \tag{7.2}$$

where π_7 is the orthogonal projection $\pi_7 : \Omega^2 \to \Omega^2_7$ with respect to the G_2-structure φ. The way to understand where the above $\mathcal{S}_\varphi$ comes from is to consider tangent vectors to the orbit of Diff_0 at φ. Such a tangent vector is of the form

$$\frac{d}{dt}\bigg|_{t=0} h_t^* \varphi = \mathcal{L}_X \varphi = \mathrm{d}(X \lrcorner \varphi)$$

where h_t is the flow of a smooth vector field X on M. By the description (4.16), the tangent space at φ of the orbit of Diff_0 is thus the space $\mathrm{d}(\Omega^2_7)$. It thus makes sense to define

$$\mathcal{S}_\varphi = \{\widetilde{\varphi} \in \mathcal{T} \mid \langle \widetilde{\varphi} - \varphi, \mathrm{d}(X \lrcorner \varphi)\rangle_{L^2} = 0 \, \forall X \in \Gamma(TM)\}, \tag{7.3}$$

because for $\widetilde{\varphi}$ close to φ, the condition of L^2-orthogonality to the tangent spaces of the orbit of Diff_0 through φ would ensure local transversality. Since φ is torsion-free, we have $\mathrm{d}^* \varphi = 0$. Thus integration by parts shows that (7.3) is equivalent to (7.2).

It still remains to explain why $\mathcal{S}_\varphi$ is a smooth manifold of dimension b^3. Given $\widetilde{\varphi} \in \mathcal{T}$, by Hodge theory with respect to g_φ we can write $\widetilde{\varphi} = \varphi + \xi + \mathrm{d}\eta$ for some $\xi \in \mathcal{H}^3$ and some $\eta \in \mathrm{d}^*(\Omega^3)$. For $\widetilde{\varphi}$ sufficiently close to φ in the C^0 norm, Joyce shows that

$$\widetilde{\varphi} \in \mathcal{S}_\varphi \quad \Longleftrightarrow \quad \Delta_{\mathrm{d}} \eta = \star \mathrm{d}\big(\mathcal{Q}(\xi, \mathrm{d}\eta)\big) \tag{7.4}$$

where $\mathcal{Q}$ is a nonlinear expression that is at least order two in $\mathrm{d}\eta$. This is a fully nonlinear elliptic equation for η given any $\xi \in \mathcal{H}^3$. Using the Banach Space Implicit Function Theorem, Joyce shows that the space of solutions (ξ, η) to the right hand side of (7.4) is a smooth manifold of dimension b^3. The complete details can be found in [23, Sect. 10.4].

Remark 7.5 A consequence of the fact from Theorem 7.4 that the period map $\mathcal{P} : \mathcal{M} \to H^3(M, \mathbb{R})$ is a local diffeomorphism is the following. The manifold $\mathcal{M}$ has a natural *affine structure*, that is a covering by coordinate charts whose transition functions are affine maps. In Karigiannis–Leung [28] this affine structure is exploited to study special structures on $\mathcal{M}$, including a natural *Hessian metric* and a *symmetric cubic form* called the "Yukawa coupling". This Hessian metric is obtained from the *Hitchin volume functional* defined in Sect. 7.3 below. ▲

We know *very little* about the *global structure* of $\mathcal{M}$. (But see the survey article by Crowley–Goette–Nordström [11] in the present volume for some recent progress on the (dis-)connectedness of $\mathcal{M}$.

7.3 A Variational Characterization of Torsion-Free G_2-Structures

It is the case that some natural geometric structures can be given a *variational interpretation*. That is, they can be characterized as critical points of a certain natural geometric functional, which means that they are solutions to the associated Euler-Lagrange equations for this functional. Some examples of such geometric structures and their associated functionals are:

- minimal submanifolds (the volume functional),
- harmonic maps (the energy functional),
- Einstein metrics (the Einstein–Hilbert functional),
- Yang–Mills connections (the Yang–Mills functional).

It was an important observation of Hitchin [19] that torsion-free G_2-structures on compact manifolds can be given a variational interpretation. The setup is as follows. Let M^7 be compact, and as usual, let Ω^3_+ be the set of G_2-structures on M. Given $\varphi \in \Omega^3_+$, we get a metric g_φ, a Riemannian volume vol_φ, and a dual 4-form $\star_\varphi \varphi$.

Definition 7.6 The *Hitchin functional* is defined to be the map $\mathcal{F} : \Omega^3_+ \to \mathbb{R}$ given by

$$\mathcal{F}(\varphi) = \int_M \varphi \wedge \star_\varphi \varphi = 7 \int_M \mathsf{vol}_\varphi = 7\,\mathrm{Vol}(M, g_\varphi), \tag{7.5}$$

where we have used the fact that $|\varphi|^2_{g_\varphi} = 7$ from (4.2). Thus, up to a positive factor, $\mathcal{F}(\varphi)$ is the total volume of M with respect to the metric g_φ. ▲

Hitchin's observation was to restrict $\mathcal{F}$ to a *cohomology class* containing a closed G_2-structure. That is, suppose φ_0 is a *closed* G_2-structure on M, and let

$$\mathcal{C}_\varphi = \Omega^3_+ \cap [\varphi] = \{\widetilde{\varphi} \in \Omega^3_+ \mid \mathrm{d}\widetilde{\varphi} = 0, [\widetilde{\varphi}] = [\varphi] \in H^3(M, \mathbb{R})\}.$$

In [19], Hitchin proved the following.

Theorem 7.7 (Hitchin's variational characterization) *Let φ be a closed G_2-structure on M, and consider the* restriction *of $\mathcal{F}$ to the set $\mathcal{C}_\varphi$ defined above. Then φ is a* critical point *of $\mathcal{F}|_{\mathcal{C}_\varphi}$ if and only if φ is torsion-free. That is,*

$$\left.\frac{d}{dt}\right|_{t=0} \mathcal{F}(\varphi + t\mathrm{d}\eta) = 0 \iff \mathrm{d} \star_\varphi \varphi = 0.$$

Moreover, at a critical point φ, the second variation *of $\mathcal{F}|_{\mathcal{C}_\varphi}$ is nonpositive. This means that critical points are* local maxima.

The proof of Theorem 7.7 is quite straightforward given the following observation, which is quite useful itself in many other applications. Let $\varphi(t)$ be a smooth family of G_2-structures with $\left.\frac{d}{dt}\right|_{t=0} \varphi(t) = \gamma$. Then

$$\left.\frac{d}{dt}\right|_{t=0} \star_{\varphi(t)} \varphi(t) = \frac{4}{3} \star \pi_1 \gamma + \star \pi_7 \gamma - \star \pi_{27} \gamma, \tag{7.6}$$

where the orthogonal projections $\pi_k : \Omega^3 \to \Omega^3_k$ and the Hodge star $\star$ are all taken with respect to $\varphi(0)$. Two different proofs of (7.6) can be found in [19] and in [26, Remark 3.6].

The interesting observation in Theorem 7.7 that torsion-free G_2-structures are *local maxima* of $\mathcal{F}$ restricted to a cohomology class motivates the idea to try to *flow* to a torsion-free G_2-structure by taking the appropriate *gradient flow* of $\mathcal{F}$. This yields the *Laplacian flow* of closed G_2-structures. See the article by Lotay [36] in the present volume for a discussion of geometric flows of G_2-structures, including the Laplacian flow.

Acknowledgements The author would like to acknowledge Jason Lotay and Naichung Conan Leung for useful discussions on the structuring of these lecture notes. The initial preparation of these notes was done while the author held a Fields Research Fellowship at the Fields Institute. The final preparation of these notes was done while the author was a visiting scholar at the Center of Mathematical Sciences and Applications at Harvard University. The author thanks both the Fields Institute and the CMSA for their hospitality.

References

1. Agricola, I. (2008). Old and new on the exceptional group G_2. *Notices of the American Mathematical Society*, *55*, 922–929. MR2441524.
2. Agricola, I., Chiossi, S., Friedrich, T., & Höll, J. (2015). Spinorial description of SU(3)- and G_2-manifolds. *Journal of Geometry and Physics*, *98*, 535–555. MR3414976.
3. Brown, R. B., & Gray, A. (1967). Vector cross products. *Commentarii Mathematici Helvetici*, *42*, 222–236. MR0222105.
4. Bryant, R. L. (1987). Metrics with exceptional holonomy. *Annals of Mathematics*, *2*(126), 525–576. MR0916718.
5. Bryant, R. L., & Salamon, S. M. (1989). On the construction of some complete metrics with exceptional holonomy. *Duke Mathematical Journal*, *58*, 829–850. MR1016448.
6. Bryant, R. L. (2005). Some remarks on G_2-structures. *Proceedings of Gökova Geometry-Topology Conference 2005* (pp. 75–109). MR2282011. arXiv:math/0305124
7. Cheng, D. R., Karigiannis, S., & Madnick, J. (2019). Bubble tree convergence of conformally cross product preserving maps. *Asian Journal of Mathematics* (to appear). arXiv:1909.03512
8. Chiossi, S., Salamon, S. (2001). The intrinsic torsion of SU(3) and G_2 structures. In *Differential geometry, Valencia, 2001* (pp. 115–133). River Edge: World Sci. Publ. MR1922042.
9. Cleyton, R., & Ivanov, S. (2008). Curvature decomposition of G_2-manifolds. *Journal of Geometry and Physics*, *58*(2008), 1429–1449. MR2453675.
10. Corti, A., Haskins, M., Nordström, J., & Pacini, T. (2015). G_2-manifolds and associative submanifolds via semi-Fano 3-folds. *Duke Mathematical Journal*, *164*, 1971–2092. MR3369307.
11. Crowley, D., Goette, S., & Nordström, J. Distinguishing G_2-manifolds. *Lectures and surveys on G_2-manifolds and related topics*. Fields institute communications. Berlin: Springer. (The present volume).
12. de la Ossa, X., Karigiannis, S., & Svanes, E. Geometry of $U(m)$-structures: Kähler identities, the dd^c lemma, and Hodge theory. (In preparation).
13. Dwivedi, S., Gianniotis, P., & Karigiannis, S. Flows of G_2-structures, II: Curvature, torsion, symbols, and functionals. (In preparation).
14. Fernández, M., & Gray, A. (1982). Riemannian manifolds with structure group G_2. *Annali di Matematica Pura ed Applicata*, *4*(132), 19–45. MR0696037.
15. Foscolo, L., Haskins, M., & Nordström, J. Complete non-compact G_2-manifolds from asymptotically conical Calabi-Yau 3-folds. arXiv:1709.04904.
16. Foscolo, L., Haskins, M., & Nordström, J. Infinitely many new families of complete cohomogeneity one G_2-manifolds: G_2 analogues of the Taub-NUT and Eguchi-Hanson spaces. arXiv:1805.02612.

17. Harvey, R. (1990). *Spinors and calibrations*. Perspectives in Mathematics (Vol. 9). Boston: Academic Press Inc. MR1045637.
18. Harvey, R., & Lawson, H. B. (1982). Calibrated geometries. *Acta Mathematica*, *148*, 47–157. MR0666108.
19. Hitchin, N. The geometry of three-forms in six and seven dimensions. arXiv:math/0010054.
20. Hitchin, N. Stable forms and special metrics. In *Global differential geometry: the mathematical legacy of Alfred Gray (Bilbao, 2000)*, 70–89, *Contemp. Math.* **288**, Amer. Math. Soc., Providence, RI. MR1871001
21. D. Huybrechts, *Complex geometry*, Universitext, Springer-Verlag, Berlin, 2005. MR2093043
22. D.D. Joyce, "Compact Riemannian 7-manifolds with holonomy G_2. I, II", *J. Differential Geom.* **43** (1996), 291–328, 329–375. MR1424428
23. D.D. Joyce, *Compact manifolds with special holonomy*, Oxford Mathematical Monographs, Oxford University Press, Oxford, 2000. MR1787733
24. D. Joyce and S. Karigiannis, "A new construction of compact torsion-free G_2 manifolds by gluing families of Eguchi-Hanson spaces", *J. Differential Geom.*, to appear. https://arxiv.org/abs/1707.09325
25. Karigiannis, S. (2005). Deformations of G_2 and Spin(7) structures on manifolds. *Canadian Journal of Mathematics*, *57*(2005), 1012–1055. MR2164593.
26. Karigiannis, S. (2009). Flows of G_2-structures, I. *Quarterly Journal of Mathematics*, *60*, 487–522. MR2559631. arXiv:math/0702077.
27. Karigiannis, S. (2010). Some notes on G_2 and Spin(7) geometry. *Recent advances in geometric analysis*. Advanced lectures in mathematics (Vol. 11, pp. 129–146). Vienna: International Press. arXiv:math/0608618.
28. Karigiannis, S., & Leung, N. C. (2009). Hodge theory for G_2-manifolds: Intermediate Jacobians and Abel-Jacobi maps. *Proceedings of the London Mathematical Society (3)*, *99*, 297–325. MR2533667.
29. Kovalev, A. (2003). Twisted connected sums and special Riemannian holonomy. *Journal Für Die Reine und Angewandte Mathematik*, *565*, 125–160. MR2024648.
30. Kovalev, A. Constructions of compact G_2-holonomy manifolds. *Lectures and surveys on* G_2-*manifolds and related topics*. Fields institute communications. Berlin: Springer. (The present volume).
31. Lawson, H. B., & Michelsohn, M.-L. (1989). *Spin geometry*. Princeton Mathematical Series (Vol. 38). Princeton: Princeton University Press. MR1031992.
32. Leung, N. C. (2002). Riemannian geometry over different normed division algebras. *Journal of Differential Geometry*, *61*, 289–333. MR1972148.
33. Lee, J.-H., & Leung, N. C. (2008). Instantons and branes in manifolds with vector cross products. *Asian Journal of Mathematics*, *12*, 121–143. MR2415016.
34. Chan, K. F., & Leung, N. C. Calibrated submanifolds in G_2 geometry. *Lectures and surveys on* G_2-*manifolds and related topics*. Fields institute communications. Berlin: Springer. (The present volume).
35. Lotay, J. D. Calibrated submanifolds. *Lectures and surveys on* G_2-*manifolds and related topics*. Fields institute communications. Berlin: Springer. (The present volume).
36. Lotay, J. D. Geometric flows of G_2 structures. *Lectures and surveys on* G_2-*manifolds and related topics*. Fields institute communications. Berlin: Springer. (The present volume).
37. Massey, W. S. (1961). Obstructions to the existence of almost complex structures. *Bulletin of the American Mathematical Society*, *67*, 559–564. MR0133137.
38. Milnor, J. W., & Stasheff, J. D. (1974). *Characteristic classes*. Princeton: Princeton University Press. MR0440554.
39. Salamon, S. (1989). *Riemannian geometry and holonomy groups*. Pitman research notes in mathematics series (Vol. 201). Harlow: Longman Scientific & Technical. MR1004008.
40. Salamon, D. A., & Walpuski, T. Notes on the octonions. In *Proceedings of the Gökova Geometry-Topology Conference 2016* (pp. 1–85). Gökova Geometry/Topology Conference (GGT), Gökova. MR3676083.

Constructions of Compact G_2-Holonomy Manifolds

Alexei Kovalev

Abstract We explain the constructions for two geometrically different classes of examples of compact Riemannian 7-manifolds with holonomy G_2. One method uses resolutions of singularities of appropriately chosen 7-dimensional orbifolds, with the help of asymptotically locally Euclidean spaces. Another method uses the gluing of two asymptotically cylindrical pieces and requires a certain matching condition for their cross-sections 'at infinity'.

1 Introduction

The Lie group G_2 occurs as an exceptional case in Berger's classification of the Riemannian holonomy groups, in dimension 7. Riemannian manifolds with holonomy G_2 are Ricci-flat and admit parallel spinor fields. The purpose of these notes is to give an introduction to two methods of producing examples of compact Riemannian 7-manifolds with holonomy group G_2.

For a detailed introduction to G_2-structures on 7-manifolds and the G_2 holonomy group we refer to [25, Chap. 11], [14, Chap. 10] and the article by Karigiannis in this volume. Here we briefly recall the foundational results that we need.

The Lie group G_2 may be defined as the stabilizer, in the action of $GL(7, \mathbb{R})$, of the 3-form [4, p. 539]

$$\varphi_0 = dx^{123} + dx^{145} + dx^{167} + dx^{246} - dx^{257} - dx^{347} - dx^{356} \in \Lambda^3(\mathbb{R}^7)^*, \quad (1)$$

where x^k are the standard coordinates on $\mathbb{R}^7$ and $dx^{klm} = dx^k \wedge dx^l \wedge dx^m$. Every linear isomorphism of $\mathbb{R}^7$ preserving φ_0 also preserves the Euclidean metric

A. Kovalev (✉)
DPMMS, University of Cambridge, Cambridge, UK
e-mail: a.g.kovalev@dpmms.cam.ac.uk

S. Karigiannis et al. (eds.), *Lectures and Surveys on G_2-Manifolds and Related Topics*, Fields Institute Communications 84, https://doi.org/10.1007/978-1-0716-0577-6_2

$\sum_{i=1}^{7}(dx^i)^2$ and orientation of $\mathbb{R}^7$, thus G_2 is a subgroup of $SO(7)$. The $GL(7,\mathbb{R})$-orbit of φ_0 is open in $\Lambda^3(\mathbb{R}^7)^*$.

Let M be a 7-dimensional manifold. Then every G_2-structure on M is induced by a choice of a smooth differential 3-form φ such that for each $p \in M$ there is a linear isomorphism $\iota_p : \mathbb{R}^7 \to T_pM$ with $\iota_p^*(\varphi(p)) = \varphi_0$. We say a 3-form φ is *positive* when φ satisfies the latter condition and denote by $\Omega^3_+(M) \subset \Omega^3(M)$ the subset of all positive 3-forms on M. Note that for a compact M the subset $\Omega^3_+(M)$ is open in the uniform norm topology. We shall sometimes, slightly informally, say that a differential form $\varphi \in \Omega^3_+(M)$ *is* a G_2-structure on M.

We can see from the above that every G_2-structure $\varphi \in \Omega^3_+(M)$ determines on M a metric $g(\varphi)$ and an orientation, hence also the Hodge star $*_\varphi$.

Theorem 1 (cf. [9]) *Let M be a 7-manifold endowed with a G_2-structure $\varphi \in \Omega^3_+(M)$. Then the following are equivalent.*

(a) The holonomy of the metric $g(\varphi)$ is contained in G_2.
(b) $\nabla\varphi = 0$, where ∇ is the Levi–Civita connection of $g(\varphi)$.
(c)

$$d\varphi = 0, \qquad d*_\varphi\varphi = 0, \tag{2}$$

(d) The intrinsic torsion of the G_2-structure φ vanishes.

Note that the second equation in (2) is non-linear because $*_\varphi$ depends non-linearly on φ.

We say that (M, φ) is a *G_2-manifold* if φ is a positive 3-form satisfying (2). If, in addition, the holonomy of $g(\varphi)$ is all of G_2, then we shall call (M, φ) an *irreducible G_2-manifold.*

Proposition 2 ([14, Proposition 10.2.2]) *A compact G_2-manifold is irreducible if and only if $\pi_1(M)$ finite.*

A key idea in the known methods of constructing irreducible G_2-manifolds is that one first achieves on M a G_2-structure φ which is, in some sense, an 'approximate' solution of (2) with $d\varphi = 0$ and $d*_\varphi\varphi$ having a small norm, in a suitable Banach space. In more geometric terms, the G_2-structure φ then has small torsion. Then one uses perturbative analysis to obtain a correction term $d\eta$, for a 2-form η small in the C^1 norm, so that $\varphi + d\eta$ is a valid G_2-structure and a solution of (2).

We shall explain methods of finding the desired approximate solutions of (2) by building compact Riemannian manifolds from 'simpler pieces'. These will be non-compact or singular G_2-manifolds whose metrics are flat or have holonomy $SU(2)$ or $SU(3)$, which are subgroups of G_2. These latter metrics can be obtained by using the Calabi–Yau analysis or written explicitly. The manifolds are patched together in a 'compatible' way to achieve, on the resulting compact manifolds, G_2-structures with arbitrarily small torsion.

More precisely, one obtains 1-dimensional families of metrics depending on a certain 'gluing parameter' taking values in a semi-closed interval. The limits of these families may be interpreted as boundary points in a 'partial compactification' of the

G_2 moduli space. (It is known that the moduli space of torsion-free G_2-structures on a compact 7-manifold M is a smooth manifold of dimension the third Betti number $b^3(M)$.)

In these notes, we shall explain two ways of implementing the above strategy with different respective limits in the boundary of the G_2 moduli space.

Recently, Joyce and Karigiannis [16] developed a new method of constructing holonomy G_2 manifolds using analysis on families of Eguchi–Hanson spaces. This construction is not reviewed here. It includes an application of perturbative methods for G_2-structures with small torsion but also requires significant additional methods to achieve a suitable small torsion.

2 Construction by Resolutions of Singularities

The method explained in this section was historically the first construction of compact 7-manifolds with holonomy G_2. It is due to Joyce [13, 14].

Joyce's method produces one-parameter families of holonomy G_2 metrics g_s, $0 < s \leq \varepsilon$. The limits of these families as $s \to 0$ can be interpreted as boundary points in the G_2-moduli space and are given by flat orbifolds. In particular, the limit spaces are *compact, connected* and *singular*.

More precisely, the construction proceeds via the following steps.

1. (a) Let $T^7 = \mathbb{R}^7/\mathbb{Z}^7$ be the 7-torus with a flat G_2-structure $\varphi_0 \in \Omega^3_+(T^7)$ induced from the standard G_2-structure (1) on the Euclidean $\mathbb{R}^7$. Choose a finite group Γ of affine transformations of $\mathbb{R}^7$ which preserve φ_0 and descend to diffeomorphisms of T^7. The quotient space (T^7/Γ) is an orbifold with a torsion-free G_2-structure, still denoted by φ_0, and a flat orbifold metric g_0 induced by φ_0.
   (b) For suitable choices of Γ, all the singularities of T^7/Γ are locally modeled on $\mathbb{R}^3 \times (\mathbb{C}^2/G)$ or $\mathbb{R} \times (\mathbb{C}^3/G)$, for G a finite subgroup of respectively $SU(2)$ or $SU(3)$, and can be resolved using methods of complex algebraic geometry. Perform the resolutions to obtain a smooth compact 7-manifold M together with a resolution map $\pi : M \to T^7/\Gamma$.
2. (a) On M, one can 'explicitly' define a 1-parameter family of closed positive 3-forms $\varphi_s \in \Omega^3_+(M)$, with $d\varphi_s = 0$ for $0 < s \leq \varepsilon$, such that the G_2-structures φ_s have small torsion. The forms φ_s converge as $s \to 0$ to $\pi^*\varphi_0$ (respectively, the induced metrics $g(\varphi_s)$ converge to $\pi^* g_0$). One may also say that the Riemannian manifolds $(M, g(\varphi_s))$ converge in the Gromov–Hausdorff sense to the flat orbifold $(T^7/\Gamma, g_0)$ as $s \to 0$.
   (b) Apply perturbative analysis (more precisely, construct a convergent sequence of iterations) to show that for every small $s > 0$, the G_2-structure φ_s can be deformed to a nearby torsion-free G_2-structure $\tilde{\varphi}_s$. If $\pi_1(M)$ is finite, then the holonomy of the induced metric $\tilde{g}_s = g(\tilde{\varphi}_s)$ is precisely the group G_2, i.e. $(M, \tilde{\varphi}_s)$ is an irreducible G_2-manifold.

We illustrate this method with an example taken from [14, Sect. 12.2] (cf. also [13]) where some technical details are relatively simple. Consider the group Γ generated by

$$\begin{aligned}
\alpha &: (x_1,\dots,x_7) \mapsto (\ \ x_1,\ \ x_2,\ \ x_3, -x_4,\ -x_5,\ -x_6,\ -x_7),\\
\beta &: (x_1,\dots,x_7) \mapsto (\ \ x_1, -x_2, -x_3,\ \ x_4,\ \ x_5, \tfrac{1}{2}-x_6,\ -x_7),\\
\gamma &: (x_1,\dots,x_7) \mapsto (-x_1,\ \ x_2, -x_3,\ \ x_4, \tfrac{1}{2}-x_5,\ \ x_6, \tfrac{1}{2}-x_7).
\end{aligned}$$

The maps α, β, γ commute and each has order 2, thus Γ is isomorphic to $\mathbb{Z}_2^3$. The elements of Γ descend to T^7 and preserve φ_0, making the quotient T^7/Γ into a G_2-orbifold.

One can further check that the only elements of Γ having fixed points are α, β, γ, each fixes 16 copies of T^3 and these are all disjoint. The subgroup generated by β, γ acts freely on the 16 tori fixed by α, so these correspond to 4 copies of T^3 in the singular locus of T^7/Γ. Similar properties hold for the tori fixed by β and by γ. Thus the singular locus S of T^7/Γ is 12 disjoint copies of T^3. A neighbourhood of each 3-torus component of S is diffeomorphic to $T^3 \times (\mathbb{C}^2/\{\pm 1\})$.

The blow-up $\sigma : Y \to \mathbb{C}^2/\{\pm 1\}$ at the origin resolves the singularity with a complex surface Y biholomorphic to $T^*\mathbb{C}P^1$, with the exceptional divisor $E = \sigma^{-1}(0) \cong \mathbb{C}P^1$ corresponding to the zero section of $T^*\mathbb{C}P^1$. The canonical bundle of Y is trivial and there is a family of Ricci-flat Kähler metrics h_s on Y with holonomy equal to $SU(2)$ depending on a real parameter $s > 0$. The Kähler form of the metric h_s may be written as $\omega_s = \sigma^*(i\partial\bar{\partial} f_s)$, where

$$f_s = \sqrt{r^4 + s^4} + 2s^2 \log r - s^2 \log(\sqrt{r^4 + s^4} + s^2),$$

$r^2 = z_1\bar{z}_1 + z_2\bar{z}_2$ and $(z_1, z_2) \in \mathbb{C}^2$. The radius function r makes sense as a smooth function on $Y \setminus E$ and the values of this function near E can be interpreted as the distance to E in the metric h_s. The forms ω_s extend smoothly over the exceptional divisor $E \subset Y$, thus the metrics h_s are well-defined on Y. These are the well-known Eguchi–Hanson metrics [8].

Comparing, for each $s > 0$, the Kähler potential f_s of h_s with the Kähler potential r^2 of the Euclidean metric h_0 on $\mathbb{C}^2$ we see that

$$\nabla^k (h_s - h_0) = O(r^{-4-k}) \quad \text{as } r \to \infty, \quad \text{for all } k = 0, 1, 2, \dots , \tag{3}$$

which means that h_s is an asymptotically locally Euclidean (ALE) metric on Y.

For each $\lambda > 0$, the dilation map $Y \to Y$ induced by $(z_1, z_2) \mapsto \lambda(z_1, z_2)$ pulls back ω_s to $\lambda^2\omega_{\lambda s}$. It follows that s is proportional to the diameter of the exceptional divisor. One can further check that the injectivity radius of the Eguchi–Hanson metric h_s is proportional to s and that the uniform norm of the Riemannian curvature is proportional to s^{-2}.

Every Ricci-flat Kähler metric h on a complex surface is in fact hyper-Kähler: in addition to the original complex structure I there are (integrable) complex structures

J and K satisfying quaternionic relations $IJ = -JI = K$. For each $p \in Y$, there is an $\mathbb{R}$-linear isomorphism $\mathbb{R}^4 \to T_pY$ such that the linear maps $I(p)$, $J(p)$, $K(p)$ correspond to multiplication by the unit quaternions i, j, k via the standard identification $\mathbb{R}^4 \cong \mathbb{H} = \langle 1, i, j, k\rangle$ with the algebra of quaternions. Also, the metric h is Kähler with respect to each I, J, K. We shall denote by κ_I, κ_J, κ_K the respective Kähler forms.

For a 3-torus T^3 with coordinates x_1, x_2, x_3, with a flat metric $dx_1^2 + dx_2^2 + dx_3^2$ and a hyper-Kähler 4-manifold Y as above, the Riemannian product $T^3 \times Y$ has holonomy in $SU(2)$. The product metric is induced by a torsion-free G_2 structure on $T^3 \times Y$, which is

$$\varphi_{SU(2)} = dx_1 \wedge dx_2 \wedge dx_3 + dx_1 \wedge \kappa_I + dx_2 \wedge \kappa_J - dx_3 \wedge \kappa_K. \tag{4}$$

We now define, for every small $\varepsilon > 0$, a smooth compact 7-manifold $M = M_\varepsilon$ by replacing a neighbourhood $T^3 \times \{r < 2\varepsilon\}$ of each 3-torus component in the singular locus of T^7/Γ by $T^3 \times U$, where $U = \sigma^{-1}(r < 2\varepsilon) \subset Y$ is a neighbourhood of the exceptional divisor on Y. (Note that the manifolds M_ε are diffeomorphic to each other.)

On each $T^3 \times U$ in M, we smoothly interpolate, for $\varepsilon < r < 2\varepsilon$, between the flat G_2-structure φ_0 induced from T^7/Γ on the complement of the regions $T^3 \times U$ and the product G_2-structure arising as in (4) from the appropriately rescaled Eguchi–Hanson hyper-Kähler h_s on $\sigma^{-1}(r < \varepsilon) \subset U$. The ALE property of the Eguchi–Hanson metric allows to take the product G_2-structure on $T^3 \times Y$ to be asymptotic to the flat G_2-structure on $T^3 \times (\mathbb{C}^2/\{\pm 1\})$. We can obtain, for each sufficiently small $s > 0$, a well-defined positive 3-form φ_s on M noting also that $\Omega^3_+(M)$ is an open subset of 3-forms in the uniform norm. Furthermore, we can choose these G_2 3-forms on M to be *closed*, $d\varphi_s = 0$. Thus the G_2-structure φ_s is torsion-free away from the interpolation region $\{\varepsilon < r < 2\varepsilon\}$ but φ_s is not co-closed in that region.

The positive 3-forms φ_s are intended as 'approximate solutions' of the torsion-free equations (2), as $s \to 0$. The parameter s may be interpreted geometrically as the maximal diameter of the pre-image of a singular point in T^7/Γ under the resolution map $M \to T^7/\Gamma$. We would like to perturb φ_s to actual solutions on M. To this end, the following two conditions satisfied by φ_s are important, cf. [14, Theorem 11.5.7].

Condition (i) One can construct a smooth 3-form ψ_s on M such that $d^*\varphi_s = d^*\psi_s$ and

$$\|\psi_s\|_{L^2} < A_1 s^4,\ \|\psi_s\|_{C^0} < A_1 s^3 \text{ and } \|d^*\psi_s\|_{L^{14}} < A_1 s^{16/7}. \tag{5}$$

Condition (ii) The injectivity radius $\delta(g_s)$ and the Riemann curvature $R(g_s)$ of the metric $g_s = g(\varphi_s)$ on M satisfy the estimates

$$\delta(g_s) > A_2\, s, \qquad \|R(g_s)\|_{C^0} < A_3 s^{-2}. \tag{6}$$

The construction of ψ_s exploits the asymptotic and scaling properties of the G_2-structure (4) on $T^3 \times U$ 'approximating' the flat G_2-structure on $T^3 \times \mathbb{C}^2/\{\pm 1\}$.

The estimates (6) follow from the properties of the metric h_s around the exceptional divisor on Y, which give the dominant contributions for small s. In the conditions (i) and (ii) the norms and the formal adjoint d^* are taken with respect to the metric $g_s = g(\varphi_s)$. The constants $A_1,\ A_2,\ A_3$ are independent of s.

We can now state the existence result for torsion-free G_2-structures.

Theorem 3 (cf. [14, Theorem 11.6.1]) *Let M be a compact 7-dimensional manifold and $\varphi_s \in \Omega^3_+(M)$, $0 < s \le s_0$, a family of G_2-structures such that $d\varphi_s = 0$ and the conditions (i) and (ii) above hold for all s.*

Then there is an $\varepsilon_0 > 0$ so that for each s with $0 < s \le \varepsilon_0$ the manifold M admits a torsion-free G_2 structure $\tilde{\varphi}_s \in \Omega^3_+(M)$ in the same cohomology class as φ_s and satisfying $\|\tilde{\varphi}_s - \varphi_s\|_{C^0} < Ks^{1/2}$ with some constant K independent of s.

We next outline the Proof of Theorem 3 following [15, pp. 236–237], dropping the subscripts s to ease the notation. The desired torsion-free G_2-structure $\tilde{\varphi} = \tilde{\varphi}_s$ will be obtained in the form $\tilde{\varphi} = \varphi + d\eta$, where $d\eta$ has a small uniform norm, so $\tilde{\varphi}$ is a closed positive 3-form. We then need to satisfy the co-closed condition $d^*_{\tilde{\varphi}}\tilde{\varphi} = 0$ and this amounts to solving for a 2-form η a non-linear elliptic PDE which may be written as

$$d^*d\eta = -d^*\psi + d^*F(d\eta) \tag{7}$$

where F satisfies a quadratic estimate. A solution of (7) is achieved by using iterations to construct a sequence $\{\eta_j\}_{j=0}^{\infty}$ with $\eta_0 = 0$ and

$$d^*d\eta_{j+1} = -d^*\psi + d^*F(d\eta_j), \qquad d^*\eta_{j+1} = 0.$$

One first argues that the sequence η_j converges.

The proof of convergence is based on the following inductive estimates (all the constants C_i below are independent of s)

$$\|d\eta_{j+1}\|_{L^2} \le \|\psi\|_{L^2} + C_1\|d\eta_j\|_{L^2}\|d\eta_j\|_{C^0}, \tag{8a}$$

$$\|\nabla d\eta_{j+1}\|_{L^{14}} \le C_2(\|d^*\psi\|_{L^{14}} + \|\nabla d\eta_j\|_{L^{14}}\|d\eta_j\|_{C^0} + s^{-4}\|d\eta_{j+1}\|_{L^2}), \tag{8b}$$

$$\|d\eta_j\|_{C^0} \le C_3(s^{1/2}\|\nabla d\eta_j\|_{L^{14}} + s^{-7/2}\|d\eta_j\|_{L^2}). \tag{8c}$$

The estimate (8a) is proved by taking the L^2 product of both sides with η_{j+1} and integrating by parts, noting also the condition (i) above. The proof of (8b) uses an elliptic regularity estimate for the operator $d + d^*$ considered for 3-forms on small balls on M with radius of order s. The condition (ii) is also required here and in (8c) which uses the Sobolev embedding of L^{14}_1 in C^0 in dimension 7 and is again achieved by working on small balls with radius of order s.

For every sufficiently small s, we deduce from (8) that if $d\eta_j$ satisfies

$$\|d\eta_j\|_{L^2} \le C_4 s^4, \quad \|\nabla d\eta_j\|_{L^{14}} \le C_5, \quad \|d\eta_j\|_{C^0} \le Ks^{1/2}, \tag{9}$$

then these latter estimates hold for $d\eta_{j+1}$ and, by induction, for all j. Thus $d\eta_j$ is a bounded sequence in the L^{14}_1 norm on $\Lambda^3 T^*M$ and one can further show that $d\eta_j$ is a Cauchy sequence. Further, we are free to assume that the forms η_j are in the L^2-orthogonal complement $\mathcal{H}^\perp$ of harmonic forms. As the elliptic operator $d + d^*$ is bounded below on $\mathcal{H}^\perp$ it follows that the sequence η_j converges in the L^{14}_2 norm. In particular, the last inequality of (9) holds for the limit η.

Finally, a careful elliptic regularity argument shows that η is in fact a smooth solution of (7), thus completing the Proof of Theorem 3.

The metrics on M induced by φ_s have holonomy in G_2 and it remains to verify that the holonomy does not reduce further to a subgroup of G_2. In the present case, the orbifold T^7/Γ is simply-connected, therefore M is so, by the properties of the blow-up. Thus (M, φ_s) is an irreducible G_2-manifold by Proposition 2.

The discussed example may be considered as a generalization of the Kummer construction of hyper-Kähler metrics of holonomy $SU(2)$ on K3 surfaces.

It is convenient to obtain the Betti numbers of M; these are determined by $b^2(M)$ and $b^3(M)$. By considering the Γ-invariant classes in $H^*_{\mathrm{dR}}(T^7/\Gamma)$ we obtain $b^2(T^7/\Gamma) = 0$ and $b^3(T^7/\Gamma) = 7$. When resolving the singularities, we replaced a deformation retract of T^3 with $T^3 \times Y$ which is homotopy equivalent to $T^3 \times \mathbb{C}P^1$. Let S denote the singular locus of T^7/Γ. Comparing the cohomology long exact sequence for the pairs $(T^7/\Gamma, S)$ and $(M, \sqcup_{i=1}^{12}(T^3 \times U))$, we find that each of the 12 instances of a resolution adds $b^i(T^3 \times Y) - b^i(T^3)$ to the ith Betti number of M. Thus $b^2(M) = 12 \cdot 1$ and $b^3(M) = 7 + 12 \cdot 3 = 43$.

Further examples of irreducible G_2-manifolds arise by using the above method with different choices of finite groups Γ and different choices of resolutions of singularities of T^7/Γ. If every component of the singular locus of T^7/Γ has a neighbourhood diffeomorphic to $T^3 \times (\mathbb{C}^2/G)$ for a finite subgroup G of $SU(2)$ or to $S^1 \times (\mathbb{C}^3/G)$ for a finite subgroup G of $SU(3)$ acting freely on $\mathbb{C}^3 \setminus \{0\}$, then it is known from complex algebraic geometry that one can find crepant resolutions, $\sigma_2 : Y_2 \to \mathbb{C}^2/G$ or $\sigma_3 : Y_3 \to \mathbb{C}^3/G$ respectively, with the canonical bundle of Y_i holomorphically trivial.

The Ricci-flat Kähler (thus hyper-Kähler) metrics on the complex surfaces Y_2 asymptotic to $\mathbb{C}^2/G$ in the sense of (3), for each G, were constructed by Kronheimer [21] using hyper-Kähler quotients.

In complex dimension 3, the existence of ALE Ricci-flat holonomy $SU(3)$ metrics on Y_3 asymptotic to $\mathbb{C}^3/G$ follows from the solution of ALE version of the Calabi conjecture, see [14, Chap. 8] and references therein. The asymptotic rate for the metrics h is given by

$$\nabla^k(h - h_0) = O(r^{-6-k}) \quad \text{as } r \to \infty, \quad \text{for all } k = 0, 1, 2, \dots ,$$

where h_0 is the pull-back of the Euclidean metric on $\mathbb{C}^3/G$. The Kähler forms of h and h_0 satisfy

$$\omega - \omega_0 = i\partial\bar{\partial}u, \quad \nabla^k u = O(r^{-4-k}) \quad \text{as } r \to \infty$$

(cf. [14, Theorem 8.2.3]). The holonomy being $SU(3)$ means there is a choice of nowhere vanishing (3, 0)-form Ω on Y_3 (sometimes called a holomorphic volume form), such that $\omega^3/3! = (i/2)^3\Omega \wedge \bar{\Omega}$. A torsion-free G_2-structure on $S^1 \times Y_3$ defined by

$$\varphi_{SU(3)} = dx \wedge \omega + \mathrm{Re}\Omega, \tag{10}$$

induces a product metric corresponding to $(dx)^2$ and ω, where x is the usual coordinate on $S^1 = \mathbb{R}/\mathbb{Z}$.

The singularities of T^7/Γ can be resolved with copies of $T^3 \times U_2$ or $S^1 \times U_3$ (where U_i is a neighbourhood of $\sigma_i^{-1}(0)$ in Y_i) in a manner similar to the example above. One obtains compact smooth 7-manifolds M and closed positive 3-forms φ_s on M satisfying the hypotheses of Theorem 3. More generally, the method extends to situations when the singularities of T^7/Γ are only *locally* modeled on $\mathbb{R}^3 \times (\mathbb{C}^2/G)$ or $\mathbb{R} \times (\mathbb{C}^3/G)$. In the latter case, G need not act freely on $\mathbb{C}^3 \setminus \{0\}$ resulting in a more complicated singular locus of T^7/Γ.

Joyce found a large number of orbifolds T^7/Γ with suitable resolutions of singularities. In particular, 252 examples of topologically distinct compact 7-manifolds admitting holonomy G_2 metrics are worked out in [14, Chap. 12], including some manifolds with non-trivial fundamental group. The Betti numbers of these examples are in the range $0 \le b^2 \le 28$ and $4 \le b^3 \le 215$. There is a evidence that many more further topological types can be constructed by the same method.

3 Construction by Generalized Connected Sums

The method of constructing compact holonomy G_2 manifolds discussed in this section is sometimes called a 'twisted connected sum'. The construction was originally developed by the author in [17] and included an important idea due to Donaldson. Generalizations and many new examples appeared in [5, 6, 18, 24].

The connected sum construction produces one-parameter families of holonomy G_2 metrics g_T, $T_0 \le T < \infty$, on compact manifolds with 'long necks'. The parameter T here is asymptotic, as $T \to \infty$, to the diameter of the metric g_T. We may think of the respective families of torsion-free G_2-structures as paths in the G_2 moduli space, going to the boundary as one 'stretches the neck', the limit boundary point corresponding to the disjoint union of the initial two asymptotically cylindrical pieces. So, in this construction, the limit spaces are *disconnected, non-compact* and *smooth.*

A twisted connected sum is an instance of generalized connected sum of a pair of asymptotically cylindrical Riemannian manifolds which, in the present case, are G_2-manifolds. The asymptotically cylindrical G_2-manifolds we require are Riemannian products $W \times S^1$, where W is a Ricci-flat Kähler manifold with cylindrical end asymptotic to a Riemannian product $D \times S^1 \times [0, \infty)$ with D a K3 surface with a hyper-Kähler metric. For certain pairs of the K3 surfaces D_1, D_2 there is a way to 'join' the two latter asymptotically cylindrical manifolds at their ends. We obtain a compact simply-connected manifold M and a G_2-structure with small torsion on M to which a perturbative analysis can be applied.

We now describe the key steps in the construction in more detail, starting with the asymptotically cylindrical Calabi–Yau threefolds W.

Theorem 4 ([11, 17, 27]) *Let $\overline{W}$ be a compact Kähler threefold with Kähler form $\overline{\omega}$ and suppose that a K3 surface $D \in |-K_{\overline{W}}|$ is an anticanonical divisor on $\overline{W}$ with holomorphically trivial normal bundle $N_{D/\overline{W}}$. Denote by z a complex coordinate around D vanishing to order one precisely on D. Suppose that $\overline{W}$ is simply-connected and the fundamental group of $W = \overline{W} \setminus D$ is finite.*

Then W admits a complete Ricci-flat Kähler metric, with holonomy $SU(3)$, with Kähler form ω and a non-vanishing holomorphic $(3,0)$-form Ω. These are asymptotic to the product cylindrical Ricci-flat Kähler structure on $D \times S^1 \times \mathbb{R}_{>0}$

$$\omega = \kappa_I + dt \wedge d\theta + d\psi,$$
$$\Omega = (\kappa_J + i\,\kappa_K) \wedge (dt + id\theta) + d\Psi,$$

where $\exp(-t - i\theta) = z$, for $(\theta, t) \in S^1 \times \mathbb{R}_{>0}$ and the forms ψ, Ψ exponentially decay as $t \to \infty$. Also κ_I is the Ricci-flat Kähler metric on D in the class $[\overline{\omega}|_D]$ and $\kappa_J + i\kappa_K$ is a non-vanishing holomorphic $(2,0)$-form on D.

Remark Any threefold $\overline{W}$ satisfying the hypotheses of Theorem 4 is necessarily projective and algebraic [18, Proposition 2.2]. The holomorphic coordinate z extends to a meromorphic function $\overline{W} \to \mathbb{C}P^1$ vanishing precisely on D.

Theorem 4 extends to higher dimensions $m \geq 3$ with D replaced by a compact simply-connected Calabi–Yau $(m-1)$-fold. The result may be regarded as a solution of an 'asymptotically cylindrical version' of the Calabi conjecture.

It will be convenient to extend the parameter t along the cylindrical end in Theorem 4 to a smooth function t defined on all of W with $t < 0$ away from a tubular neighbourhood of D. We shall also assume that the holomorphic 2-form on a Kähler K3 surface D is normalized so that $\kappa_I^2 = \kappa_J^2 = \kappa_K^2$, with the implied normalization of a holomorphic 3-form Ω on W. The Ricci-flat Kähler (hyper-Kähler) structure on D is in fact determined by the triple $\kappa_I, \kappa_J, \kappa_K$ (cf. [10, p. 91]).

The following relation between K3 surfaces is crucial for the connected sum construction of G_2-manifolds.

Definition 1 We say that two Ricci-flat Kähler K3 surfaces $(D_1, \kappa'_I, \kappa'_J + i\kappa'_K)$, $(D_2, \kappa''_I, \kappa''_J + i\kappa''_K)$ satisfy the *Donaldson matching condition* if there exists an isometry of lattices $h : H^2(D_2, \mathbb{Z}) \to H^2(D_1, \mathbb{Z})$, so that the $\mathbb{R}$-linear extension of h satisfies

$$h : [\kappa''_I] \mapsto [\kappa'_J], \qquad [\kappa''_J] \mapsto [\kappa'_I], \qquad [\kappa''_K] \mapsto [-\kappa'_K]. \tag{11}$$

It follows, by application of the Torelli theorem for K3 surfaces, that there is a smooth map

$$f : D_1 \to D_2, \text{ such that } h = f^*.$$

Note that f is *not* a holomorphic map between D_1 and D_2 (with complex structures I), though f is an isometry of the underlying Riemannian 4-manifolds. In particular, the pull back f^* rotates the 2-forms of the hyper-Kähler triple (not just their cohomology classes), $\kappa''_I \mapsto \kappa'_J$, $\kappa''_J \mapsto \kappa'_I$, $\kappa''_K \mapsto -\kappa'_K$.

Now if (W, ω, Ω) is an asymptotically cylindrical Calabi–Yau manifold given by Theorem 4, then $W \times S^1$ has a torsion-free G_2-structure given by (10)

$$\varphi_W = d\tilde{\theta} \wedge \omega + \mathrm{Re}\Omega,$$

where $\tilde{\theta}$ is the standard coordinate on the S^1 factor. The form φ_W is asymptotic to a cylindrical product torsion-free G_2-structure φ_∞ on the cylindrical end $D \times [0, \infty) \times S^1 \times S^1 \subset W \times S^1$,

$$\varphi_\infty = dt \wedge d\theta \wedge d\tilde{\theta} + d\tilde{\theta} \wedge \kappa_I + dt \wedge \kappa_J - d\theta \wedge \kappa_K.$$

corresponding to the hyper-Kähler structure $(\kappa_I, \kappa_J, \kappa_K)$ on D (cf. (4)).

For $i = 1, 2$ and $T > 0$, let $W_{i,T}$ be a compact manifold with boundary obtained by truncating W_i at $t_i = T + 1$ (where t_i is the parameter along the cylindrical end as in Theorem 4). We can smoothly cut off each φ_{W_i} to obtain on $W_{i,T}$ a closed G_2-structure $\varphi_{W,T}$ so that $\varphi_{W_i,T}$ equals its cylindrical asymptotic model φ_∞ on a collar neighbourhood $D_i \times S^1 \times S^1 \times [T, T+1]$ of the boundary.

Suppose that D_1 and D_2 satisfy the Donaldson matching condition. Then we can define a compact 7-manifold

$$M = M_T = (W_{1,T+1} \times S^1) \cup_F (W_{2,T+1} \times S^1) \tag{12}$$

by identifying the collar neighbourhoods of the boundaries using a map

$$\begin{aligned} F : D_1 \times S^1 \times S^1 \times [T, T+1] &\to D_2 \times S^1 \times S^1 \times [T, T+1], \\ (y, \theta, \tilde{\theta}, T+t) &\mapsto (f(y), \tilde{\theta}, \theta, T+1-t). \end{aligned} \tag{13}$$

The form $\varphi_\infty|_{[T,T+1]}$ is preserved by F, so the G_2-structures $\varphi_{i,T}$ agree on the overlap and patch together to a well-defined closed 3-form φ_T on M. It is easy to see that φ_T is a well-defined G_2-structure on M for every large T.

Another important property of the map F is that F identifies the S^1 factor in $W_{1,T+1} \times S^1$ with a circle around the divisor on the other threefold W_2 and vice versa. This eliminates the possibility of an infinite fundamental group of M, in particular, M will be simply-connected when the threefolds W_1 and W_2 are so.

The G_2-structure form on M satisfies $d\varphi_T = 0$, one of the two equations in (2), but the co-derivative $d *_T \varphi_T$ in general will not vanish. The cut-off functions introduce 'error terms' which depend on the difference between the $SU(3)$-structures on the end of W_i and on its cylindrical asymptotic model, and can be estimated as

$$\|d *_T \varphi_T\|_{L^p_k} < C_{p,k} e^{-\lambda T},$$

with $\lambda > 0$. Here $*_T$ denotes the Hodge star of the metric $g(\varphi_T)$.

The next result shows that for a sufficiently long neck the G_2-structure φ_T on M can be made torsion-free by adding a small correction term.

Theorem 5 *Suppose that each of $\overline{W}_1, D_1$ and $\overline{W}_2, D_2$ satisfies the hypotheses of Theorem 4 and the $K3$ surfaces $D_j \in |-K_{\overline{W}_j}|$ satisfy the Donaldson matching condition. Let M be the compact 7-manifold M defined in* (12) *with a closed G_2-structure φ_T induced from $\varphi_{W_1}, \varphi_{W_2}$.*

Then M has finite fundamental group. Furthermore, there exists $T_0 \in \mathbb{R}$ and for every $T \geq T_0$ a unique smooth 2-form η_T on M, orthogonal to the closed forms, so that the following holds.
(a) $\|\eta_T\|_{C^1} < A \cdot e^{-\mu T}$, for some constants $A, \mu > 0$ independent of T, where the norm is defined using the metric $g(\varphi_T)$. In particular, $\varphi_T + d\eta_T$ is a valid G_2-structure on M.
(b) The closed 3-form $\varphi_T + d\eta_T$ satisfies

$$d *_{\varphi_T + d\eta_T} (\varphi_T + d\eta_T) = 0. \tag{14}$$

and so $\varphi_T + d\eta_T$ defines a metric with holonomy G_2 on M.

As discussed in the previous section, the perturbative problem (14) can be rewritten as a non-linear elliptic PDE for the 2-form η. When η has a small norm this PDE takes the form $a(\eta) = a_0 + A\eta + Q(\eta) = 0$, where $a_0 = d *_T \varphi_T$, the linear elliptic operator $A = A_T$ is a compact perturbation of the Hodge Laplacian of the form $dd^* + d^*d + O(e^{-\varepsilon T})$, $\varepsilon > 0$ and $Q(\eta)$ satisfies a quadratic estimate in $d\eta$.

One can use elliptic theory for manifolds with cylindrical ends and the gluing analysis for the problem at hand is then simplified, compared to the general situation of Theorem 3. The central idea in the proof of Theorem 5 may be informally described as follows. For small η, the map $a(\eta)$ is approximated by its linearization and so there would be a unique small solution η to the equation $a(\eta) = 0$, for every small a_0 in the range of A. This perturbative approach requires the invertibility of A and a suitable upper bound on the operator norm $\|A_T^{-1}\|$, as $T \to \infty$. This bound determines what is meant by 'small a_0' above.

As we actually need the value of $d\eta$ rather than η we may consider the equation for η in the orthogonal complement of harmonic 2-forms on M where the Laplacian is invertible. We use the technique similar to [20, Sect. 4.1] based on Fredholm theory for the asymptotically cylindrical manifolds and weighted Sobolev spaces to find an upper bound $\|A_T^{-1}\| < Ge^{\delta T}$. Here the constant G is independent of T and $\delta > 0$ can be taken arbitrarily small. So, for large T, the growth of $\|A_T^{-1}\|$ is negligible compared to the decay of $\|d *_T \varphi_T\|$ and the 'inverse function theorem' strategy applies to give the required small solution η_T in a (appropriately chosen) Sobolev space. Standard elliptic methods show that this η_T is in fact smooth.

3.1 Some Examples and Further Results

In order to make irreducible G_2-manifolds using the connected sum construction, we require pairs $\overline{W}_1$, $\overline{W}_2$ of complex algebraic threefolds with matching anticanonical K3 divisors $D_i \subset \overline{W}_i$. We begin with an example based on some classical algebraic geometry.

Example 1 The intersection of three generically chosen quadric hypersurfaces in $\mathbb{C}P^6$ defines a smooth Kähler threefold X_8. It is simply-connected and the characteristic class $c_1(X_8)$ of its anticanonical bundle is the pull-back to X_8 of the positive generator of the cohomology ring $H^*(\mathbb{C}P^6)$. This tells us that the anticanonical bundle $K_{X_8}^{-1}$ is the restriction to X_8 of the tautological line bundle $\mathcal{O}(1)$ over $\mathbb{C}P^6$. It follows that any anticanonical divisor D on X_8 is obtained by taking an intersection $D = X_8 \cap H$ with a hyperplane H in $\mathbb{C}P^6$. A generic such hyperplane section D is a complex surface, isomorphic to a smooth complete intersection of three quadrics in $\mathbb{C}P^5$. This is a well-known example of a K3 surface.

Conversely, starting from a smooth intersection D of three quadrics in $\mathbb{C}P^5$ we can write down a smooth threefold $X_8 \subset \mathbb{C}P^6$ as above containing the K3 surface D as an anticanonical divisor.

Consider another anticanonical divisor $D' = X_8 \cap H'$ and let $\tilde{X}_8 \to X_8$ be the blow-up of the second hyperplane section $C = D \cap D' = X_8 \cap H \cap H'$. (It is convenient, though not strictly necessary, to choose D' so that C is a non-singular connected complex curve.) The pencil defined by D and D' lifts, via the proper transform, to a pencil consisting of the fibres of a holomorphic map $\tilde{X}_8 \to \mathbb{C}P^1$. In particular, the K3 divisor D lifts to an isomorphic K3 surface $\tilde{D}$ which is an anticanonical divisor on $\tilde{X}_8$ and has trivial normal bundle. Moreover, a Kähler metric on $\tilde{X}_8$ may be chosen so that $\tilde{D}$ and D are isometric Kähler K3 surfaces.

It is not difficult to check that $\tilde{X}_8 \setminus \tilde{D}$ is simply-connected, noting that $\tilde{D}$ and X_8 are so and considering an exceptional curve in the blow up $\tilde{X}_8$. The pair $\tilde{X}_8$, $\tilde{D}$ thus satisfies all the hypotheses of Theorem 4, and so the quasiprojective threefold $W = \tilde{X}_8 \setminus \tilde{D}$ admits an asymptotically cylindrical Ricci-flat Kähler metric with holonomy $SU(3)$. Note that the cylindrical asymptotic model for this metric is determined by the Ricci-flat Kähler structure in the Kähler class of D in X_8.

We would like to choose two octic threefolds $X_8^{(i)}$, $i = 1, 2$ and a K3 surface D_i in each, so as to satisfy the Donaldson matching condition. We do this by applying some general theory of K3 surfaces and their moduli (see [2, Chap. VIII]). The key point is that one can determine a Ricci-flat Kähler K3 surface D, up to isomorphism, by a data of the integral second cohomology $H^2(D, \mathbb{Z})$.

Recall that all K3 surfaces are diffeomorphic and the intersection form makes $H^2(D, \mathbb{Z})$ into a lattice. There is an isomorphism, called a marking, $p : H^2(D, \mathbb{Z}) \to L$ to a fixed non-degenerate even unimodular lattice L with signature (3, 19). We shall refer to L as the *K3 lattice*; its bilinear form is given by the orthogonal direct sum $L = 3H \oplus 2(-E_8)$ of 3 copies of the hyperbolic plane lattice $H = \left(\begin{smallmatrix} 0 & 1 \\ 1 & 0 \end{smallmatrix}\right)$ and 2 copies of the negative definite root lattice $-E_8$ of rank 8. Now if $D \subset \mathbb{C}P^5$ is an octic K3 surface, then the image $p(\kappa_I)$ of the Kähler class of D is primitive (non-divisible in L by an integer > 1) and $p(\kappa_I) \cdot p(\kappa_I) = 8$, computed in the bilinear form of L. The images $p(\kappa_J)$, $p(\kappa_K)$ span a positive 2-plane P orthogonal to $p(\kappa_I)$ in the real vector space $L \otimes \mathbb{R}$. Conversely, the positive 2-planes P arising in this way form a dense open set in the Grassmannian of positive 2-planes orthogonal to $p(\kappa_I)$ in $L \otimes \mathbb{R}$.

It is known that the group of lattice isometries of L acts transitively on the set of all primitive vectors with a fixed value of $v \cdot v$ in L. We can therefore choose two octic K3 surfaces with hyper-Kähler structures (in the respective Kähler classes) $(D_1; \kappa'_I, \kappa'_J, \kappa'_K)$, $(D_2; \kappa''_I, \kappa''_J, \kappa''_K)$ and the markings p_1, p_2 with $p_1(\kappa'_I) = p_2(\kappa''_J)$, $p_1(\kappa'_J) = p_2(\kappa''_I)$ in L, and $p_1(\kappa'_K) = -p_2(\kappa''_K)$ in $L \otimes \mathbb{R}$ thereby achieving a matching.

Choosing the ambient octic threefolds X'_8, X''_8 for the latter D_1, D_2, blowing up these threefolds to obtain asymptotically cylindrical Ricci-flat threefolds by Theorem 4, and applying Theorem 5 to the respective connected sum, we obtain a simply-connected compact 7-manifold M with a metric of holonomy G_2.

We may consider in a very similar way, in place of one of both X_8's above, a smooth intersection X_6 of a quadric and a cubic in $\mathbb{C}P^5$. The respective K3 divisor then is an intersection of a quadric and a cubic in $\mathbb{C}P^4$ and the image of the Kähler class of this divisor has square 6 in the bilinear form L.

More generally, it was shown in [17, Sect. 6,7] that in place of X_8, X_6 in the above example we can consider any non-singular *Fano threefold* V, i.e. a projective complex 3-dimensional manifold such that the image of the first Chern class $c_1(V)$ in the de Rham cohomology can be represented by some Kähler form on V. Equivalently, the anticanonical bundle K_V^{-1} is ample. Smooth Fano threefolds are completely classified; up to deformations, there are 105 algebraic types [12, 22].

Every Fano threefold V is simply-connected and a generic anticanonical divisor D on V is a (smooth) K3 surface [26]. A threefold $\overline{W}$ is obtained by blowing up a connected complex curve representing the self-intersection cycle $D \cdot D$ (in the sense of the Chow ring). Then $\overline{W}$ and the proper transform of D satisfy the hypotheses of Theorem 4. Alternatively, if $D \cdot D$ is represented by a finite sequence of curves, then $\overline{W}$ may be defined by successively blowing up these curves. We shall refer to any such threefold $\overline{W}$ to be of *Fano type*.

A Kähler K3 surface D and its proper transform in $\overline{W}$ can be assumed isomorphic by choosing an appropriate Kähler metric on $\overline{W}$. Then the cylindrical asymptotic model for W is determined by the K3 surface D with the Kähler metric restricted from V.

For a general Fano V, the class of anticanonical K3 surfaces D arising in the deformations of V will correspond to an open dense subset of *lattice-polarized* K3 surfaces. This latter class is defined by the condition that the Picard lattice $H^{1,1}(D, \mathbb{R}) \cap H^2(D, \mathbb{Z})$ contains a sublattice isomorphic to a fixed lattice N and this sublattice contains a class of some Kähler form. In the case of algebraic K3 surfaces of a fixed degree, as in the example above, N is generated by the Kähler form κ_I induced from the embedding of D in the projective space. In general, N

arises as $\iota^* H^2(V, \mathbb{Z})$ from the embedding $\iota : D \to V$. The rank of N is the Betti number $b^2(V)$ as ι^* in injective by the Lefschetz hyperplane theorem.

Another source of examples for Theorem 4 was given by Lee and the author in [18]. The construction uses K3 surfaces S with *non-symplectic involution*, a holomorphic map $\rho : S \to S$, such that ρ^* restricts to -1 on $H^{2,0}(S)$. The K3 surfaces of this type were completely classified up to deformation by Nikulin [1], who determined the complete system of invariants and fixed point set of ρ for each deformation family. We require the fixed point set of ρ to be non-empty; this occurs in all but one of the 75 deformation families.

Let $\psi : \mathbb{C}P^1 \to \mathbb{C}P^1$ denote the holomorphic involution $\psi(z_0 : z_1) = z_1 : z_0$ fixing exactly two points. The quotient $Z = (S \times \mathbb{C}P^1)/(\rho, \psi)$ is then an orbifold whose singular locus is a disjoint union of smooth curves. The desired 3-fold $\overline{W}$ is defined by the resolution of singularities diagram for Z,

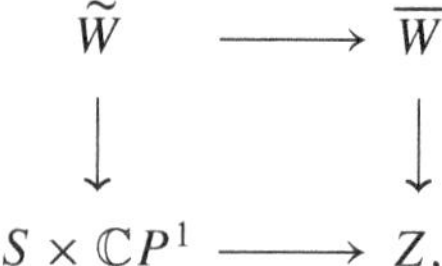

where the vertical arrows correspond to blowing up the fixed locus of (ρ, ψ) in $S \times \mathbb{C}P^1$ and the singular locus of Z and the horizontal arrows are the quotient maps.

The anticanonical divisor D on $\overline{W}$ arises as the (pre-)image of $S \times \{p\}$, via the above diagram, where $\psi(p) \neq p$. Such D is clearly isomorphic to the Kähler K3 surface S and has trivial normal bundle in $\overline{W}$. It can be checked that $\overline{W}$ and $W = \overline{W} \setminus D$ are simply-connected (the condition that ρ have fixed points is needed here). Thus W has an asymptotically cylindrical Ricci-flat Kähler metric by Theorem 4.

The pull-back $\iota^* : H^2(\overline{W}, \mathbb{Z}) \to H^3(D, \mathbb{Z})$ defined by the embedding of D makes D into a lattice polarized K3 surface with N corresponding to the sublattice of all classes fixed by ρ^* in $H^2(D, \mathbb{Z})$. On the other hand, ι^* has a kernel of dimension at least 2. A threefold $\overline{W}$ obtained from K3 surface with non-symplectic involution is therefore *never* deformation equivalent to any threefold of Fano type (assuming $D \cdot D$ in the latter threefold was represented by a single curve).

The matching problem in all the examples becomes entirely a consideration on the K3 lattice L, as illustrated by the example in the beginning of this subsection. In general, the argument is more technical and requires results on the lattice embeddings [23].

One simple sufficient (though not necessary) condition for the existence of the Donaldson matching for representatives in the two classes of lattice polarized K3 surfaces is that the rank of each polarizing lattice N_i is ≤ 5.

All the irreducible G_2-manifolds M constructed from threefolds in the above examples are simply-connected. The cohomology of compact irreducible G_2-manifolds M coming from the connected sum construction may be determined by application of the Mayer–Vietoris exact sequence and generally depends on the choice of matching. However, the sum of the Betti numbers

$$b^2(M) + b^3(M) = b^3(\overline{W}_1) + b^3(\overline{W}_2) + 2d_1 + 2d_2 + 23, \tag{15}$$

for any matching, depends only on the threefolds $\overline{W}_i$ and the dimensions d_i of the kernel of $\iota^* : H^2(\overline{W}_i, \mathbb{R}) \to H^3(D_i, \mathbb{R})$. The quantities in (15) can be determined by standard methods (adjunction formula, Lefschetz–Bott hyperplane theorem) from known algebraic invariants of Fano threefolds or, respectively, of non-symplectic involutions.

In particular, the Fano threefold X_8 discussed in Example 1 above has $b^2(X_8) = 1$, $b^3(X_8) = 28$ and its blow-up has $b^2(\overline{W}) = 2$, $b^3(\overline{W}) = 38$. An irreducible compact G_2-manifold M constructed from a pair of X_8's has $b^2(M) = 0$ and then $b^3(M) = 99$ as d_l vanish in this case. This irreducible G_2-manifold is topologically distinct from the examples given by Joyce via resolution of singularities; the only irreducible G_2-manifold in [14] with $b^2 = 0$ has $b_3 = 215$. The property $b^2(M) = 0$ holds in many other examples coming from pairs of threefolds of Fano type and these latter G_2-manifolds typically have smaller b^2 and larger b^3 than the examples given by Joyce. (Note also that every compact irreducible G_2-manifold M must have $b^1(M) = 0$ by Proposition 2 but $b^3(M)$ cannot vanish as the G_2 3-form φ on M is harmonic.)

Corti, Haskins, Nordström and Pacini [5, 6] generalized the class of threefolds of Fano type by considering *weak Fano* threefolds V whose anticanonical bundle K_V^{-1} is only required to be big and nef. (Every such V may be obtained as a resolution of an appropriate singular Fano threefold.) They identified a large subclass called semi-Fano threefolds and generalized for this class the properties required in the construction of G_2-manifolds from threefolds W of Fano type. This generalization dramatically increased the number of examples of connected sum G_2-manifolds. Some of the examples were shown to be 2-connected which allows to determine their diffeomorphism type by computing certain standard invariants.

More recently, Braun [3] gave a toric geometry construction, from certain lattice polytops, of examples of pairs $\overline{W}$, D defining asymptotically cylindrical Calabi–Yau threefolds by Theorem 4. Useful invariants of $\overline{W}$ e.g. the Hodge numbers can be computed by combinatorial formulae.

Nordström [24] gave an interesting generalization of the connected sum construction, by replacing (11) with a different 'hyper-Kähler rotation' and taking finite quotients of asymptotically cylindrical Calabi–Yau threefolds W. Applications of the construction include topologically new examples of compact irreducible G_2-manifolds some of which have a non-trivial finite fundamental group.

In conclusion, we mention two works which contain results concerning relations between the two types of construction of G_2-manifolds discussed in these notes.

Nordström and the author identified in [19] an example of a compact irreducible G_2-manifold given by Joyce [14] where the underlying 7-manifold is diffeomorphic to one obtainable from the construction in [18]. Further, the two respective families of G_2-metrics on this manifold are connected in the G_2-moduli space.

On the other hand, some of the G_2-manifolds given by Joyce cannot possibly be obtained by the connected sum construction. The result is due to Crowley and Nordstrom [7] who constructed an invariant of G_2-structures which is equal to 24 for each connected sum (12) but is odd for some examples in [14].

Acknowledgements These notes are an expanded version of lectures given in the Minischool on G_2 manifolds at the Fields Institute, Toronto, in August 2017. I would like to thank the organizers for inviting me.

References

1. Alexeev, V., & Nikulin, V. V. (2006). *Del Pezzo and K3 surfaces*. MSJ memoirs (Vol. 15). Tokyo: Mathematical Society of Japan.
2. Barth, W., Hulek, K., Peters, C., & van de Ven, A. (2004). *Compact complex surfaces*. Berlin: Springer.
3. Braun, A. (2017). Tops as building blocks for G_2 manifolds. *Journal of High Energy Physics, 10*, 083.
4. Bryant, R. L. (1987). Metrics with exceptional holonomy. *Annals of Mathematics (2), 126*, 525–576.
5. Corti, A., Haskins, M., Nordström, J., & Pacini, T. (2013). Asymptotically cylindrical Calabi-Yau 3-folds from weak Fano 3-folds. *Geometry and Topology, 17*, 1955–2059.
6. Corti, A., Haskins, M., Nordström, J., & Pacini, T. (2015). G_2-manifolds and associative submanifolds via semi-Fano 3-folds. *Duke Mathematical Journal, 164*, 1971–2092.
7. Crowley, D., & Nordström, J. (2015). New invariants of G_2-structures. *Geometry and Topology, 19*, 2949–2992.
8. Eguchi, T., & Hanson, A. J. (1978). Asymptotically flat self-dual solutions to Euclidean gravity. *Physics Letters B, 74*, 249–251.
9. Fernández, M., & Gray, A. (1982). Riemannian manifolds with structure group G_2. *Annali di Matematica Pura ed Applicata (4), 132*, 19–45.
10. Hitchin, N. (1987). Self-duality equations on Riemann surface. *Proceedings of the London Mathematical Society, 55*, 59–126.
11. Haskins, M., Hein, H.-J., & Nordström, J. (2015). Asymptotically cylindrical Calabi-Yau manifolds. *Journal of Differential Geometry, 101*, 213–265.
12. Iskovskih, V. A. (1977). Fano threefolds. I. II. *Izv. Akad. Nauk SSSR Ser. Mat., 41*, 516–562, 717 and *42*, 506–549 (1978). English translation: *Math. USSR Izvestia, 11*, 485–527 (1977), and *12*, 469–506 (1978).
13. Joyce, D. D. (1996). Compact Riemannian 7-manifolds with holonomy G_2. I, II. *Journal of Differential Geometry, 43*, 291–328, 329–375.
14. Joyce, D. D. (2000). *Compact manifolds with special holonomy*. Oxford: Oxford University Press.
15. Joyce, D. D. (2007). *Riemannian holonomy groups and calibrated geometry*. Oxford: Oxford University Press.
16. Joyce, D. D., & Karigiannis, S. A new construction of compact torsion-free G_2-manifolds by gluing families of Eguchi-Hanson spaces. *Journal of Differential Geometry*, to appear.
17. Kovalev, A. G. (2003). Twisted connected sums and special Riemannian holonomy. *Journal für die Reine und Angewandte Mathematik, 565*, 125–160.
18. Kovalev, A. G., & Lee, N.-H. (2011). $K3$ surfaces with non-symplectic involution and compact irreducible G_2-manifolds. *Mathematical Proceedings of the Cambridge Philosophical Society, 151*, 193–218.

19. Kovalev, A. G., & Nordström, J. (2010). Asymptotically cylindrical 7-manifolds of holonomy G_2 with applications to compact irreducible G_2-manifolds. *Annals of Global Analysis and Geometry*, *38*, 221–257.
20. Kovalev, A. G., & Singer, M. A. (2001). Gluing theorems for complete anti-self-dual spaces. *Geometric and Functional Analysis*, *11*, 1229–1281.
21. Kronheimer, P. B. (1989). The construction of ALE spaces as hyper-Kähler quotients. *Journal of Differential Geometry*, *29*, 665–683.
22. Mori, S., & Mukai, S. (1981/1982). Classification of Fano 3-folds with $B_2 \geq 2$. *Manuscripta Mathematica*, *36*, 147–162. Erratum: Classification of Fano 3-folds with $B_2 \geq 2$. *Manuscripta Mathematica*, *110*, 407 (2003).
23. Nikulin, V. V. (1979). Integer symmetric bilinear forms and some of their geometric applications. *Izv. Akad. Nauk SSSR Ser. Mat.*, *43*, 111–177, 238. English translation: *Math. USSR Izvestia*, *14*, 103–167 (1980).
24. Nordström, J. Extra-twisted connected sum G_2-manifolds. arXiv:1809.09083.
25. Salamon, S. M. (1989). *Riemannian geometry and holonomy groups*. Pitman research notes in mathematics (Vol. 201). Harlow: Longman.
26. Shokurov, V. V. (1979). Smoothness of a general anticanonical divisor on a Fano variety. *Izv. Akad. Nauk SSSR Ser. Mat.*, *43*, 430–441. English translation: *Math. USSR-Izv.*, *14*, 395–405 (1980).
27. Tian, G., & Yau, S.-T. (1990). Complete Kähler manifolds with zero Ricci curvature. I. *Journal of the American Mathematical Society*, *3*, 579–609.

Calibrated Submanifolds

Jason D. Lotay

Abstract We provide an introduction to the theory of calibrated submanifolds through the key examples related with special holonomy. We focus on calibrated geometry in Calabi–Yau, G_2 and Spin(7) manifolds, and describe fundamental results and techniques in the field.

1 Introduction

A key aspect of mathematics is the study of variational problems. These can vary from the purely analytic to the very geometric. A classic geometric example is the study of geodesics, which are critical points for the length functional on curves. As we know, understanding the geodesics of a given Riemannian manifold allows us to understand some of the ambient geometry, for example the curvature. The higher dimensional analogue would be to study critical points for the volume functional, and we would hope (and it indeed turns out to be the case) that these critical points, called *minimal submanifolds*, encode crucial aspects of the geometry of the manifold.

Just like the geodesic equation, we would expect (and it is true) that minimal submanifolds are defined by a (nonlinear) second order partial differential equation. Such equations are very difficult to solve in general, so a key idea is to find a special class of minimal submanifolds, called *calibrated submanifolds*, which are instead defined by a first order partial differential equation. The definition of calibrated submanifolds is motivated by the properties of complex submanifolds in Kähler manifolds, and turns out to be useful in finding minimizers for the volume functional rather than just critical points. However, finding examples outside the classical complex setting turns out to be difficult, leading to important methods coming from a variety of sources, as well as motivating the study of the deformation theory of these objects.

J. D. Lotay (✉)
University of Oxford, Oxford, UK
e-mail: lotay@maths.ox.ac.uk

S. Karigiannis et al. (eds.), *Lectures and Surveys on G_2-Manifolds and Related Topics*, Fields Institute Communications 84, https://doi.org/10.1007/978-1-0716-0577-6_3

Calibrated submanifolds naturally arise when the ambient manifold has *special holonomy*, including holonomy G_2. In this situation, we would hope that the calibrated submanifolds encode even more, finer, information about the ambient manifold, potentially leading to the construction of new invariants. In this setting, there is also a relationship between calibrated submanifolds and gauge theory: specifically, connections whose curvature satisfies a natural constraint determined by the special holonomy group (so-called *instantons*). For these reasons, calibrated submanifolds form a hot topic in current research, especially in the G_2 setting.

Note These notes are primarily based on a lecture course the author gave at the LMS–CMI Research School "An Invitation to Geometry and Topology via G_2" at Imperial College London in July 2014.

2 Minimal Submanifolds

We start by analysing the submanifolds which are critical points for the volume functional. Let N be a submanifold (without boundary) of a Riemannian manifold (M, g) and let $F : N \times (-\epsilon, \epsilon) \to M$ be a variation of N with compact support; i.e. $F = \text{Id}$ outside a compact subset $\overline{S}$ of N with S open and $F(p, 0) = p$ for all $p \in N$. The vector field $X = \frac{\partial F}{\partial t}|_N$ is called the variation vector field (which will be zero outside of $\overline{S}$). We then have the following definition.

Definition 2.1 N is *minimal* if $\frac{\mathrm{d}}{\mathrm{d}t}\text{Vol}(F(S,t))|_{t=0} = 0$ for all variations F with compact support $\overline{S}$ (depending on F).

Remark Notice that we do not ask for N to minimize volume: it is only stationary for the volume. It could even be a maximum!

Example A plane in $\mathbb{R}^n$ is minimal since any small variation will have larger volume.

Example Geodesics are locally length minimizing, so geodesics are minimal. However, as an example, the equator in $\mathcal{S}^2$ is minimal but not length minimizing since we can deform it to a shorter line of latitude.

For simplicity let us suppose that N is compact. We wish to calculate $\frac{\mathrm{d}}{\mathrm{d}t}\text{Vol}(F(N,t))|_{t=0}$. Given local coordinates x_i on N we know that

$$\text{Vol}(F(N,t)) = \int_N \sqrt{\det\left(g\left(\frac{\partial F}{\partial x_i}, \frac{\partial F}{\partial x_j}\right)\right)}\, \text{vol}_N .$$

Let $p \in N$ and choose our coordinates x_i to be normal coordinates at p: i.e. so that $\frac{\partial F}{\partial x_i}(p,t) = e_i(t)$ satisfy $g(e_i(0), e_j(0)) = \delta_{ij}$. If $g_{ij}(t) = g(e_i(t), e_j(t))$ and $(g^{ij}(t))$ denotes the inverse of the matrix $(g_{ij}(t))$ then we know that

$$\frac{\mathrm{d}}{\mathrm{d}t}\sqrt{\det(g_{ij}(t))}|_{t=0} = \frac{1}{2}\frac{\sum_{i,j} g^{ij}(t)g'_{ij}(t)}{\sqrt{\det(g_{ij}(t))}}|_{t=0} = \frac{1}{2}\sum_i g'_{ii}(0).$$

Now, if we let ∇ denote the Levi-Civita connection of g, then

$$\begin{aligned}\frac{1}{2}\sum_i g'_{ii}(0) &= \frac{1}{2}\sum_i \frac{\mathrm{d}}{\mathrm{d}t} g\left(\frac{\partial F}{\partial x_i}, \frac{\partial F}{\partial x_i}\right)|_{t=0} \\ &= \sum_i g(\nabla_X e_i, e_i) \\ &= \sum_i g(\nabla_{e_i} X, e_i) = \mathrm{div}_N(X)\end{aligned}$$

since $[X, e_i] = 0$ (i.e. the t and x_i derivatives commute). Moreover, we see that

$$\begin{aligned}\mathrm{div}_N(X) = \sum_i g(\nabla_{e_i} X, e_i) &= \mathrm{div}_N(X^{\mathrm{T}}) - \sum_i g(X^{\perp}, \nabla_{e_i} e_i) \\ &= \mathrm{div}_N(X^{\mathrm{T}}) - g(X, H)\end{aligned}$$

(since $\nabla_{e_i}\big(g(X^{\perp}, e_i)\big) = 0$) where ^{T} and $^{\perp}$ denote the tangential and normal parts and

$$H = \sum_i \nabla^{\perp}_{e_i} e_i$$

is the *mean curvature vector*. Overall we have the following.

Theorem 2.2 *The first variation formula is*

$$\frac{\mathrm{d}}{\mathrm{d}t}\mathrm{Vol}(F(N,t))|_{t=0} = \int_N \mathrm{div}_N(X)\,\mathrm{vol}_N = -\int_N g(X, H)\,\mathrm{vol}_N .$$

Remark The $\mathrm{div}_N(X^{\mathrm{T}})$ term does not appear in the first variation formula because its integral vanishes by the divergence theorem as N is compact without boundary. In general, it will still vanish since we assume for our variations that there exists a compact submanifold of N with boundary which contains the support of X^{T} and so that X^{T} vanishes on the boundary.

We deduce the following.

Definition 2.3 N is a *minimal submanifold* if and only if $H = 0$.

The equation $H = 0$ is a *second order nonlinear PDE*. We can see this explicitly in the following simple case. For a function $f : U \subseteq \mathbb{R}^{n-1} \to \mathbb{R}$ where $\overline{U}$ is compact, we see that if $N = \mathrm{Graph}(f) \subseteq \mathbb{R}^n$ then the volume of N is given by

$$\mathrm{Vol}(N) = \int_U \sqrt{1 + |\nabla f|^2}\,\mathrm{vol}_U .$$

Any sufficiently small variation can be written $F(N,t) = \text{Graph}(f + th)$ for some $h : U \to \mathbb{R}$, so we can compute

$$\begin{aligned}\frac{\mathrm{d}}{\mathrm{d}t}\operatorname{Vol}(F(N,t))|_{t=0} &= \frac{\mathrm{d}}{\mathrm{d}t}|_{t=0}\int_U \sqrt{1+|\nabla f + t\nabla h|^2}\,\mathrm{vol}_U \\ &= \int_U \frac{\mathrm{d}}{\mathrm{d}t}|_{t=0}\sqrt{1+|\nabla f|^2 + 2t\langle\nabla f,\nabla h\rangle + t^2|\nabla h|^2}\,\mathrm{vol}_U \\ &= \int_U \frac{\langle\nabla f,\nabla h\rangle}{\sqrt{1+|\nabla f|^2}}\,\mathrm{vol}_U \\ &= -\int_U h \operatorname{div}\left(\frac{\nabla f}{\sqrt{1+|\nabla f|^2}}\right)\mathrm{vol}_U\,.\end{aligned}$$

We therefore see that N is minimal if and only if this vanishes for all h. Hence, $\text{Graph}(f)$ is minimal in $\mathbb{R}^n$ if and only if

$$\operatorname{div}\left(\frac{\nabla f}{\sqrt{1+|\nabla f|^2}}\right) = 0.$$

We see that we can write this equation as $\Delta f + Q(\nabla f, \nabla^2 f) = 0$ where Q consists of nonlinear terms (but linear in $\nabla^2 f$). Hence, if we linearise this equation we just get $\Delta f = 0$, so f is harmonic. In other words, the minimal submanifold equation is a nonlinear equation whose linearisation is just Laplace's equation: this is an example of a nonlinear *elliptic* PDE, which we shall discuss further later.

Example A plane in $\mathbb{R}^n$ is trivially minimal because if X, Y are any vector fields on the plane then $\nabla_X^\perp Y = 0$ as the second fundamental form of a plane is zero.

Example For curves γ, $H = 0$ is equivalent to the geodesic equation $\nabla_{\dot\gamma}\dot\gamma = 0$.

The most studied minimal submanifolds (other than geodesics) are minimal surfaces in $\mathbb{R}^3$, since here the equation $H = 0$ becomes a scalar equation on a surface, which is the simplest to analyse. In general we would have a system of equations, which is more difficult to study.

Example The helicoid $M = \{(t\cos s, t\sin s, s) \in \mathbb{R}^3 : s, t \in \mathbb{R}\}$ is a complete embedded minimal surface, discovered by Meusnier in 1776.

Example The catenoid $M = \{(\cosh t\cos s, \cosh t\sin s, t) \in \mathbb{R}^3 : s, t, \in \mathbb{R}\}$ is a complete embedded minimal surface, discovered by Euler in 1744 and shown to be minimal by Meusnier in 1776. The catenoid is another explicit example which is a critical point for volume but not minimizing.

In fact the helicoid and the catenoid are locally isometric, and there is a 1-parameter family of locally isometric minimal surfaces deforming between the catenoid and helicoid: see, for example, [18, Theorem 16.5] for details.

It took about 70 years to find the next minimal surface, but now we know many examples of minimal surfaces in $\mathbb{R}^3$, as well as in other spaces by studying the nonlinear elliptic PDE given by the minimal surface equation. The amount of literature in the area is vast, with key results including the proofs of the Lawson [1], Willmore [63] and Yau [29, 64, 77] Conjectures, and minimal surfaces have applications to major problems in geometry including the Positive Mass Theorem [75, 76], Penrose Inequality [24] and Poincaré Conjecture [74].

3 Introduction to Calibrations

As we have seen, minimal submanifolds are extremely important. However there are two key issues.

- Minimal submanifolds are defined by a second order nonlinear PDE system—therefore they are hard to analyse.
- Minimal submanifolds are only critical points for the volume functional, but we are often interested in minima for the volume functional—we need a way to determine when this occurs.

We can help resolve these issues using the notion of calibration and calibrated submanifolds, introduced by Harvey–Lawson [20] in 1982.

Definition 3.1 A differential k-form η on a Riemannian manifold (M, g) is a *calibration* if

- $\mathrm{d}\eta = 0$ and
- $\eta(e_1, \ldots, e_k) \leq 1$ for all unit tangent vectors $e_1, \ldots, e_k$ on M.

Example Any non-zero form with constant coefficients on $\mathbb{R}^n$ can be rescaled so that it is a calibration with at least one plane where equality holds.

This example shows that there are many calibrations η, but the interesting question is: for which oriented planes $P = \mathrm{Span}\{e_1, \ldots, e_k\}$ does $\eta(e_1, \ldots, e_k) = 1$? More importantly, can we find submanifolds N so that this equality holds on each tangent space? This motivates the next definition.

Definition 3.2 Let η be a calibration k-form on (M, g). An oriented k-dimensional submanifold N of (M, g) is *calibrated* by η if $\eta|_N = \mathrm{vol}_N$, i.e. if for all $p \in N$ we have $\eta(e_1, \ldots, e_k) = 1$ for an oriented orthonormal basis $e_1, \ldots, e_k$ for T_pN.

Example Any oriented plane in $\mathbb{R}^n$ is calibrated. If we change coordinates so that the plane P is $\{x \in \mathbb{R}^n : x_{k+1} = \cdots = x_n = 0\}$ (with the obvious orientation) then $\eta = \mathrm{d}x_1 \wedge \cdots \wedge \mathrm{d}x_k$ is a calibration and P is calibrated by η.

Notice that the calibrated condition is now an algebraic condition on the tangent vectors to N, so being calibrated is a *first order nonlinear PDE*. We shall motivate these definitions further later, but for now we make the following observation.

Theorem 3.3 *Let N be a calibrated submanifold. Then N is minimal and, moreover, if F is any variation with compact support $\overline{S}$ then* $\mathrm{Vol}(F(S,t)) \geq \mathrm{Vol}(S)$*; i.e. N is volume-minimizing. In particular, if N is compact then N is volume-minimizing in its homology class.*

Proof Suppose that N is calibrated by η and suppose for simplicity that N is compact. We will show that N is homologically volume-minimizing.

Suppose that N' is homologous to N. Then there exists a compact K with boundary $-N \cup N'$ and, since $\mathrm{d}\eta = 0$, we have by Stokes' Theorem that

$$0 = \int_K \mathrm{d}\eta = \int_{N'} \eta - \int_N \eta.$$

We deduce that

$$\mathrm{Vol}(N) = \int_N \eta = \int_{N'} \eta \leq \mathrm{Vol}(N').$$

We then have the result by the definition of minimal submanifold. □

We conclude this introduction with the following elementary result.

Proposition 3.4 *There are no compact calibrated submanifolds in $\mathbb{R}^n$.*

Proof Suppose that η is a calibration and N is compact and calibrated by η. Then $\mathrm{d}\eta = 0$ so by the Poincaré Lemma $\eta = \mathrm{d}\zeta$, and hence

$$\mathrm{Vol}(N) = \int_N \eta = \int_N \mathrm{d}\zeta = 0$$

by Stokes' Theorem. □

Although there are many calibrations, having calibrated submanifolds greatly restricts the calibrations you want to consider. The calibrations which have calibrated submanifolds have special significance and there is a particular connection with special holonomy, due to the following observations.

Let G be the holonomy group of a Riemannian metric g on an n-manifold M. Then G acts on the k-forms on $\mathbb{R}^n$, so suppose that η_0 is a G-invariant k-form. We can always rescale η_0 so that $\eta_0|_P \leq \mathrm{vol}_P$ for all oriented k-planes P and equality holds for at least one P. Since η_0 is G-invariant, if P is calibrated then so is $\gamma \cdot P$ for any $\gamma \in G$, which usually means we have quite a few calibrated planes. We know by the *holonomy principle* (see, for example, [42, Proposition 2.5.2]) that we then get a parallel k-form η on M which is identified with η_0 at every point. Since $\nabla\eta = 0$, we have $\mathrm{d}\eta = 0$ and hence η is a calibration. Moreover, we have a lot of calibrated tangent planes on M, so we can hope to find calibrated submanifolds.

4 Complex Submanifolds

We would now like to address the question: where does the calibration condition come from? The answer is from *complex geometry*. On $\mathbb{R}^{2n} = \mathbb{C}^n$ with coordinates $z_j = x_j + iy_j$, we have the complex structure J and the distinguished Kähler 2-form

$$\omega = \sum_{j=1}^{n} \mathrm{d}x_j \wedge \mathrm{d}y_j = \frac{i}{2} \sum_{j=1}^{n} \mathrm{d}z_j \wedge \mathrm{d}\bar{z}_j.$$

More generally we can work with a *Kähler manifold* (M, J, ω). Our first key result is the following.

Theorem 4.1 *On a Kähler manifold* (M, J, ω), $\frac{\omega^k}{k!}$ *is a calibration whose calibrated submanifolds are the complex k-dimensional submanifolds: i.e. submanifolds N such that $J(T_pN) = T_pN$ for all $p \in N$.*

Since $\mathrm{d}\omega^k = k\mathrm{d}\omega \wedge \omega^{k-1} = 0$, Theorem 4.1 follows immediately from the following result.

Theorem 4.2 (Wirtinger's inequality) *For any unit vectors* $e_1, \ldots, e_{2k} \in \mathbb{C}^n$,

$$\frac{\omega^k}{k!}(e_1, \ldots, e_{2k}) \leq 1$$

with equality if and only if $\mathrm{Span}\{e_1, \ldots, e_{2k}\}$ *is a complex k-plane in $\mathbb{C}^n$.*

Before proving this we make the following observation.

Lemma 4.3 *If η is a calibration and $*\eta$ is closed then $*\eta$ is a calibration. Moreover an oriented tangent plane P is calibrated by η if and only if there is an orientation on the orthogonal complement $P^\perp$ so that it is calibrated by $*\eta$.*

Proof Suppose that η is a calibration k-form on (M, g) with $\mathrm{d}*\eta = 0$. Let $p \in M$. Take any $n - k$ orthonormal tangent vectors $e_{k+1}, \ldots, e_n$ at p. Then there exist $e_1, \ldots, e_k \in T_pM$ so that $\{e_1, \ldots, e_n\}$ is an oriented orthonormal basis for T_pM. Since $\{e_1, \ldots, e_n\}$ is an oriented orthonormal basis, we can use the definition of the Hodge star to calculate

$$*\eta(e_{k+1}, \ldots, e_n) = \eta(e_1, \ldots, e_k) \leq 1.$$

Hence $*\eta$ is a calibration by Definition 3.1. Moreover, the oriented plane $P = \mathrm{Span}\{e_{k+1}, \ldots, e_n\}$ is calibrated by $*\eta$ if and only if there is an orientation on $\mathrm{Span}\{e_1, \ldots, e_k\} = P^\perp$ so that it is calibrated by η, since $\eta(e_1, \ldots, e_k) = \pm *\eta(e_{k+1}, \ldots, e_n) = \pm 1$. □

We can now prove Wirtinger's inequality.

Proof of Theorem 4.2 We see that $|\frac{\omega^k}{k!}|^2 = \frac{n!}{k!(n-k)!}$ and $\mathrm{vol}_{\mathbb{C}^n} = \frac{\omega^n}{n!}$ so $*\frac{\omega^k}{k!} = \frac{\omega^{n-k}}{(n-k)!}$. Hence, by Lemma 4.3, it is enough to study the case where $k \leq \frac{n}{2}$.

Let P be any $2k$-plane in $\mathbb{C}^n$ with $2k \leq n$. We shall find a canonical form for P. First consider $\langle Ju, v\rangle$ for orthonormal vectors $u, v \in P$. This must have a maximum, so let $\cos\theta_1 = \langle Ju, v\rangle$ be this maximum realised by some orthonormal vectors $u, v \in P$, where $0 \leq \theta_1 \leq \frac{\pi}{2}$.

Suppose that $w \in P$ is a unit vector orthogonal to $\mathrm{Span}\{u, v\}$, where $\cos\theta_1 = \langle Ju, v\rangle$. The function

$$f_w(\theta) = \langle Ju, \cos\theta v + \sin\theta w\rangle$$

has a maximum at $\theta = 0$ so $f'_w(0) = \langle Ju, w\rangle = 0$. Similarly we have that $\langle Jv, w\rangle = 0$, and thus $w \in \mathrm{Span}\{u, v, Ju, Jv\}^\perp$.

We then have two cases. If $\theta_1 = 0$ then $v = Ju$ so we can set $u = e_1, v = Je_1$ and see that $P = \mathrm{Span}\{e_1, Je_1\} \times Q$ where Q is a $2(k-1)$-plane in $\mathbb{C}^{n-1} = \mathrm{Span}\{e_1, Je_1\}^\perp$. If $\theta_1 \neq 0$ we have that $v = \cos\theta_1 Ju + \sin\theta_1 w$ where w is a unit vector orthogonal to u and Ju, so we can let $u = e_1$, $w = e_2$ and see that $P = \mathrm{Span}\{e_1, \cos\theta_1 Je_1 + \sin\theta_1 e_2\} \times Q$ where Q is a $2(k-1)$-plane in $\mathbb{C}^{n-2} = \mathrm{Span}\{e_1, Je_1, e_2, Je_2\}^\perp$.

Proceeding by induction we see that we have an oriented basis $\{e_1, Je_1, \ldots, e_n, Je_n\}$ for $\mathbb{C}^n$ so that

$$P = \mathrm{Span}\{e_1, \cos\theta_1 Je_1 + \sin\theta_1 e_2, \ldots, e_{2k-1}, \cos\theta_k Je_{2k-1} + \sin\theta_k e_{2k}\},$$

where $0 \leq \theta_1 \leq \cdots \leq \theta_{k-1} \leq \frac{\pi}{2}$ and $\theta_{k-1} \leq \theta_k \leq \pi - \theta_{k-1}$.

Since we can write $\omega = \sum_{j=1}^n e^j \wedge Je^j$ we see that $\frac{\omega^k}{k!}$ restricts to P to give a product of $\cos\theta_j$ which is certainly less than or equal to 1. Moreover, equality holds if and only if all of the $\theta_j = 0$ which means that P is complex. □

Putting together Theorems 4.1 and 3.3 yields the following.

Corollary 4.4 *Compact complex submanifolds of Kähler manifolds are homologically volume-minimizing.*

We know that complex submanifolds are defined by holomorphic functions; i.e. solutions to the Cauchy–Riemann equations, which are a first-order PDE system, as one would expect for calibrated submanifolds.

Example $N = \{(z, \frac{1}{z}) \in \mathbb{C}^2 : z \in \mathbb{C} \setminus \{0\}\}$ is a complex curve in $\mathbb{C}^2$, and thus is calibrated.

Example An important non-trivial example of a Kähler manifold is $\mathbb{CP}^n$, where the zero set of a system of polynomial equations defines a (possibly singular) complex submanifold.

5 Special Lagrangians

Complex submanifolds are very familiar, but can we find any other interesting classes of calibrated submanifolds? The answer is that indeed we can, particularly when the manifold has special holonomy. We begin with the case of holonomy $\mathrm{SU}(n)$—so-called *Calabi–Yau manifolds*. The model example for Calabi–Yau manifolds is $\mathbb{C}^n$ with complex structure J, Kähler form ω and holomorphic volume form

$$\Upsilon = \mathrm{d}z_1 \wedge \cdots \wedge \mathrm{d}z_n,$$

if $z_1, \ldots, z_n$ are complex coordinates on $\mathbb{C}^n$.

Theorem 5.1 *Let M be a Calabi–Yau manifold with holomorphic volume form Υ. Then* $\mathrm{Re}(e^{-i\theta}\Upsilon)$ *is a calibration for any $\theta \in \mathbb{R}$.*

Since $\mathrm{d}\Upsilon = 0$, the result follows immediately from the following.

Theorem 5.2 *On $\mathbb{C}^n$, $|\Upsilon(e_1, \ldots, e_n)| \leq 1$ for all unit vectors $e_1, \ldots, e_n$ with equality if and only if $P = \mathrm{Span}\{e_1, \ldots, e_n\}$ is a Lagrangian plane, i.e. P is an n-plane such that $\omega|_P \equiv 0$.*

Proof Let $e_1, \ldots, e_n$ be the standard basis for $\mathbb{R}^n$ and let P be an n-plane in $\mathbb{C}^n$. There exists $A \in \mathrm{GL}(n, \mathbb{C})$ so that $f_1 = Ae_1, \ldots, f_n = Ae_n$ is an orthonormal basis for P. Then $\Upsilon(Ae_1, \ldots, Ae_n) = \det_{\mathbb{C}}(A)$ so

$$\begin{aligned} |\Upsilon(f_1, \ldots, f_n)|^2 &= |\det{}_{\mathbb{C}}(A)|^2 \\ &= |\det{}_{\mathbb{R}}(A)| \\ &= |f_1 \wedge Jf_1 \wedge \cdots \wedge f_n \wedge Jf_n| \leq |f_1||Jf_1| \cdots |f_n||Jf_n| = 1 \end{aligned}$$

with equality if and only if $f_1, Jf_1, \ldots, f_n, Jf_n$ are orthonormal. However, this is exactly equivalent to the Lagrangian condition, since $\omega(u, v) = g(Ju, v)$ so $\omega|_P \equiv 0$ if and only if $JP = P^{\perp}$. □

Definition 5.3 A submanifold N of M calibrated by $\mathrm{Re}(e^{-i\theta}\Upsilon)$ is called *special Lagrangian* with phase $e^{i\theta}$. If $\theta = 0$ we say that N is simply special Lagrangian. By Theorem 5.2, we see that N is special Lagrangian if and only if $\omega|_N \equiv 0$ (i.e. N is Lagrangian) and $\mathrm{Im}\,\Upsilon|_N \equiv 0$ (up to a choice of orientation so that $\mathrm{Re}\,\Upsilon|_N > 0$).

Example Consider $\mathbb{C} = \mathbb{R}^2$ with coordinates $z = x + iy$, complex structure J given by $Jw = iw$, Kähler form $\omega = \mathrm{d}x \wedge \mathrm{d}y = \frac{i}{2}\mathrm{d}z \wedge \mathrm{d}\bar{z}$ and holomorphic volume form $\Upsilon = \mathrm{d}z = \mathrm{d}x + i\mathrm{d}y$. We want to consider the special Lagrangians in $\mathbb{C}$, which are 1-dimensional submanifolds or curves N in $\mathbb{C} - \mathbb{R}^2$.

Since ω is a 2-form, it vanishes on any curve in $\mathbb{C}$. Hence every curve in $\mathbb{C}$ is Lagrangian. For N to be special Lagrangian with phase $e^{i\theta}$ we need that

$$\mathrm{Re}(e^{-i\theta}\Upsilon) = \cos\theta \mathrm{d}x + \sin\theta \mathrm{d}y$$

is the volume form on N, or equivalently that

$$\operatorname{Im}(e^{-i\theta}\Upsilon) = \cos\theta \mathrm{d}y - \sin\theta \mathrm{d}x$$

vanishes on N. This means that $\cos\theta\partial_x + \sin\theta\partial_y$ is everywhere a unit tangent vector to N, so N is a straight line given by $N = \{(t\cos\theta, t\sin\theta) \in \mathbb{R}^2 : t \in \mathbb{R}\}$ (up to translation), so it makes an angle θ with the x-axis, hence motivating the term "phase $e^{i\theta}$".

Notice that this result is compatible with the fact that special Lagrangians are minimal, and hence must be geodesics in $\mathbb{R}^2$; i.e. straight lines.

Example Consider $\mathbb{C}^2 = \mathbb{R}^4$. We know that $\omega = \mathrm{d}x_1 \wedge \mathrm{d}y_1 + \mathrm{d}x_2 \wedge \mathrm{d}y_2$. Since $\Upsilon = \mathrm{d}z_1 \wedge \mathrm{d}z_2 = (\mathrm{d}x_1 + i\mathrm{d}y_1) \wedge (\mathrm{d}x_2 + i\mathrm{d}y_2)$, we also know that $\operatorname{Re}\Upsilon = \mathrm{d}x_1 \wedge \mathrm{d}x_2 + \mathrm{d}y_2 \wedge \mathrm{d}y_1$, which looks somewhat similar. In fact, if we let J' denote the complex structure given by $J'(\partial_{x_1}) = \partial_{x_2}$ and $J'(\partial_{y_2}) = \partial_{y_1}$, then $\operatorname{Re}\Upsilon = \omega'$, the Kähler form corresponding to the complex structure J'. Hence special Lagrangians in $\mathbb{C}^2$ are complex curves for a different complex structure.

In fact, we have a hyperkähler triple of complex structures J_1, J_2, J_3, where $J_1 = J$ is the standard one and $J_3 = J_1J_2 = -J_2J_1$ so that $J_1 = J_2J_3 = -J_3J_2$ and $J_2 = J_3J_1 = -J_1J_3$, and the corresponding Kähler forms are $\omega = \omega_1, \omega_2, \omega_3$ which are orthogonal and the same length with $\Upsilon = \omega_2 + i\omega_3$.

This shows we should only consider complex dimension 3 and higher to find new calibrated submanifolds.

Example Let $f : \mathbb{R}^n \to \mathbb{R}^n$ be a smooth function and let $N = \operatorname{Graph}(f) \subseteq \mathbb{R}^{2n} = \mathbb{C}^n$. We want to see when N is special Lagrangian. We see that tangent vectors to N are given by

$$e_1 + i\nabla_{e_1} f, \ldots, e_n + i\nabla_{e_n} f.$$

Hence N is Lagrangian if and only if

$$\omega(e_j + i\nabla_{e_j} f, e_k + i\nabla_{e_k} f) = \nabla_{e_k} f_j - \nabla_{e_j} f_k = 0$$

for all j, k. Since $\mathbb{R}^n$ is simply connected, this occurs if and only if there exists F such that $f_j = \nabla_{e_j} F$; i.e. $f = \nabla F$.

Recall that $\Upsilon = \mathrm{d}z_1 \wedge \cdots \wedge \mathrm{d}z_n$. We know that N is special Lagrangian if and only if N is Lagrangian and $\operatorname{Im}\Upsilon$ vanishes on N. Now

$$\Upsilon(a_1 + ib_1, \ldots, a_n + ib_n) = \det{}_{\mathbb{C}}(A + iB)$$

where A, B are the matrices with columns a_i, b_j respectively. Hence

$$\Upsilon(e_1 + i\nabla_{e_1}\nabla F, \ldots, e_n + i\nabla_{e_n}\nabla F) = \det{}_{\mathbb{C}}(I + i\operatorname{Hess} F),$$

where $\operatorname{Hess} F = (\frac{\partial^2 F}{\partial x_i \partial x_j})$.

Therefore $N = \text{Graph}(f)$ is special Lagrangian (up to a choice of orientation) if and only if $f = \nabla F$ and

$$\text{Im}\,\det_{\mathbb{C}}(I + i\,\text{Hess}\,F) = 0.$$

If $n = 2$,

$$I + i\text{Hess}\,F = \begin{pmatrix} 1 + iF_{xx} & iF_{xy} \\ iF_{yx} & 1 + iF_{yy} \end{pmatrix}.$$

Therefore, the determinant gives

$$1 - F_{xx}F_{yy} + F_{xy}^2 + i(F_{xx} + F_{yy}),$$

then the imaginary part is $F_{xx} + F_{yy}$. Therefore, N is special Lagrangian if and only if $\Delta F = 0$.

As we know, a graph in $\mathbb{C}^2$ of $f = u + iv : \mathbb{C} \to \mathbb{C}$ is a complex surface if and only if $u + iv$ is holomorphic, which implies that u, v are harmonic. We know that special Lagrangians in $\mathbb{C}^2$ are complex surfaces for a different complex structure, so this is expected.

If $n = 3$,

$$I + i\text{Hess}\,F = \begin{pmatrix} 1 + iF_{xx} & iF_{xy} & iF_{xz} \\ iF_{yx} & 1 + iF_{yy} & iF_{yz} \\ iF_{zx} & iF_{zy} & 1 + iF_{zz} \end{pmatrix}.$$

Hence,

$$\begin{aligned} \text{Im}\,\det_{\mathbb{C}}(I + i\,\text{Hess}\,F) &= F_{xx} + F_{yy} + F_{zz} \\ &\quad - F_{xx}(F_{yy}F_{zz} - F_{yz}^2) - F_{xy}(F_{yz}F_{zx} - F_{xy}F_{zz}) - F_{zx}(F_{xy}F_{yz} - F_{yy}F_{zx}). \end{aligned}$$

Therefore, N is special Lagrangian if and only if

$$\begin{aligned} -\Delta F &= F_{xx} + F_{yy} + F_{zz} \\ &= F_{xx}(F_{yy}F_{zz} - F_{yz}^2) - F_{xy}(F_{xy}F_{zz} - F_{yz}F_{zx}) + F_{zx}(F_{xy}F_{yz} - F_{yy}F_{zx}) \\ &= \det \text{Hess}\,F. \end{aligned}$$

We now wish to describe some very important examples of special Lagrangians, which are asymptotic to pairs of planes.

Example $\text{SU}(n)$ acts transitively on the space of special Lagrangian planes with isotropy $\text{SO}(n)$. So any special Lagrangian plane is given by $A \cdot \mathbb{R}^n$ for $A \in \text{SU}(n)$ where $\mathbb{R}^n$ is the standard real $\mathbb{R}^n$ in $\mathbb{C}^n$.

Given $\theta = (\theta_1, \ldots, \theta_n)$ we can define a plane

$$P(\theta) = \{(e^{i\theta_1}x_1, \ldots, e^{i\theta_n}x_n) \in \mathbb{C}^n \;:\; (x_1, \ldots, x_n) \in \mathbb{R}^n\}$$

(where we can swap orientation). We see that $P(\theta)$ is special Lagrangian if and only if $\mathrm{Re}\,\Upsilon|_P = \pm\cos(\theta_1 + \cdots + \theta_n) = 1$ so that $\theta_1 + \cdots + \theta_n \in \pi\mathbb{Z}$. Given any $\theta_1, \ldots, \theta_n \in (0, \pi)$ with $\theta_1 + \cdots + \theta_n = \pi$, there exists a special Lagrangian N (called a *Lawlor neck*) asymptotic to $P(0) \cup P(\theta)$: see, for example, [42, Example 8.3.15] or Sect. 9 for details. It is diffeomorphic to $\mathcal{S}^{n-1} \times \mathbb{R}$. By rotating coordinates we have a special Lagrangian with phase i asymptotic to $P(-\frac{\theta}{2}) \cup P(\frac{\theta}{2})$.

The simplest case is when $\theta_1 = \cdots = \theta_n = \frac{\pi}{n}$: here N is called the *Lagrangian catenoid*. When $n = 2$, under a coordinate change the Lagrangian catenoid becomes the complex curve $\{(z, \frac{1}{z}) \in \mathbb{C}^2 : z \in \mathbb{C} \setminus \{0\}\}$ that we saw before. When $n = 3$, the only possibilities for the angles are $\sum_i \theta_i = \pi, 2\pi$, but if $\sum_i \theta_i = 2\pi$ we can rotate coordinates and change the order of the planes so that $P(0) \cup P(\theta)$ becomes $P(0) \cup P(\theta')$ where $\sum_i \theta'_i = \pi$. Hence, given any pair of transverse special Lagrangian planes in $\mathbb{C}^3$, there exists a Lawlor neck asymptotic to their union.

Remark Using complex geometry it is easy to classify all of the smooth special Lagrangians in $\mathbb{C}^2$ asymptotic to a pair of transverse planes, and one sees that the Lawlor necks in $\mathbb{C}^2$ are the unique exact special Lagrangians with this property. It is now known that the Lawlor necks are the unique smooth exact special Lagrangians asymptotic to a pair of planes in all dimensions [25].

We can find special Lagrangians in Calabi–Yau manifolds using the following easy result.

Proposition 5.4 *Let (M, ω, Υ) be a Calabi–Yau manifold and let $\sigma : M \to M$ be such that $\sigma^2 = \mathrm{Id}$, $\sigma^*(\omega) = -\omega$, $\sigma^*(\Upsilon) = \overline{\Upsilon}$. Then Fix($\sigma$) is special Lagrangian, if it is non-empty.*

Example Let $X = \{[z_0, \ldots, z_4] \in \mathbb{CP}^4 : z_0^5 + \cdots + z_4^5 = 0\}$ (the *Fermat quintic*) with its Calabi–Yau structure (which exists by Yau's solution of the Calabi conjecture since the first Chern class of X vanishes). Let σ be the restriction of complex conjugation on $\mathbb{CP}^4$ to X. Then the fixed point set of σ, which is the real locus in X, is a special Lagrangian 3-fold (if it is non-empty). (There is a subtlety here: σ is certainly an anti-holomorphic isometric involution for the induced metric on X, but this is *not* the same as the Calabi–Yau metric on X. Nevertheless, it is the case that σ satisfies the conditions of Proposition 5.4.)

Example There exists a Calabi–Yau metric on $T^*\mathcal{S}^n$ (the Stenzel metric [78]) so that the base $\mathcal{S}^n$ is special Lagrangian: When $n = 2$ this is a hyperkähler metric called the Eguchi–Hanson metric [11].

6 Constructing Calibrated Submanifolds

It is easy to construct complex submanifolds in Kähler manifolds algebraically. Constructing other calibrated submanifolds is much more challenging because one needs to solve a nonlinear PDE, even in Euclidean space. There are approaches in Euclidean space and other simple spaces which have involved reducing the problem to ODEs or other problems which do not require PDE (for example, algebraic methods). For example, we have the following methods, which you can find out more about in [42] or the references provided.

- Symmetries/evolution equations [17, 20, 21, 28, 30, 31, 33, 34, 39–41, 52, 54].
- Use of integrable systems to study calibrated cones [8, 9, 22, 43, 65].
- Calibrated cones and ruled smoothings of these cones [2, 4, 13, 14, 32, 52, 53, 59].
- Vector sub-bundle constructions [27, 45, 46].
- Classification of calibrated submanifolds satisfying pointwise constraints on their second fundamental form [4, 12, 26, 59, 60].

However, an important direction which has borne fruit in calibrated geometry and special holonomy recently has been to study the nonlinear PDE head on, especially by perturbative and gluing methods.

We want to solve nonlinear PDE, so how do we tackle this? The idea is to use the linear case to help. Suppose we are on a compact manifold N and recall the theory of linear *elliptic* operators L of order l on N, including:

- the definition of ellipticity of L via the *principal symbol* σ_L (which encodes the highest order derivatives in the operator) being an isomorphism;
- the use of *Hölder spaces* $C^{k,a}$ to give elliptic regularity theory (so-called *Schauder theory*), namely that if $w \in C^{k,a}$ and $Lv = w$ then $v \in C^{k+l,a}$ and there is a universal constant C so that
$$\|v\|_{C^{k+l,a}} \leq C(\|Lv\|_{C^{k,a}} + \|v\|_{C^0})$$
(and we can drop the $\|v\|_{C^0}$ term if v is orthogonal to $\operatorname{Ker} L$);
- the adjoint operator L^* and that $\sigma_{L^*} = (-1)^l \sigma_L^*$ so that L^* is elliptic if and only if L is elliptic; and
- the Fredholm theory of L, namely that $\operatorname{Ker} L$ (and hence $\operatorname{Ker} L^*$) is finite-dimensional, and we can solve $Lv = w$ if and only if $w \in (\operatorname{Ker} L^*)^\perp$.

We shall discuss this in a model example which we shall use throughout this section.

Example The Laplacian on functions is given by $\Delta f = \mathrm{d}^*\mathrm{d} f$ which in normal coordinates at a point is given by $f \mapsto -\sum_i \frac{\partial^2 f}{\partial x_i^2}$, so it is a linear second order differential operator. We see that its principal symbol is $\sigma_\Delta(x, \xi) f = -|\xi|^2 f$ which is an isomorphism for $\xi \in T_x^* N \setminus \{0\}$, so Δ is elliptic. We therefore have that if $h \in C^{k,a}(N)$ and $\Delta f = h$ then $f \in C^{k+2,a}(N)$, and we have an estimate
$$\|f\|_{C^{k+2,a}} \leq C(\|\Delta f\|_{C^{k,a}} + \|f\|_{C^0}).$$

We also know that $\Delta^* = \Delta$ and Ker Δ is given by the constant functions (since if $f \in \text{Ker}\,\Delta$ then

$$0 = \langle f, \Delta f\rangle_{L^2} = \langle f, \mathrm{d}^*\mathrm{d}f\rangle_{L^2} = \|\mathrm{d}f\|^2_{L^2}$$

so $\mathrm{d}f = 0$). Hence, we can solve $\Delta f = h$ if and only if h is orthogonal to the constants, i.e. $\int_N h\,\mathrm{vol}_N = 0$.

The operator defining the minimal graph equation for a hypersurface is

$$P(f) = -\,\mathrm{div}\left(\frac{\nabla f}{\sqrt{1+|\nabla f|^2}}\right),$$

which is a nonlinear second order operator whose linearisation L_0P at 0 is Δ. Thus P is a nonlinear elliptic operator at 0. If we linearise P at f_0 we find a more complicated expression depending on f_0, but it is still a perturbation of the Laplacian.

Suppose we are on a compact manifold N and we want to solve $P(f) = 0$ where P is the minimal graph operator on functions f. Let us consider regularity for f. We can re-arrange $P(f) = 0$ by taking all of the second derivatives to one side as:

$$R(x, \nabla f(x))\nabla^2 f(x) = E(x, \nabla f(x))$$

where $x \in N$. Since $L_0P = \Delta$ is elliptic and ellipticity is an open condition we know that the operator L_f (depending on f) given by

$$L_f(h)(x) = R(x, \nabla f(x))\nabla^2 h(x)$$

is a *linear* elliptic operator whenever $\|\nabla f\|_{C^0}$ is small, in particular if $\|f\|_{C^{1,a}}$ is sufficiently small. The operator L_f does not have smooth coefficients, but if $f \in C^{k,a}$ then the coefficients $R \in C^{k-1,a}$.

Suppose that $f \in C^{1,a}$ and $\|f\|_{C^{1,a}}$ is small with $P(f) = 0$. Then $L_f(f) = E(f)$ and L_f is a linear *second order* elliptic operator with coefficients in $C^{0,a}$ and $E(f)$ is in $C^{0,a}$. So by elliptic regularity we can deduce that $f \in C^{2,a}$. We have gained one degree of regularity, so we can "bootstrap", i.e. proceed by induction and deduce that any $C^{1,a}$ solution to $P(f) = 0$ is smooth.

Example $C^{1,a}$*-minimal submanifolds (and thus calibrated submanifolds) are smooth.*

Remark More sophisticated techniques can be used to deduce that C^1-minimal submanifolds are real analytic [69]. Notice that elliptic regularity results are *not* valid for C^k spaces, so this result is not obvious.

We can also arrange our simple equation $P(f) = 0$ as $\Delta f + Q(\nabla f, \nabla^2 f) = 0$, where Q is nonlinear but linear in $\nabla^2 f$. If we know that $\int_N P(f)\,\mathrm{vol}_N = 0$, i.e. that $P(f)$ is orthogonal to the constants, then we can always solve $\Delta f_0 = -Q(\nabla f, \nabla^2 f)$. We do know that $\int_N P(f)\,\mathrm{vol}_N = 0$ since P has a divergence form. This means we are in the setting for implementing the Implicit Function Theorem for Banach spaces to conclude that we can always solve $P(f) = 0$ for some f near 0, and f will be smooth by our regularity argument above. In general, we will use the following.

Theorem 6.1 (Implicit Function Theorem) *Let X, Y be Banach spaces, let $U \ni 0$ be open in X, let $P : U \to Y$ with $P(0) = 0$ and $L_0 P : X \to Y$ surjective with finite-dimensional kernel K.*

Then for some U, $P^{-1}(0) = \{u \in U : P(u) = 0\}$ is a manifold of dimension $\dim K$. *Moreover, if we write $X = K \oplus Z$, $P^{-1}(0) = \text{Graph}\, G$ for some map G from an open set in K to Z with $G(0) = 0$.*

This gives us a way to describe all perturbations of a given calibrated submanifold, as we now see in the special Lagrangian case, due to McLean [66].

Theorem 6.2 *Let N be a compact special Lagrangian in a Calabi–Yau manifold M. Then the moduli space of deformations of N is a smooth manifold of dimension $b^1(N)$.*

Remark One should compare this result to the deformation theory for complex submanifolds in Kähler manifolds. There, one does not get that the moduli space is a smooth manifold: in fact, it can be singular, and one has *obstructions* to deformations. It is somewhat remarkable that special Lagrangian calibrated geometry enjoys a much better deformation theory than this classical calibrated geometry. The deformation theory of embedded compact complex submanifolds in Calabi–Yau manifolds has recently been revisited using analytic techniques [67].

Proof The tubular neighbourhood theorem gives us a diffeomorphism $\exp : S \subseteq \nu(N) \to T \subseteq M$ which maps the zero section to N; in other words, we can write any nearby submanifold to N as the graph of a normal vector field on N. We know that N is Lagrangian, so the complex structure J gives an isomorphism between $\nu(N)$ and TN and the metric gives an isomorphism between TN and T^*N: $v \mapsto g(Jv, .) = \omega(v, .) = \alpha_v$. Therefore any deformation of N in T is given as the graph of a 1-form. In fact, using the Lagrangian neighbourhood theorem, we can arrange that any $N' \in T$ is the graph of a 1-form α, so that if $f_\alpha : N \to N_\alpha$ is the natural diffeomorphism then

$$f_\alpha^*(\omega) = \mathrm{d}\alpha \quad \text{and} \quad -*f_\alpha^*(\operatorname{Im}\Upsilon) = F(\alpha, \nabla\alpha) = \mathrm{d}^*\alpha + Q(\alpha, \nabla\alpha),$$

where the second formula follows from a calculation using the special Lagrangian condition on N and the fact that the ambient structure is Calabi–Yau. Hence, N_α is special Lagrangian if and only if $P(\alpha) = (F(\alpha, \nabla\alpha), \mathrm{d}\alpha) = 0$. This means that infinitesimal special Lagrangian deformations are given by closed and coclosed 1-forms, which is the kernel of $L_0 P$.

Since $\operatorname{Im}\Upsilon = 0$ on N we have that $[\operatorname{Im}\Upsilon] = 0$ on N_α, which means that $f_\alpha^*(\operatorname{Im}\Upsilon)$ is exact. Thus $F(\alpha, \nabla\alpha) = - * f_\alpha^*(\operatorname{Im}\Upsilon)$ is coexact and so

$$P : C^\infty(S) \to \mathrm{d}^*(C^\infty(T^*N)) \oplus \mathrm{d}(C^\infty(T^*N)) \subseteq C^\infty(\Lambda^0 T^*N \oplus \Lambda^2 T^*N).$$

If we let $X = C^{1,a}(T^*N)$, $Y = \mathrm{d}^*(C^{1,a}(T^*N)) \oplus \mathrm{d}(C^{1,a}(T^*N))$ and $U = C^{1,a}(S)$ we can apply the Implicit Function Theorem if we know that

$$L_0P : \alpha \in X \mapsto (\mathrm{d}^*\alpha, \mathrm{d}\alpha) \in Y$$

is surjective, i.e. given $\mathrm{d}\beta + \mathrm{d}^*\gamma \in Y$ does there exist α such that $\mathrm{d}\alpha = \mathrm{d}\beta$ and $\mathrm{d}^*\alpha = \mathrm{d}^*\gamma$? If we let $\alpha = \beta + \mathrm{d}f$ then we need $\Delta f = \mathrm{d}^*\mathrm{d}f = \mathrm{d}^*(\gamma - \beta)$. Since

$$\int_N \mathrm{d}^*(\gamma - \beta)\,\mathrm{vol}_N = \pm \int_N \mathrm{d} * (\gamma - \beta) = 0$$

we can solve the equation for f, and hence L_0P is surjective.

Therefore $P^{-1}(0)$ is a manifold of dimension $\dim \operatorname{Ker} L_0P = b^1(N)$ by Hodge theory. Moreover, if $P(\alpha) = 0$ then N_α is special Lagrangian, hence minimal and since $\alpha \in C^{1,a}$ we deduce that α is in fact smooth. □

Example The special Lagrangian $\mathcal{S}^n$ in $T^*\mathcal{S}^n$ has $b^1 = 0$ and so is rigid.

Observe that if we have a special Lagrangian T^n in M then $b^1(T^n) = n$ and, if the torus is close to flat then its deformations locally foliate M (as there will be n nowhere vanishing harmonic 1-forms), so we can hope to find special Lagrangian torus fibrations. This cannot happen in compact manifolds without singular fibres, but still motivates the SYZ conjecture in Mirror Symmetry. The deformation result also motivates the following theorem [3].

Theorem 6.3 *Every compact oriented real analytic Riemannian 3-manifold can be isometrically embedded in a Calabi–Yau 3-fold as the fixed point set of an involution.*

Remark Theorem 6.2 has also been extended to certain non-compact, singular and boundary settings, for example in [6, 36, 72].

Another well-known way to get a solution of a linear PDE from two solutions is simply to add them. However, for a nonlinear PDE $P(v) = 0$ this will not work. Intuitively, we can try to add two solutions to give us a solution v_0 for which $P(v_0)$ is small. Then we may try to perturb v_0 by v to solve $P(v + v_0) = 0$.

Geometrically, this occurs when we have two calibrated submanifolds N_1, N_2 and then glue them together to give a submanifold N which is "almost" calibrated, then we deform N to become calibrated. If the two submanifolds N_1, N_2 are glued using a very long neck then one can imagine that N is almost the disjoint union of N_1, N_2 and so close to being calibrated. If instead one scales N_2 by a factor t and then glues it into a singular point of N_1, we can again imagine that as t becomes very small N resembles N_1 and so again is close to being calibrated. These two examples are in fact related, because if we rescale the shrinking N_2 to fixed size, then we get a long neck between N_1 and N_2 of length of order $-\log t$. However, although these pictures are appealing, they also reveal the difficulty in this approach: as t becomes small, N becomes more "degenerate", giving rise to analytic difficulties which are encoded in the geometry of N_1, N_2 and N.

These ideas are used extensively in geometry, and particularly successfully in calibrated geometry e.g. [7, 23, 35, 37, 38, 51, 57, 62, 73]. A particular simple case is the following, which we will describe to show the basic idea of the gluing method.

Theorem 6.4 *Let N be a compact connected 3-manifold and let $i : N \to M$ be a special Lagrangian immersion with transverse self-intersection points in a Calabi–Yau manifold M. Then there exist embedded special Lagrangians N_t such that $N_t \to N$ as $t \to 0$.*

Remark One might ask about the sense of convergence here: for definiteness, we can say that N_t converges to N in the sense of currents; that is, if we have any compactly supported 3-form χ on M then $\int_{N_t} \chi \to \int_N \chi$ as $t \to 0$. However, all sensible notions of convergence of submanifolds will be true in this setting.

Proof Here we only provide a sketch of the proof: see, for example, [35, Sect. 9] for a detailed proof.

At each self-intersection point of N the tangent spaces are a pair of transverse 3-planes, which we can view as a pair of tranverse special Lagrangian 3-planes P_1, P_2 in $\mathbb{C}^3$. Since we are in dimension 3, we know that there exists a (unique up to scale) special Lagrangian Lawlor neck L asymptotic to $P_1 \cup P_2$. We can then glue tL into N near each intersection point to get a compact embedded submanifold $S_t = N\#tL$ (if we glue in a Lawlor neck for every self-intersection point). We can also arrange that S_t is Lagrangian, i.e. that it is a Lagrangian connect sum.

Now we want to perturb S_t to be special Lagrangian. Since S_t is Lagrangian, by the deformation theory we can write any nearby submanifold as the graph of a 1-form α, and this graph will be special Lagrangian if and only if (using the same notation as in our deformation theory discussion)

$$P_t(\alpha) = (- * f_\alpha^*(\mathrm{Im}\,\Upsilon), f_\alpha^*(\omega)) = 0.$$

Since S_t is Lagrangian but not special Lagrangian we have that

$$f_\alpha^*(\omega) = \mathrm{d}\alpha \quad \text{and} \quad - * f_\alpha^*(\mathrm{Im}\,\Upsilon) = P_t(0) + \mathrm{d}_t^*\alpha + Q_t(\alpha, \nabla\alpha)$$

where $P_t(0) = - * \mathrm{Im}\,\Upsilon|_{S_t}$ and $\mathrm{d}_t^* = L_0 P_t$, which is a perturbation of the usual d^* since we are no longer linearising at a point where $P_t(0) = 0$. By choosing $\alpha = \mathrm{d}f$, we then have to solve

$$\Delta_t f = -P_t(0) - Q_t(\nabla f, \nabla^2 f)$$

where Δ_t is a perturbation of the Laplacian.

For simplicity, let us suppose that Δ_t is the Laplacian on S_t. The idea is to view our equation as a fixed point problem. We know that if we let $X^k = \{f \in C^{k,a}(N) : \int_N f \,\mathrm{vol}_N = 0\}$ then $\Delta_t : X^{k+2} \to X^k$ is an isomorphism so it has an inverse G_t. We know by our elliptic regularity result that there exists a constant $C(\Delta_t)$ such that

$$\|f\|_{C^{k+2,a}} \leq C(\Delta_t)\|\Delta_t f\|_{C^{k,a}} \Leftrightarrow \|G_t h\|_{C^{k+2,a}} \leq C(\Delta_t)\|h\|_{C^{k,a}}$$

for any $f \in X^{k+2}$, $h \in X^k$.

We thus see that $P_t(f) = 0$ for $f \in X^{k+2}$ if and only if

$$f = G_t(-P_t(0) - Q_t(f)) = F_t(f).$$

The idea is now to show that F_t is a contraction sufficiently near 0 for all t small enough. Then it will have a (unique) fixed point near 0, which will also be smooth because it satisfies $P_t(f) = 0$ and hence defines a special Lagrangian as the graph of $\mathrm{d}f$ over S_t.

We know that $F_t : X^{k+2} \to X^{k+2}$ with

$$\|F_t(f_1) - F_t(f_2)\|_{C^{k+2,a}} = \|G_t(Q_t(f_1) - Q_t(f_2))\|_{C^{k+2,a}} \leq C(\Delta_t)\|Q_t(f_1) - Q_t(f_2)\|_{C^{k,a}}.$$

Since Q_t and its first derivatives vanish at 0 we know that

$$\|Q_t(f_1) - Q_t(f_2)\|_{C^{k,a}} \leq C(Q_t)\|f_1 - f_2\|_{C^{k+2,a}}(\|f_1\|_{C^{k+2,a}} + \|f_2\|_{C^{k+2,a}}).$$

We deduce that

$$\|F_t(f_1) - F_t(f_2)\|_{C^{k+2,a}} \leq C(\Delta_t)C(Q_t)\|f_1 - f_2\|_{C^{k+2,a}}(\|f_1\|_{C^{k+2,a}} + \|f_2\|_{C^{k+2,a}})$$

and

$$\|F_t(0)\|_{C^{k+2,a}} = \|G_t(P_t(0))\|_{C^{k+2,a}} \leq C(\Delta_t)\|P_t(0)\|_{C^{k,a}}.$$

Hence, F_t is a contraction on $\overline{B_{\epsilon_t}(0)} \subseteq X^{k+2}$ if we can choose ϵ_t so that

$$2C(\Delta_t)\|P_t(0)\|_{C^{k,a}} \leq \epsilon_t \leq \frac{1}{4C(\Delta_t)C(Q_t)}.$$

(This also proves Theorem 6.2, where we used the Implicit Function Theorem, by hand since there $P_t(0) = P(0) = 0$ so we just need to take ϵ_t small enough.) In other words, we need that

- $P_t(0)$ is small, so S_t is "close" to being calibrated and is a good approximation to $P_t(f) = 0$;
- $C(\Delta_t)$, $C(Q_t)$, which are determined by the linear PDE and geometry of N, L and S_t, are well-controlled as $t \to 0$.

The statement of the theorem is then that there exists t sufficiently small and ϵ_t so that the contraction mapping argument works.

This is a delicate balancing act since as $t \to 0$ parts of the manifold are collapsing, so the constants $C(\Delta_t)$, $C(Q_t)$ above (which depend on t) can and typically do blow-up as $t \to 0$. To control this, we need to understand the Laplacian on N, L and S_t and introduce "weighted" Banach spaces so that tL gets rescaled to constant size (independent of t), and S_t resembles the union of two manifolds with a cylindrical neck (as we described earlier). It is also crucial to understand the relationship between the kernels and cokernels of the Laplacian on the *non-compact* N (with the intersection points removed), L and compact S_t: here is where connectedness is important so that the kernel and cokernel of the Laplacian is 1-dimensional. □

Remark In more challenging gluing problems it is not possible to show that the relevant map is a contraction, but rather one can instead appeal to an alternative theorem (e.g. Schauder fixed point theorem) to show that it still has a fixed point.

7 Associative and Coassociative Submanifolds

We now want to introduce our calibrated geometry associated with G_2 holonomy. The first key result is the following.

Theorem 7.1 *Let (M^7, φ) be a G_2 manifold (so φ is a closed and coclosed positive 3-form). Then φ and $*\varphi$ are calibrations.*

Proof Let u, v, w be oriented orthonormal vectors in $\mathbb{R}^7$. There exists an element A of G_2 so that $Au = e_1$. The subgroup of G_2 fixing e_1 is isomorphic to $SU(3)$, and we know from the proof of Wirtinger's inequality (Theorem 4.2) there exists a (special) unitary transformation so that $v = e_2$ and $w = \cos\theta e_3 + \sin\theta v$ for some θ and v orthogonal to e_1, e_2, e_3. Since $\varphi(e_1, e_2, .) = dx_3$ by the formula below, we see that $\varphi(u, v, w) = \cos\theta$. Hence, since φ is closed, φ is a calibration and the calibrated planes are given by $A.\mathrm{Span}\{e_1, e_2, e_3\}$ for $A \in G_2$.

By Lemma 4.3, $*\varphi$ is also a calibration. □

Let us look at the calibrated planes and start with φ, which we take to be the following on $\mathbb{R}^7$:

$$\varphi = dx_{123} + dx_{145} + dx_{167} + dx_{246} - dx_{257} - dx_{347} - dx_{356},$$

where we use the short-hand notation $dx_{ij\ldots k} = dx_i \wedge dx_j \wedge \cdots \wedge dx_k$. Hence, $*\varphi$ on $\mathbb{R}^7$ is given by:

$$*\varphi = dx_{4567} + dx_{2367} + dx_{2345} + dx_{1357} - dx_{1346} - dx_{1256} - dx_{1247}.$$

If u, v, w are unit vectors in $\mathbb{R}^7 \cong \mathrm{Im}\,\mathbb{O}$ (the imaginary octonions), then $\varphi(u, v, w) = \langle u \times v, w\rangle = 1$ if and only if $w = u \times v$, so $P = \mathrm{Span}\{u, v, w\}$ is a copy of $\mathrm{Im}\,\mathbb{H}$ in $\mathrm{Im}\,\mathbb{O}$; in other words, $\mathrm{Span}\{1, u, v, w\}$ is an associative subalgebra of $\mathbb{O}$. Moreover, suppose we define a vector-valued 3-form χ on $\mathbb{R}^7$ by

$$\chi(u, v, w) = [u, v, w] = u(vw) - (uv)w,$$

known as the *associator*. Then we observe the following.

Lemma 7.2 *A 3-plane P in $\mathbb{R}^7$ satisfies $\chi|_P \equiv 0$ if and only if P admits an orientation so that it is calibrated by φ.*

Proof Since the associator is clearly invariant under G_2 we can put any plane P in standard position using G_2, i.e. as in the proof of Theorem 7.1, we can write $P = \mathrm{Span}\{e_1, e_2, \cos\theta e_3 + \sin\theta v\}$ for some v orthogonal to e_1, e_2, e_3. We can calculate that $[e_1, e_2, e_3] = 0$ whereas $[e_1, e_2, v] \neq 0$ for any v orthogonal to e_1, e_2, e_3. Moreover, P is calibrated by φ if and only if $\theta = 0$. We thus see that P is calibrated by φ (up to a choice a orientation) if and only if $\chi|_P \equiv 0$. □

Hence we call the φ-calibrated planes *associative*. In general on a G_2 manifold we can define a 3-form χ with values in TM using the pointwise formula above.

For $*\varphi$ we see that $*\varphi|_P = \mathrm{vol}_P$ for a plane P if and only if $\varphi|_{P^\perp} = \mathrm{vol}_{P^\perp}$. Hence the planes calibrated by $*\varphi$ are the orthogonal complements of the associative planes, so we call them *coassociative*. We have a similar alternative characterisation for 4-planes calibrated by $*\varphi$.

Lemma 7.3 *A 4-plane P in $\mathbb{R}^7$ satisfies $\varphi|_P \equiv 0$ if and only if P admits an orientation so that it is calibrated by $*\varphi$.*

Proof We know that given a 4-plane P we can choose coordinates such that $P^\perp = \mathrm{Span}\{e_1, e_2, \cos\theta e_3 + \sin\theta(a_4e_4 + a_5e_5 + a_6e_6 + a_7e_7)\}$ where $\sum_j a_j^2 = 1$. Then

$$\begin{aligned} P = \mathrm{Span}\{ &-\sin\theta e_3 + \cos\theta(a_je_j), a_5e_4 - a_4e_5 + a_7e_6 - a_6e_7, \\ &a_6e_4 - a_7e_5 - a_4e_6 + a_5e_7, a_7e_4 + a_6e_5 - a_5e_6 - a_4e_7\}. \end{aligned}$$

We can then see directly that $*\varphi|_P = \cos\theta$. We also have $\varphi(e_i, e_j, e_k) = 0$ for $i, j, k \in \{4, 5, 6, 7\}$ and $e_3 \lrcorner \varphi = -\mathrm{d}x_{47} - \mathrm{d}x_{56}$, so that $\varphi(-\sin\theta e_3 + \cos\theta(a_je_j), v, w)$ is a non-zero multiple of $\sin\theta$ for some $v, w \in P$. Hence $\varphi|_P = 0$ if and only if $\theta = 0$, which is if and only if P is calibrated by $*\varphi$ (again up to a choice of orientation). □

We thus can define our calibrated submanifolds.

Definition 7.4 Submanifolds in (M^7, φ) calibrated by φ are called *associative* 3-folds. Moreover, N is associative if and only if $\chi|_N \equiv 0$ (up to a choice of orientation).

Submanifolds in (M^7, φ) calibrated by $*\varphi$ are called *coassociative* 4-folds. Moreover, N is coassociative if and only if $\varphi|_N \equiv 0$ (up to a choice of orientation).

It is instructive to see the form that the associative or coassociative condition takes by studying associative or coassociative graphs in $\mathbb{R}^7$: see [20] for details.

A simple way to get associative and coassociative submanifolds is by using known geometries.

Proposition 7.5 *Let $x_1, \ldots, x_7$ be coordinates on $\mathbb{R}^7$ and let $z_j = x_{2j} + ix_{2j+1}$ be coordinates on $\mathbb{C}^3$ so that $\mathbb{R}^7 = \mathbb{R} \times \mathbb{C}^3$.*

(a) $N = \mathbb{R} \times S \subseteq \mathbb{R} \times \mathbb{C}^3$ *is associative or coassociative if and only if S is a complex curve or a special Lagrangian 3-fold with phase* $-i$, *respectively.*
(b) $N \subseteq \{0\} \times \mathbb{C}^3$ *is associative or coassociative if and only if N is a special Lagrangian 3-fold or a complex surface, respectively.*

Proof Recall the Kähler form ω and holomorphic volume form Υ on $\mathbb{C}^3$. We can write

$$\varphi = dx_1 \wedge \omega + \operatorname{Re} \Upsilon \quad \text{and} \quad *\varphi = \frac{1}{2}\omega^2 - dx_1 \wedge \operatorname{Im} \Upsilon.$$

For associatives, we see that $\varphi|_{\mathbb{R}\times S} = dx_1 \wedge \operatorname{vol}_S$ if and only if $\omega|_S = \operatorname{vol}_S$ and $\varphi|_N = \operatorname{Re} \Upsilon|_N$ for $N \subseteq \mathbb{C}^3$. For coassociatives, we see that $*\varphi|_{\mathbb{R}\times S} = dx_1 \wedge \operatorname{vol}_S$ if and only if $-\operatorname{Im} \Upsilon|_S = \operatorname{vol}_S$ and $*\varphi|_N = \frac{1}{2}\omega^2|_N$ for $N \subseteq \mathbb{C}^3$. The results quickly follow. □

We can also produce examples in G_2 manifolds with an isometric involution.

Proposition 7.6 *Let* (M, φ) *be a* G_2 *manifold with an isometric involution* $\sigma \neq \mathrm{id}$ *such that* $\sigma^*\varphi = \varphi$ *or* $\sigma^*\varphi = -\varphi$. *Then Fix*$(\sigma)$ *is an associative or coassociative submanifold in M respectively, if it is non-empty.*

We also have explicit examples of associatives and coassociatives.

Example The first explicit examples of associatives in $\mathbb{R}^7$ not arising from other geometries are given in [52] from symmetry and evolution equation considerations.

The first explicit non-trivial examples of coassociatives in $\mathbb{R}^7$ are given in [20]. There are two dilation families: one which has one end asymptotic to a cone C on a non-round $\mathcal{S}^3$, and one which has two ends asymptotic to $C \cup \mathbb{R}^4$. The cone C was discovered earlier by Lawson–Osserman [50] and was a key example of a volume-minimizing submanifold which is not smooth (it is Lipschitz but not C^1).

Example In the Bryant–Salamon holonomy G_2 metric on the spinor bundle of $\mathcal{S}^3$ [5], the base $\mathcal{S}^3$ is associative. In the Bryant–Salamon holonomy G_2 metrics on $\Lambda^2_+ T^*\mathcal{S}^4$ and $\Lambda^2_+ T^*\mathbb{CP}^2$ [5], the bases $\mathcal{S}^4$ and $\mathbb{CP}^2$ are coassociative.

We now want to understand deformations of associatives and coassociatives, from which perturbation or gluing results will follow. We begin with associatives.

Notice that if P is an associative plane, $u \in P$ and $v \in P^\perp$ then $u \times v \in P^\perp$ since $\varphi(w, u, v) = g(w, u \times v) = g(v, w \times u) = 0$ for all $w \in P$ since $w \times u \in P$. Thus, if N is associative, cross product gives a (Clifford) multiplication $m : C^\infty(T^*N \otimes \nu(N)) \to C^\infty(\nu(N))$ (viewing tangent vectors as cotangent vectors via the metric). Hence, using the normal connection $\nabla^\perp : C^\infty(\nu(N)) \to C^\infty(T^*N \otimes \nu(N))$ on $\nu(N)$ we get a linear operator

$$D\!\!\!/ = m \circ \nabla^\perp : C^\infty(\nu(N)) \to C^\infty(\nu(N)).$$

We call $D\!\!\!/$ the *Dirac operator*. We see that its principal symbol is given by $\sigma_{D\!\!\!/}(x, \xi)v = i\xi \times v$, so $D\!\!\!/$ is elliptic, and we also have that $D\!\!\!/^* = D\!\!\!/$.

Remark Since a 3-manifold is always spin, we have a spinor bundle $\mathbb{S}$ on N, a connection $\nabla : C^\infty(\mathbb{S}) \to C^\infty(T^*M \otimes \mathbb{S})$ (a lift of the Levi-Civita connection) and we have Clifford multiplication $m : C^\infty(T^*M \otimes \mathbb{S}) \to C^\infty(\mathbb{S})$ given by $m(\xi, v) = \xi \cdot v$. Hence we have a composition $\not{D} = m \circ \nabla : C^\infty(\mathbb{S}) \to C^\infty(\mathbb{S})$, which is a first order linear differential operator called the Dirac operator. Locally it is given by $\not{D}v = \sum_i e_i \cdot \nabla_{e_i} v$, so we have that $\sigma_{\not{D}}(\xi, v) = i\xi \cdot v$. Hence $\not{D}$ is elliptic. Moreover $\not{D}$ is self-adjoint.

In fact, it is possible (see e.g. [66]) to see that the complexified normal bundle $\nu(N) \otimes \mathbb{C} = \mathbb{S} \otimes V$ for a $\mathbb{C}^2$-bundle V over N, so that the Dirac operator on $\nu(N)$ is just a "twist" of the usual Dirac operator on $\mathbb{S}$.

Consider a compact associative N. We want to describe the associative deformations of N, just as in the case of special Lagrangians above. To be consistent with that previous setting, we will now use P to denote a nonlinear deformation map: we trust that this will not cause confusion given the context.

We know that $\exp_v(N) = N_v$, which is the graph of v, is associative for a normal vector field v if and only if $*\exp_v^*(\chi) \in C^\infty(TM|_N)$ is 0. In fact, it turns out that $P(v) = *\exp_v^*(\chi) \in C^\infty(\nu(N))$ since N is associative and

$$L_0 P(v) = *\mathrm{d}(v \lrcorner \chi) = \not{D}v.$$

Here $L_0 P$ is not typically surjective so we cannot apply our Implicit Function Theorem, except when $\operatorname{Ker} \not{D} = \operatorname{Ker} \not{D}^* = \{0\}$. However, we can still say something in these circumstances, for which we make a small digression to a more general situation.

Suppose X, Y are Banach spaces. Let $U \subseteq X$ be an open set with $0 \in U$ and let $P : U \to Y$ be a smooth map with $P(0) = 0$ such that $L_0 P : X \to Y$ is Fredholm.

Let $\mathcal{I} = \operatorname{Ker} L_0 P$ and let $\mathcal{O}$ be such that $Y = L_0 P(X) \oplus \mathcal{O}$, which exists and is finite-dimensional by the assumption that $L_0 P$ is Fredholm. We then let $Z = X \oplus \mathcal{O}$ and define $F : U \oplus \mathcal{O} \to Y$ by $F(u, y) = P(u) + y$. We see that $L_0 F : X \oplus \mathcal{O} \to Y = L_0 P(X) \oplus \mathcal{O}$ is given by $L_0 F(x, y) = L_0 P(x) + y$ which is surjective and $L_0 F(x, y) = 0$ if and only if $L_0 P(x) = 0$ and $y = 0$, thus $\operatorname{Ker} L_0 F = \operatorname{Ker} L_0 P \times \{0\}$.

There exists $W \subseteq X$ such that $\operatorname{Ker} L_0 P \oplus W = X$. Applying the Implicit Function Theorem, there exist open sets $U_1 \subseteq \operatorname{Ker} L_0 P$ containing 0, $U_2 \subseteq W$ containing 0 and $U_3 \subseteq \mathcal{O}$ containing 0 and smooth maps $G_2 : U_1 \to U_2$, $G_3 : U_1 \to U_3$ such that

$$F^{-1}(0) \cap U_1 \times U_2 \times U_3 = \{(u, G_2(u), G_3(u)) \, : \, u \in U_1\}.$$

We also know that $P(x) = 0$ if and only if $F(x, y) = 0$ and $y = 0$. Hence

$$P^{-1}(0) \cap U_1 \times U_2 = \{(u, G_2(u)) \, : \, u \in G_3^{-1}(0)\}.$$

Let $\mathcal{U} = U_1$ and define $\pi : \mathcal{U} \to \mathcal{O}$ by $\pi(u) = G_3(u)$. Then $P^{-1}(0) \cap U_1 \times U_2$ is a graph over $\pi^{-1}(0)$, and hence $P^{-1}(0)$ is locally homeomorphic to $\pi^{-1}(0)$.

Sard's Theorem says that generically $\pi^{-1}(y)$ is a smooth manifold of dimension $\dim \mathcal{I} - \dim \mathcal{O} = \dim \operatorname{Ker} L_0 P - \dim \operatorname{Coker} L_0 P$, which is the index of $L_0 P$. Hence, the expected dimension of $P^{-1}(0)$ is the index of $L_0 P$.

In the associative setting we have that the linearisation is $\not{D}$, which is elliptic and thus Fredholm, and we know that index $\not{D} = \dim \operatorname{Ker} \not{D} - \dim \operatorname{Ker} \not{D}^* = 0$. We deduce the following [66].

Theorem 7.7 *The expected dimension of the moduli space of deformations of a compact associative 3-fold N in a G_2 manifold is 0 and infinitesimal deformations of N are given by the kernel of $\not{D}$ on $\nu(N)$. Moreover, if* $\operatorname{Ker} \not{D} = \{0\}$ *then N is rigid.*

Remark The dimension of the kernel of $\not{D}$ typically depends on the metric on N rather than just the topology, so it is usually difficult to determine. However, there are some cases where one can ensure the moduli space is smooth cf. [15].

Example For the associative $N = \mathcal{S}^3$ in $\mathbb{S}(\mathcal{S}^3)$, $\nu(N) = \mathbb{S}(\mathcal{S}^3)$ so $\not{D}$ is just the usual Dirac operator. A theorem of Lichnerowicz states that $\operatorname{Ker} \not{D} = \{0\}$ as $\mathcal{S}^3$ has positive scalar curvature so N is rigid.

Example Corti–Haskins–Nordström–Pacini construct rigid associative $\mathcal{S}^1 \times \mathcal{S}^2$s in compact holonomy G_2 twisted connected sums [10].

For coassociatives, the deformation theory is much better behaved, like for special Lagrangians [66].

Theorem 7.8 *Let N be a compact coassociative in a G_2 manifold (or just a 7-manifold with closed G_2 structure). The moduli space of deformations of N is a smooth manifold of dimension $b^2_+(N)$.*

Proof Since N is coassociative the map $v \mapsto v \lrcorner \varphi = \alpha_v$ defines an isomorphism from $\nu(N)$ to a rank 3 vector bundle on N, which is $\Lambda^2_+ T^*N$, the 2-forms on N which are self-dual (so $*\alpha = \alpha$). We can therefore view nearby submanifolds to N as graphs of self-dual 2-forms.

We know that $N_v = \exp_v(N)$ is coassociative if and only if $\exp_v^*(\varphi) = 0$. We see that

$$\frac{\mathrm{d}}{\mathrm{d}t} \exp_{tv}^*(\varphi)|_{t=0} = \mathcal{L}_v \varphi = \mathrm{d}(v \lrcorner \varphi) = \mathrm{d}\alpha_v.$$

Hence nearby coassociatives N' to N are given by the zeros of $P(\alpha) = \mathrm{d}\alpha + Q(\alpha, \nabla\alpha)$. Moreover, since $\varphi = 0$ on N, $[\varphi] = 0$ on N' and hence $P : C^\infty(\Lambda^2_+ T^*N) \to \mathrm{d}(C^\infty(\Lambda^2 T^*N))$.

Here P is not elliptic, but $L_0 P = \mathrm{d}$ has finite-dimensional kernel, the closed self-dual 2-forms, since $\mathrm{d}\alpha = 0$ implies that $\mathrm{d}^*\alpha = - * \mathrm{d} * \alpha = 0$ so α is harmonic. Moreover, $L_0 P$ has injective symbol so it is overdetermined elliptic, which means that elliptic regularity still holds. Another way to deal with this is to consider $F(\alpha, \beta) = P(\alpha) + \mathrm{d}^*\beta$ for β a 4-form. Now $F^{-1}(0)$ is the disjoint union of $P^{-1}(0)$

and multiples of the volume form, as exact and coexact forms are orthogonal. Moreover, $L_0F(\alpha,\beta)=\mathrm{d}\alpha+\mathrm{d}^*\beta$ is now elliptic. Overall, we can apply our standard Implicit Function Theorem if we know that

$$\mathrm{d}(C^{k+1,a}(\Lambda^2_+T^*N))=\mathrm{d}(C^{k+1,a}(\Lambda^2T^*N)).$$

This is true because by Hodge theory if α is a 2-form, we can write $\alpha=\mathrm{d}^*\beta+\gamma$ for a 3-form β and a closed form γ, so $\mathrm{d}\alpha=\mathrm{d}\mathrm{d}^*\beta=\mathrm{d}(\mathrm{d}^*\beta+*\mathrm{d}^*\beta)$ and $\mathrm{d}^*\beta+*\mathrm{d}^*\beta$ is self-dual. □

Example The $\mathcal{S}^4$ and $\mathbb{CP}^2$ in the Bryant–Salamon metrics on $\Lambda^2_+T^*\mathcal{S}^4$ and $\Lambda^2_+T^*\mathbb{CP}^2$ have $b^2_+=0$ and so are rigid.

For a K3 surface and T^4 we have $b^2_+=3$ and Λ^2_+ is trivial, so we can hope to find coassociative K3 and T^4 fibrations of compact G_2 manifolds. There is a programme [47] for constructing a coassociative K3 fibration (with singular fibres). Towards completing this programme, the first examples of compact coassociative 4-folds with conical singularities in compact holonomy G_2 twisted connected sums were constructed in [61].

Again, we have a similar isometric embedding result for coassociative 4-folds, motivated by the deformation theory result [3].

Theorem 7.9 *Any compact oriented real analytic Riemannian 4-manifold whose bundle of self-dual 2-forms is trivial can be isometrically embedded in a* G_2 *manifold as the fixed points of an isometric involution.*

Remark The deformation theory results for compact associative and coassociative submanifolds have been extended to certain non-compact, singular and boundary settings, for example in [16, 44, 48, 55, 56, 58].

8 Cayley Submanifolds

We now discuss our final class of calibrated submanifolds.

Theorem 8.1 *On a* $\mathrm{Spin}(7)$ *manifold* (M^8,Φ) *(so* Φ *is a closed admissible form),* Φ *is a calibration.*

Proof Let P be a plane in $\mathbb{R}^8\cong\mathbb{C}^4$. Since $\mathrm{SU}(4)\subseteq\mathrm{Spin}(7)$, by the proof of Wirtinger's inequality (Theorem 4.2), we can choose $A\in\mathrm{Spin}(7)$ so that $A(P)$ is spanned by

$$\{e_1,\cos\theta_1e_2+\sin\theta_1e_3,e_5,\cos\theta_2e_6+\sin\theta_2e_7\}.$$

We take the standard $\mathrm{Spin}(7)$ form Φ on $\mathbb{R}^8$ to be:

$$\Phi = \mathrm{d}x_{1234} + \mathrm{d}x_{1256} + \mathrm{d}x_{1278} + \mathrm{d}x_{1357} - \mathrm{d}x_{1367} - \mathrm{d}x_{1458} - \mathrm{d}x_{1467}$$
$$+ \mathrm{d}x_{5678} + \mathrm{d}x_{3478} + \mathrm{d}x_{3456} + \mathrm{d}x_{2468} - \mathrm{d}x_{2457} - \mathrm{d}x_{2367} - \mathrm{d}x_{2358},$$

again using the notation $\mathrm{d}x_{ij\ldots k} = \mathrm{d}x_i \wedge \mathrm{d}x_j \wedge \cdots \wedge \mathrm{d}x_k$. Then $\Phi|_P = (\cos\theta_1 \cos\theta_2 + \sin\theta_1 \sin\theta_2)\,\mathrm{vol}_P = \cos(\theta_1 - \theta_2)\,\mathrm{vol}_P$. Hence Φ is a calibration as it is closed. □

We can thus define our calibrated submanifolds in Spin(7) manifolds.

Definition 8.2 Submanifolds in (M^8, Φ) calibrated by Φ are called Cayley 4-folds.

Remark The name Cayley submanifolds is because of the relation between the submanifolds and the octonions or Cayley numbers $\mathbb{O}$.

We can relate Cayley submanifolds to all of the other calibrated geometries we have seen.

Proposition 8.3 *(a) Complex surfaces and special Lagrangian 4-folds in $\mathbb{C}^4$ are Cayley in $\mathbb{R}^8 = \mathbb{C}^4$.*
(b) Write $\mathbb{R}^8 = \mathbb{R} \times \mathbb{R}^7$. Then $\mathbb{R} \times S$ is Cayley if and only if S is associative in $\mathbb{R}^7$ and $N \subseteq \mathbb{R}^7$ is Cayley in $\mathbb{R}^8$ if and only if N is coassociative in $\mathbb{R}^7$.

Proof Recall the Kähler form ω and holomorphic volume form Υ on $\mathbb{C}^4$ and the G_2 3-form φ on $\mathbb{R}^7$.

Part (a) is immediate from the formula $\Phi = \frac{1}{2}\omega^2 + \mathrm{Re}\,\Upsilon$, since complex surfaces are calibrated by $\frac{1}{2}\omega^2$, special Lagrangians are calibrated by $\mathrm{Re}\,\Upsilon$, Υ vanishes on complex surfaces and ω vanishes on special Lagrangians.

Part (b) follows immediately from the formula $\Phi = \mathrm{d}x_1 \wedge \varphi + *\varphi$. □

We can also use an isometric involution to construct Cayley submanifolds as in our previous calibrated geometries.

Proposition 8.4 *Let (M, Φ) be a* Spin(7) *manifold and let $\sigma \neq$* id *be an isometric involution with $\sigma^*\Phi = \Phi$. Then Fix(σ) is Cayley submanifold, if it is non-empty.*

Example The first interesting explicit examples of Cayleys in $\mathbb{R}^8$ not arising from other geometries were given in [53] and are asymptotic to cones.

Example The base S^4 in the Bryant–Salamon holonomy Spin(7) metric on $\mathbb{S}_+(S^4)$ [5] is Cayley.

We now discuss deformations of a compact Cayley N, for which we need some discussion of algebra related to Spin(7). Since $\Lambda^2(\mathbb{R}^8)^*$ is 28-dimensional and the 21-dimensional Lie algebra of Spin(7) sits inside the space of 2-forms, we must have a distinguished 7-dimensional subspace Λ^2_7 of 2-forms on $\mathbb{R}^8$. So what is this subspace? Let $u, v \in \mathbb{R}^8$. Then we can construct a 2-form $u \wedge v$, viewing u, v as cotangent vectors. We can also construct a 2-form from u, v by considering $\Phi(u, v, ., .)$. It is then true that

$$\Lambda^2_7 = \{u \wedge v + \Phi(u, v, ., .) \,:\, u, v \in \mathbb{R}^8\}.$$

When P is a Cayley plane and $u, v \in P$ are orthogonal we see that $\Phi(u, v, ., .) = *_P(u \wedge v)$ so that $u \wedge v + \Phi(u, v, ., .)$ is self-dual on P. Since $\Lambda^2_+ P^*$ is 3-dimensional, we see that there must be a 4-dimensional space E of 2-forms on P such that $\Lambda^2_7|_P = \Lambda^2_+ P^* \oplus E$. Moreover, if $u \in P$ and $v \in P^\perp$ then $m(u, v) = u \wedge v + \Phi(u, v, ., .) \in E$ and the map $m : P \times P^\perp \to E$ is surjective.

Now let us move to a Cayley submanifold N in a $\mathrm{Spin}(7)$ manifold (M, Φ). On M we have a rank 7 bundle Λ^2_7 of 2-forms and we have that $\Lambda^2_7|_N = \Lambda^2_+ T^*N \oplus E$ for some rank 4 bundle E over N. The map m above defines a (Clifford) multiplication $m : C^\infty(T^*N \otimes \nu(N)) \to C^\infty(E)$ (viewing tangent vectors as cotangent vectors via the metric), and thus using the normal connection $\nabla^\perp : C^\infty(\nu(N)) \to C^\infty(T^*N \otimes \nu(N))$ we get a linear first order differential operator

$$\not{D}_+ = m \circ \nabla^\perp : C^\infty(\nu(N)) \to C^\infty(E).$$

Again this is an elliptic operator called the *positive Dirac operator*, but it is not self-adjoint: its adjoint is the negative Dirac operator from E to $\nu(N)$.

Remark If N is spin, the spinor bundle $\mathbb{S}$ splits as $\mathbb{S}_+ \oplus \mathbb{S}_-$, and the Dirac operator $\not{D}$ splits into $\not{D}_\pm$ from $\mathbb{S}_\pm$ to $\mathbb{S}_\mp$ so that $\not{D}(v_+, v_-) = (\not{D}_- v_-, \not{D}_+ v_+)$. Hence $\not{D}^* = \not{D}$ says that $\not{D}^*_\pm = \not{D}_\mp$.

It turns out (see, for example, [66]) that there exists a $\mathbb{C}^2$-bundle V on N so that $\nu(N) \otimes \mathbb{C} = \mathbb{S}_+ \otimes V$, $E \otimes \mathbb{C} = \mathbb{S}_- \otimes V$ and $\not{D}_+$ on $\nu(N)$ is a "twist" of the usual positive Dirac operator. However, not every 4-manifold is spin, so we cannot always make this identification.

On $\mathbb{O}$ there exists a 4-fold cross product, whose real part gives Φ and whose imaginary part we call τ. Perhaps unsurprisingly, we have the following result, which we will leave as an exercise for the reader.

Lemma 8.5 *A 4-plane P in $\mathbb{R}^8$ satisfies $\tau|_P \equiv 0$ if and only if it admits an orientation so that it is calibrated by Φ.*

We can extend τ to a $\mathrm{Spin}(7)$ manifold, except that we need a rank 7 vector bundle on M in which τ takes values: we have one, namely Λ^2_7. So we have the following alternative characterisation of Cayley 4-folds.

Lemma 8.6 *A submanifold N in a* $\mathrm{Spin}(7)$ *manifold is Cayley (up to a choice of orientation) if and only if $\tau \in C^\infty(\Lambda^4 T^*M; \Lambda^2_7)$ vanishes on N.*

Now suppose that N is a compact Cayley 4-fold. Then the zeros of the equation $F(v) = *\exp_v^*(\tau)$ for $v \in C^\infty(\nu(N))$ define Cayley deformations (as the graph of v). We know that F takes values in $\Lambda^2_7|_N = \Lambda^2_+ T^*N \oplus E$ and it turns out that

$$L_0 F(v) = *\mathrm{d}(v \lrcorner \tau) = \not{D}_+ v$$

since N is Cayley. So, we potentially have a problem because F does not necessarily take values only in E (and in general it will not just take values in E). However, the Cayley condition on N means that $F(v) = 0$ if and only $P(v) = \pi_E F(v) = 0$, where π_E is the projection onto E (again, we are using P to denote the nonlinear deformation map as in our previous discussion, and we expect it will not cause confusion given the context). Then the operator $P : C^\infty(\nu(N)) \to C^\infty(E)$ and $L_0 P = \not{D}_+$ is elliptic.

Again, we cannot say that $L_0 P$ is surjective, so we have the following using the same argument as in the lead up to Theorem 7.7, cf. [66].

Theorem 8.7 *The expected dimension of the moduli space of deformations of a compact Cayley 4-fold N in a* Spin(7) *manifold is* ind $\not{D}_+ = \dim \text{Ker } \not{D}_+ - \dim \text{Ker } \not{D}_+^*$ *with infinitesimal deformations given by* Ker $\not{D}_+$ *on $\nu(N)$. Moreover,*

$$\text{ind } \not{D}_+ = \frac{1}{2}\sigma(N) + \frac{1}{2}\chi(N) - [N].[N], \tag{1}$$

where $\sigma(N) = b^2_+(N) - b^2_-(N)$ (the signature of N), $\chi(N) = 2b^0(N) - 2b^1(N) + b^2(N)$ (the Euler characteristic of N) and $[N].[N]$ is the self-intersection of N, which is the Euler number of $\nu(N)$.

Example For the Cayley $N = \mathcal{S}^4$ in $\mathbb{S}_+(\mathcal{S}^4)$, $\nu(N) = \mathbb{S}_+(\mathcal{S}^4)$ and $\not{D}_+$ is the usual positive Dirac operator. Again, since N has positive scalar curvature, we see that Ker $\not{D}_\pm = \{0\}$ so N is rigid.

Remark Theorem 8.7 has been extended to various other non-compact, singular and boundary settings, for example in [68, 70, 71].

9 The Angle Theorem

To conclude these notes, we now discuss a very natural and elementary problem in Euclidean geometry where calibrations play a major, and perhaps unexpected, role.

If one takes two lines in $\mathbb{R}^2$ intersecting transversely, then their union is never length-minimizing. A natural question to ask is: does this persist in higher dimensions? In other words, when is the union of two transversely intersecting n-planes in $\mathbb{R}^{2n}$ volume-minimizing? Two such planes are determined by the n angles between them as follows.

Lemma 9.1 *Let P, Q be oriented n-planes in $\mathbb{R}^{2n}$. There exists an orthonormal basis $e_1, \ldots, e_{2n}$ for $\mathbb{R}^{2n}$ such that $P = \text{Span}\{e_1, \ldots, e_n\}$ and*

$$Q = \text{Span}\{\cos\theta_1 e_1 + \sin\theta_1 e_{n+1}, \ldots, \cos\theta_n e_n + \sin\theta_n e_{2n}\}$$

where $0 \leq \theta_1 \leq \cdots \leq \theta_{n-1} \leq \frac{\pi}{2}$ and $\theta_{n-1} \leq \theta_n \leq \pi - \theta_{n-1}$. These angles are called the characterising angles *of P, Q.*

Proof The proof is very similar to the argument in the proof of Wirtinger's inequality (Theorem 4.2). Choose unit $e_1 \in P$ and maximise $\langle e_1, u_1\rangle$ for $u_1 \in Q$, and let $e_{n+1} \in P^\perp$ be defined by $u_1 = \cos\theta_1 e_1 + \sin\theta_1 e_{n+1}$. Now choose $e_2 \in P \cap e_1^\perp$ and maximise $\langle e_2, u_2\rangle$ for $u_2 \in Q \cap u_1^\perp$, then proceed by induction. □

If the characterising angles of P, Q are $\theta_1, \ldots, \theta_n$, then the characterising angles of $P, -Q$ are $\psi_1, \ldots, \psi_n$ where $\psi_j = \theta_j$ for $j = 1, \ldots, n-1$ and $\psi_n = \pi - \theta_n$.

The idea of the following theorem is that the union of $P \cup Q$ is area-minimizing if $P, -Q$ are not too close together [49].

Theorem 9.2 (Angle Theorem) *Let P, Q be oriented transverse n-planes in $\mathbb{R}^{2n}$ and let $\psi_1, \ldots, \psi_n$ be the characterising angles between $P, -Q$. Then $P \cup Q$ is volume-minimizing if and only if $\psi_1 + \cdots + \psi_n \geq \pi$.*

Notice that this criteria is impossible to fulfill in 1 dimension.

Proof We will sketch the proof which involves calibrations in a fundamental way in both directions. For details, we recommend looking at [19].

First if $P \cup Q$ does not satisfy the angle condition, we can choose coordinates by Lemma 9.1 so that $P = P(-\frac{\psi}{2})$ and $-Q = P(\frac{\psi}{2})$ where $\psi = (\psi_1, \ldots, \psi_n)$ and $P(\psi) = \{(e^{i\psi_1}x_1, \ldots, e^{i\psi_n}x_n) : (x_1, \ldots, x_n) \in \mathbb{R}^n\}$ as given earlier. We know that we have a special Lagrangian Lawlor neck N asymptotic to $P(-\frac{\psi'}{2}) \cup P(\frac{\psi'}{2})$ for any ψ' where $\sum_{i=1}^n \psi_i' = \pi$. The claim is then that since $\sum \psi_i < \pi$ we can find ψ' so that $\sum \psi_i' = \pi$ and $N \cap P(\pm\frac{\psi'}{2})$ is non-empty (in fact, an ellipsoid). This is actually a way to characterise N. Hence since N is calibrated by $\mathrm{Im}\,\Upsilon$ and $\mathrm{Im}\,\Upsilon|_{P\cup Q} < \mathrm{vol}_{P\cup Q}$ by the condition on the characterising angles, $P \cup Q$ cannot be volume-minimizing by the usual Stokes' Theorem argument for calibrated submanifolds.

We now provide a few extra details, for which we need to describe N. For maps $z_1, \ldots, z_n : \mathbb{R} \to \mathbb{C}$ define

$$N = \{(t_1 z_1(s), \ldots, t_n z_n(s)) \in \mathbb{C}^n \ : s \in \mathbb{R},\ t_1, \ldots, t_n \in \mathbb{R},\ \sum_{j=1}^n t_j^2 = 1\}.$$

It is not difficult to calculate that N is special Lagrangian with phase i (so calibrated by $\mathrm{Im}\,\Upsilon$) if and only if

$$\overline{z_j}\frac{\mathrm{d}z_j}{\mathrm{d}s} = i f_j \overline{z_1 \ldots z_n}$$

for positive real functions f_j.

Suppose that $f_j = 1$ for all j. Write $z_j = r_j e^{i\theta_j}$, let $\theta = \sum_{j=1}^n \theta_j$ and suppose that $z_j(0) = c_j > 0$. From the differential equation, one quickly sees that $r_j^2 = c_j^2 + u$ for some function u with $u(0) = 0$ and $r_1 \ldots r_n \cos\theta = c_1 \ldots c_n$.

If we now suppose that $u = t^2$, we see that

$$\theta_j(t) = \int_0^t \frac{a_j \mathrm{d}t}{(1 + a_j t^2)\sqrt{\frac{1}{t^2}((1 + a_1 t^2) \ldots (1 + a_n t^2) - 1)}}$$

where $a_j = c_j^{-2}$. We observe that $\theta \to \pm\frac{\pi}{2}$ as $t \to \pm\infty$ and hence N, which is a Lawlor neck, is asymptotic to a pair of planes where the sum of the angles is $\pm\frac{\pi}{2}$.

Now fix $t > 0$ and define

$$f : X = \{(a_1, \dots, a_n) \in \mathbb{R}^n \ : \ a_j \geq 0\}$$
$$\to Y = \{(\theta_1, \dots, \theta_n) \in \mathbb{R}^n \ : \ \theta_j \geq 0, \sum_{j=1}^{n} \theta_j < \frac{\pi}{2}\}$$

by $f(a_1, \dots, a_n) = (\theta_1, \dots, \theta_n)$ where

$$\theta_i = \int_0^t \frac{a_j \mathrm{d}t}{(1 + a_j t^2)\sqrt{\frac{1}{t^2}\big((1 + a_1 t^2) \dots (1 + a_n t^2) - 1\big)}}.$$

It is clear that if $n = 1$, $f : X \to Y$ is surjective. We want to show it is surjective for all n. For $\theta \in (0, \frac{\pi}{2})$ define $H_\theta = \{(\theta_1, \dots, \theta_n) \in Y \ : \ \sum_{j=1}^{n} \theta_j = \theta\}$. By our discussion above we see that

$$f^{-1}(H_\theta) \subseteq S_\theta = \{(a_1, \dots, a_n) \in X \ : \ (1 + a_1 t^2) \dots (1 + a_n t^2) = \cos^{-2} t\}.$$

Notice that if the degree of $f : \partial S_\theta \to \partial H_\theta$ is 1 then the degree of $f : S_\theta \to H_\theta$ is 1. Thus, by induction on n, we see that $f : S_\theta \to H_\theta$ is surjective.

Now, given any plane $\{(e^{i\theta_1} x_1, \dots e^{i\theta_n} x_n) \ : \ (x_1, \dots, x_n) \in \mathbb{R}^n\}$ where $(\theta_1, \dots, \theta_n) \in Y$, $\theta_j \neq 0$ for all j, we see that we can choose a Lawlor neck N which intersects the plane in a hypersurface as claimed.

We now move to the other direction in the statement of the Angle Theorem. If $P \cup Q$ does satisfy the angle condition, then (by choosing coordinates so that $P = \mathbb{R}^n$ and Q is in standard position) we claim that it is calibrated by a so-called *Nance calibration*:

$$\eta(u_1, \dots, u_n) = \mathrm{Re}\,\big((\mathrm{d}x_1 + u_1 \mathrm{d}y_1) \wedge \dots \wedge (\mathrm{d}x_n + u_n \mathrm{d}y_n)\big)$$

where $u_1, \dots, u_n \in \mathcal{S}^2 \subseteq \mathrm{Im}\,\mathbb{H}$. If $u_m = i$ for all m then $\eta = \mathrm{Re}\,\Upsilon$, so it is believable that it is a calibration in general, but we now show that it is indeed true.

Let $x_1, y_1, \dots, x_n, y_n$ be coordinates on $\mathbb{R}^{2n}$. We call an n-form η on $\mathbb{R}^{2n}$ a torus form if η lies in the span of forms of type

$$\mathrm{d}x_{i_1} \wedge \dots \wedge \mathrm{d}x_{i_k} \wedge \mathrm{d}y_{j_1} \wedge \dots \wedge \mathrm{d}y_{j_l}$$

where $\{i_1, \dots, i_k\} \cap \{j_1, \dots, j_l\} = \emptyset$ and $\{i_1, \dots, i_k\} \cup \{j_1, \dots, j_l\} = \{1, \dots, n\}$. We now claim that a torus form η is a calibration if and only if

$$\eta(\cos\theta_1 e_1 + \sin\theta_1 e_{n+1}, \dots, \cos\theta_n e_n + \sin\theta_n e_{2n}) \leq 1$$

for all $\theta_1, \ldots, \theta_n \in \mathbb{R}$.

For $n = 1$, $\eta = dx_1 \wedge dy_1$ which is a calibration. Suppose that the result holds for $n = k$. Let η be a torus form on $\mathbb{R}^{2(k+1)}$ and rescale η so that the maximum of η is 1 and is attained at some plane. The idea is to show using the argument in the proof of Wirtinger's inequality to put planes in standard position that we can write $\eta = e_1 \wedge \eta_1 + e_2 \wedge \eta_2$ where e_1, e_2 are orthonormal and span an $\mathbb{R}^2$ and η_1, η_2 are torus forms on $\mathbb{R}^{2k}$. The claim then follows by induction on n.

Hence, the Nance calibration η above is a calibration and moreover we know $P(\theta)$ is calibrated by $\eta(u)$ if and only if

$$\prod_{j=1}^{n}(\cos\theta_j + \sin\theta_j u_j) = 1.$$

We then just need to find the u_j determined by θ_j. Notice that the condition that $\psi_1 + \cdots + \psi_n \geq \pi$ holds if and only if $\theta_n \leq \theta_1 + \cdots + \theta_{n-1}$. If we write $\cos\theta_j + \sin\theta_j u_j = w_j \overline{w}_{j+1}$ where $w_{n+1} = w_1$ and w_j are unit imaginary quaternions, then the product condition is satisfied and we just need $\langle w_j, w_{j+1}\rangle = \cos\theta_j$, which is equivalent to finding n points on the unit 2-sphere so that $d(w_j, w_{j+1}) = \theta_j$, where $\theta_n \leq \theta_1 + \cdots + \theta_{n-1}$. This is indeed possible, by considering an n-sided spherical polygon. □

References

1. Brendle, S. (2013). Embedded minimal tori in S^3 and the Lawson conjecture. *Acta Mathematica, 211*(2), 177–190.
2. Bryant, R. L. (1982). Submanifolds and special structures on the octonians. *Journal of Differential Geometry, 17*, 185–232.
3. Bryant, R. L. (2000). Calibrated embeddings in the special Lagrangian and coassociative cases. *Annals of Global Analysis and Geometry, 18*(3–4), 405–435.
4. Bryant, R. L. (2006). Second order families of special Lagrangian 3-folds. *Perspectives in Riemannian geometry* (Vol. 40, pp. 63–98)., CRM proceedings & lecture notes Providence: American Mathematical Society.
5. Bryant, R. L., & Salamon, S. (1989). On the construction of some complete metrics with exceptional holonomy. *Duke Mathematical Journal, 58*, 829–850.
6. Butscher, A. (2002). Deformations of minimal Lagrangian submanifolds with boundary. *Proceedings of the American Mathematical Society, 131*, 1953–1964.
7. Butscher, A. (2004). Regularizing a singular special Lagrangian variety. *Communications in Analysis and Geometry, 12*, 733–791.
8. Carberry, E. (2010). Associative cones in the imaginary octonions. *Riemann surfaces, harmonic maps and visualization* (Vol. 3, pp. 249–263), OCAMI Stud. Osaka: Osaka Municipal University Press.
9. Carberry, E., & McIntosh, I. (2004). Minimal Lagrangian 2-tori in $\mathbb{CP}^2$ come in real families of every dimension. *Journal of the London Mathematical Society (2), 69*(2), 531–544.
10. Corti, A., Haskins, M., Nordström, J., & Pacini, T. (2015). G_2-manifolds and associative submanifolds via semi-Fano 3-folds. *Duke Mathematical Journal, 164*(10), 1971–2092.

11. Eguchi, T., & Hanson, A. J. (1978). Asymptotically flat self-dual solutions to Euclidean gravity. *Physics Letters*, *74B*(3), 249–251.
12. Fox, D. (2005). *Second order families of coassociative 4-folds*. Thesis (Ph.D.) Duke University, ProQuest LLC, Ann Arbor, MI.
13. Fox, D. (2007). Coassociative cones ruled by 2-planes. *Asian Journal of Mathematics*, *11*(4), 535–553.
14. Fox, D. (2008). Cayley cones ruled by 2-planes: Desingularization and implications of the twistor fibration. *Communications in Analysis and Geometry*, *16*(5), 937–968.
15. Gayet, D. (2014). Smooth moduli spaces of associative submanifolds. *The Quarterly Journal of Mathematics*, *65*(4), 1213–1240.
16. Gayet, D., & Witt, F. (2011). Deformations of associative submanifolds with boundary. *Advances in Mathematics*, *226*(3), 2351–2370.
17. Goldstein, E. (2001). Calibrated fibrations on noncompact manifolds via group actions. *Duke Mathematical Journal*, *110*(2), 309–343.
18. Gray, A., Abbena, E., & Salamon, S. (2006). *Modern differential geometry of curves and surfaces with Mathematica®* (3rd ed.), Studies in advanced mathematics. Boca Raton: Chapman & Hall/CRC.
19. Harvey, R. (1990). *Spinors and calibrations* (Vol. 9)., Perspectives in mathematics Boston: Academic Press Inc.
20. Harvey, R., & Lawson, H. B. (1982). Calibrated geometries. *Acta Mathematica*, *148*, 47–157.
21. Haskins, M. (2004). Special Lagrangian cones. *American Journal of Mathematics*, *126*(4), 845–871.
22. Haskins, M. (2004). The geometric complexity of special Lagrangian T^2-cones. *Inventiones Mathematicae*, *157*(1), 11–70.
23. Haskins, M., & Kapouleas, N. (2007). Special Lagrangian cones with higher genus links. *Inventiones Mathematicae*, *167*, 223–294.
24. Huisken, G., & Ilmanen, T. (2001). The inverse mean curvature flow and the Riemannian Penrose inequality. *Journal of Differential Geometry*, *59*(3), 353–437.
25. Imagi, Y., Joyce, D. D., & Oliveira dos Santos, J. (2016). Uniqueness results for special Lagrangians and Lagrangian mean curvature flow expanders in $\mathbb{C}^m$. *Duke Mathematical Journal*, *165*(5), 847–933.
26. Ionel, M. (2003). Second order families of special Lagrangian submanifolds in $\mathbb{C}^4$. *Journal of Differential Geometry*, *65*(2), 211–272.
27. Ionel, M., Karigiannis, S., & Min-Oo, M. (2005). Bundle constructions of calibrated submanifolds in $\mathbb{R}^7$ and $\mathbb{R}^8$. *Mathematical Research Letters*, *12*(4), 493–512.
28. Ionel, M., & Min-Oo, M. (2008). Cohomogeneity one special Lagrangian 3-folds in the deformed and the resolved conifolds. *Illinois Journal of Mathematics*, *52*(3), 839–865.
29. Irie, K., Marques, F., & Neves, A. (2018). Density of minimal hypersurfaces for generic metrics. *Annals of Mathematics (2)*, *187*(3), 963–972.
30. Joyce, D. D. (2001). Evolution equations for special Lagrangian 3-folds in $\mathbb{C}^3$. *Annals of Global Analysis and Geometry*, *20*(4), 345–403.
31. Joyce, D. D. (2001). Constructing special Lagrangian m-folds in $\mathbb{C}^m$ by evolving quadrics. *Mathematische Annalen*, *320*(4), 757–797.
32. Joyce, D. D. (2002). Ruled special Lagrangian 3-folds in $\mathbb{C}^3$. *Proceedings of the London Mathematical Society (3)* *85*(1), 233–256.
33. Joyce, D. D. (2002). Special Lagrangian m-folds in $\mathbb{C}^m$ with symmetries. *Duke Mathematical Journal*, *115*(1), 1–51.
34. Joyce, D. D. (2003). $U(1)$-invariant special Lagrangian 3-folds in $\mathbb{C}^3$ and special Lagrangian fibrations. *Turkish Journal of Mathematics*, *27*(1), 99–114.
35. Joyce, D. D. (2003). Special Lagrangian submanifolds with isolated conical singularities. V. Survey and applications. *Journal of Differential Geometry*, *63*, 279–347.
36. Joyce, D. D. (2004). Special Lagrangian submanifolds with isolated conical singularities. II. Moduli spaces. *Annals of Global Analysis and Geometry*, *25*, 301–352.

37. Joyce, D. D. (2004). Special Lagrangian submanifolds with isolated conical singularities. III. Desingularization, the unobstructed case. *Annals of Global Analysis and Geometry*, *26*, 1–58.
38. Joyce, D. D. (2004). Special Lagrangian submanifolds with isolated conical singularities. IV. Desingularization, obstructions and families. *Annals of Global Analysis and Geometry*, *26*, 117–174.
39. Joyce, D. D. (2005). $U(1)$-invariant special Lagrangian 3-folds. I. Nonsingular solutions. *Advances in Mathematics*, *192*(1), 35–71.
40. Joyce, D. D. (2005). $U(1)$-invariant special Lagrangian 3-folds. II. Existence of singular solutions. *Advances in Mathematics*, *192*(1), 72–134.
41. Joyce, D. D. (2005). $U(1)$-invariant special Lagrangian 3-folds. III. Properties of singular solutions. *Advances in Mathematics*, *192*(1), 135–182.
42. Joyce, D. D. (2007). *Riemannian holonomy groups and calibrated geometry* (Vol. 12), Oxford graduate texts in mathematics. Oxford: Oxford University Press.
43. Joyce, D. D. (2008). Special Lagrangian 3-folds and integrable systems. *Surveys on geometry and integrable systems* (Vol. 51, pp. 189–233), Advanced studies in pure mathematics. Tokyo: Mathematical Society of Japan.
44. Joyce, D. D., & Salur, S. (2005). Deformations of asymptotically cylindrical coassociative submanifolds with fixed boundary. *Geometry and Topology*, *9*, 1115–1146.
45. Karigiannis, S., & Leung, N. C.-H. (2012). Deformations of calibrated subbundles of Euclidean spaces via twisting by special sections. *Annals of Global Analysis and Geometry*, *42*(3), 371–389.
46. Karigiannis, S., & Min-Oo, M. (2005). Calibrated subbundles in noncompact manifolds of special holonomy. *Annals of Global Analysis and Geometry*, *28*(4), 371–394.
47. Kovalev, A. G. Coassociative $K3$ fibrations of compact G_2-manifolds. arXiv:math/0511150.
48. Kovalev, A., & Lotay, J. D. (2009). Deformations of compact coassociative 4-folds with boundary. *Journal of Geometry and Physics*, *59*, 63–73.
49. Lawlor, G. (1989). The angle criterion. *Inventiones Mathematicae*, *95*, 437–446.
50. Lawson, H. B., & Osserman, R. (1977). Non-existence, non-uniqueness and irregularity of solutions to the minimal surface system. *Acta Mathematica*, *139*, 1–17.
51. Lee, Y.-I. (2003). Embedded special Lagrangian submanifolds in Calabi-Yau manifolds. *Communications in Analysis and Geometry*, *11*, 391–423.
52. Lotay, J. D. (2005). Constructing associative 3-folds by evolution equations. *Communications in Analysis and Geometry*, *13*, 999–1037.
53. Lotay, J. D. (2006). 2-ruled calibrated 4-folds in $\mathbb{R}^7$ and $\mathbb{R}^8$. *Journal of the London Mathematical Society*, *74*, 219–243.
54. Lotay, J. D. (2007). Calibrated submanifolds of $\mathbb{R}^7$ and $\mathbb{R}^8$ with symmetries. *Quarterly Journal of Mathematics*, *58*, 53–70.
55. Lotay, J. D. (2007). Coassociative 4-folds with conical singularities. *Communications in Analysis and Geometry*, *15*, 891–946.
56. Lotay, J. D. (2009). Deformation theory of asymptotically conical coassociative 4-folds. *Proceedings of the London Mathematical Society*, *99*, 386–424.
57. Lotay, J. D. (2009). Desingularization of coassociative 4-folds with conical singularities. *Geometric and Functional Analysis*, *18*, 2055–2100.
58. Lotay, J. D. (2011). Asymptotically conical associative 3-folds. *Quarterly Journal of Mathematics*, *62*, 131–156.
59. Lotay, J. D. (2011). Ruled Lagrangian submanifolds of the 6-sphere. *Transactions of the American Mathematical Society*, *363*, 2305–2339.
60. Lotay, J. D. (2012). Associative submanifolds of the 7-sphere. *Proceedings of the London Mathematical Society*, *105*, 1183–1214.
61. Lotay, J. D. (2012). Stability of coassociative conical singularities. *Communications in Analysis and Geometry*, *20*, 803–867.
62. Lotay, J. D. (2014). Desingularization of coassociative 4-folds with conical singularities: Obstructions and applications. *Transactions of the American Mathematical Society*, *366*, 6051–6092.

63. Marques, F. C., & Neves, A. (2014). Min-max theory and the Willmore conjecture. *Annals of Mathematics (2)*, *179*(2), 683–782.
64. Marques, F. C., & Neves, A. (2017). Existence of infinitely many minimal hypersurfaces in positive Ricci curvature. *Inventiones Mathematicae*, *209*(2), 577–616.
65. McIntosh, I. (2003). Special Lagrangian cones in $\mathbb{C}^3$ and primitive harmonic maps. *Journal of the London Mathematical Society (2)*, *67*(3), 769–789.
66. McLean, R. C. (1998). Deformations of calibrated submanifolds. *Communications in Analysis and Geometry*, *6*, 705–747.
67. Moore, K. (2019). Cayley deformations of compact complex surfaces. *Journal of the London Mathematical Society, 100*, 668–691.
68. Moore, K. (2019). Deformations of conically singular Cayley submanifolds. *Journal of Geometric Analysis, 29*, 2147–2216.
69. Morrey, C. B. (1966). *Multiple integrals in the calculus of variations* (Vol. 130)., Grundlehren series Berlin: Springer.
70. Ohst, M. Deformations of compact Cayley submanifolds with boundary. arXiv:1405.7886.
71. Ohst, M. Deformations of asymptotically cylindrical Cayley submanifolds. arXiv:1506.00110.
72. Pacini, T. (2013). Special Lagrangian conifolds, I: Moduli spaces. *Proceedings of the London Mathematical Society (3) 107*, 198–224.
73. Pacini, T. (2013). Special Lagrangian conifolds, II: Gluing constructions in $\mathbb{C}^m$. *Proceedings of the London Mathematical Society (3)*, *107*(2), 225–266.
74. Perelman, G. Finite extinction time for the solutions to the Ricci flow on certain three-manifolds. arXiv:math/0307245.
75. Schoen, R., & Yau, S. T. (1979). On the proof of the positive mass conjecture in general relativity. *Communications in Mathematical Physics*, *65*(1), 45–76.
76. Schoen, R., & Yau, S. T. Positive scalar curvature and minimal hypersurface singularities. arXiv:1704.05490.
77. Song, A. Existence of infinitely many minimal hypersurfaces in closed manifolds. arXiv:1806.08816.
78. Stenzel, M. B. (1993). Ricci-flat metrics on the complexification of a compact rank one symmetric space. *Manuscripta Mathematica*, *80*(2), 151–163.

Calibrated Submanifolds in G_2 Geometry

Ki Fung Chan and Naichung Conan Leung

This article summarizes the lecture given by the second author at the *Minischool on "G$_2$ manifolds and related topics"* which was held at the Fields Institute in August 2017, and is based on notes made by the first author.

All the content in this note is well-known, and the authors make no claim of originality. In this note, the symbols $\mathbb{R}$, $\mathbb{C}$, $\mathbb{H}$, and $\mathbb{O}$ denote the real numbers, the complex numbers, the quaternions, and the octonions, respectively, equipped with the standard Euclidean inner product.

1 Vector Cross Products

We start by introducing a general notion: vector cross products.

Definition 1.1 Let $V, \langle \cdot, \cdot \rangle$ be a finite dimensional (positive definite) inner product space, and let r be a non-negative integer. An r-fold vector cross product on V is an r-linear map

$$P : V^{\otimes r} \to V$$

satisfying the axioms

(1) $\langle P(u_1, \ldots, u_r), u_i \rangle = 0$ for all $u_1, \ldots u_r \in V$ and $0 \leq i \leq r$.
(2) $\|P(u_1, \ldots, u_r)\|^2 = \det(\langle u_i, u_j \rangle)$ for all $u_1, \ldots u_r \in V$.

▲

Using the inner product, it is easy to see that prescribing an r-fold vector cross product P is equivalent to prescribing the associated $(r+1)$-linear map $\Omega_P : (u_1, \ldots, u_{r+1}) \mapsto \langle P(u_1, \ldots, u_r), u_{r+1} \rangle$. Condition (1) is the same as saying that

K. F. Chan · N. C. Leung (✉)
Chinese University of Hong Kong, Sha Tin, Hong Kong
e-mail: leung@math.cuhk.edu.hk

S. Karigiannis et al. (eds.), *Lectures and Surveys on G$_2$-Manifolds and Related Topics*, Fields Institute Communications 84, https://doi.org/10.1007/978-1-0716-0577-6_4

Ω_P is alternating, and condition (2) implies that for orthonormal vectors $e_1, \ldots, e_{r+1} \in V$, we have $\Omega_P(e_1, \ldots, e_{r+1}) \leq 1$. This will be important when we talk about calibration forms.

Example 1.2 A 0-fold vector cross product is just a unit vector $v \in V$, and the corresponding 1-form Ω_v is $\langle \cdot, v \rangle$. ▲

Example 1.3 A 1-fold vector cross product is a linear map $J : V \to V$. Condition (2) says J is an isometry, and we write ω for Ω_J. Suppose $u, v \in V$ with $Ju = v$. By the skew-symmetry of ω, we have $\langle Jv, u \rangle = -\langle Ju, v \rangle = -\|v\|^2 = -\|Jv\| \cdot \|u\|$, which implies $Jv = -u$, and this proves $J^2 = -\text{id}$. That is, J is an orthogonal complex structure. Conversely, suppose $J : V \to V$ is a linear isometry satisfying $J^2 = -\text{id}$, then $\langle Ju, v \rangle = \langle J^2 u, Jv \rangle = -\langle Jv, u \rangle$, and hence J is a 1-fold vector cross product. This in particular says that an inner product space V admits a 1-fold vector cross product if and only if dim V is even. ▲

Example 1.4 Let P be a 2-fold vector cross product on V. We will write $u \times v$ for $P(u, v)$. Condition (2) says that $|u \times v|$ equals the area of the parallelogram spanned by u and v. Now, we define a ring R which is isomorphic to $\mathbb{R} \oplus V$ as a vector space, whose product structure is given by extending linearly the following relations:

$$a \cdot b = ab$$
$$a \cdot u = au$$
$$u \cdot v = -\langle u, v \rangle + u \times v$$

for $a, b \in \mathbb{R}$ and $u, v \in V$. The ring R so defined is a (not necessarily associative) division $\mathbb{R}$ algebra. Because, if $a + u \neq 0$, then $(a+u)(a-u) = (a-u)(a+u) = a^2 + \langle u, u \rangle > 0$ is a unit in $\mathbb{R}$. By the theorem of Frobenius on classification of finite dimensional (non-associative) division $\mathbb{R}$ algebras, we must have $R \cong \mathbb{R}, \mathbb{C}, \mathbb{H}$ or $\mathbb{O}$. In the two former cases, the vector cross product must vanish, therefore, up to isomorphism, there are only two non-trivial examples of 2-fold vector cross products. ▲

Example 1.5 On Im $\mathbb{H} \cong \mathbb{R}^3$, the cross product is defined by

$$u \times v = \text{Im}\,(uv).$$

One can check that this vector cross product can be identified with the usual cross product on $\mathbb{R}^3$ under the isomorphism $(a, b, c)^t \mapsto ai + bj + ck$. ▲

Example 1.6 On Im $\mathbb{O} = \text{Im}\,\mathbb{H} \oplus e\mathbb{H} \cong \mathbb{R}^7$, the cross product is similarly defined by

$$u \times v = \text{Im}\,(uv).$$

Under the isomorphism Im $\mathbb{O} \cong \text{Im}\,\mathbb{H} \oplus \mathbb{H}$, if ϕ_i, $i = 1, 2, 3$ are coordinates of the first factor Im $\mathbb{H}$ and x_i, $i = 1, 2, 3, 4$ are coordinates of the second factor $e\mathbb{H}$, the associated three form $\Omega_\times$ has the explicit formula:

$$\Omega_\times = d\phi_1 d\phi_2 d\phi_3 + d\phi_1 \wedge (dx_1dx_2 + dx_3dx_4) + d\phi_2 \wedge (dx_1dx_3 - dx_2dx_4) + d\phi_3 \wedge (dx_1dx_4 + dx_2dx_3)$$

This vector product is the basis for our definition of G_2 structure. We will write $(\mathrm{Im}\,\mathbb{O}, \langle\cdot,\cdot\rangle, \Omega_\times)$ to refer to this standard model. ▲

Definition 1.7 The group G_2 is the subgroup of $O(\mathrm{Im}\,\mathbb{O}) \cong O(7,\mathbb{R})$ that preserves $\Omega_\times$. ▲

Definition 1.8 We make the following two definitions.

(a) A G_2 structure on a 7-dimensional Riemannian manifold (M, g) is a smoothly varying (2-fold) vector cross product $\times$ defined on each tangent space. Equivalently, a G_2 structure on (M, g) is a differential 3-form Ω such that at each $p \in M$, $(T_pM, g|_{T_pM}, \Omega|_{T_pM})$ is isomorphic to our standard model $(\mathrm{Im}\,\mathbb{O}, \langle\cdot,\cdot\rangle, \Omega_\times)$.
(b) Furthermore, if $\nabla\Omega = 0$ for the Levi-Civita connection ∇ of g, then (M, g, Ω) is said to be a G_2 manifold. This is the same as saying $\mathrm{Hol}(g) \subset G_2$.

▲

The following is a useful property of the vector cross product.

Lemma 1.9 *Let $(V, \langle\cdot,\cdot\rangle)$ be an inner product space equipped with a 2-fold vector cross product $\times$ and let Ω be the associated 3-form. Suppose $u, v \in V$ are orthonormal. If $w = u \times v$, then $w \times u = v$.*

Proof From condition (2) of a vector cross product, we see that w is of unit length, and hence $\Omega(u, v, w) = \langle u \times v, w\rangle = \langle w, w\rangle = 1$. But this implies $\langle w \times u, v\rangle = \Omega(w, u, v) = 1$ as well. Now $w \times u$ is a unit vector whose orthogonal projection to the direction of v is v, so we must have $w \times u = v$ as claimed. □

This lemma allow us to define linear complex structures on the hyperplanes of V. More precisely, let $u \in V$ be any normal vector. Then $u \times \cdot$ defines a complex structure on the orthogonal complement $u^\perp$ of u. For if v is an element in $u^\perp$, then $\langle u \times v, u\rangle = 0$ by property (1). Using the notations in the lemma. we have $u \times (u \times v) = u \times w = -v$.

2 Calibrations

Definition 2.1 We make the following two definitions.

(a) Let (X, g) be a Riemannian manifold, and $\phi \in \Omega^k(X)$. Suppose $d\phi = 0$ and $\phi(e_1, e_2, \ldots, e_k) \le 1$ whenever $e_1, e_2, \ldots, e_k$ form an orthonormal subset in T_xX. Then ϕ is called a calibration form.

(b) Let ϕ be a calibration form, and let C be an oriented k-dimensional submanifold of X. Then C is said to be calibrated by ϕ if $\phi|_C = \mathrm{vol}_C$. (Here vol_C is the volume form of C as an oriented Riemannian submanifold of X.)

▲

By the discussion after Definition 1.1, we see that the associated $(r+1)$-form of an r-fold vector cross product is a calibration form.

Example 2.2 A 1-fold vector cross product on X is a vector field $\mathfrak{X}$. Its associated 1-form $\alpha = \langle \cdot, \mathfrak{X} \rangle$ is a calibration form if and only it is closed. In this case, an oriented 1-dimensional submanifold $C \subset X$ is calibrated by α if and only it is a flow line of $\mathfrak{X}$. ▲

Example 2.3 Let X be a Kähler manifold. A 2-dimensional oriented submanifold C of M is calibrated by the Kähler form $\omega = \langle J(\cdot), \cdot \rangle$ if and only if for each $p \in C$, and each unit vector $u \in T_pC$, the pair $\{u, Ju\}$ forms an orthonormal basis for T_pC. That is, each T_pC is preserved by J, so C is a complex curve in X. Similarly, a $2k$-dimensional oriented submanifold M of X is calibrated by the k-form $\frac{\omega^k}{k!}$ if and only if M is a complex submanifold of X. ▲

Example 2.4 If (X, g, Ω) is a G_2 manifold, then Ω is a calibration form, and the oriented submanifolds calibrated by Ω are called associative submanifolds of M. Equivalently, an oriented 3-dimensional submanifold $M \subset X$ is an associative submanifold if TM is preserved by the vector cross product $\times$. ▲

Example 2.5 If an 8-dimensional manifold X admits a closed 3-fold vector cross product, then the corresponding calibrated submanifolds are called Cayley submanifolds. ▲

Example 2.6 We may regard $\operatorname{Im}\mathbb{O} = \operatorname{Im}\mathbb{H} \oplus e\mathbb{H}$ as a Riemannian manifold with the canonical vector cross product structure. One can check easily that $\operatorname{Im}\mathbb{H} \subset \operatorname{Im}\mathbb{O}$ is an associative submanifold. ▲

Now, we come to an easy but important theorem.

Theorem 2.7 *If C is calibrated by some form ϕ, and C' is another k-dimensional submanifold of X such that $[C] = [C']$ in $H_k(X)$, then* $\mathrm{vol}(C) \leq \mathrm{vol}(C')$.

Proof We have $\mathrm{vol}(C) = \int_C \phi = \int_{C'} \phi \leq \int_{C'} \mathrm{vol}_{C'} = \mathrm{vol}(C')$. □

We may also want to extend this theorem to the case when the calibrated submanifold has a boundary. Let X^{2n} be a Kähler manifold, and let C be a 2-dimensional submanifold (possibly with boundary) calibrated by the Kähler form ω. Suppose the boundary C lies in some submanifold L. In order to repeat the above proof to show that C attains the absolute minimum volume in $H_k(X, L)$, we ought to have $\omega|_L = 0$. By simple linear algebra, we must have $\dim L \leq n$.

Definition 2.8 A submanifold $L \subset (X^{2n}, \omega)$ is called a Lagrangian submanifold if

$$(i) \ \dim L = n, \qquad \text{and} \qquad (ii) \ \omega|_L = 0.$$

▲

Remark 2.9 We have

$$\mathcal{N}_{L/X} = T^*L.$$

The reason for this is as follows. We have the following commutative diagram:

$$\begin{array}{ccccccccc} 0 & \longrightarrow & TL & \xrightarrow{i} & TX & \xrightarrow{\pi} & \mathcal{N}_{L/X} & \longrightarrow & 0 \\ & & \Big\downarrow{\scriptstyle \alpha} & & \Big\downarrow{\scriptstyle \cong \ \beta = \lrcorner\omega} & & \Big\downarrow{\scriptstyle \gamma} & & \\ 0 & \longrightarrow & \mathcal{N}^* & \xrightarrow{\pi^*} & T^*X & \xrightarrow{i^*} & T^*L & \longrightarrow & 0 \end{array}$$

By the non-degeneracy of ω, the middle map β is an isomorphism. We have the composite map $i^* \circ \beta \circ i : TL \to TX \to T^*X \to T^*L$. Condition (ii) of the Lagrangian condition says this composition vanishes, and this defines the maps α and γ. The map γ is surjective since $\gamma \circ \pi = i^* \circ \beta$ is, and α is injective since $\pi^* \circ \alpha = \beta \circ i$ is. By dimension count, both maps must be isomorphisms.

Suppose L is a manifold, and $X = T^*L$ is its cotangent bundle. We have a canonical symplectic form ω on X defined by $\omega(p, q) = dp \wedge dq$. The inclusion $L \subset X$ makes L a Lagrangian submanifold. The above says any Lagrangian submanifold L of X looks like the inclusion $L \subset X$ in a tubular neighbourhood of L. In fact, this identification can be made into a symplectomorphism, by the Weinstein neighborhood theorem.

Now we return to the G_2 case. Let (X, g, Ω) be a G_2 manifold, and C a submanifold with boundary. Suppose the boundary of C lies in some submanifold $K \subset X$. As before, in order for C to achieve absolute minimal volume in $H_3(X, K)$, we require $\Omega|_K = 0$. It is also not hard to show that we must have $\dim K \leq 4$. To see this, suppose W is a subspace of our standard model $(\mathrm{Im}\,\mathbb{O}, \Omega_\times)$, and $\dim W = s$. Suppose $\Omega_\times|_W = 0$. Then for any non-zero $w \in W$, we must have $(w \times W) \cap W = \{0\}$, and hence $(s-1) + s \leq 7$ and our claim follows.

Definition 2.10 A submanifold $K \subset (X, g, \Omega)$ is called a coassociative submanifold if

$$(i) \ \dim K = 4, \qquad \text{and} \qquad (ii) \ \Omega|_K = 0.$$

One may also check that the condition $\Omega|_K = 0$ is equivalent to the condition that K is calibrated by the 4-form $*\Omega$. ▲

Let W be a coassociative subspace of $(\mathrm{Im}\,\mathbb{O}, \Omega_\times)$. That is, $\dim W = 4$ and $\Omega_\times|_W = 0$. By a change of basis, we may write $\mathrm{Im}\,\mathbb{O} = \mathrm{Im}\,\mathbb{H} \oplus e\mathbb{H}$, where W under this identification is isomorphic to $e\mathbb{H}$. Let $\{e_1, e_2, e_3\}$ be an orthonormal subset of W.

One can then check that $\{e_1, e_2, e_3, (e_1 \times e_2) \times e_3\}$ forms an oriented orthonormal basis of W. Suppose $u \in \operatorname{Im}\mathbb{H}$ is a unit vector, we may assume $u = e_1 \times e_2$ without loss of generality. We can form interior product of u with $\Omega_\times$ and project it to a 2-form $\Omega_u \in \Lambda^2 W^*$ with $\Omega_u = \Omega_\times(u, \cdot, \cdot)$. Then one can check that

$$\Omega_u^2(e_1, e_2, e_3, (e_1 \times e_2) \times e_3) = \Omega(e_1 \times e_2, e_1, e_2)\Omega(u, e_3, u \times e_3) = 1.$$

Recall that in a 4-dimensional inner product space $(V, \langle \cdot, \cdot \rangle)$ with an orientation $\mathsf{vol} \in \Lambda^4 V^*$, we have the Hodge dual map $* : \Lambda^2 V^* \to \Lambda^2 V^*$ defined by $\langle a, b \rangle \mathsf{vol} = a \wedge *b$. We have $*^2 = \mathrm{id}$, and therefore we can decompose $\Lambda^2 V^*$ into the direct sum $\Lambda_+^2 V^* \oplus \Lambda_-^2 V^*$ of eigenspaces of $*$ with eigenvalues 1 and -1 respectively. The paragraph above simply says that the assignment $u \mapsto \Omega_u$ is an injective map from $W^\perp$ into $\Lambda_+^2 W^*$. Due to dimension reason, this must be an isomorphism. Globalizing this observation, we deduce the following.

Theorem 2.11 *Let (X, g, Ω) be a* G_2 *manifold, and let $K \subset X$ a coassociative submanifold. Then we have*

$$\mathcal{N}_{K/X} \overset{\lrcorner\Omega}{\cong} \Lambda_+^2 T^* K.$$

Observe that we have also established an isomorphism $\operatorname{Im}\mathbb{O} = \operatorname{Im}\mathbb{H} \oplus e\mathbb{H} \cong \Lambda_+^2(\mathbb{H}^*)^4 \times \mathbb{H}^4$, given by $\lrcorner\Omega|_\times \oplus (-e)$. Let x_i for $i = 1, 2, 3, 4$ be the standard coordinates of $\mathbb{H}$. Then $\Lambda_+^2(\mathbb{H}^*)^4$ has a basis formed by $e_1 = \mathrm{d}x_1\mathrm{d}x_2 + \mathrm{d}x_3\mathrm{d}x_4$, $e_2 = \mathrm{d}x_1\mathrm{d}x_3 + \mathrm{d}x_4\mathrm{d}x_2$, and $e_3 = \mathrm{d}x_1\mathrm{d}x_4 + \mathrm{d}x_2\mathrm{d}x_3$. Let $\{\phi_1, \phi_2, \phi_3\}$ be its dual basis. With this identification, the three form $\Omega|_\times$ is now given by

$$\Omega|_\times = \mathrm{d}\phi_1 \wedge \mathrm{d}\phi_2 \wedge \mathrm{d}\phi_3 + \mathrm{d}\phi_1 \wedge e_1 + \mathrm{d}\phi_2 \wedge e_2 + \mathrm{d}\phi_3 \wedge e_3.$$

We can easily see that each fibre $\Lambda_+^2(\mathbb{R}^*)^4 \times \{x\}$ is an associated submanifold.

Although the theory is good for finding volume minimizing submanifolds, in general it is difficult to find forms that restrict to volume forms.

Remark 2.12 We outline some similarities between Kähler manifold and G_2 manifold:

(a) When X is a Kähler manifold, $C^{2k} \subset X$ is calibrated by $\frac{\omega^k}{k!}$ if and only if C is a complex submanifold of X. When X is a G_2 manifold, $M^3 \subset X$ is calibrated by Ω if and only if M is an associative submanifold of X.

(b) When X is a Kähler manifold, $C^{2k} \subset X$ is calibrated by $\frac{\omega^k}{k!}$ if and only if TC is preserved by the 1-fold vector cross product J. When X is a G_2 manifold, $M^3 \subset X$ is calibrated by Ω if and only if TM is preserved by the 2-fold vector cross product $\times$.

(c) In a Kähler manifold, both $\phi = \frac{\omega^k}{k!}$ and $*\phi = \frac{\omega^{n-k}}{(n-k)!}$ are closed. In a G_2 manifold, both Ω and $*\Omega$ are closed.

(d) In a Kähler manifold, we have the notion of Lagrangian submanifold, defined as the largest submanifold on which the Kähler form ω restricts to zero. In a G_2

manifold, we have the notion of coassociated submanifold, defined as the largest submanifold on which the G_2 form Ω restricts to zero.

Example 2.13 Let (M^7, g, Ω) be a G_2 manifold. Note that the group G_2 acts on $\mathrm{Im}\,\mathbb{O}$, and for any unit element $u \in \mathrm{Im}\,\mathbb{O}$, the subgroup $\{g \in \mathrm{G}_2 | gu = u\}$ of G_2 is isomorphic to the group SU(3). We may consider $J = u \times \cdot : \langle u \rangle^\perp \to \langle u \rangle^\perp$ as an almost complex structure. We have $\Omega = u^\sharp \wedge \alpha + \beta$, and $*\Omega = u^\sharp \wedge \gamma + \delta$, where α is a 2-form, β, γ are 3-forms, and δ is a 4-form.

We can also write $\Omega = u^\sharp \wedge \omega + \mathrm{Re}\,\Omega_{CY}$ and $*\Omega = u^\sharp \wedge \mathrm{Im}\,\Omega_{CY} + \frac{1}{2}\omega^2$, where $\Omega_{CY} = \mathrm{Re}\,\Omega_{CY} + i\mathrm{Im}\,\Omega_{CY} \in \Omega^{3,0}(\langle u \rangle^\perp)$ is a holomorphic volume form for a Calabi-Yau manifold.

By the above, we may locally express M^7 as the product $Y^6 \times \mathbb{R}$, where Y^6 is a Calabi-Yau 3-fold. Then Ω can be generated by $\Omega_{CY} \in \Omega^{3,0}(Y)$ and $\omega \in \Omega^{1,1}$. Explicitly we have $\Omega = dt \wedge \omega + \mathrm{Re}\,\Omega_{CY}$, and $*\Omega = u^\sharp \wedge \mathrm{Im}\,\Omega_{CY} + \frac{1}{2}\omega^2$. Thus G_2 geometry is locally Calabi-Yau geometry.

Now let $M = Y \times \mathbb{R}$. Then a calibrated submanifold must be of the form $C = C \times \{\mathrm{pt}\}$ or $C = D \times \mathbb{R}$.

For an associative submanifold, it may be $C \times \{\mathrm{pt}\}$, where C is a special Lagrangian submanifold of Y calibrated by $\mathrm{Re}\,\Omega_{CY}$; or $D \times \mathbb{R}$, where D is a complex curve in Y calibrated by ω.

For a coassociative submanifold, it may be $C \times \{\mathrm{pt}\}$, where C is a complex submanifold in Y calibrated by $\frac{1}{2}\omega^2$; or $D \times \mathbb{R}$, where D is a special Lagrangian submanifold in Y calibrated by $\mathrm{Im}\,\Omega_{CY}$. ▲

3 Two Theorems

Suppose (L, g) is a Riemannian manifold, and $X = T^*L$ is its cotangent bundle. Then X has a natural Riemannian metric inherited from L, which we also denote by g. We can also define a canonical 1-form α on X as follows. Let p be a point on L, let q be a covector in T_p^*L, and let (U, V) be a tangent vector in $T_{(p,q)}X$, where $U \in T_pL$ and V is tangent to the fibre T_p^*L. Then $\alpha_{(p,q)}(U, V)$ is defined by

$$\alpha_{(p,q)}(U, V) = q(U).$$

In local coordinates, we have $\alpha = q\mathrm{d}p$, and $\omega = -\mathrm{d}\alpha = \mathrm{d}p \wedge \mathrm{d}q$ is a symplectic form on X. An almost complex structure J_g on X can thus be defined by the formula:

$$\omega(U, V) = g(J_gU, V).$$

The Weinstein neighbourhood theorem says that any Lagrangian submanifold L of a symplectic manifold has a tubular neighbourhood isomorphic to the inclusion $L \subset T^*L$, so we may reduce the general case to this special situation.

As we have seen before, the zero section L is a Lagrangian submanifold of X. Generally, any section $s : L \to T^*L$ can be regarded as the graph of a 1-form α_s on L, and we have the relations $s^*\alpha = \alpha_s$ and $s^*\omega = -\mathrm{d}\alpha_s$. In particular, a section $s : L \to T^*L$ defines a Lagrangian submanifold if and only if the associated 1-form α_s is a closed form.

Now, suppose β is a closed 1-form on L. For each $t \in \mathbb{R}$, the graph L_t of $t\beta$ is a Lagrangian submanifold in X. Suppose further that for some small positive ϵ, there exists a family of J_g-holomorphic curves A_t bounding $L \cup L_t$ for $t \in [0, \epsilon]$.

Since β is closed, when it is restricted to a smaller neighbourhood, we may assume that $\beta = \mathrm{d}f$ is exact. Let ∇f be the gradient of f defined by $g(\nabla f, \cdot) = \mathrm{d}f$. If we embed TL and T^*L into to TX in the obvious way, then we have the equality $\nabla f = -J(\mathrm{d}f)$. As the J_g holomorphic curves preserve the J_g structure, one may well think that $A_t \cap L$ is a gradient flow line of f in the limiting case. In fact, we have the following result of Fukaya–Oh.

Theorem 3.1 *With the same notation as above, there exists a constant $\epsilon > 0$, such that for any $t \in (0, \epsilon]$, there is a one-to-one correspondence between J_g-holomorphic curves bounding $L \cup L_t$ and the gradient flow lines of α on L.*

We also have a similar theorem for G_2 manifolds.

Suppose $\{C_t\}_{t\in[0,\epsilon]}$ is a family of coassociative submanifolds in a G_2 manifold (M, g, Ω), regarded as a deformation of $C = C_0$ along the normal vector field $n = \frac{\mathrm{d}C_t}{\mathrm{d}t}|_{t=0}$. Then $\iota_n\Omega$ is a self-dual harmonic 2-form on C. We further impose the condition that n is nowhere vanishing on C. Then $\omega_n = \iota_n\Omega$ defines a symplectic structure on C as $\omega_n^2 = |\omega_n|^2\mathrm{vol}_C$ is nowhere zero. Moreover, $J_n = \frac{n}{\|n\|} \times \cdot$ defines a compatible almost complex structure on (C, ω_n).

Theorem 3.2 *Suppose that (M, Ω) is a G_2 manifold and that C_t is a 1-parameter smooth family of coassociative submanifolds in M. When $\iota_n\Omega \in \Omega^+(C_0)$ is non-vanishing, then*

(a) If A_t is any one-parameter family of associative submanifolds in M satisfying

$$\partial A_t \subset C_t \cup C_0, \text{ and } \lim_{t\to 0} = \Sigma_0 \text{ in the } C^1\text{-topology,}$$

then Σ_0 is a J_n-holomorphic curve in C_0.

(b) Conversely, every regular J_n-holomorphic curve Σ_0 in C_0 is the limit of a family of associative submanifolds A_t as described above.

References

1. Bott, R., & Tu, L. W. (1982). *Differential forms in algebraic topology* (Vol. 82), Graduate texts in mathematics. New York: Springer.
2. Harvey, R., & Lawson, H. B. (1982). Calibrated geometries. *Acta Mathematica, 148*, 47–157.

3. Huybrechts, D. (2005). *Complex geometry*, Universitext. Berlin: Springer.
4. Joyce, D. D. (2000). *Compact manifolds with special holonomy*. Oxford: Oxford University Press.
5. Karigiannis, S. (2010). Some notes on G_2 and Spin(7) geometry. *Recent advances in geometric analysis* (Vol. 11, pp. 129–146), Advanced lectures in mathematics. Vienna: International Press. https://arxiv.org/abs/math/0608618.
6. Karigiannis, S., Lin, C., & Loftin, J. Octonionic-algebraic structure and curvature of the Teichmüller space of G_2manifolds, in preparation.
7. Kolář, I., Michor, P. W., & Slovák, J. (1993). *Natural operations in differential geometry*. Berlin: Springer.
8. Lawson, H. B., & Michelsohn, M. L. (1989). *Spin geometry*. Princeton: Princeton University Press.
9. Lee, J.-H., & Leung, N. C. (2008). Instantons and branes in manifolds with vector cross products. *Asian Journal of Mathematics*, *12*, 121–144.
10. Leung, N. C., Wang, X., & Zhu, K. (2013). Thin instantons in G2-manifolds and Seiberg-Witten invariants. *Journal of Differential Geometry*, *95*, 419–481.

Geometric Flows of G_2 Structures

Jason D. Lotay

Abstract Geometric flows have proved to be a powerful geometric analysis tool, perhaps most notably in the study of 3-manifold topology, the differentiable sphere theorem, Hermitian–Yang–Mills connections and canonical Kähler metrics. In the context of G_2 geometry, there are several geometric flows which arise. Each flow provides a potential means to study the geometry and topology associated with a given class of G_2 structures. We will introduce these flows, and describe some of the key known results and open problems in the field.

1 Introduction

Our understanding of G_2 structures, and particularly the question of when a G_2 structure can be deformed to become torsion-free, is very limited. It is therefore useful to look to new tools to tackle open problems in the area. An obvious avenue of attack is to use geometric flows, given their success in other geometric contexts: for example in analysing Hermitian connections (via Yang–Mills flow), convex hypersurfaces (via mean curvature flow) and perhaps most notably 3-manifolds and $\frac{1}{4}$-pinched Riemannian manifolds (via Ricci flow).

The goal of these notes is to explain some of the basics behind the geometric flow approach to studying G_2 structures and give a brief overview of what is known. It is important to note that several different flows of G_2 structures have been studied, based on various well-founded motivations. We shall attempt to give a brief description of each of these flows, the reasons behind them and some of the pros and cons in their study.

As well as giving this brief survey of the landscape in geometric flows of G_2 structures, we will provide some indication of some key open questions that we believe are worthy of further exploration.

Note These notes are based primarily on a lecture given at a Minischool on "G_2 Manifolds and Related Topics" at the Fields Institute, Toronto in August 2017.

J. D. Lotay (✉)
University of Oxford, Oxford, UK
e-mail: lotay@maths.ox.ac.uk

S. Karigiannis et al. (eds.), *Lectures and Surveys on G_2-Manifolds and Related Topics*, Fields Institute Communications 84, https://doi.org/10.1007/978-1-0716-0577-6_5

2 Geometric Flows

What is a geometric flow? Informally, it is a mechanism for "simplifying" or "decomposing" a given geometric structure into one or several "canonical" or "special" pieces. Thus, the primary goals of geometric flows are to show the existence of special geometric objects and to determine which geometric objects can be deformed to special ones. By answering these questions, one can then hope to understand large classes of geometric structures by understanding the much smaller class of canonical ones.

2.1 Heat Flow

Functions. The motivation for geometric flows comes from the heat flow for functions f on a Riemannian manifold:

$$\left(\frac{\partial}{\partial t}+\Delta\right)f=0. \tag{2.1}$$

(Here, and throughout, we will use the geometer's convention that the Laplacian Δ is a non-negative operator, so $\Delta = \mathrm{d}^*\mathrm{d}$ on functions.) The heat flow is parabolic, which means that if we consider (2.1) on a compact manifold M, then a short time solution to (2.1) is guaranteed to exist and the equation is "regularizing" (a notion we shall clarify in a moment).

We now make an elementary but fundamental observation.

Proposition 2.1 *The heat flow is the negative gradient flow for the Dirichlet energy:*

$$\frac{1}{2}\int_M |\mathrm{d}f|^2\,\mathrm{vol}_M \geq 0. \tag{2.2}$$

Proof We see that for any t-dependent family of functions we have

$$\frac{\partial}{\partial t}\frac{1}{2}\int_M |\mathrm{d}f|^2\,\mathrm{vol}_M = \langle\frac{\partial}{\partial t}\mathrm{d}f,\mathrm{d}f\rangle_{L^2} = \langle\mathrm{d}\frac{\partial}{\partial t}f,\mathrm{d}f\rangle_{L^2} = \langle\frac{\partial f}{\partial t},\Delta f\rangle_{L^2}. \tag{2.3}$$

□

Thus (2.2) will decrease fastest along the heat flow (2.1), and the critical points for the Dirichlet energy (which are exactly the stationary points for (2.1)) are given by the constant functions (i.e. $\mathrm{d}f = 0$) which are the absolute minimizers for the energy.

From the gradient flow point of view we should expect that given any smooth function, by following the heat flow we should be able to deform it into a critical point for (2.2), i.e. a constant function. We now show that this is the case.

Theorem 2.2 *Suppose that $f = f(x,t)$ solves the heat equation* (2.1) *on a compact manifold M and $f(x,0)$ is smooth. Then $f(x,t)$ exists for all $t > 0$ and $f(x,t) \to c \in \mathbb{R}$ smoothly as $t \to \infty$, where*

$$c = \frac{1}{\mathrm{Vol}(M)} \int_M f(x,0)\,\mathrm{vol}_M\,.$$

Proof If we suppose that f is smooth at $t = 0$, then it is smooth for all $t > 0$ as well and the eigenfunctions of Δ span $L^2(M)$, so at each time t we can write

$$f(x,t) = \sum_{\lambda} c_\lambda(t) f_\lambda(x) \tag{2.4}$$

for functions c_λ of time t and functions f_λ on M, where $\Delta f_\lambda = \lambda f_\lambda$ for $\lambda \geq 0$ and the f_λ form a complete orthonormal system for $L^2(M)$. It quickly follows from inserting (2.4) in (2.1) that

$$c_\lambda(t) = c_\lambda(0) e^{-\lambda t}, \tag{2.5}$$

and so the solution of (2.1) actually exists for all time $t > 0$. Moreover, the solution converges as $t \to \infty$ to

$$c_0 f_0 = \frac{1}{\mathrm{Vol}(M)} \int_M f(x,0)\,\mathrm{vol}_M, \tag{2.6}$$

the "average value" of f at time 0. □

Thus, the heat flow "regularizes" the function f in that it simplifies it as much as possible (it turns it into a constant) and we see that the higher the frequency (i.e. eigenvalue) of the eigenfunction of Δ in the expansion (2.4), the faster that component of f decays under the flow by (2.5). The f_λ for high λ correspond to higher "oscillations" of f, and so these "wiggles" in f get smoothed out by (2.1), eventually giving a constant. In terms of the Dirichlet energy functional (2.2), it shows that every function can be deformed to a minimizer (so the space of smooth functions retracts onto the constant functions, which are the critical points of the functional), and the minimizer we find is determined by the average value of f.

Notice that our analysis in the proof of Theorem 2.2 implies the following.

Lemma 2.3 *The integral of f is constant along* (2.1).

We are therefore free to modify (2.1) and consider

$$\left(\frac{\partial}{\partial t} + \Delta - \lambda_1\right) f = 0, \tag{2.7}$$

for functions f with $\int_M f = 0$, where λ_1 is the first positive eigenvalue of Δ on M. It is easy to see that (2.7) is still parabolic and that if $\int_M f = 0$ initially then it stays zero for all t under (2.7). However, under (2.7), we see that the flow no longer converges to a constant, but instead to the projection of f to the λ_1-eigenspace of Δ (which may now have several components if λ_1 is not a simple eigenvalue). Again, this flow "regularizes" f, throwing away all of the higher eigenmodes of Δ in the limit.

Forms. We can also consider the heat flow on differential k-forms α on a compact manifold M:

$$\left(\frac{\partial}{\partial t}+\Delta\right)\alpha=0, \tag{2.8}$$

where Δ is the Hodge Laplacian

$$\Delta=\mathrm{dd}^*+\mathrm{d}^*\mathrm{d}. \tag{2.9}$$

The flow (2.8) is now the gradient flow for the Dirichlet energy

$$\frac{1}{2}\int_M |\mathrm{d}\alpha|^2+|\mathrm{d}^*\alpha|\,\mathrm{vol}_M\geq 0 \tag{2.10}$$

by a similar argument as before. Again decomposing $\alpha(t)$ at each time t using eigenforms for Δ, we have the following.

Theorem 2.4 *The heat equation* (2.8) *for* $\alpha(t)$ *on a compact manifold starting at a smooth form* $\alpha(0)$ *exists for all time and converges smoothly to the projection of* $\alpha(0)$ *to the* 0*-eigenforms for* Δ*, i.e. the harmonic* k*-forms*

$$\mathrm{d}\alpha=\mathrm{d}^*\alpha=0. \tag{2.11}$$

The harmonic forms are precisely the critical points of (2.10) and are clearly absolute minimizers as they are zeros for the energy functional.

Now, the harmonic forms are only a finite-dimensional space in the space of k-forms, so given any initial k-form it could well be that the heat flow will just send it to zero, which is clearly a legitimate critical point for the flow (though not an interesting one!). For example if $\alpha(0)$ is exact or coexact (or the sum of forms of this type), the heat flow will just go to 0.

To ensure that we find a non-trivial critical point, we could restrict attention to closed k-forms:

$$\mathrm{d}\alpha=0. \tag{2.12}$$

Notice that this is preserved by (2.8) since in this case we have

$$\frac{\partial}{\partial t}\alpha=-\Delta\alpha=-(\mathrm{dd}^*+\mathrm{d}^*\mathrm{d})\alpha=-\mathrm{dd}^*\alpha, \tag{2.13}$$

so in fact we have that $\alpha(t)$ lies in the fixed cohomology class $[\alpha(0)]$ for all time as the right-hand side of (2.13) is exact. Therefore, if we have that $[\alpha(0)]\neq 0$ is a non-trivial cohomology class, we know that (2.13) will exist for all time and converge to the non-zero harmonic representative of that class (which we know exists and is unique by Hodge theory).

We could also equally well have restricted to coclosed k-forms

$$\mathrm{d}^*\alpha = 0, \tag{2.14}$$

since this is also preserved by a similar argument. This time $*\alpha(t)$ will lie in the fixed cohomology class $[*\alpha(0)]$ for all time and the flow will converge to the Hodge dual of the harmonic representative of $[*\alpha(0)]$.

We summarize these findings.

Proposition 2.5 *Suppose that $\alpha(t)$ is a family of k-forms on a compact n-manifold M solving* (2.8).

(a) If $\mathrm{d}\alpha(0) = 0$, $\alpha(t)$ exists for all $t > 0$ satisfying $\mathrm{d}\alpha(t) = 0$ for all t and converges smoothly to the unique harmonic k-form in $[\alpha(0)] \in H^k(M)$.

(b) If $\mathrm{d}^*\alpha(0) = 0$, $\alpha(t)$ exists for all $t > 0$ satisfying $\mathrm{d}^*\alpha(t) = 0$ for all t and converges smoothly to the Hodge dual of the unique harmonic $(n-k)$-form in $[*\alpha(0)] \in H^{n-k}(M)$.

We might hope, at least naively, that we could have similar good behaviour in geometric flows as for the heat flow, and thus obtain ways to canonically represent classes of geometric structures, just as harmonic forms uniquely represent all cohomology classes.

2.2 *Ricci Flow and Mean Curvature Flow*

Geometric flows aim to act on the same principle as the heat flow, two canonical examples being Ricci flow on metrics g and mean curvature flow on immersions F into a Riemannian manifold:

$$\frac{\partial}{\partial t}g = -2\,\mathrm{Ric}(g) \quad \text{and} \quad \frac{\partial}{\partial t}F = H, \tag{2.15}$$

where $\mathrm{Ric}(g)$ denotes the Ricci curvature tensor of g and H denotes the mean curvature vector of the immersion F. (Two other key examples of geometric flows of significant interest where many results have been obtained are the harmonic map heat flow and Yang–Mills flow, but we do not discuss them here.) Under suitable choices of coordinates, (2.15) can be seen as "heat flows", however this time the Laplacian depends on the metric or immersion respectively, and so the flows are nonlinear.

Parabolicity. The flows (2.15) are not parabolic due to geometric invariance in the problem: in Ricci flow this is diffeomorphism invariance, and in mean curvature flow this is invariance under reparametrisation. However, once one kills this geometric invariance, one obtains a parabolic equation.

For example, in Ricci flow, one can apply the so-called DeTurck's trick:

$$\frac{\partial}{\partial t}h = -2\,\mathrm{Ric}(h) + \mathcal{L}_{X(h)}h, \tag{2.16}$$

where $X(h)$ is a suitably chosen vector field (depending on the metric h) which ensures that (2.16) is parabolic and so has a short time solution which is regularizing. We can get a solution to Ricci flow from a solution to (2.16) by considering $g = \Phi^*h$, where Φ are diffeomorphisms defined by

$$\frac{\partial}{\partial t}\Phi = -X(h) \quad \text{and} \quad \Phi(0) = \mathrm{id}\,. \tag{2.17}$$

Proposition 2.6 *Suppose that h are metrics satisfying* (2.16) *and Φ are diffeomorphisms satisfying* (2.17). *Then $g = \Phi^*h$ satisfies the Ricci flow in* (2.15).

Proof By (2.16) and (2.17),

$$\frac{\partial}{\partial t}g = \frac{\partial}{\partial t}\Phi^*h = \Phi^*\frac{\partial}{\partial t}h - \Phi^*\mathcal{L}_{X(h)}h = -2\Phi^*\,\mathrm{Ric}(h) = -2\,\mathrm{Ric}(g). \tag{2.18}$$

□

This result is great but, it is natural to ask: what is a good choice of $X(h)$? The idea is, given h, for any symmetric 2-tensor k to consider the "gravitational tensor"

$$G(k) = k - \frac{1}{2}(\mathrm{tr}\,k)h, \tag{2.19}$$

whose divergence is given by the 1-form:

$$\mathrm{div}\,G(k) = \mathrm{div}(k) + \frac{1}{2}\mathrm{d}(\mathrm{tr}\,k). \tag{2.20}$$

(Here, by the divergence we mean the formal adjoint of the map $X^\flat \mapsto \frac{1}{2}\mathcal{L}_X h$, so that $\mathrm{div}(k)$ is the negative of the trace on the first two indices of ∇k; i.e. $\mathrm{div}(k)_j = -\nabla_i k_{ij}$. The musical isomorphisms, ∇ and trace are all defined by h.) If k is a *fixed* Riemannian metric then, using h, we can view k as an invertible map on 1-forms and so

$$X(h) = (k^{-1}\,\mathrm{div}\,G(k))^\sharp, \tag{2.21}$$

where the musical isomorphism is again given by h, is a well-defined vector field.

Theorem 2.7 *If we choose the vector field $X(h)$ as in* (2.21)*, the Ricci–DeTurck flow* (2.16) *is parabolic.*

Proof It is straightforward to compute that, with the choice of $X(h)$ in (2.21), the linearisation of (2.16) is simply

$$\frac{\partial}{\partial t}h = -\Delta h, \tag{2.22}$$

the heat equation on h. □

Moreover, (2.17) is harmonic map flow, which is parabolic, so we can solve (2.16) and (2.17) uniquely for short time by parabolic theory, and hence obtain a unique short time solution to the Ricci flow by Proposition 2.6.

Now, by analogy with the heat flow, in Ricci flow the "eigenmodes" we need to consider are solutions to

$$\text{Ric}(g) = \lambda g; \tag{2.23}$$

in other words, Einstein metrics. Here the "eigenvalues" λ have no distinguished sign and so, by analogy with the heat flow analysis above, we cannot say what will happen along the flow in general. However, if the flow exists for all time and converges, then it must tend to a Ricci-flat metric—the "zero mode" in the expansion of g in Einstein metrics, if you will. Similarly, for mean curvature flow, if the flow exists for all time and converges, we would obtain a minimal immersion (and the "expansion" of the immersion should be into constant mean curvature immersions). We see in both cases that we are breaking up our geometric object into pieces of significant interest. Moreover, the long-time existence and convergence of the flow allows us both to find the special object (a Ricci-flat metric or minimal immersion) and, at the same time, show that our initial geometric object can be deformed smoothly into the special one, which again is an important and challenging problem to solve.

Compact surfaces. To see the power of geometric flows it is instructive to look at Ricci flow in dimension 2. Here, Ricci curvature is just the Gauss curvature of the surface (up to a multiple) and there are three possibilities on a compact orientable surface.

- The flow exists for all time and converges. This means that the surface has a flat metric, and so must have genus 1 (by Gauss–Bonnet).
- The flow exists for all time but does not converge. In this case, just as when we perturbed the heat flow in (2.7), we can modify the Ricci flow and show that this modified flow exists for all time and converges to a hyperbolic metric. Thus the surface must have genus at least 2.
- The flow exists for only a finite time. This does not have a direct heat flow analogue, but one can again modify the Ricci flow as in the previous case (now by adding a term with the opposite sign), and show that this converges now to a constant positive curvature metric, which means that the surface must be a sphere. This is the difficult case in the analysis of the Ricci flow and this is typical of geometric flows: the case corresponding to "negative eigenvalues" (which do not happen for the Laplacian in the heat flow) is the most challenging to understand.

Thus, the Ricci flow in dimension 2 gives an alternative means to prove the uniformization theorem. In particular, that constant curvature metrics exist on any compact orientable surface, and the topology of the surface uniquely determines the sign of the constant curvature.

Gradient flow. Finally, one can also interpret (2.15) as gradient flows: in the case of mean curvature flow this is nothing but the negative gradient flow of the volume functional on immersions, but for Ricci flow the gradient flow interpretation is more subtle and involved so we shall not describe it here.

Needless to say, the fact that they are gradient flows is very helpful, since then one has some expectation of what one might hope to happen along the flow, as one has a monotone quantity (the analogue of the Dirichlet energy) along the flow which is trying to reach a critical value for the functional.

However, even with the gradient flow point-of-view, the nonlinearity of the problem and the potential complexity of the topology of the space of geometric objects we are considering means that we cannot always hope for the analysis of our flow to be straightforward and to go as expected. For example, the Ricci flow and mean curvature flow have special features (both good and bad) due to nonlinearity which simply cannot possibly occur in the standard heat flow.

2.3 *Singularities*

A singularity in a geometric flow is a point where the flow cannot be continued, because some quantity blows up to infinity. We already saw this in the Ricci flow in dimension 2, where there is always a singularity in finite time if we work on a sphere. Singularities may sound bad, and they definitely can be, but they can also be very helpful because they may tell you that you need to break up your geometric object into several pieces to get canonical objects. This happens for example in 3-dimensional Ricci flow, where singularities can be used to decide how to break up the 3-manifold according to Thurston's Geometrization Conjecture (now a theorem by Perelman's work).

The question is: what happens at a singularity geometrically? In good situations the singularity will be modelled on a special solution to the flow called a soliton. By "modelled on" we mean that by appropriately rescaling the flow around the singular point, both in space and time, in the limit we should see a soliton.

Definition 2.8 A soliton is a solution to the flow which is "self-similar", meaning that it moves very simply under the flow, just under rigid motions and diffeomorphisms or reparametrisations (or whatever notion of invariance is present in the problem).

Solitons which just move under diffeomorphisms are called steady, those which rescale getting smaller are called shrinking, and those which rescale getting larger are called expanding.

That is usually all of them (like in Ricci flow), but in mean curvature flow a soliton can also just translate, which is, rather unimaginatively, called a translating soliton.

Simple examples of solitons in Ricci flow are given by constant curvature metrics: flat metrics are steady (in fact critical points), constant positive curvature metrics are shrinking (like standard round spheres) and constant negative curvature metrics are expanding (like hyperbolic space).

From this point of view we see that shrinking solitons are the ones we should be most concerned with, since they will become singular in finite time, shrinking away. However, steady and expanding solitons also play an important role.

Steady solitons are potentially where the flow can get "stuck" going round and round under diffeomorphisms and never converging. Non-stationary compact examples of steady solitons are ruled out if you have a standard gradient flow interpretation of the flow, since the corresponding functional would be constant for steady solitons, which is a contradiction unless they are stationary. This shows one of the benefits of knowing that your geometric flow is the gradient flow of some functional.

On the other hand, expanding solitons give a potential mechanism to escape from a singularity, since they expand away from a singular geometric object.

It is therefore clear that understanding singularities and solitons is an important part of the study of any geometric flow.

3 G_2 Structures

Given this discussion of geometric flows, we are now motivated to ask the question: are there (useful) geometric flows of G_2 structures on a (compact) 7-manifold M and what do they want to achieve? We should perhaps not expect there to be just one useful flow to consider: for immersions, both mean curvature flow and inverse mean curvature flow have important geometric uses, for example. We therefore need to think about what are the important classes of G_2 structures that we want to analyse and what we expect to be "canonical" representatives for these classes. (For details about G_2 structures, which are equivalent to positive 3-forms, see for example [21].)

3.1 Torsion-Free and Torsion Forms

Clearly the most important class of G_2 structures are the torsion-free ones, given by positive 3-forms φ on M satisfying

$$\nabla_\varphi \varphi = 0 \quad \Leftrightarrow \quad \mathrm{d}\varphi = \mathrm{d}^*_\varphi \varphi = 0 \quad \Leftrightarrow \quad \mathrm{Hol}(g_\varphi) \subseteq \mathrm{G}_2\,. \tag{3.1}$$

(We are being slightly sloppy in the last equivalence, since given a metric there are infinitely many G_2 structures inducing that metric, so we mean that the holonomy $\mathrm{Hol}(g_\varphi)$ of g_φ is contained in G_2 if and only if there is some G_2 structure φ inducing g_φ which is closed and coclosed.) We also know that we may equivalently define G_2 structures on oriented, spin, Riemannian 7-manifolds using unit spinors σ, and the condition for the G_2 structure to be torsion-free is that σ is parallel with respect to the spin connection :

$$\nabla\sigma = 0. \tag{3.2}$$

(We can recall the relationship between unit spinors and positive 3-forms:

$$4\sigma \otimes \sigma = 1 + \varphi + *_\varphi\varphi + \mathrm{vol}_\varphi, \tag{3.3}$$

up to appropriate normalizations and sign conventions.)

Closed and coclosed G_2 structures. However, there are other obvious (potentially) important classes of G_2 structures, for example closed G_2 structures

$$d\varphi = 0, \tag{3.4}$$

or coclosed G_2 structures

$$d^*_\varphi\varphi = 0. \tag{3.5}$$

It is worth noting that, on the face of it, (3.4) is much stronger than (3.5): the first is a condition on a 4-form in 7-dimensions (so 35 equations at each point), whereas the second is a condition on a 5-form in 7-dimensions (so 21 equations at each point).

Both conditions (3.4) and (3.5) can be satisfied independently on any open 7-manifold admitting a G_2 structure by a straightforward h-principle argument; thus one can say (in some sense) that these conditions are only truly meaningful on compact 7-manifolds. In fact, (3.5) can always be satisfied on a compact 7-manifold admitting a G_2 structure also by an h-principle [7], but it is currently unknown whether the same is true for condition (3.4) or not: this again reflects the fact that (3.4) is a stronger condition than (3.5).

Theorem 3.1 *Let φ be a G_2 structure on M.*

(a) If M is open, then there exists a G_2 structure $\tilde{\varphi}$ C^0-close to φ satisfying (3.4).
(b) If M is either open or compact, then there exists a G_2 structure $\tilde{\varphi}$ C^0-close to φ satisfying (3.5).

One can interpret the h-principle result for coclosed G_2 structures as both positive and negative. On the one hand, it is good because we can always assume the condition (3.5) holds for φ if we want, though we have very little control on the φ produced by the h-principle: we can assume it is C^0-close to our original G_2 structure but the method only produces φ which will be very far away in the C^1-topology. On the other hand, it says that the condition (3.5) is, in a sense, meaningless and that talking about coclosed G_2 structures is the same as talking about all G_2 structures, which becomes a topological rather than a geometric question.

Torsion forms. One could also conceivably look at other special torsion classes by setting various combinations of the intrinsic torsion forms to vanish, recalling that these are given by

$$d\varphi = \tau_0 * \varphi + 3\tau_1 \wedge \varphi + *_\varphi \tau_3 \quad \text{and} \quad d *_\varphi \varphi = 4\tau_1 \wedge *_\varphi \varphi + \tau_2 \wedge \varphi, \tag{3.6}$$

where $\tau_0 \in C^\infty(M)$, $\tau_1 \in \Omega^1(M)$, $\tau_2 \in \Omega^2_{14}(M)$, $\tau_3 \in \Omega^3_{27}(M)$, with the standard notation referring to the type decomposition of forms determined by φ (see, for example, [5]). Recall that $\beta \in \Omega^2_{14}(M)$ if and only if

$$\beta \wedge \varphi = - *_\varphi \beta \quad \Leftrightarrow \quad \beta \wedge *_\varphi \varphi = 0, \tag{3.7}$$

and that $\gamma \in \Omega^3_{27}(M)$ if and only if

$$\gamma \wedge \varphi = \gamma \wedge *_\varphi \varphi = 0. \tag{3.8}$$

Moreover, recall that we have an isomorphism $i_\varphi : S^2T^*M = \mathrm{Span}\{g_\varphi\} \oplus S^2_0T^*M \to \Lambda^3_1 \oplus \Lambda^3_{27}$ given on decomposable elements $\alpha \circ \beta$ by

$$i_\varphi(\alpha \circ \beta) = \alpha \wedge *_\varphi(\beta \wedge *_\varphi \varphi) + \beta \wedge *_\varphi(\alpha \wedge *_\varphi \varphi).$$

We also have an explicit way to invert i_φ using $j_\varphi : \Lambda^3T^*M \to S^2T^*M$ given by

$$j_\varphi(\gamma)(u, v) = *_\varphi(u \lrcorner \varphi \wedge v \lrcorner \varphi \wedge \gamma).$$

Notice that $i_\varphi(g_\varphi) = 6\varphi$, $j_\varphi(\varphi) = 6g_\varphi$ and Ker $j_\varphi = \Lambda^3_7$.

Other classes. A particular class of G_2 structures one could consider are the nearly parallel G_2 structures

$$d\varphi = \tau_0 *_\varphi \varphi \tag{3.9}$$

for a constant τ_0. These structures define Einstein metrics with non-negative scalar curvature, and so there is a potential relation between these structures and our discussion of the Ricci flow above.

One could also view matters in terms of spinors, and study geometric flows of unit spinors. One can then try studying parallel spinors or, more generally, Killing spinors, as well as other special types of spinors (e.g. twistor spinors).

3.2 General Flows

Based on this discussion, it is clear that there are many possible geometric flows one could write down, and each one could potentially tackle different open problems in G_2 geometry.

That said, one can describe how various key quantities vary under a general flow of G_2 structures (see [5, 22]). By the type decomposition of 3-forms, any geometric flow of G_2 structures can be written

$$\frac{\partial}{\partial t}\varphi = 3f_0\varphi + *_\varphi(f_1\wedge\varphi) + f_3 = i_\varphi(h) + X\lrcorner *_\varphi \varphi \tag{3.10}$$

where $f_0 \in C^\infty(M)$, $f_1 \in \Omega^1(M)$ and $f_3 \in \Omega^3_{27}(M)$, $h \in C^\infty(S^2T^*M)$ and $X \in C^\infty(TM)$ at each time t. From this one can see that the metric and 4-form evolve as follows.

Proposition 3.2 *Along* (3.10) *we have*

$$\frac{\partial}{\partial t}g_\varphi = 2f_0 g + \frac{1}{2}j_\varphi(f_3) = 2h \tag{3.11}$$

and

$$\frac{\partial}{\partial t} *_\varphi\varphi = 4f_0 * \varphi + f_1\wedge\varphi - *_\varphi f_3. \tag{3.12}$$

In particular, along any flow (3.10), the evolution of the metric is independent of the vector field X, and the volume form evolves as follows.

Proposition 3.3 *Along* (3.10) *we have*

$$\frac{\partial}{\partial t}\operatorname{vol}_\varphi = 7f_0\operatorname{vol}_\varphi = \frac{1}{3}\frac{\partial}{\partial t}\varphi\wedge *_\varphi\varphi. \tag{3.13}$$

This formula is useful to study Hitchin's volume functional [17] on a compact manifold M. (Note that the published version [18] of [17] omits the material on G_2 structures.)

Proposition 3.4 *Along* (3.10), *the volume functional*

$$\operatorname{Vol}(\varphi) = \frac{1}{7}\int_M \varphi\wedge *\varphi = \int_M \operatorname{vol}_\varphi = \operatorname{Vol}(M, g_\varphi). \tag{3.14}$$

satisfies (by (3.13)*)*

$$\frac{\partial}{\partial t}\operatorname{Vol}(\varphi) = 7\int_M f_0\operatorname{vol}_\varphi = \frac{1}{3}\langle\frac{\partial}{\partial t}\varphi, \varphi\rangle_{L^2}. \tag{3.15}$$

This shows in particular that $\operatorname{Vol}(\varphi)$ will be monotone along any flow for which f_0 has a sign, and the critical points of the functional will be characterised by the G_2 structures for which $f_0 = 0$. It also shows that the obvious gradient flow for $\operatorname{Vol}(\varphi)$ is

$$\frac{\partial}{\partial t}\varphi = \lambda\varphi \tag{3.16}$$

for some $\lambda > 0$ (or $\lambda < 0$ for the negative gradient flow). This is clearly useless since all it does is rescale φ! Therefore, if one wants to think about making use of the volume functional for a gradient flow, we should consider restricting the class of G_2 structures we work with.

4 Laplacian Flow

The geometric flow of G_2 structures that has received the most attention is the Laplacian flow due to Bryant [5].

Definition 4.1 The Laplacian flow is given by

$$\frac{\partial}{\partial t}\varphi = \Delta_\varphi \varphi, \tag{4.1}$$

where

$$\Delta_\varphi = \mathrm{d}\mathrm{d}^*_\varphi + \mathrm{d}^*_\varphi \mathrm{d} \tag{4.2}$$

is the Hodge Laplacian.

On a compact manifold, we see that

$$\Delta_\varphi \varphi = 0 \quad \Leftrightarrow \quad \mathrm{d}\varphi = \mathrm{d}^*_\varphi \varphi = 0 \tag{4.3}$$

by integration by parts, and so torsion-free G_2 structures will be the critical points of (4.1).

4.1 *Closed G_2 Structures*

Bryant's suggestion is to restrict (4.1) to closed G_2 structures φ as in (3.4). A key motivation is the usual one in G_2 geometry: namely that the torsion-free condition naturally splits into a linear condition (3.4) and a nonlinear condition (3.5). Thus, it is useful to assume the linear condition is satisfied then try to solve the nonlinear one. This strategy is the only one that has proved to be successful, by the work of Joyce [21, Chap. 11]. Hence, it clearly makes sense to follow the same approach in a geometric flow.

It will turn out that when (4.1) exists the closed condition is preserved. In that case

$$\frac{\partial}{\partial t}\varphi = \mathrm{d}\mathrm{d}^*_\varphi \varphi, \tag{4.4}$$

so in fact (4.1) stays within a fixed cohomology class $[\varphi(0)]$.

Proposition 4.2 *If $\varphi(t)$ satisfies (4.1) on M and $\mathrm{d}\varphi(t) = 0$ then $\varphi(t) \in [\varphi(0)] \in H^3(M)$ for all t for which the flow exists.*

This is perhaps reminiscent of the Kähler–Ricci flow on a manifold with $c_1 = 0$, which starts with a Kähler form and stays within the Kähler class, but it is not clear at all whether such an analogy is pertinent or a red herring.

When we restrict to closed G_2 structures we can decompose

$$\Delta_\varphi\varphi = \frac{1}{7}|\tau_2|^2\varphi + f_3. \tag{4.5}$$

We can see this because

$$\mathrm{d}*_\varphi\varphi = \tau_2 \wedge \varphi = -*_\varphi\tau_2 \tag{4.6}$$

by (3.6) and (3.7), and therefore

$$\Delta_\varphi\varphi = \mathrm{d}\mathrm{d}^*_\varphi\varphi = \mathrm{d}\tau_2 \tag{4.7}$$

and differentiating (4.6) gives

$$0 = \mathrm{d}\tau_2 \wedge \varphi = \Delta_\varphi\varphi \wedge \varphi = 0. \tag{4.8}$$

This means that $f_1 = 0$ in (3.10). Moreover, (4.6) and (3.7) imply that

$$\mathrm{d}\tau_2 \wedge *_\varphi\varphi = \mathrm{d}(\tau_2 \wedge *_\varphi\varphi) - \tau_2 \wedge \mathrm{d}*_\varphi\varphi = -\tau_2 \wedge \tau_2 \wedge \varphi = |\tau_2|^2 \operatorname{vol}_\varphi. \tag{4.9}$$

Hence, for closed G_2 structures, (4.5) implies the following.

Lemma 4.3 *For a closed* G_2 *structure* φ, *we have*

$$\Delta_\varphi\varphi = 0 \quad \Leftrightarrow \quad \mathrm{d}^*_\varphi\varphi = 0. \tag{4.10}$$

Therefore, the critical points of (4.1) on closed G_2 structures are precisely the torsion-free G_2 structures, without assuming compactness. This might seem like a minor point (since we will mainly only care about the Laplacian flow on compact manifolds) but it appears to hint at the special character of the Laplacian flow because it is restricted to closed G_2 structures.

Moreover, we can show that

$$\Delta_\varphi\varphi = i_\varphi\Big(-\operatorname{Ric}(g_\varphi) + \frac{4}{21}|\tau_2|^2 g_\varphi + \frac{1}{8} j_\varphi\big(*_\varphi (\tau_2 \wedge \tau_2)\big)\Big) \tag{4.11}$$

so that by (3.11) we have the following.

Proposition 4.4 *Along* (4.1) *we have*

$$\frac{\partial}{\partial t} g_\varphi = -2\operatorname{Ric}(g_\varphi) + \frac{8}{21}|\tau_2|^2 g_\varphi + \frac{1}{4} j_\varphi\big(*_\varphi (\tau_2 \wedge \tau_2)\big). \tag{4.12}$$

4.2 Volume Functional

Lemma 4.5 *Along* (4.1) *for closed* G_2 *structures, the volume functional* $\operatorname{Vol}(\varphi)$ *is monotonically increasing.*

Proof We can already see from (4.5) that, in terms of (3.10),

$$f_0 = \frac{1}{7}|\tau_2|^2 \geq 0. \tag{4.13}$$

Hence, by (3.15), the volume functional $\mathrm{Vol}(\varphi)$ is monotonically increasing along the Laplacian flow. □

This allows us to give a very quick alternative proof of the main result of [27].

Proposition 4.6 *Let $\varphi(t)$ be a steady or shrinking soliton to the Laplacian flow* (4.1) *for closed* G_2 *structures on a compact manifold. Then $\varphi(t)$ is stationary, i.e. $\varphi(t) = \varphi(0)$ is torsion-free for all t.*

Proof If $\varphi(t)$ is steady or shrinking then

$$\frac{\partial}{\partial t}\mathrm{Vol}(\varphi(t)) \leq 0. \tag{4.14}$$

Therefore, by Lemma 4.5, $\mathrm{Vol}(\varphi(t))$ is constant. Hence, by (3.15) and (4.13), we must have that

$$f_0 = \frac{1}{7}|\tau_2|^2 = 0, \tag{4.15}$$

which means $\tau_2 = 0$, as required. □

It is important to note that this result fails in the non-compact setting: Lauret [26] has constructed examples of non-compact shrinking and steady solitons which are not stationary. The proof of Proposition 4.6 is not valid here since the volume is not well-defined for these non-compact examples.

We now show that the Laplacian flow for closed G_2 structures is actually the gradient flow for $\mathrm{Vol}(\varphi)$ where, from now on, we only consider the volume functional restricted to a given cohomology class.

Proposition 4.7 *The Laplacian flow* (4.1) *for closed* G_2 *structures φ is the gradient flow for the volume functional* $\mathrm{Vol}(\varphi)$ *in* (3.14) *restricted to* $[\varphi]$.

Proof We know that the flow stays within a given cohomology class, so we can write

$$\varphi(t) = \varphi(0) + \mathrm{d}\eta(t) \tag{4.16}$$

for some 2-forms η and the Laplacian flow is really, in some sense, a flow on 2-forms. (Again, this is reminiscent of Kähler–Ricci flow with $c_1 = 0$ as the flow becomes a flow on Kähler potentials, but this analogy is made with the usual caveats.) Then (3.15) gives us that

$$\frac{\partial}{\partial t}\mathrm{Vol}(\varphi) = \frac{1}{3}\langle\frac{\partial}{\partial t}\varphi, \varphi\rangle_{L^2} = \frac{1}{3}\langle \mathrm{d}\frac{\partial}{\partial t}\eta, \varphi\rangle_{L^2} = \frac{1}{3}\langle\frac{\partial}{\partial t}\eta, \mathrm{d}^*_\varphi\varphi\rangle_{L^2}. \tag{4.17}$$

Hence, the gradient flow for $\mathrm{Vol}(\varphi)$ is

$$\frac{\partial}{\partial t}\eta = \mathrm{d}^*_\varphi\varphi \quad \Rightarrow \quad \frac{\partial}{\partial t}\varphi = \mathrm{d}\frac{\partial}{\partial t}\eta = \mathrm{d}\mathrm{d}^*_\varphi\varphi = \Delta_\varphi\varphi, \tag{4.18}$$

the Laplacian flow. (We have ignored the factor $\frac{1}{3}$ which amounts to rescaling time t.) □

Proposition 4.7 immediately yields the following.

Proposition 4.8 *A closed* G_2 *structure on a compact manifold is a critical point of* $\mathrm{Vol}(\varphi)$ *within a fixed cohomology class if and only if* φ *is torsion-free.*

We can even say more about the critical points of the volume functional. If we look at the second derivative at a critical point $\varphi(0)$, then by (3.15) along any variation $\varphi(s) = \varphi(0) + \mathrm{d}\eta(s)$ in the cohomology class

$$\frac{\partial^2}{\partial s^2}\mathrm{Vol}(\varphi)|_{s=0} = \frac{1}{3}\int_M \frac{\partial}{\partial s}\varphi \wedge \frac{\partial}{\partial s} *_\varphi\varphi|_{s=0}, \tag{4.19}$$

since

$$\frac{1}{3}\int_M \frac{\partial^2}{\partial s^2}\varphi \wedge *_\varphi\varphi|_{s=0} = \frac{1}{3}\langle \mathrm{d}\frac{\partial^2}{\partial s^2}\eta, *_\varphi\varphi\rangle_{L^2}|_{s=0} = \frac{1}{3}\langle\frac{\partial^2}{\partial s^2}\eta, \mathrm{d}^*_\varphi\varphi\rangle_{L^2}|_{s=0} = 0 \tag{4.20}$$

as $\mathrm{d}^*_\varphi\varphi = 0$ at $s = 0$ by assumption. Now if we write the variation of $\varphi(s)$ at $s = 0$ as in formula (3.10) then $*_\varphi\varphi(s)$ varies by (3.12) and so we see that

$$\int_M \frac{\partial}{\partial s}\varphi \wedge \frac{\partial}{\partial s} *_\varphi\varphi|_{s=0} = c_0\|f_0\|^2_{L^2} + c_1\|f_1\|^2_{L^2} - \|f_3\|^2_{L^2} \tag{4.21}$$

for some positive constants c_0, c_1. If the variation $\varphi(s)$ is orthogonal to the action by diffeomorphisms (meaning that $\frac{\partial\varphi}{\partial s}|_{s=0}$ and the tangent directions to the diffeomorphism orbit through $\varphi(0)$ are orthogonal), then a slice theorem argument forces $f_0 = f_1 = 0$ (see [17]). In other words, the slice condition

$$\mathrm{d}^*_\varphi\mathrm{d}\eta \in \Omega^2_{14}(M) \quad \Rightarrow \quad \mathrm{d}\eta \in \Omega^3_{27}(M), \tag{4.22}$$

so $f_3 = \mathrm{d}\eta(0)$. Putting this observation together with (4.19) and (4.21), we see that, orthogonal to the action by diffeomorphisms, we have

$$\frac{\partial^2}{\partial s^2}\mathrm{Vol}(\varphi)|_{s=0} = -\|\mathrm{d}\eta(0)\|^2_{L^2} \le 0. \tag{4.23}$$

Thus, we have the following.

Theorem 4.9 *Critical points of* $\mathrm{Vol}(\varphi)$ *on* $[\varphi]$ *are strict local maxima (modulo the action of diffeomorphisms).*

This suggests that the gradient flow of the volume functional (i.e. (4.1)) could be well-behaved since its only critical points are maxima.

4.3 Short-Time Existence

A key issue we have avoided in our discussion thus far is the question of whether the Laplacian flow exists or not. Certainly, if we look at (4.1) and compare it to (2.1) we would seem to have the wrong sign! In general, (4.1) does not seem to be parabolic in any sense, which is very bad news analytically.

However, again the fact that we are restricting to closed G_2 structures comes to our rescue. In this case, we have already seen in (4.12) that the metric evolves by Ricci flow plus lower order terms and so its flow is parabolic modulo diffeomorphisms.

If we do DeTurck's trick for the Laplacian flow, using $d\varphi = 0$:

$$\frac{\partial}{\partial t}\varphi = \Delta_\varphi \varphi + \mathcal{L}_{X(\varphi)}\varphi = \Delta_\varphi \varphi + d(X(\varphi) \lrcorner \varphi), \tag{4.24}$$

(as in the Ricci flow case, and with the same vector field X given in (2.21), in fact) then we might hope that we end up with a genuine parabolic equation in (4.24). However, this is not the case!

In fact, (4.24) is only parabolic in the direction of closed forms, so one has to consider the restricted flow in order to prove short-time existence. This is a little bit tricky but was done by Bryant–Xu [6]. Their paper, which definitely gives a correct result that is fundamental to the subject, has never been published, so we give an account of the proof here, which is essentially the same as in [6].

Theorem 4.10 *Let φ_0 be a smooth closed G_2 structure on a compact manifold M. There exists $\epsilon > 0$ so that a unique solution $\varphi(t)$ to the Laplacian flow (4.1) with $\varphi(0) = \varphi_0$ and $d\varphi(t) = 0$ exists for all $t \in [0, \epsilon]$, where ϵ depends on φ_0.*

Proof We know that if (4.1) exists then it will stay in the cohomology class $[\varphi_0]$ by Proposition 4.2. Therefore, we could write

$$\varphi(t) = \varphi_0 + d\eta(t) \tag{4.25}$$

for a family of exact 3-forms $d\eta(t)$, and (4.1) with the initial condition $\varphi(t) = \varphi_0$ is equivalent to

$$\frac{\partial}{\partial t} d\eta = \Delta_{\varphi_0 + d\eta} d\eta \quad \text{and} \quad d\eta(0) = 0. \tag{4.26}$$

Let $X(\varphi)$ be the vector field given by $X(g_\varphi)$ in (2.21), where we can choose the fixed background metric $k = g_{\varphi_0}$ for example. Suppose we can solve

$$\frac{\partial}{\partial t} d\eta(t) = \Delta_{\varphi_0 + d\eta} d\eta + d(X(\varphi_0 + d\eta) \lrcorner d\eta) \quad \text{and} \quad d\eta(0) = 0 \tag{4.27}$$

uniquely for short time. Then we can find a unique family of diffeomorphisms Φ solving (2.17) (since this is the harmonic map heat flow). Therefore, just as in Proposition 2.6, it follows that $\varphi = \Phi^*(\varphi_0 + \mathrm{d}\eta)$ satisfies (4.1) with $\varphi(0) = \varphi_0$, $\mathrm{d}\varphi(t) = 0$ for short time, and the solution is unique. (We have used here that the condition for a 3-form to be positive is open, so for t sufficiently small, $\varphi_0 + \mathrm{d}\eta(t)$ will be a positive 3-form.) We are therefore left to show that (4.27) has a unique short-time solution.

We know from (4.11) and [28, Lemma 9.3] that

$$\Delta_\varphi \varphi + \mathcal{L}_{X(\varphi)}\varphi = \frac{1}{2} i_\varphi\big(- 2\,\mathrm{Ric}(g_\varphi) + \mathcal{L}_{X(g_\varphi)} g_\varphi + \frac{2}{21}|\tau_2|^2 g_\varphi + \frac{1}{4} j_\varphi(*_\varphi(\tau_2 \wedge \tau_2))\big) + \frac{1}{2}\big(\mathrm{d}^*(X(\varphi) \lrcorner \varphi)\big)^\sharp \lrcorner *_\varphi \varphi. \tag{4.28}$$

Given that the terms with τ_2 in them in (4.28) are lower order, and the Ricci–DeTurck flow (2.16) is parabolic, it would seem likely that the linearisation of (4.28) is parabolic when restricted to closed forms, and hence that (4.27) is parabolic. In fact, one may explicitly compute as in [6] that the linearisation of (4.27) in the direction of exact forms is

$$\frac{\partial}{\partial t}\mathrm{d}\zeta = -\Delta_{\varphi_0+\mathrm{d}\eta}\mathrm{d}\zeta + \mathrm{d}\big(Q_{\varphi_0+\mathrm{d}\eta}(\mathrm{d}\zeta)\big), \tag{4.29}$$

where $Q_{\varphi_0+\mathrm{d}\eta}(\mathrm{d}\zeta)$ is order zero in $\mathrm{d}\zeta$ (meaning it depends on just $\mathrm{d}\zeta$ and not its derivatives). Somewhat surprisingly, the sign has switched and (4.29) is manifestly parabolic.

However, we have only shown that (4.27) is parabolic in the direction of exact forms, so we cannot apply standard parabolic theory. Instead we will invoke the Nash–Moser Inverse Function Theorem (see [16] for a detailed discussion of this theorem).

We start by setting up the notation. We let

$$\mathcal{X} = \mathrm{d}\big(C^\infty([0,\epsilon] \times M, \Lambda^2 T^* M)\big) \quad \text{and} \quad \mathcal{Y} = \mathrm{d}\Omega^2(M), \tag{4.30}$$

and

$$\mathcal{U} = \{\mathrm{d}\eta \in \mathcal{X} \,:\, \varphi_0 + \mathrm{d}\eta(t) \text{ is a } \mathrm{G}_2 \text{ structure for all } t\}, \tag{4.31}$$

which is an open set in $\mathcal{X}$ containing 0. We define $\mathcal{F} : \mathcal{U} \to \mathcal{X} \times \mathcal{Y}$ by

$$\mathcal{F}(\mathrm{d}\eta) = \left(\frac{\partial}{\partial t}\mathrm{d}\eta(t) - \Delta_{\varphi_0+\mathrm{d}\eta}\mathrm{d}\eta - \mathrm{d}(X(\varphi_0 + \mathrm{d}\eta) \lrcorner \mathrm{d}\eta), \mathrm{d}\eta(0)\right), \tag{4.32}$$

so that $\mathcal{F}(\mathrm{d}\eta) = (0,0)$ if and only if $\mathrm{d}\eta$ solves (4.27). If we can show that $\mathcal{F}$ is locally invertible near 0, we have that (4.27) has a unique short-time solution and so the proof is complete.

By (4.29), the linearisation of $\mathcal{F}$ at $\mathrm{d}\eta \in \mathcal{U}$ is given by

$$\mathrm{d}\mathcal{F}|_{\mathrm{d}\eta}(\mathrm{d}\zeta) = \left(\frac{\partial}{\partial t}\mathrm{d}\zeta + \Delta_{\varphi_0+\mathrm{d}\eta}\mathrm{d}\zeta - \mathrm{d}\big(\mathcal{Q}_{\varphi_0+\mathrm{d}\eta}(\mathrm{d}\zeta)\big), \mathrm{d}\zeta(0)\right). \tag{4.33}$$

Since (4.29) is parabolic, we see that if $\mathrm{d}\mathcal{F}|_{\mathrm{d}\eta}(\mathrm{d}\zeta) = (0,0)$ then $\mathrm{d}\zeta = 0$ as this is the unique solution to the linear parabolic equation (4.29) with zero initial condition. Further, given $(\mathrm{d}\xi, \mathrm{d}\xi_0) \in \mathcal{X} \times \mathcal{Y}$, we see that $\mathrm{d}\mathcal{F}|_{\mathrm{d}\eta}(\mathrm{d}\zeta) = (\mathrm{d}\xi, \mathrm{d}\xi_0)$ if and only if

$$\frac{\partial}{\partial t}\mathrm{d}\zeta = -\Delta_{\varphi_0+\mathrm{d}\eta}\mathrm{d}\zeta + \mathrm{d}\big(\mathcal{Q}_{\varphi_0+\mathrm{d}\eta}(\mathrm{d}\zeta)\big) + \mathrm{d}\xi \quad \text{and} \quad \mathrm{d}\zeta(0) = \mathrm{d}\xi_0, \tag{4.34}$$

which can be solved by linear parabolic theory.

We therefore have that $\mathrm{d}\mathcal{F}|_{\mathrm{d}\eta} : \mathcal{X} \to \mathcal{X} \times \mathcal{Y}$ is invertible for all $\mathrm{d}\eta \in \mathcal{U}$. However, this is not yet enough to show that $\mathcal{F}$ is locally invertible. Notice that the family of inverses provides a map $\mathcal{G} : \mathcal{U} \times \mathcal{X} \times \mathcal{Y} \to \mathcal{X}$ given by

$$\mathcal{G}(\mathrm{d}\eta, \mathrm{d}\xi, \mathrm{d}\xi_0) = \mathrm{d}\mathcal{F}|_{\mathrm{d}\eta}^{-1}(\mathrm{d}\xi, \mathrm{d}\xi_0). \tag{4.35}$$

To deduce that $\mathcal{F}$ is locally invertible, as we said, we want to invoke the Nash–Moser Inverse Function Theorem which means that we need the following:

- $\mathcal{X}$ and $\mathcal{Y}$ are tame Fréchet spaces and $\mathcal{F}$ is a smooth tame map;
- $\mathrm{d}\mathcal{F}|_{\mathrm{d}\eta}$ is invertible for all $\mathrm{d}\eta \in \mathcal{U}$ and $\mathcal{G}$ is a smooth tame map.

The fact that $C^\infty([0,\epsilon] \times M, \Lambda^2 T^*M)$ and $\Omega^2(M)$ are naturally tame Fréchet spaces is standard, and it therefore quickly follows (from Hodge theory) that $\mathcal{X}$ and $\mathcal{Y}$ are also tame Fréchet spaces. Smooth partial differential operators are smooth tame maps, so $\mathcal{F}$ is a smooth tame map.

The last thing we need to show is that $\mathcal{G}$ is a smooth tame map, but this follows immediately from a general result about the family of inverses given by solutions of a smooth family of parabolic partial differential equations, due to Hamilton [15].

Thus, the Nash–Moser Inverse Function Theorem applies to $\mathcal{F}$ and so it is locally invertible as desired. □

Actually, before DeTurck's trick the same method of Bryant–Xu was used by Hamilton [15] to prove short-time existence of the Ricci flow. The reason is that the Ricci flow is not a flow amongst all symmetric 2-tensors really, since the Ricci tensor always satisfies the contracted Bianchi identity. Therefore, one could consider the flow restricted to those which satisfy this identity. To prove rigorously that one can do this, one must employ the Nash–Moser Inverse Function Theorem, as Hamilton did. This is not needed as we have said for the Ricci flow, since DeTurck's trick already removes the issue caused by the Bianchi identity there, but it is needed currently for the Laplacian flow.

4.4 Results and Questions

Results. There are several important areas in the Laplacian flow for closed G_2 structures where progress has been made.

- Long-time existence criteria based on curvature and torsion estimates along the flow, uniqueness and compactness theory, and real analyticity of the flow [28, 30].
- Stability of the critical points [29].
- Non-collapsing under assumption of bounded torsion [8].
- Explicit study of the flow in homogeneous situations and other symmetric cases such as nilmanifolds and warped products with a circle [9, 11, 25, 26, 31, 33].
- Examples and non-existence results for solitons [25–28].
- Eternal solutions for the flow arising from extremally Ricci pinched G_2 structures [12].
- Reduction of the flow to 4 dimensions, with improved long-time existence criteria [10] and analysis of the 4-torus case [19].
- Reduction of the flow to 3 dimensions, with striking long-time existence and convergence results [24].

It is worth remarking that the scalar curvature here is given by

$$R(g_\varphi) = -\frac{1}{2}|\tau_2|^2, \tag{4.36}$$

so having a bound on torsion is equivalent to a bound on scalar curvature.

Questions. There are many open problems in the area.

- Does the flow exist as long as the torsion is bounded?
- Can a volume bound be used to control the flow?
- Are there any compact examples which develop a singularity in finite time?
- Is there a relationship between the flow and calibrated submanifolds, specifically coassociative submanifolds?

For the last two points, there is an example due to Bryant [5] which shows that singularities can happen at infinite time (i.e. the flow exists for all time but does not converge), and that the singularity is related to coassociative geometry.

We can also ask whether the Laplacian flow is potentially useful to study other classes of G_2 structures. For example, naively if we assume φ is coclosed and the flow exists then it should stay coclosed since then

$$\Delta_\varphi \varphi = d_\varphi^* d\varphi. \tag{4.37}$$

So, an obvious question is: does this flow exist? Gavin Ball has informed the author that, in general, the Laplacian flow will *not* preserve the coclosed condition, so the answer would appear to be negative.

5 Laplacian Coflow

Another approach to studying G_2 structures was introduced in [23].

Definition 5.1 The Laplacian coflow for G_2 structures is given by:

$$\frac{\partial}{\partial t} *_{\varphi}\varphi = \Delta_{*_{\varphi}\varphi} *_{\varphi}\varphi, \tag{5.1}$$

where $\Delta_{*_{\varphi}\varphi}$ is the Hodge Laplacian of the metric determined by $*_{\varphi}\varphi$. (Actually, in [23], they introduced (5.1) with a minus sign on the right-hand side by analogy with the heat equation (2.1), but as we shall see below the "correct" sign in the equation is that given, just as in (4.1).)

Here one has to be a little careful since the 4-form $*_{\varphi}\varphi$ is not quite equivalent to the 3-form φ. In particular, the 4-form does not determine the orientation, but we can assume we have an initial orientation which stays fixed along the flow.

Again by integration by parts it is easy to see that on a compact manifold

$$\Delta_{*_{\varphi}\varphi} *_{\varphi}\varphi = 0 \quad \Leftrightarrow \quad \mathrm{d}\varphi = \mathrm{d}^*_{\varphi}\varphi = 0, \tag{5.2}$$

so the critical points are again the torsion-free G_2 structures.

5.1 Coclosed G_2 Structures

The proposal in [23] is to restrict (5.1) to closed 4-forms (so coclosed G_2 structures). If the flow exists, meaning it preserves closed forms as in the Laplacian flow setting, we would have that:

$$\frac{\partial}{\partial t} *_{\varphi}\varphi = \mathrm{d}\mathrm{d}^*_{\varphi} *_{\varphi}\varphi. \tag{5.3}$$

Thus, again, the flow will stay in the given cohomology class $[*_{\varphi}\varphi(0)]$ as long as it exists. Therefore, the Laplacian coflow can be seen as a possible means to deform $*_{\varphi}\varphi$ in its cohomology class so that it becomes torsion-free.

Proposition 5.2 *If $*_{\varphi}\varphi(t)$ satisfies (5.1) on M and $\mathrm{d}*_{\varphi}\varphi(t) = 0$ then $*_{\varphi}\varphi \in [*_{\varphi}\varphi(0)] \in H^4(M)$ for all t for which the flow exists.*

We can easily modify our discussion of the volume functional $\mathrm{Vol}(\varphi)$, given in (3.14), and the Laplacian flow in Proposition 4.7 and Theorem 4.9 to show the following.

Theorem 5.3 *The flow (5.1) for coclosed G_2 structures $*_{\varphi}\varphi$ is the gradient flow of the volume functional in (3.14) restricted to $[*_{\varphi}\varphi]$ and the critical points are strict local maxima for the volume functional (modulo diffeomorphisms).*

At this point, things are looking quite good in the study of the Laplacian coflow.

However, it is now worth going back to the earlier discussion of the coclosed condition (3.5). Taking a positive interpretation of the h-principle result (Theorem 3.1), we can always assume that our G_2 structure is coclosed and therefore (5.1) potentially allows us to study the space of all G_2 structures whilst restricting to the coclosed ones. A more negative outlook is to say that studying the Laplacian coflow means that we are effectively studying all G_2 structures, which only has topological rather than geometric content, and so perhaps it is unrealistic to expect the flow to be well-behaved in such generality.

5.2 *Short-Time Existence: A Modification*

Given this discussion it is worth confronting the key issue of whether the flow (5.1) even exists. To tackle this, we would employ the DeTurck's trick and hope to show that (5.1) is parabolic, possibly only in the direction of closed forms, just as in the case of (4.1) as in Theorem 4.10. Unfortunately, this does not work! Therefore, we cannot currently say (regardless of which sign we choose in (5.1)) that the Laplacian coflow even exists. Nonetheless, one can find solutions to it in special cases [23], so we can continue to ask the question: does (5.1) exist restricted to coclosed G_2 structures?

An approach taken in [13] is to modify (5.1) to get a family of flows depending on an arbitrary constant c.

Definition 5.4 The modified Laplacian coflow(s) (recalling the torsion forms in (3.6)) for $c \in \mathbb{R}$ is defined by

$$\frac{\partial}{\partial t} *_\varphi \varphi = \Delta_{*_\varphi \varphi} *_\varphi \varphi + \mathrm{d}\left(\left(c - \frac{7}{2}\tau_0\right)\varphi\right), \tag{5.4}$$

again restricted to coclosed G_2 structures.

This again moves in the cohomology class and has the added benefit of defining a flow which is parabolic in the direction of closed forms (modulo diffeomorphisms), and therefore (5.4) is guaranteed to exist on a compact 7-manifold by essentially the same proof as Theorem 4.10.

Theorem 5.5 *Let φ_0 be a smooth coclosed G_2 structure on a compact manifold M. There exists $\epsilon > 0$ so that a unique solution $*_\varphi\varphi(t)$ to the modified Laplacian coflow (5.4) with $*_\varphi\varphi(0) = *_{\varphi_0}\varphi_0$ and $\mathrm{d} *_\varphi \varphi(t) = 0$ exists for all $t \in [0, \epsilon]$.*

Critical points. The short-time existence is of course important, but (5.4) is no longer obviously a gradient flow and there is also no reason why the critical points of the flow are torsion-free G_2 structures. It is clear that if φ is torsion-free then the right-hand side of (5.4) vanishes (since $\tau_0 = 0$ and $\mathrm{d}\varphi = 0$). However, suppose we choose a nearly parallel G_2 structure φ as in (3.9), recalling that τ_0 is constant in that case. Then we see that

$$\Delta_{*_\varphi\varphi} *_\varphi \varphi = dd^*_\varphi *_\varphi \varphi = d*_\varphi d\varphi = d*_\varphi \tau_0 *_\varphi \varphi = \tau_0 d\varphi = \tau_0^2 *_\varphi \varphi \tag{5.5}$$

and

$$d\left(\left(c - \frac{7}{2}\tau_0\right)\varphi\right) = \tau_0\left(c - \frac{7}{2}\tau_0\right) *_\varphi \varphi. \tag{5.6}$$

Hence,

$$\Delta_{*_\varphi\varphi} *_\varphi \varphi + d\left(\left(c - \frac{7}{2}\tau_0\right)\varphi\right) = \tau_0\left(c - \frac{5}{2}\tau_0\right) *_\varphi \varphi, \tag{5.7}$$

which vanishes if $c = \frac{5}{2}\tau_0$. Therefore, we will also get certain nearly parallel G_2 structures as critical points. Notice here that we can change τ_0 by rescaling our nearly parallel G_2 structure, so the flow will distinguish a certain scale for the nearly parallel G_2 structures. For example, the 7-sphere has a canonical nearly parallel G_2 structure, and only the 7-sphere of a certain size (depending on a choice of positive c) will be a critical point whereas others will not.

Altogether, this is a rather strange situation, which shows that the modified Laplacian coflow, though parabolic, has some potentially undesirable properties.

5.3 *Results and Questions*

The Laplacian coflow (5.1) and its modification (5.4) have so far received rather little attention, but the key results in the area include the following.

- Soliton solutions arising from warped products and symmetries [14, 23].
- Explicit study of the flow for symmetric situations [3, 4, 31].
- Long-time existence criteria based on curvature and torsion estimates along the modified Laplacian coflow, and non-collapsing for (5.4) under assumption of bounded scalar curvature [8].

There are also many open problems.

- Does the Laplacian coflow exist (possibly under some further assumptions)?
- What are the critical points of the modified Laplacian coflow?
- What do dimensional reductions of the Laplacian coflow or its modification look like?

6 Dirichlet Energies and Spinorial Flows

The final set of flows we will discuss are explicitly constructed as gradient flows. We have already seen gradient flows of the volume functional (3.14) for G_2 structures (when restricted to closed or coclosed G_2 structures) but these are actually atypical examples of gradient flows. The reason is that $\mathrm{Vol}(\varphi)$ is of order 0 in φ. Typically, gradient flows are with respect to functionals which are first order in the geometric quantity in question. For example, mean curvature flow is the gradient flow for the volume functional which is first order in the immersion.

6.1 Dirichlet Energies

In light of (3.1), there are two clear choices for possible functionals on a compact 7-manifold to consider.

Definition 6.1 We let

$$\mathcal{C}(\varphi)=\frac{1}{2}\int_M |\nabla_\varphi\varphi|^2_{g_\varphi}\,\mathrm{vol}_\varphi\geq 0\quad\text{and}\quad \mathcal{D}(\varphi)=\frac{1}{2}\int_M |\mathrm{d}\varphi|^2_{g_\varphi}+|\mathrm{d}^*_\varphi\varphi|^2_{g_\varphi}\,\mathrm{vol}_\varphi\geq 0. \tag{6.1}$$

Observe that $\mathcal{D}$ and $\mathcal{C}$ differ by the total scalar curvature functional (see [37]):

$$\mathcal{D}(\varphi)-\mathcal{C}(\varphi)=\int_M R(g_\varphi)\,\mathrm{vol}_\varphi\,. \tag{6.2}$$

The functionals $\mathcal{C}$ and $\mathcal{D}$ are both what one might call “Dirichlet energies”, by analogy with (2.2). Therefore, one could potentially call the gradient flows of these quantities “heat flows”. There is also the possibility to generalise slightly and consider (in the notation of (3.6)):

$$\mathcal{D}_\nu(\varphi)=\sum_i\frac{\nu_i}{2}\int_M|\tau_i|^2_{g_\varphi}\,\mathrm{vol}_\varphi\geq 0 \tag{6.3}$$

for positive constants ν_i: this encompasses $\mathcal{C}$ and $\mathcal{D}$ for appropriate choices of ν_i.

All of these functionals are considered in [36, 37] where the authors show the following.

Theorem 6.2 *The critical points of the Dirichlet energies $\mathcal{D}_\nu$ in (6.3) are the torsion-free G_2 structures, which are the absolute minimizers for the functionals (since they are precisely zero at those points).*

They have also shown short-time existence of the gradient flows of the functionals $\mathcal{D}_\nu$, since they are parabolic (modulo diffeomorphisms) and so the standard DeTurck's trick approach can be used.

The key main results are the following (again in [36, 37]).

- Stability of the critical points.
- Some simple examples of solitons.

In [37] the following observation is made.

Proposition 6.3 *The volume functional* Vol(φ) *in* (3.14) *is monotone decreasing along the gradient flow of* $\mathcal{C}$ *in* (6.1). *In fact, it is a convex function along the flow.*

This is contrary to our earlier results for the Laplacian flow and coflow which viewed Vol(φ) as having strict maxima (modulo diffeomorphisms). In our view, this indicates a key drawback in this Dirichlet energy approach which we shall return to later.

6.2 Spinorial Flow

Up to a constant multiplicative factor, the functional $\mathcal{C}$ can also be written in terms of unit spinors σ on a 7-manifold as

$$\mathcal{E}(\sigma) = \frac{1}{2}\int_M |\nabla_g \sigma|_g^2 \operatorname{vol}_\sigma \geq 0. \tag{6.4}$$

This formulation of the Dirichlet energy is something which can clearly be extended beyond G_2 geometry.

Definition 6.4 Define the following functional on pairs of metrics and unit spinors on a compact oriented spin manifold M:

$$\mathcal{E}(g, \sigma) = \frac{1}{2}\int_M |\nabla_g \sigma|_g^2 \operatorname{vol}_g \geq 0. \tag{6.5}$$

The gradient flow of $\mathcal{E}$ is called the spinorial or spinor flow.

Here, unlike the G_2 case, one can vary the metric and spinor independently, with the caveat that the spinor must remain a g-spinor and be unit length.

In [1] the authors show that the critical points of $\mathcal{E}(g, \sigma)$, when the dimension of the manifold M is at least 3, are given by parallel unit spinors σ and so the metric g is Ricci-flat of special holonomy. They also show that the associated gradient flow (the spinorial flow) is parabolic modulo diffeomorphisms and so has short-time existence. This time the analysis is more involved because the space of unit spinors varies as the metric varies, but despite these complications the final result is as one would expect.

- All of the ingredients enable one to prove that the critical points of (6.5) are stable under its gradient flow, as shown in [34].
- The special case of the spinorial flow on Berger spheres is studied in detail in [38].
- The full analysis of the 2-dimensional case, which has special features not covered in [1], is part of work in progress at the time of writing.

6.3 *Questions*

Morse functionals. The motivation for studying gradient flows of functionals is that one would hope (at least formally) that the functional is Morse (or Morse–Bott) on the space of geometric objects in question. Therefore, the critical points would be encoded by the topology of the space of geometric objects. Very often making this formal picture rigorous is very challenging, but still motivational. For example, in the study of surfaces in 3-manifolds (or, more generally, hypersurfaces in n-manifolds), since the topology of the space of such surfaces is infinite and the volume functional is (in some sense) formally a Morse function on this space, one might hope to construct infinitely many minimal surfaces in any 3-manifold: this is a conjecture due to Yau from 1982, which has been proved when the Ricci curvature of the 3-manifold is positive or generic [20, 32], and recently claimed (at the time of writing) for all metrics on 3-manifolds in [35].

Here, the Dirichlet functionals are defined on the space of G_2 structures (modulo diffeomorphism). However, unlike the case of the volume functional on hypersurfaces, the only critical points of the functionals are absolute minimizers. Therefore, if the functional is a Morse function then the best we can hope for is that the space of G_2 structures on our given manifold could be contractible onto the torsion-free G_2 structures.

Scaling. More than that, just as we saw with the volume functional in (3.14), the best way to reduce the Dirichlet energy is to send the 3-form to zero by scaling, which is clearly useless. The same thing of course happens when studying hypersurfaces under mean curvature flow, but we can stop the hypersurface from being contracted to a point by simply choosing a nontrivial homology class for our initial hypersurface, which then obviously cannot contain the “zero” hypersurface. The same happens in the Laplacian flow and coflow: the cohomology class is fixed so that one kills the action of rescaling.

Unfortunately, when studying the Dirichlet energies, the class of G_2 structure is not preserved and so all one can do is look at the homotopy class of the initial G_2 structure φ: this class is always homotopic to 0 in the space of G_2 structures just by rescaling (although 0 is, of course, not a G_2 structure). Therefore, one might expect for generic initial conditions that the Dirichlet energy gradient flows just send the 3-form to 0, which is certainly an absolute minimizer of the energy, but does not appear to provide any meaningful content.

This discussion leads to the following question: is there a way to modify or restrict the gradient flows of (6.1) to ensure that the 3-form does not go to 0?

7 Conclusions

The study of geometric flows of G_2 structures has seen some important progress, but it is fair to say that at the time of writing the subject is still in its relative infancy. The flows we have described have both pros and cons, and seek to tackle different problems, so it is potentially interesting to further investigate all of them to see what we can learn about G_2 structures. There are also potentially further flows of G_2 structures that could be useful, such as the flow of isometric G_2 structures introduced in [2]. The field is clearly vibrant and wide open for discovery and progress.

In particular, it we would be very exciting if by studying geometric flows we can uncover a new criteria (geometric or topological) for the existence or otherwise of torsion-free G_2 structures. Whilst this is an ambitious goal, by seeking to solve it we may well acquire a much better understanding of the space of G_2 structures.

References

1. Amman, B., Weiss, H., & Witt, F. (2016). A spinorial energy functional: Critical points and gradient flow. *Mathematische Annalen*, *365*, 1559–1602.
2. Bagaglini, L. (2019). The energy functional of G_2structures compatible with a background metric. *The Journal of Geometric Analysis*.
3. Bagaglini, L., Fernández, M., & Fino, A. Laplacian coflow on the 7-dimensional Heisenberg group. *Asian Journal of Mathematics*.
4. Bagaglini, L., & Fino, A. (2018). The Laplacian coflow on almost-abelian Lie groups. *Annali di Matematica Pura ed Applicata*, *197*, 1855–1873.
5. Bryant, R. L. (2005). Some remarks on G_2-structures. In *Proceedings of Gökova Geometry-Topology Conference* (pp. 75–109).
6. Bryant, R. L., & Xu, F. Laplacian flow for closed G_2-structures: Short time behavior. arXiv:1101.2004.
7. Crowley, D., & Nordström, J. (2015). New invariants of G_2 structures. *Geometry & Topology*, *19*, 2949–2992.
8. Chen, G. (2018). Shi-type estimates and finite time singularities of flows of G_2 structures. *The Quarterly Journal of Mathematics*, *69*, 779–797.
9. Fernández, M., Fino, A., & Manero, V. (2016). Laplacian flow of closed G_2-structures inducing nilsolitons. *The Journal of Geometric Analysis*, *26*, 1808–1837.
10. Fine, J., & Yao, C. (2018). Hypersymplectic 4-manifolds, the G_2-Laplacian flow and extension assuming bounded scalar curvature. *Duke Mathematical Journal, 167*(18), 3533–3589.
11. Fino, A., & Raffero, A. (2020). Closed warped G_2-structures evolving under the Laplacian flow. *Annali della Scuola Normale Superiore di Pisa, Classe di Scienze, 5*(20), 315–348.
12. Fino. A., & Raffero, A. A class of eternal solutions to the G_2-Laplacian flow. arXiv:1807.01128.
13. Grigorian, S. (2013). Short-time behaviour of a modified Laplacian coflow of G_2-structures. *Advances in Mathematics*, *248*, 378–415.
14. Grigorian, S. (2016). Modified Laplacian coflow of G_2-structures on manifolds with symmetry. *Differential Geometry and Its Applications*, *46*, 39–78.
15. Hamilton, R. S. (1982). Three-manifolds with positive Ricci curvature. *Journal of Differential Geometry*, *17*, 255–306.
16. Hamilton, R. S. (1982). The inverse function theorem of Nash and Moser. *Bulletin of the American Mathematical Society*, *7*(1), 65–222.
17. Hitchin, N. The geometry of three-forms in six and seven dimensions. arXiv:math/0010054.
18. Hitchin, N. (2000). The geometry of three-forms in six dimensions. *Journal of Differential Geometry*, *55*(3), 547–576.
19. Huang, H., Wang, Y., & Yao, C. (2018). Cohomogeneity-one G_2-Laplacian flow on 7-torus. *Journal of the London Mathematical Society, 98*(2), 349–368.
20. Irie, K., Marques, F., & Neves, A. (2018). Density of minimal hypersurfaces for generic metrics. *Annals of Mathematics* (2) **187**(3), 963–972.

21. Joyce, D. D. (2000). *Compact manifolds with special holonomy*. Oxford: OUP.
22. Karigiannis, S. (2009). Flows of G_2-structures. I. *Quarterly Journal of Mathematics* **60**(4), 487–522.
23. Karigiannis, S., McKay, B., & Tsui, M.-P. (2012). Soliton solutions for the Laplacian coflow of some G_2-structures with symmetry. *Differential Geometry and Its Applications*, *30*, 318–333.
24. Lambert, B., & Lotay, J. D. (2019). Spacelike mean curvature flow. *The Journal of Geometric Analysis*.
25. Lauret, J. (2017). Laplacian flow of homogeneous G_2 structures and their solitons. *Proceedings of the London Mathematical Society, 114*(3), 527–560.
26. Lauret, J. (2017). Laplacian flow: Questions and homogeneous examples. *Differential Geometry and Its Applications*, *54*, 345–360.
27. Lin, C. (2013). Laplacian solitons and symmetry in G_2-geometry. *Journal of Geometry and Physics*, *64*, 111–119.
28. Lotay, J. D., & Wei, Y. (2017). Laplacian flow for closed G_2 structures: Shi-type estimates, uniqueness and compactness. *Geometric and Functional Analysis*, *27*, 165–233.
29. Lotay, J. D., & Wei, Y. (2019). Stability of torsion-free G_2 structures along the Laplacian flow. *Journal of Differential Geometry, 111*(3), 495–526.
30. Lotay, J. D., & Wei, Y. (2019). Laplacian flow for closed G_2 structures: Real analyticity. *Communications in Analysis and Geometry, 27*, 73–109.
31. Manero, V., Otal, A., & Villacampa, R. Solutions of the Laplacian flow and coflow of a locally conformal parallel G_2-structure. arXiv:1711.08644.
32. Marques, F. C., & Neves, A. (2017). Existence of infinitely many minimal hypersurfaces in positive Ricci curvature. *Inventiones Mathematicae*, *209*, 577–616.
33. Nicolini, M. (2018). Laplacian solitons on nilpotent Lie groups. *Bulletin of the Belgian Mathematical Society Simon Stevin*, *25*, 183–196.
34. Schiemanowski, L. Stability of the spinor flow. arXiv:1706.09292.
35. Song, A. Existence of infinitely many minimal hypersurfaces in closed manifolds. arXiv:1806.08816.
36. Weiss, H., & Witt, F. (2012). A heat flow for special metrics. *Advances in Mathematics*, *231*, 3288–3322.
37. Weiss, H., & Witt, F. (2012). Energy functionals and soliton equations for G_2-forms. *Annals of Global Analysis and Geometry*, *42*, 585–610.
38. Wittmann, J. (2016). The spinorial energy functional: Solutions of the gradient flow on Berger spheres. *Annals of Global Analysis and Geometry*, *49*, 329–348.

Surveys

Distinguishing G_2-Manifolds

Diarmuid Crowley, Sebastian Goette, and Johannes Nordström

Abstract In this survey, we describe invariants that can be used to distinguish connected components of the moduli space of holonomy G_2 metrics on a closed 7-manifold, or to distinguish G_2-manifolds that are homeomorphic but not diffeomorphic. We also describe the twisted connected sum and extra-twisted connected sum constructions used to realise G_2-manifolds for which the above invariants differ.

2000 Mathematics Subject Classification Primary: 57R20 · Secondary: 53C29 · 58J28

1 Introduction

This is a survey of recent results on the topology of closed Riemannian 7-manifolds with holonomy G_2 and their G_2-structures. Among the highlights are examples of

- closed 7-manifolds whose moduli space of holonomy G_2 metrics is disconnected, i.e. the manifold admits a pair of G_2-metrics that cannot be connected by a path of G_2-metrics (even after applying a diffeomorphism to one of them)
- pairs of closed 7-manifolds that both admit holonomy G_2 metrics, which are homeomorphic but not diffeomorphic.

The key ingredients are

- invariants that can distinguish homeomorphic closed 7-manifolds up to diffeomorphism, or G_2-structures on 7-manifolds up to homotopy and diffeomorphism

---

D. Crowley (✉)
School of Mathematics and Statistics, University of Melbourne, Parkville, VIC 3010, Australia
e-mail: dcrowley@unimelb.edu.au

S. Goette
Mathematisches Institut, Universität Freiburg, Ernst-Zermelo Str. 1, 79104 Freiburg, Germany
e-mail: sebastian.goette@math.uni-freiburg.de

J. Nordström
Department of Mathematical Sciences, University of Bath, Bath BA2 7AY, UK
e-mail: j.nordstrom@bath.ac.uk

S. Karigiannis et al. (eds.), *Lectures and Surveys on G_2-Manifolds and Related Topics*, Fields Institute Communications 84, https://doi.org/10.1007/978-1-0716-0577-6_6

- classification theorems for smooth 7-manifolds or G_2-structures on smooth 7-manifolds, at least for 7-manifolds that are 2-connected
- a method for producing many examples of closed G_2-manifolds, many of which are 2-connected, and for which the above invariants can be computed.

By a homotopy of G_2-structures we simply mean a continuous path of G_2-structures on a fixed manifold. The relevance is that metrics with holonomy G_2 are essentially equivalent to *torsion-free* G_2-structures. If two G_2-metrics on M are in the same component of the moduli space, then their associated G_2-structures are certainly related by homotopy and diffeomorphism. However, studying homotopy classes of G_2-structures is essentially a topological problem, and avoids considering the complicated partial differential equation of torsion-freeness.

This survey will concentrate on describing the invariants and the constructions, while only stating the classification results.

1.1 Invariants and Classification Results for 2-Connected 7-Manifolds

By Poincaré duality, all information about the cohomology of a closed 2-connected 7-manifold M is captured by $H^4(M)$. For simplicity, let us from now on assume that $H^4(M)$ is torsion-free. Then in particular the data about the cohomology reduces to the integer $b_3(M)$.

A 2-connected manifold has a unique spin structure. The only interesting relevant characteristic class of M is the spin characteristic class $p_M \in H^4(M)$ (which determines the first Pontrjagin class by $p_1(M) = 2p_M$). Since we assume $H^4(M)$ to be torsion-free, the data of p_M amounts to specifying the greatest integer d that divides p_M in $H^4(M)$ (where we set $d := 0$ if $p_M = 0$). This d is in fact even, see Sect. 2.2.

Theorem 1.1 ([31, Theorem 3]) *Closed 2-connected 7-manifolds with torsion-free cohomology are classified up to homeomorphism by* (b_3, d).

If $p_M = 0$, or more generally, if p_M is a torsion class, then M admits 28 diffeomorphisms classes of smooth structures, distinguished by the diffeomorphism invariant of Eells and Kuiper [15]. A generalisation of this invariant to the case when p_M is non-torsion was introduced in [13]. Under the simplifying assumption that $H^4(M)$ is torsion-free, this generalised Eells–Kuiper invariant is a constant

$$\mu(M) \in \mathbb{Z}/\gcd(28, \tfrac{\tilde{d}}{4}) ,$$

where

$$\tilde{d} := \mathrm{lcm}(4, d) .$$

Theorem 1.2 ([13, Theorem 1.3]) *Closed 2-connected 7-manifolds with torsion-free cohomology are classified up to diffeomorphism by* (b_3, d, μ).

In particular, the number of diffeomorphism classes of smooth structures on a 2-connected M is exactly $\gcd(28, \frac{\tilde{d}}{4})$.

Given a G_2-structure on M, [11] defines two further invariants ν and ξ. The first is simply a constant $\nu \in \mathbb{Z}/48$. Under the assumption that $H^4(M)$ is torsion-free, ξ is also a constant $\xi(M) \in \mathbb{Z}/3\tilde{d}$. Both are invariant not only under diffeomorphisms but also under homotopies of G_2-structures. They satisfy the relations

$$\nu = \sum_{i=0}^{3} b_i(M) \mod 2 \tag{1a}$$

$$12\mu = \xi - 7\nu \mod \gcd(12 \cdot 28, 3\tilde{d}) . \tag{1b}$$

Theorem 1.3 ([11, Theorem 6.9]) *Closed 2-connected 7-manifolds with torsion-free cohomology equipped with a G_2-structure are classified up to diffeomorphism and homotopy by (b_3, d, ν, ξ).*

In particular, the number of classes of G_2-structures modulo homotopy and diffeomorphism on a fixed smooth 2-connected M is determined by computing the number of pairs (ν, ξ) that satisfy (1); for each of the 24 values of ν allowed by the parity constraint (1a), there are $\mathrm{Num}\left(\frac{d}{112}\right)$ values for ξ that satisfy (1b), so the number of classes is $24\,\mathrm{Num}\left(\frac{d}{112}\right)$. Here, "Num" denotes the numerator of a fraction written in lowest terms. We can say that ν on its own always distinguishes at least 24 classes of G_2-structures on any fixed M, and if d divides 112 then it determines the classes completely.

The invariants μ, ν and ξ are all defined as "coboundary defects" of characteristic class formulas valid for closed 8-manifolds. The definitions of ν and ξ rely on interpreting G_2-structures in terms of non-vanishing spinor fields.

The ν-invariant is a bit more robust than the other two, in that its range does not depend on d. If the G_2-structure is torsion-free, it is also possible to define a closely related invariant $\bar{\nu} \in \mathbb{Z}$ in terms of spectral invariants of the metric induced by the G_2-structure, which satisfies

$$\nu = \bar{\nu} + 24\,(1 + b_1(M)) \mod 48 ,$$

see Corollary 4.3. The analytic refinement is invariant under diffeomorphisms, but *not* under arbitrary homotopies of G_2-structures. However, $\bar{\nu}$ *is* invariant under homotopies through torsion-free G_2-structures. Therefore $\bar{\nu}$ is capable of distinguishing components of the moduli space of G_2-metrics on a manifold M, even when the associated G_2-structures are homotopic.

1.2 Twisted Connected Sums

The source of examples that we use is the twisted connected sum construction pioneered by Kovalev [22] and studied further in [9], and the "extra-twisted" generalisation from [10, 28]. Let us first outline the original version of the construction.

Suppose that V_+ and V_- is a pair of asymptotically cylindrical Calabi–Yau 3-folds: Ricci-flat Kähler 3-folds with an asymptotic end exponentially close to a product cylinder $\mathbb{R}^+ \times U$. We require the asymptotic cross-section $U_\pm$ of $V_\pm$ to be of the form $S^1 \times \Sigma_\pm$ where $\Sigma_\pm$ is a K3 surface. Then $M_\pm := S^1 \times V_\pm$ is an asymptotically cylindrical (ACyl) G_2-manifold with asymptotic cross-section $Y_\pm = T_\pm \times \Sigma_\pm$, where the 2-torus $T_\pm$ is a product of an 'internal' circle factor from the asymptotic cross-section of $V_\pm$ and the 'external' circle factor in the definition of $M_\pm$. Let $\mathfrak{t} : T_+ \to T_-$ be an orientation-reversing isometry that swaps the internal and external circle directions. We call $\mathfrak{r} : \Sigma_+ \to \Sigma_-$ a *hyper-Kähler rotation* if the product map

$$(-1) \times \mathfrak{t} \times \mathfrak{r} : \mathbb{R} \times T_+ \times \Sigma_+ \to \mathbb{R} \times T_- \times \Sigma_- \tag{2}$$

is an isomorphism of the asymptotic limits of the torsion-free G_2-structures of M_+ and M_- (see Definition 3.4). Given a hyper-Kähler rotation $\mathfrak{r}$ and a sufficiently large 'neck length' parameter ℓ, we can truncate the cylinders of $M_\pm$ at distance ℓ, form a closed 7-manifold M_ℓ by gluing the boundaries using $\mathfrak{t} \times \mathfrak{r}$ and patch the torsion-free G_2-structures from the halves to a closed G_2-structure φ with small torsion on M_ℓ. By Kovalev [22, Theorem 5.34] or the more general results of Joyce [20, Theorem 11.6.1], φ_ℓ can be perturbed to a torsion-free G_2-structure $\bar{\varphi}$.

The cohomology of the twisted connected sum M_ℓ can be computed from the cohomology of V_+ and V_- using Mayer–Vietoris, given some data about the action of $\mathfrak{r}$ on cohomology. It is convenient to describe the latter piece of data in terms of what we call the *configuration* of $\mathfrak{r}$. Call the image $N_\pm \subset H^2(\Sigma_\pm; \mathbb{Z})$ of the restriction map $H^2(V_\pm; \mathbb{Z}) \to H^2(\Sigma_\pm; \mathbb{Z})$ the *polarising lattice* of $V_\pm$. If L is an even unimodular lattice of signature (3, 19), then $H^2(\Sigma_\pm; \mathbb{Z}) \cong L$ by the classification of lattices, so we can identify N_+ and N_- with sublattices of L, each well-defined up to the action of the isometry group $O(L)$. Given $\mathfrak{r}$ we can instead consider the pair of embeddings $N_+, N_- \hookrightarrow L$ as well-defined up to the action of $O(L)$, and we call such a pair a configuration of the polarising lattices.

In a similar way, p_M can be computed from data about V_+ and V_- together with the configuration. But even without considering that data, there are some strong general restrictions on the possible values of the greatest divisor d of p_M. It follows from [20, Proposition 10.2.7] that p_M is rationally non-trivial, so $d > 0$. Note also that M always contains a K3 surface with trivial normal bundle. Since $p_{K3} \in H^4(K3; \mathbb{Z}) \cong \mathbb{Z}$ corresponds to 24, d must always divide 24. As explained in Sect. 2.2, d is always even, so a priori the only possible values for d are 2, 4, 6, 8, 12 and 24.

As explained in Sect. 3.4, it is easier to find examples of pairs of ACyl Calabi–Yau 3-folds V_+, V_- with a hyper-Kähler rotation of their asymptotic K3s where the configuration is 'perpendicular' (in the sense that every element of N_+ is perpendicular to every element of N_- in L) than where it is not. In [9] it is shown that there are at least 10^8 pairs V_+, V_- with a perpendicular matching, which are 2-connected with $H^4(M)$ torsion-free. Computing b_3 and d shows that many of the resulting twisted connected sums are homeomorphic. However, these examples on their own turn out to be insufficient for addressing the questions above.

If M is 2-connected with torsion-free $H^4(M)$ and $d = 2, 4, 6$ or 12 then M has a unique smooth structure (up to diffeomorphism), while if $d = 8$ or 24 then M admits exactly two classes of smooth structure distinguished by the generalised Eells–Kuiper invariant $\mu \in \mathbb{Z}/2$. For twisted connected sums, μ is computed in [12] in terms of the same data used to determine p_M. It turns out that $\mu = 0$ for any twisted connected sum with perpendicular configuration. However, [12] studies the problem of finding twisted connected sums with non-perpendicular configuration, and thereby also produces some examples with $\mu = 1$.

Example 1.4 The smooth 2-connected 7-manifolds with torsion-free $H^4(M)$ and $(b_3, d, \mu) = (101, 8, 0)$ and $(101, 8, 1)$ both admit metrics with holonomy G_2 (see Example 3.8); they form a pair of G_2-manifolds that are homeomorphic but not diffeomorphic.

Turning to the G_2-moduli space of twisted connected sums, we find that if one attempts to distinguish components of the moduli space using the ν-invariant, the ν-invariant of a twisted connected sum turns out to always take the same value. Indeed, it was computed in [11] in terms of a spin cobordism that all twisted connected sums have $\nu = 24$, regardless of the ACyl Calabi–Yaus used or the configuration. Considering the analytic refinement $\bar{\nu}$ does not help either.

Theorem 1.5 ([10, Corollary 3]) $\bar{\nu} = 0$ *for any twisted connected sum.*

If we want to use this circle of ideas to exhibit examples of closed 7-manifolds with disconnected moduli space of G_2 metrics, we are left with two possible approaches. One is to make use of the ξ-invariant, and this approach has very recently been successfully followed by Wallis [30]. His computation of the ξ-invariant for twisted connected sums shows that, like μ, it is uninteresting whenever the configuration is perpendicular. In particular, none of the examples found in [9] can be distinguished using ξ. However, [30, Examples 1.6 & 1.7] provide twisted connected sums with non-perpendicular configuration and $d = 6$ or 24, where ξ does distinguish components of the moduli space.

The other approach available for disconnecting the G_2-moduli space is to consider a more general class of examples and we review recent work along this direction in the following subsection. Among these new examples we will also find G_2-manifolds that are not G_2-nullbordant, see Remark 1.10 below, so G_2-bordism presents no obstruction against holonomy G_2.

1.3 Extra-Twisted Connected Sums

Our generalisation of the twisted connected sum construction relies on using ACyl Calabi–Yau manifolds $V_\pm$ with automorphism groups $\Gamma_\pm \cong \mathbb{Z}/k_\pm$. The action of $\Gamma_\pm$ on the asymptotic cross-section $S^1 \times \Sigma_\pm$ of $V_\pm$ is required to be trivial on $\Sigma_\pm$ and free on the 'internal' S^1 factor. If we also let $\Gamma_\pm$ act freely on the 'external' S^1 factor of $S^1 \times V_\pm$, then the quotient $M_\pm := (S^1 \times V_\pm)/\Gamma_\pm$ is a smooth ACyl G_2-manifold. The asymptotic cross-section is of the form $T_\pm \times \Sigma_\pm$, for $T_\pm := (S^1 \times S^1)/\Gamma_\pm$. Note that the 2-torus $T_\pm$ need not be a metric product of two circles; on the other hand it could be even if $k_\pm > 1$, depending on the choice of circumferences of the internal and external circles.

Suppose we have arranged the circumferences of the circles in such a way that there exists an orientation-reversing isometry $\mathrm{t} : T_+ \to T_-$. A key parameter of t is the angle ϑ between the external circle directions. For a diffeomorphism $\mathrm{r} : \Sigma_+ \to \Sigma_-$, the condition that (2) be an isomorphism of the asymptotic limits of M_+ and M_- depends on ϑ (see Definition 3.4). Given such a ϑ-hyper-Kähler rotation, we can proceed to glue M_+ and M_- similarly as before to form a closed manifold M with a torsion-free G_2-structure. We assume that ϑ is not a multiple of π, so that M has finite fundamental group, and thus holonomy exactly G_2 (otherwise the external circles are aligned, so that M has an S^1 factor). Note that if $k_+ = k_- = 1$ then ϑ is forced to be a right angle and we recover the ordinary twisted connected sum construction from the previous subsection.

Unlike for ordinary twisted connected sums, the analytic invariant of an extra-twisted connected sum *is* affected by the configuration. In fact, when both $k_\pm \le 2$, the only contributions to $\bar{\nu}$ come from $\rho = \pi - 2\vartheta$, and the invariant $m_\rho(L; N_+, N_-) \in \mathbb{Z}$ of the configuration defined in (27) (see [10, Def 2.5]).

If $k_\pm \ge 3$, there are two more contributions. The generalised Dedekind sum $D_{\gamma_\pm}(V_\pm) \in \mathbb{Q}$ defined in (28) (see [17]) depends on the action of $\Gamma_\pm$ on $V_\pm$. It vanishes if no element $\gamma \in \Gamma_\pm$ has isolated fixed points. On an odd-dimensional Calabi–Yau manifold, no structure preserving involution can have isolated fixed points, so this contribution vanishes if $k_\pm \le 2$.

Finally, (29) defines a number $F_\pm \in \mathbb{R}$ that depends on the circumferences of the internal and external S^1 and the $\Gamma_\pm$-action on their product (see [17]). It vanishes in the rectangular case ($k_\pm = 1$) and in the rhombic case ($k_\pm = 2$). While it is hard to compute $F_\pm$ individually in general, we sketch ways to determine $F_+ + F_-$ in Sects. 4.3 and 4.4 below.

Theorem 1.6 ([10, Thm 1], see also [17]) *Let (M, g) be an extra-twisted connected sum. Let ϑ be the gluing angle, and let $m_\rho(L; N_+, N_-)$, $D_{\gamma_\pm}(V_\pm)$ and $F_\pm$ be as above. Then*

$$\bar{\nu}(M) = D_{\gamma_+}(V_+) + D_{\gamma_-}(V_-) + F_+ + F_- - 72\frac{\rho}{\pi} + 3m_\rho(L; N_+, N_-)\,.$$

Example 1.7 The smooth 2-connected 7-manifold with torsion-free $H^4(M)$ and $(b_3, d, \mu) = (97, 2, 0)$ admits two torsion-free G_2-structures with $\bar{\nu} = 0$ and $\bar{\nu} = -36$, see [10, Ex 3.7]. Hence these torsion-free G_2-structures are not homotopic, so the corresponding holonomy G_2 metrics must lie in different components of the G_2 moduli space. One of the two G_2-structures comes from an extra-twisted connected sum with gluing angle $\vartheta = \frac{\pi}{4}$, while the other is a rectangular twisted connected sum.

Example 1.8 The smooth 2-connected 7-manifold with torsion-free $H^4(M)$ and $(b_3, d, \mu) = (109, 2, 0)$ admits two torsion-free G_2-structures with $\bar{\nu} = 0$ and $\bar{\nu} = -48$, see [10, Ex 3.11]. Both have $\nu = 24$, and because d divides 112, the underlying G_2-structures are homotopic (after choosing the diffeomorphism appropriately) by Theorem 1.3. Nevertheless, the analytic invariant $\bar{\nu}$ shows that the corresponding holonomy G_2 metrics are in different components of the G_2 moduli space. One of the two G_2-structures comes from an extra-twisted connected sum with gluing angle $\frac{\pi}{6}$, while the other is a rectangular twisted connected sum.

Remark 1.9 The examples above all have $d = \tilde{d} = 4$ and $12|\nu$, so they have $\xi = 0$ by (1b). On the other hand, the examples found by Wallis all have $\bar{\nu} = 0$ by Theorem 1.5, demonstrating that neither ξ nor $\bar{\nu}$ is a complete invariant of the connected components of the G_2-moduli space. (Nor do we have any reason to believe that these methods can give a complete set of invariants of the connected components.)

Remark 1.10 The ν-invariants in Theorem 1.6 are always divisible by 3 if $k_\pm \leq 2$. If $k_\pm > 2$, it is possible to construct examples where this is no longer the case, see Example 4.4. Indeed, we expect that ν can attain all values in $\mathbb{Z}/48$ satisfying the parity constraint (1a).

This is significant because a topological G_2-structure is trivial in G_2-bordism if and only if $3|\nu$ (or equivalently, $3|\xi$), see Remark 2.1. Hence we see that G_2-bordism does not give an obstruction against the existence of torsion-free G_2-structures.

Remark 1.11 The fact that $\bar{\nu}(M)$ is always an integer poses interesting restrictions on the possible asymptotically cylindrical pieces $V_\pm$, the groups $\Gamma_\pm$, the torus matchings, and the matchings of K3 surfaces. For $k_\pm \geq 3$, the values of $F_\pm$ and $\frac{\rho}{\pi}$ can be irrational (see Fig. 4). Using elementary hyperbolic geometry, one can prove that the linear combination of these terms that occurs in Theorem 1.6 is always rational.

The generalised Dedekind sums $D_{\gamma_\pm}(V_\pm)$ are always rational, too. The fractional part of their sum is determined by the remaining terms in Theorem 1.6. As an example, if an asymptotically cylindrical Calabi–Yau manifold $V_\pm$ with an action of $\Gamma_\pm \cong \mathbb{Z}/5\mathbb{Z}$ occurs in a matching, then the action of $\Gamma_\pm$ on that space must have isolated fixed points.

Remark 1.12 Finally, one may wonder if it is worthwhile to extend the construction above by allowing group actions of $\Gamma_\pm$ on $V_\pm$ that do not necessarily act trivially on the K3-factor $\Sigma_\pm$. While on one hand it may be difficult to provide such examples, it turns out that on the other hand this will only give quotients of the examples we can produce by our methods above.

To understand this, let $\Gamma_{0,\pm} \subset \Gamma_\pm$ be the normal subgroup of $\Gamma_\pm$ that fixes $\Sigma_\pm$ pointwise. The cross-section of $(V_\pm \times S^1)/\Gamma_\pm$ at infinity can be regarded as the total space of a singular fibration

$$(S^1 \times S^1 \times \Sigma_\pm)/\Gamma_\pm \longrightarrow \Sigma_\pm/(\Gamma_\pm/\Gamma_{0,\pm})$$

that is locally of product geometry, and whose regular fibres are all isometric to $(S^1 \times S^1)/\Gamma_{0,\pm}$.

Now assume that we can glue $(S^1 \times V_-)/\Gamma_-$ to $(S^1 \times V_+)/\Gamma_+$, obtaining a G_2-manifold M. Then the isometry of the cross-sections at infinity on both sides lifts to an isometry

$$(S^1 \times S^1)/\Gamma_{0,-} \times K_- \xrightarrow{\cong} (S^1 \times S^1)/\Gamma_{0,+} \times K_+$$

by the de Rham decomposition theorem. Hence, we may write $M = \tilde{M}/(\Gamma_\pm/\Gamma_{0,\pm})$, where $\tilde{M}$ is an extra-twisted connected sum as considered above, using only the subgroups $\Gamma_{0,\pm} \subset \Gamma_\pm$.

1.4 Further Questions

Several questions not answered above are

- What are μ and ξ for extra-twisted connected sums? While ν was originally defined as a coboundary defect, our computation for extra-twisted connected sums was analytic. We do not know suitable coboundaries for extra-twisted connected sums that could be used to compute μ and ξ.
- There are now more than 10^8 different constructions of G_2-manifolds. Only a small number of these constructions give $\bar{\nu} \neq 0$, even fewer give $3 \nmid \bar{\nu}$. In case the number of possible G_2-manifolds up to deformation is finite, it would be interesting to know if $\bar{\nu} = 0$ or $3|\nu$ is preferred, or if all values occur roughly equally often.
- Is there a 2-connected 7-manifold that admits a torsion-free G_2-structure in every homotopy class of topological G_2-structures?

2 Coboundary Defects

Suppose that there is a formula valid for closed n-manifolds with a certain structure, such that each term is well-defined also for n-manifolds with boundary. If each term is additive under gluing along boundary components, then the failure of the formula to hold for manifolds with boundary can be interpreted as an invariant of the boundary itself (with relevant induced structure). We explain how combinations of the Hirzebruch signature theorem, the Atiyah–Singer theorem for the index of the Dirac operator and a relation for the Euler class of the positive spinor bundle of closed spin 8-manifolds lead to the definitions of the invariants μ, ν and ξ.

2.1 Prototypical Example

The first example of such a "coboundary defect" invariant is Milnor's λ-invariant of a closed oriented 7-manifold M with $p_1(M) = 0$. The starting point in this case is the Hirzebruch signature theorem for a closed oriented 8-manifold X:

$$\sigma(X) = \frac{7p_2(X) - p_1(X)^2}{45} \tag{3}$$

Here we have implicitly identified $p_2(X), p_1(X)^2 \in H^8(X) \cong \mathbb{Z}$ by evaluation on the fundamental class.

Now consider instead a compact 8-manifold W with boundary M. Let $H_0^4(W)$ be the image of the push-forward $H^4(W, M) \to H^4(W)$. For elements $x, y \in H_0^4(W)$ we can define a product $xy \in \mathbb{Z}$ by picking a pre-image $\bar{x} \in H^4(W, M)$ of x and setting xy to be $\bar{x}y \in H^8(W, M) \cong \mathbb{Z}$. One makes sense of the signature $\sigma(W)$ as the signature of this intersection form on $H_0^4(W)$. If we impose the condition that $p_1(M) = 0$, then $p_1(W) \in H_0^4(W)$, so $p_1(W)^2 \in \mathbb{Z}$ is well-defined.

According to Novikov additivity [2, 7.1], the signature is additive under gluing boundary components: if $X^8 = W_0 \cup_M W_1$ for manifolds W_i with $\partial W_i = M$ (but opposite orientations), then $\sigma(X) = \sigma(W_0) + \sigma(W_1)$. The integral of p_1^2 is additive in the same sense.

While there is no good way to interpret $p_2(W)$ under these conditions, we can eliminate the corresponding term from (3) by reducing modulo 7:

$$45\sigma(X) + p_1(X)^2 \equiv 0 \mod 7 \tag{4}$$

for any closed oriented 8-manifold X. The consequence is that if M is a smooth oriented 7-manifold with $p_1(M) = 0$ and W is an oriented coboundary, then

$$\lambda(M) := 45\sigma(W) + p_1(W)^2 \mod 7$$

in fact depends only on M, and not on W (Milnor [24, Theorem 1]). Because the oriented bordism group Ω_7^{SO} is trivial, this allows us to define $\lambda(M) \in \mathbb{Z}/7$ for any oriented M with $p_1(M) = 0$.

2.2 The Spin Characteristic Class

Since we are interested in invariants of spin manifolds, it will be important to summarise some properties of the generator of $H^4(B\mathrm{Spin}; \mathbb{Z})$. This corresponds to a degree 4 characteristic class $p(E)$ of spin vector bundles E. It is related to the first Pontrjagin class by $p_1(E) = 2p(E)$, while its mod 2 reduction is the 4th Stiefel–Whitney class $w_4(E)$.

For a manifold M, we will abbreviate $p(TM)$ as p_M. For a closed spin manifold of $\dim M = n$, Wu's formula [26, Theorem 11.14] implies that $w_4(M)$ coincides with the 4th Wu class $v_4(M)$, i.e. the Poincaré dual to the Steenrod square $Sq^4 : H^{n-4}(M; \mathbb{Z}/2) \to H^n(M; \mathbb{Z}/2)$.

If $\dim M \leq 7$ then $v_4(M) = 0$, so p_M is even.

If X is closed of dimension 8 then the definition of $v_4(X)$ means it is a characteristic element for the intersection form on $H^4(X; \mathbb{Z}/2)$, that is

$$p_X x = x^2 \mod 2 \quad \text{for any } x \in H^4(X; \mathbb{Z}). \tag{5}$$

The van der Blij lemma (see Milnor–Husemöller [25, Chap. II, Lemma 5.2]) implies in turn that

$$p_X^2 = \sigma(X) \mod 8. \tag{6}$$

One can in fact deduce that (5) and (6) remain valid also if X is compact with boundary (taking $x \in H_0^4(X; \mathbb{Z}/2)$ in (5)). See [13, Sect. 2.1] for further details.

2.3 The Eells–Kuiper Invariant and Its Generalisation

In the context of closed 7-manifolds that are spin, there is another relevant formula for closed 8-manifolds in addition to the signature theorem (3). A closed spin X^8 has a Dirac operator D_X, and by the Atiyah–Singer theorem its index is computed by the $\widehat{A}$-genus of X:

$$\operatorname{ind} D_X = \frac{7p_1(X)^2 - 4p_2(X)}{45 \cdot 2^7} \tag{7}$$

While it is possible to define an index of the Dirac operator on a manifold with boundary, it is not a topological invariant. While we make use of that below, in the context of defining defect invariants we will need to eliminate this term.

To understand how to extract coboundary defect invariants from (3) and (7), it is helpful to rearrange them as

$$\begin{aligned} 7p_2(X) &= 4p_X^2 + 45\sigma(X) \\ 45 \cdot 2^5 \operatorname{ind} D_X + p_2(X) &= 7p_X^2; \end{aligned} \tag{8}$$

we have put the terms that have useful interpretations for manifolds with boundary on the right and the ones that do not on the left, and we also used $p_1(X) = 2p_X$ to simplify slightly. Clearly we cannot completely eliminate both $p_2(X)$ and ind D_X using these two equations. But if we eliminate the $p_2(X)$ term, then we are left with

$$7 \cdot 45 \cdot 2^5 \operatorname{ind} D_X = 45(p_X^2 - \sigma).$$

Clearly we can eliminate a common factor of 45. However, in view of (6) it is more natural to reduce to

$$28 \operatorname{ind} D_X = \frac{p_X^2 - \sigma(X)}{8}.$$

Thus, if for a closed spin 7 manifold M with $p_M = 0$ we define

$$\mu(M) := \frac{p_W^2 - \sigma(W)}{8} \in \mathbb{Z}/28$$

for any spin coboundary W, then this will be independent of the choice of W. This is (up to normalisation) the invariant of Eells and Kuiper [15]. It is the best possible defect invariant that can be extracted from (8) in the following sense:

- Even given the constraint (6), μ can take any value in $\mathbb{Z}/28$.
- Given $(p_W^2, \sigma(W))$ satisfying (6), there is a solution $(\operatorname{ind} D_X, p_2(X)) \in \mathbb{Z}^2$ to (8) if and only if $\mu = 0$.

If W is a compact spin manifold with boundary M but we drop the condition that $p_M = 0$, then we can no longer interpret p_W^2 as a well-defined element of $\mathbb{Z}$. However, if p_M is divisible by an integer d, then p_W mod d belongs to the image of $H^4(W, M; \mathbb{Z}/d)$, so there is a well-defined $p_W^2 \in H^8(W, M; \mathbb{Z}/d) \cong \mathbb{Z}/d$. Because d is even, there is also a well-defined Pontrjagin square in $H^8(W, M; \mathbb{Z}/2d) \cong \mathbb{Z}/2d$. But if we impose that $H^4(M)$ is torsion-free and that there exists $u \in H^4(W)$ such that $du_{|M} = p_M$, then there is a more elementary way to interpret p_W^2 even as an element of $\mathbb{Z}/2\tilde{d}$ (for $\tilde{d} = \operatorname{lcm}(4, d)$ as in the introduction): if u' is another such element then

$$(p_W - du')^2 = (p_W - du)^2 + 2dp_W(u' - u) + d^2(u' - u)^2 \in \mathbb{Z}$$

is equal to $(p_W - u)^2$ modulo $2d$ (because d is even) and also modulo 8 (because p_W is a characteristic element for the intersection form as explained in (5)), so they are equal modulo $\operatorname{lcm}(8, 2d) = 2\tilde{d}$.

If $H^4(M)$ is torsion-free then one can always find some spin coboundary W and $u \in H^4(W)$ such that $du_{|M} = p_M$. Defining

$$\mu(M) := \frac{(p_W - u)^2 - \sigma(W)}{8} \in \mathbb{Z}/\gcd(28, \tfrac{\tilde{d}}{4}) \tag{9}$$

is independent of both W and u (see [13, Definition 1.8]).

2.4 Defect Invariants of G_2-Structures

If we seek invariants of G_2-structures on 7-manifolds rather than just a spin manifold itself, then one further formula for a closed spin 8-manifold X becomes relevant. The integral of the Euler class of the tangent bundle TX is just the Euler characteristic $\chi(X)$, while the integral of the Euler class of the positive spinor bundle can be interpreted as the number of zeros (counted with signs) $n_+(X)$ of any transverse positive spinor field. They are related by

$$n_+(X) = \frac{1}{16}\left(p_1(X)^2 - 4p_2(X) + 8\chi(X)\right); \tag{10}$$

this appears to have been first established by Gray and Green [18, p. 89]. The Euler characteristic of course makes perfect sense also for manifolds with boundary, and for even-dimensional oriented manifolds it is also additive under gluing of boundary components.

On a compact spin manifold W^8 with boundary, the number of zeros of a positive spinor field is not a topological invariant. However, if we fix a non-vanishing spinor field s on the boundary M and consider transverse positive spinors $\bar{s}$ on W that restrict to s, then the number of zeros $n_+(W, s)$ does in fact depend only on s. Since a non-vanishing spinor field defines a G_2-structure, $n_+(W, s)$ is a sensible term to consider (only) in the context of manifolds with G_2-structure.

We now consider how to define defect invariants from combinations of (3), (7) and (10), which we present as

$$\begin{aligned} 7p_2(X) &= 4p_X^2 + 45\sigma(X) \\ 45 \cdot 2^5 \operatorname{ind} D_X + p_2(X) &= 7p_X^2 \\ p_2(X) &= p_X^2 + 2\chi(X) - 4n_+(X) \end{aligned} \tag{11}$$

In this case we have enough equations that we can eliminate all terms on the LHS, obtaining

$$0 = 7\chi(X) - 14n_+(X) + \frac{3p_X^2 - 45\sigma(X)}{2}$$

Thus for a G_2-structure on a closed 7-manifold M with $p_M = 0$, defined by a non-vanishing spinor field s, we can define

$$\xi(s) := 7\chi(W) - 14n_+(W, s) + \frac{3p_W^2 - 45\sigma(W)}{2} \in \mathbb{Z}$$

for any spin coboundary W.

To capture the remaining constraints from (11), there are many different ways that we could eliminate $p_2(X)$ while leaving an ind D_X term with a coefficient. If we decide to eliminate the p_X^2 term too then we obtain

$$-48 \operatorname{ind} D_X = \chi(X) - 2n_+(X) - 3\sigma(X).$$

This allows us to define an invariant of G-structures by

$$\nu(s) := \chi(W) - 2n_+(W, s) - 3\sigma(W) \in \mathbb{Z}/48. \tag{12}$$

Note that ξ and ν are *not* independent: we find using (6) that

$$\xi(s) = 7\nu(s) \mod 12 \tag{13}$$

If we were aiming to identify a "basic" set of coboundary defects from (11), we would instead be led to consider ξ together with a $\mathbb{Z}/4$-valued invariant.

The advantage of instead considering ν is that it is more robust: if we drop the condition that $p_M = 0$ then we can no longer define ξ as an integer-valued invariant, but (12) defines $\nu(s) \in \mathbb{Z}/48$ for G_2-structures on *any* closed 7-manifold. However, if we require $H^4(M)$ to be torsion-free then we can define

$$\xi(s) \in \mathbb{Z}/3\tilde{d}$$

analogously to (9).

Since we are claiming that ν and ξ capture all the coboundary-defect information that can be extracted from (11), it should also be possible to recover μ from ν and ξ. Indeed it is easy to check that

$$\frac{\xi(s) - 7\nu(s)}{12} = \mu(M) \mod \gcd\left(28, \tfrac{\tilde{d}}{4}\right).$$

Remark 2.1 In [11, Definition 1.2 and (10)], ν and ξ are initially defined in terms of Spin(7)-coboundaries of the G_2-structure, i.e. using not just a spin 8-manifold W such that $\partial W = M$, but also requiring W to admit a Spin(7)-structure whose restriction to M is the given G_2-structure. This is equivalent to requiring $n_+(s) = 0$, so leads to a slight simplification of the defining formulas. An elementary argument ([11, Lemma 3.4]) assures that Spin(7)-coboundaries exist for any G_2-structure.

One could also ask whether a given G_2-structure on a 7-manifold admits a G_2-coboundary W. Reducing the structure group of W to G_2 defines a preferred non-vanishing vector field, so forces $\chi(W) = 0$. Since $G_2 \subset \mathrm{Spin}(7)$, also $n_+(s) = 0$, so (12) implies $\nu = 0 \mod 3$. In fact, this condition is also sufficient for the existence of a G_2-coboundary [29].

3 Extra-Twisted Connected Sums

We now provide some further details regarding the constructions of twisted connected sums and extra-twisted connected sums outlined in Sects. 1.2–1.3.

3.1 ACyl Calabi–Yau Manifolds

The first step in the construction is to produce asymptotically cylindrical Calabi–Yau 3-folds, i.e. complex 3-folds with a complete Ricci-flat Kähler metric ω and a choice of (normalised) holomorphic 3-form Ω, exponentially close to a product structure $(\omega_\infty, \Omega_\infty)$ on an end $\mathbb{R}^+ \times U$.

We will only be concerned with the case when the asymptotic cross-section U is of the form $S^1 \times \Sigma$, for Σ a K3 surface. Let ζ be the circumference of the circle factor, and let u be a coordinate on S^1 with period ζ. Then there is a hyper-Kähler triple $(\omega^I, \omega^J, \omega^K)$ on Σ such that

$$\begin{aligned} \omega_\infty &= dt \wedge du + \omega^I, \\ \Omega_\infty &= (du - i dt) \wedge (\omega^J + i\omega^K) . \end{aligned} \tag{14}$$

To produce such ACyl Calabi–Yau manifolds, we use a non-compact version of Yau's solution of the Calabi conjecture. The following result from [19] is a special case of the Tian–Yau theorem, but with improved control on the asymptotics.

Theorem 3.1 *Let Z be a closed complex Kähler manifold, and Σ an anticanonical divisor with trivial normal bundle. Then $Z \setminus \Sigma$ admits ACyl Ricci-flat Kähler metrics.*

That Σ is an anticanonical divisor essentially means it is a complex submanifold Poincaré dual to $c_1(Z)$. A convenient way to produce examples of 'building blocks' Z to which Theorem 3.1 can be applied is to blow up the intersection of two anticanonical divisors in a Fano 3-fold, i.e. a closed complex 3-fold Y where $c_1(Y)$ is a Kähler class. The topology of such manifolds is well-understood, as is their deformation theory which is relevant for the matching problem discussed in Sect. 3.4.

Example 3.2 Let $Y \subset \mathbb{P}^2 \times \mathbb{P}^2$ be a smooth divisor of bidegree (2, 2). Then the anticanonical bundle $-K_Y$ is the restriction of $\mathcal{O}(1, 1)$. The intersection of two generic anticanonical divisors Σ_0, Σ_1 is a smooth curve C of genus 7. Let Z be the blow-up of Y in C. Then the proper transform of Σ_0 is an anticanonical divisor in Z with trivial normal bundle.

The 'Picard lattice' of Y is $H^2(Y)$ equipped with the bilinear form $(x, y) \mapsto xy(-K_Y)$. In the basis for $H^2(Y)$ given by the restrictions of the hyperplane classes of the $\mathbb{P}^2$ factors, this form is represented by

$$\begin{pmatrix} 2 & 4 \\ 4 & 2 \end{pmatrix} .$$

This equals the polarising lattice N of the resulting ACyl Calabi–Yau 3-folds, see Sect. 1.2.

If Z has a cyclic automorphism group Γ that fixes a smooth anticanonical divisor Σ point-wise, then the restriction of Γ to $Z \setminus \Sigma$ gives the type of automorphism we need on the ACyl Calabi–Yau. For instance, in Example 3.2 choosing Y, Σ_0 and Σ_1 to be invariant under the involution that swaps the $\mathbb{P}^2$ factors ensures that this involution lifts to Z.

Example 3.3 Let Y be a triple cover of the smooth quadric $Q \subset \mathbb{P}^4$, branched over a smooth cubic section $\Sigma_0 \subset Q$. Let $\Sigma_1 \subset Y$ be the pre-image of a generic hyperplane section of Q. Then $C := \Sigma_0 \cap \Sigma_1$ is a smooth curve of genus 4. Let Z be the blow-up of Y in C. Then the proper transform $\Sigma \subset Z$ of Σ_0 is an anticanonical divisor with trivial normal bundle, and the branch-switching automorphisms of Y lift to automorphisms of Z that fix Σ.

The Picard lattice of Y has rank 1, with a generator that squares to 6.

3.2 Gluing ACyl G_2-Manifolds

Choose $\xi > 0$ and let v be a coordinate with period ξ on S^1. Given an ACyl Calabi–Yau structure (ω, Ω) on V, the 3-form $\varphi := \operatorname{Re}\Omega + dv \wedge \omega$ defines a torsion-free ACyl G_2-structure on $S^1 \times V$ and hence a metric with holonomy contained in (but not equal to) G_2; the circumference of the 'external' S^1 factor equals ξ. If the asymptotic limit (ω, Ω) is given by (14), then φ is asymptotic to

$$\varphi_\infty = dv \wedge dt \wedge du + dv \wedge \omega^I + du \wedge \omega^J + dt \wedge \omega^K \tag{15}$$

If (V, ω, Ω) admits an isomorphic action by $\Gamma = \mathbb{Z}/k\mathbb{Z}$ with $k \geq 2$ as above, then we can extend the action to $S^1 \times V$ by making a generator act on the external S^1 factor as rotation by angle $\frac{2\pi}{k}$; let $\varepsilon \in \mathbb{Z}/k$ be the unit such that the action of that generator on the internal S^1 by $\frac{2\pi\varepsilon}{k}$. The ACyl G_2-structure φ descends to the quotient $M := (S^1 \times V)/\Gamma$. It has asymptotic limit $T^2 \times \Sigma$, where the T^2 factor is isometric to the quotient of a product of circles of circumference ζ and ξ by Γ.

More precisely, the T^2 factor could be described as the quotient of $\mathbb{C}$ by a lattice generated by ξ and $\frac{\xi + i\varepsilon\zeta}{k}$, with a complex coordinate

$$z = v + iu\,. \tag{16}$$

Given a pair $(\zeta_+, \xi_+, k_+, \varepsilon_+)$, $(\zeta_-, \xi_-, k_-, \varepsilon_-)$ of sets of data defining such tori T^2_+, T^2_- and an angle $\vartheta \neq 0$, we consider in the next subsection whether

$$\mathbb{C} \to \mathbb{C},\ z \mapsto e^{i\vartheta}\bar{z}$$

descends to a well-defined orientation reversing isometry

$$\mathfrak{t} : T^2_+ \to T^2_-\,. \tag{17}$$

If we have such a ϑ, we can attempt to find a diffeomorphism $\mathfrak{r} : \Sigma_+ \to \Sigma_-$ such that (2) is an isomorphism of G_2-structures. Let us now identify this condition in terms of the action on hyper-Kähler structures. In terms of the complex coordinate $z = v + iu$ in (16), we can rewrite (15) as

$$\varphi_\infty = \mathrm{Re}\left(dz \wedge (\omega^I - i\omega^J)\right) + dt \wedge \left(\omega^K - \tfrac{i}{2} dz \wedge d\bar{z}\right). \tag{18}$$

For (2) to be an isomorphism of cylindrical G_2-structures is thus equivalent to the following.

Definition 3.4 Given $\vartheta \in \mathbb{R}$ and hyper-Kähler structures $(\omega^I_\pm, \omega^J_\pm, \omega^K_\pm)$ on K3 surfaces $\Sigma_\pm$, call a diffeomorphism $\mathfrak{r} : \Sigma_+ \to \Sigma_-$ a ϑ-hyper-Kähler *rotation* (or simply a hyper-Kähler rotation if $\vartheta = \frac{\pi}{2}$) if

$$\begin{aligned} \mathfrak{r}^* \omega^K_- &= -\omega^K_+ \\ \mathfrak{r}^*(\omega^I_- + i\omega^J_-) &= e^{i\vartheta}(\omega^I_+ - i\omega^J_+) \,. \end{aligned} \tag{19}$$

We will now consider in turn the problems of finding suitable $\mathfrak{t} : T^2_+ \to T^2_-$ and $\mathfrak{r} : \Sigma_+ \to \Sigma_-$.

3.3 *Isometries of Tori*

Given $k_\pm$, identifying the possible data $\varepsilon_\pm$, $\xi_\pm$, $\zeta_\pm$ and ϑ for which (17) is well-defined is essentially a combinatorial problem. To study it, it is helpful to associate to such a $\mathfrak{t}$ a *gluing matrix* $G = \left(\begin{smallmatrix} m & p \\ n & q \end{smallmatrix}\right)$ such that

$$\begin{aligned} \xi_- \partial_{v_-} &= \frac{1}{k_+} d\mathfrak{t}\left(m\xi_+ \partial_{v_+} + n\zeta_+ \partial_{u_-}\right) , \\ \zeta_- \partial_{u_-} &= \frac{1}{k_+} d\mathfrak{t}\left(p\xi_+ \partial_{v_+} + q\zeta_+ \partial_{u_-}\right) . \end{aligned} \tag{20}$$

Let us write $s_\pm$ for the ratio $\frac{\xi_\pm}{\zeta_\pm}$. Amongst other relations, the matrix coefficients satisfy

$$\det\left(\begin{smallmatrix} m & p \\ n & q \end{smallmatrix}\right) = -k_- k_+ \,, \tag{21a}$$

$$mnpq \leq 0 \,, \tag{21b}$$

$$\varepsilon_+ m - n \equiv \varepsilon_+ p - q \equiv 0 \mod k_+ \,, \tag{21c}$$

$$\varepsilon_- p + m \equiv \varepsilon_- q + n \equiv 0 \mod k_- \,, \tag{21d}$$

see [17]. Because $\xi_- \partial_{v_-}$ and $\zeta_- \partial_{u_-}$ are perpendicular, there are three possibilities. If $\vartheta = 0$, then $n = p = 0$. This leads to a manifold with infinite fundamental group,

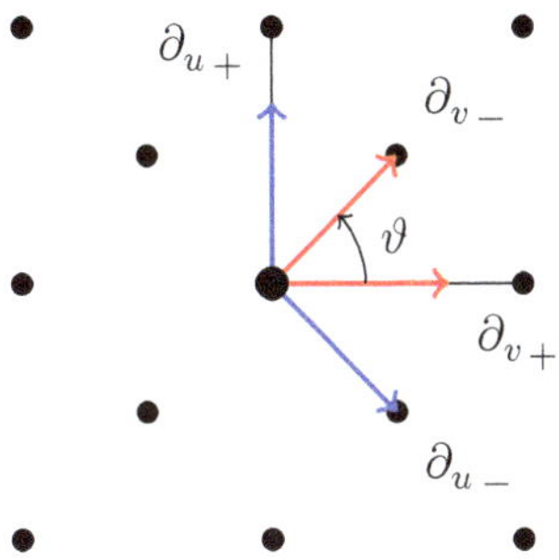

Fig. 1 $G = \begin{pmatrix} 1 & 1 \\ 1 & -1 \end{pmatrix}$, $\vartheta = \frac{\pi}{4}$

so we do not consider this case. If $\vartheta = \pm\frac{\pi}{2}$, then $m = q = 0$, and $\xi_+ = \zeta_-$ and $\zeta_+ = \xi_-$ are independent of each other. If $\vartheta \notin \frac{\pi}{2}\mathbb{Z}$, then $mnpq < 0$, and

$$s_- = \frac{\xi_-}{\zeta_-} = \sqrt{-\frac{mn}{pq}} , \qquad s_+ = \frac{\xi_+}{\zeta_+} = \sqrt{-\frac{nq}{mp}} , \tag{22a}$$

$$\vartheta = \arg\bigl(ms_+ + in\bigr) . \tag{22b}$$

For given $k_\pm$, Eqs. (21) leave only finitely many possibilities for G and $\varepsilon_\pm$. If $k_+ = k_- = 1$ then essentially the only possible gluing matrix is $G = \begin{pmatrix} 0 & 1 \\ 1 & 0 \end{pmatrix}$, leading to $\vartheta = \pm\frac{\pi}{2}$.

For $k_+ = 2$ and $k_- = 1$ there is essentially only one possibility $\begin{pmatrix} 1 & 1 \\ 1 & -1 \end{pmatrix}$. We can take $\zeta_+ = \xi_+$ and $\zeta_- = \xi_-$. That way T^2_- is a square torus, and T^2_+ is a $\mathbb{Z}/2$-quotient of a square torus that is again a square torus. If we take $\zeta_+ = \sqrt{2}\zeta_-$ then T^2_+ and T^2_- have equal size, and there is an isometry with $\vartheta = \frac{\pi}{4}$. To illustrate it we can draw a single lattice corresponding to the two tori identified by t, while adding vectors $\partial_{u_\pm}$ and $\partial_{v_\pm}$ indicating the directions of the 'internal' and 'external' circle directions of the two tori; see Fig. 1.

For $k_+ = k_- = 2$ there are more possibilities, but essentially only two that lead to simply-connected G_2-manifolds. In both of those cases, the tori T^2_+ and T^2_- are 'hexagonal'. One possibility is to take $\xi_+ = \xi_- = \sqrt{3}\zeta_+ = \sqrt{3}\zeta_-$, leading to the existence of an isometry t with $\vartheta = \frac{\pi}{3}$ illustrated in Fig. 3. The other has $\xi_+ = \zeta_- = \sqrt{3}\zeta_+ = \sqrt{3}\xi_-$ and $\vartheta = \frac{\pi}{6}$, illustrated in Fig. 2.

Once we allow k_+ or k_- to be greater than 2, the number of combinatorial possibilities increases (while the supply of examples of building blocks with the relevant symmetry decreases). For $k_+ = 3$ (and $\varepsilon_+ = -1$) and $k_- = 1$, one possibility is to take $\zeta_+ = \sqrt{2}\xi_+ = \sqrt{3}\zeta_- = \sqrt{6}\xi_-$. That way T^2_+ and T^2_- are both rectangular (with the proportions of European A4 paper), and there is an isometry t with $\cos\vartheta = \frac{1}{\sqrt{3}}$ illustrated in Fig. 4; note that $\frac{\vartheta}{\pi}$ is irrational in this case.

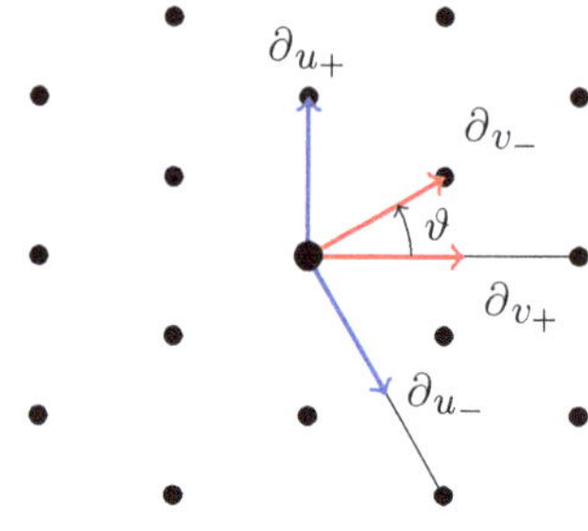

Fig. 2 $G = \left(\begin{smallmatrix} 1 & 1 \\ 1 & -3 \end{smallmatrix}\right)$, $\vartheta = \frac{\pi}{6}$

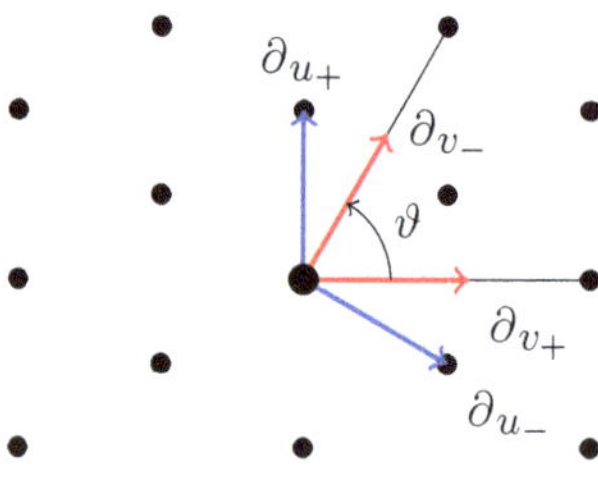

Fig. 3 $G = \left(\begin{smallmatrix} 1 & 1 \\ 3 & -1 \end{smallmatrix}\right)$, $\vartheta = \frac{\pi}{3}$

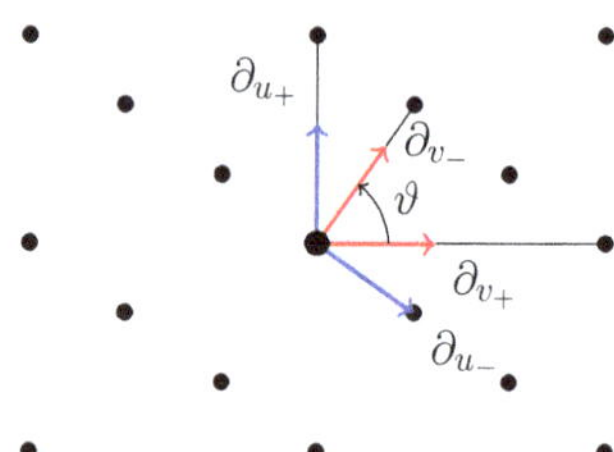

Fig. 4 $G = \left(\begin{smallmatrix} 1 & 1 \\ 2 & -1 \end{smallmatrix}\right)$, $\vartheta = \arccos \frac{1}{\sqrt{3}}$

3.4 *The Matching Problem*

If we first produce some examples of ACyl Calabi–Yau 3-folds $V_\pm$ with automorphism groups $\Gamma_\pm$ as in Sect. 3.1 and pick a compatible torus isometry $\mathfrak{t}$ as in Sect. 3.3, it is very unlikely that we will be able to find a ϑ-hyper-Kähler rotation between the asymptotic K3s (for the angle ϑ determined by $\mathfrak{t}$). A more fruitful approach is to first fix a pair $\mathcal{Z}_+$, $\mathcal{Z}_-$ of deformation families of building blocks with automorphism groups $\Gamma_\pm$, fix $\mathfrak{t}$, and *then* construct the *pair* V_+, V_- with the desired $\mathfrak{r}$ from elements of $\mathcal{Z}_\pm$.

Fixing a deformation family of blocks $\mathcal{Z}_\pm$ also fixes the polarising lattice $N_\pm$ of the resulting ACyl Calabi–Yaus. Given a ϑ-hyper-Kähler rotation $\mathfrak{r} : \Sigma_+ \to \Sigma_-$

between some pair (Z_+, Σ_+), (Z_-, Σ_-), we can identify both $H^2(\Sigma_+)$ and $H^2(\Sigma_-)$ with a fixed copy of the K3 lattice L, and hence obtain a pair of embeddings of N_+ and N_- into L. As in the introduction, Sect. 1.2, we refer to this pair as the 'configuration' of $\mathfrak{r}$, and it controls much of the topology of the resulting G_2-manifolds. It is therefore reasonable to further refine the problem to look for a $\mathfrak{r}$ compatible with a fixed configuration.

Remark 3.5 According to Nikulin [27, Theorem 1.12.4], an even indefinite lattice of rank up to 11 has essentially a unique embedding into L. As long as the ranks of N_+ and N_- are not too big, specifying a configuration is therefore essentially equivalent to describing a "push-out" lattice W that is spanned by images of isometric embeddings of N_+ and N_-.

Let us note some necessary conditions on the configuration for the existence of such a $\mathfrak{r}$. Observe that $[\omega^I_\pm]$ belongs to $N^{\mathbb{R}}_\pm := N_\pm \otimes \mathbb{R} \subset H^2(\Sigma_\pm; \mathbb{R})$, and is also the restriction of a Kähler class from $Z_\pm$. On the other hand, $[\omega^J_\pm]$ and $[\omega^K]_\pm$ are orthogonal to $N^{\mathbb{R}}_\pm$. If we let $\pi_\pm : L \to N^{\mathbb{R}}_\pm$ be the orthogonal projection, then (19) implies that $\pi_\pm[\omega^I_\mp] = (\cos\vartheta)[\omega^I_\pm]$, and hence that $[\omega^I_\pm]$ belongs to the $(\cos\vartheta)^2$-eigenspace of the self-adjoint endomorphism $\pi_+\pi_-$ on $N^{\mathbb{R}}_\pm$; let us denote that by $N^\vartheta_\pm \subseteq N^{\mathbb{R}}_\pm$.

Since the positive-definite subspace spanned by $[\omega^I_+]$ and $[\omega^I_-]$ is contained in W while $[\omega^K_+] = -[\omega^K_-]$ is perpendicular to W, we see that W must be non-degenerate of signature $(2, \mathrm{rk}\, W - 2)$.

Now let $\Lambda_\pm \subset L$ be the primitive overlattice of $N_\pm + N^{\neq\vartheta}_\mp$, where $N^{\neq\vartheta}_\mp \subset N_\mp$ is the orthogonal complement of $N^\vartheta_\mp$ in $N_\mp$. Recall that the Picard lattice of $\Sigma_\pm$ is $H^2(\Sigma_\pm; \mathbb{Z}) \cap H^{1,1}(\Sigma_\pm; \mathbb{R})$. Since $H^{1,1}(\Sigma_\pm; \mathbb{R})$ is the orthogonal complement in $H^2(\Sigma_\pm; \mathbb{R})$ to the span of $[\omega^J_\pm]$ and $[\omega^K_\pm]$, (19) further forces that $\Lambda_\pm$ is contained in the Picard lattice of $\Sigma_\pm$, so "$\Sigma_\pm$ is $\Lambda_\pm$-polarised".

In summary, given a pair of primitive embeddings $N_+, N_- \hookrightarrow L$ of the polarising lattices of a pair of deformation families $\mathcal{Z}_+, \mathcal{Z}_-$ of building blocks, three necessary conditions for finding a ϑ-hyper-Kähler rotation between asymptotic K3s in some ACyl Calabi–Yau 3-folds arising from some elements of $\mathcal{Z}_+$ and $\mathcal{Z}_-$ are that:

(i) $W := N_+ + N_-$ is non-degenerate of signature $(2, \mathrm{rk}\, W - 2)$
(ii) $N^\vartheta_\pm$ contains the restriction of some Kähler class from $Z_\pm$; in particular $N^\vartheta_\pm$ is non-trivial
(iii) there are some elements $(Z_\pm, \Sigma_\pm) \in \mathcal{Z}_\pm$ such that $\Sigma_\pm$ is $\Lambda_\pm$-polarised.

On the other hand, a combination of the Torelli theorem with a more precise statement of Theorem 3.1 turns out to show that a sufficient condition for finding a ϑ-hyper-Kähler rotation compatible with the configuration is given essentially by (i) and (ii) together with

(iii′) a *generic* element of the moduli space of $\Lambda_\pm$-polarised K3s appears as the anticanonical divisor in some element of $\mathcal{Z}_\pm$

A general principle is that a generic $N_{\pm}$-polarised K3 surface does appear as an anticanonical divisor in some element of $\mathcal{Z}_{\pm}$. For example, for blocks obtained from Fano 3-folds, as in Example 3.2, this is a consequence of the results of Beauville [3] on the deformation theory of anticanonical divisors in Fano 3-folds (see [8, Proposition 6.9]). The matching problem is therefore easiest to solve if one restricts attention to configurations where $\Lambda_{\pm} = N_{\pm}$. That is equivalent to requiring that the only configuration angles in (27) are 0 and $\pm 2\vartheta$.

3.5 Examples of Matchings

For $\vartheta = \frac{\pi}{2}$, it is very easy to produce such configurations where $\Lambda_{\pm} = N_{\pm}$: simply take the push-out W of Remark 3.5 to be the perpendicular direct sum $N_+ \perp N_-$; then (i) and (ii) are automatically satisfied too. This way one can produce literally millions of matchings, see [9]. However, there is limited diversity among the topological types realised this way, e.g. they all have $\mu = 0$ [12, Corollary 3.7].

On the other hand, if $\vartheta \neq \frac{\pi}{2}$, then for a given pair of polarising lattices N_+ and N_- there need not be any configurations at all with $N_{\pm} = \Lambda_{\pm}$. For polarising lattices of rank 1, it is not so difficult to decide whether such a configuration exists.

Example 3.6 Let $\mathcal{Z}_+$ be the deformation family of blocks with automorphism group $\Gamma_+ \cong \mathbb{Z}/3$ described in Example 3.3, and let $\mathcal{Z}_-$ be the family of blocks obtained from blow-ups of Fano 3-folds of rank 1, index 1 and degree 2 (see [9, Example 7.1$\frac{1}{2}$] in the notation used there). The relevant polarising lattices are $N_+ = (6)$ and $N_- = (2)$. By the reasoning in Remark 3.5, the matrix

$$W = \begin{pmatrix} 6 & 2 \\ 2 & 2 \end{pmatrix}$$

defines a configuration of N_+ and N_-. The angle ϑ between the basis vectors has

$$(\cos\vartheta)^2 = \frac{2^2}{2 \cdot 6} = \frac{1}{3} .$$

We can find a ϑ-hyper-Kähler rotation compatible with this configuration, and hence form an extra-twisted connected sum using the torus matching illustrated in Fig. 4.

On the other hand, for polarising lattices of higher rank the existence can be less immediately obvious.

Example 3.7 Let $\mathcal{Z}_+$ be the deformation family of blocks with automorphism group $\Gamma_+ \cong \mathbb{Z}/2$ described in Example 3.2, and let $\mathcal{Z}_-$ be the family of blocks obtained by blowing up the blow-up of $\mathbb{P}^3$ in a conic (number 30 in the Mori–Mukai classification of rank 2 Fano 3-folds, see [12, Entry 30 of Table 3]). The relevant polarising lattices are

$$N_+ = \begin{pmatrix} 2 & 4 \\ 4 & 2 \end{pmatrix}, \qquad N_- = \begin{pmatrix} 6 & 6 \\ 6 & 4 \end{pmatrix}.$$

Then

$$W = \begin{pmatrix} 2 & 4 & 3 & 4 \\ 4 & 2 & 3 & 2 \\ 3 & 3 & 6 & 6 \\ 4 & 2 & 6 & 4 \end{pmatrix}$$

defines a configuration of N_+ and N_-, such that $N_\pm^{\frac{\pi}{4}} = N_\pm$. We can find a $\frac{\pi}{4}$-hyper-Kähler rotation compatible with this configuration, and hence form an extra-twisted connected sum using the torus matching illustrated in Fig. 1.

Even if we look for configurations without the assumption that $N_\pm = \Lambda_\pm$, the conditions (i) and (ii) on their own can be still be quite restrictive. But having found such a configuration, one then has to check condition (iii′). This typically requires some detailed understanding of the particular families of building blocks involved.

Example 3.8 Take both $\mathcal{Z}_+$ and $\mathcal{Z}_-$ to be the family of blocks obtained from blowing up the blow-up of $\mathbb{P}^3$ in a twisted cubic (number 27 in the Mori–Mukai classification of rank 2 Fano 3-folds, see [12, Entry 27 in Table 3]). In this case the polarising lattices are

$$N_+ = N_- = \begin{pmatrix} 4 & 5 \\ 5 & 2 \end{pmatrix}.$$

We can define a configuration satisfying condition (i) (with $\vartheta = \frac{\pi}{2}$) and (ii) using the push-out

$$W = \begin{pmatrix} 4 & 5 & 1 & -1 \\ 5 & 2 & -1 & 1 \\ 1 & -1 & 4 & 5 \\ -1 & 1 & 5 & 2 \end{pmatrix}.$$

Now $\Lambda_\pm$ is a rank 3 overlattice of $N_\pm$, with quadratic form represented by

$$\begin{pmatrix} 4 & 5 & 16 \\ 5 & 2 & -16 \\ 16 & -16 & -272 \end{pmatrix}.$$

It is checked in [12, Lemma 7.7] that any K3 surface with Picard lattice isomorphic to that can be embedded as an anticanonical divisor in the blow-up of $\mathbb{P}^3$ in a twisted cubic, so that (iii′) holds. Thus it is possible to form a rectangular twisted connected sum of two blocks from this family. The resulting G_2-manifolds have $b_3 = 101$, $d = 8$ and $\mu = 1$, and are used in Example 1.4.

4 The Extended ν-Invariant

By definition, coboundary defect invariants for M can be computed if one knows enough about some appropriate manifold W with $\partial W = M$. For rectangular twisted connected sums, this was used in [11] to show that $\nu(s) = 24$, in [12] to compute the generalised Eells–Kuiper invariant, and recently by Wallis to compute $\xi(s)$ [30]. For extra-twisted connected sums, zero-bordisms are harder to find, and we therefore pursue a different approach to computing ν.

We rewrite the definition of $\nu(M)$ using the Atiyah–Patodi–Singer index theorem for manifolds with boundary and Mathai–Quillen currents. This yields a formula for ν in terms of η-invariants and Mathai–Quillen currents. In the case of G_2-holonomy, the Mathai–Quillen terms drop out, and the η-invariants become $\mathbb{R}$-valued rather than just $\mathbb{R}/2\mathbb{Z}$-valued. This way, the ν-invariant lifts to a $\mathbb{Z}$-valued invariant $\bar{\nu}$, the *extended ν-invariant,* that is locally constant on the moduli space of G_2-holonomy manifolds. It is possible to compute $\bar{\nu}$ for extra-twisted connected sums, see Examples 1.7 and 1.8 above.

4.1 The Analytic Description of the ν-Invariant

The definition of $\nu(s)$ in (12) involves the signature $\sigma(X)$ of an 8-manifold X, which can be written as the analytic index of the signature operator B_X on X. Implicitly, $\nu(s)$ also involves the index of the Atiyah–Singer spin Dirac operator D_X on X.

The Atiyah–Patodi–Singer index theorem allows us to write $\sigma(W)$ as an analytic index of the signature operator B_W on an 8-manifold with boundary $\partial W = M$. We assume that W has product geometry near its boundary. Let ∇^{TW} be the Levi-Civita connection, and let $L(TW, \nabla^{TW}) \in \Omega^\bullet(W)$ be the Chern–Weil representative of the L-class. If B_M denotes the odd signature operator on the boundary M, with spectrum $\cdots \leq \lambda_0 \leq \lambda_1 \leq \cdots$ counted with multiplicities, we can define its *η-invariant* by

$$\eta(B_M) = \sum_{\lambda_i \neq 0} \operatorname{sign}(\lambda_i)\ |\lambda_i|^{-s}\Big|_{s=0} = \int_0^\infty \operatorname{tr}\big(B_M\, e^{-tB_M^2}\big)\, \frac{dt}{\sqrt{\pi t}}\ .$$

The spectral expression is defined if the real part of s is sufficiently large and has a meromorphic continuation that is holomorphic at $s = 0$. The η-invariant is the value at $s = 0$, or equivalently the value of the integral on the right. The Atiyah–Patodi–Singer signature theorem [1, Thm 4.14] implies that

$$\sigma(W) = \int_W L\big(TW, \nabla^{TW}\big) - \eta(B_M)\ . \tag{23}$$

Similarly, let D_W denote the spin Dirac operator on W with the given spin structure, and let D_M denote the spin Dirac operator on M. Let ind APS$(D_W) \in \mathbb{Z}$ denote the analytic index of the spin Dirac operator with respect to the Atiyah–Patodi–Singer boundary conditions, let $\hat{A}(TW, \nabla^{TW})$ be the Chern–Weil representative of the $\hat{A}$-class, let $\eta(D_M)$ be the defined as above, and let $h(D_M) = \dim\ker(D_M)$. The Atiyah–Patodi–Singer index theorem [1, Thm 4.2] states that

$$\mathrm{ind}_{\mathrm{APS}}(D_W) = \int_W \hat{A}\big(TW, \nabla^{TW}\big) - \frac{\eta + h}{2}(D_M)\,. \tag{24}$$

The Euler class of the positive spinor bundle has to be treated differently. Let $\pi : E \to W$ be a Euclidean vector bundle with metric g^E and compatible connection ∇^E. Mathai and Quillen [23] defined a current $\psi(\nabla^E, g^E)$ on the total space E, which is singular along the zero section $W \subset TW$, such that

$$d\psi(\nabla^E, g^E) = \pi^* e\big(E, \nabla^E\big) - \delta_W\,.$$

Here, $e(E, \nabla^E)$ is the Euler class of E and δ_W denotes the Dirac delta distribution on TW along the zero section W. As a bundle E, we consider the positive spinor bundle $S^+W \to W$, so $SM = S^+W|_M$ is the spinor bundle on M. If $\bar{s} \in \Gamma(S^+W)$ extends a nowhere vanishing spinor s on M, then [23, Thm 7.6], see also [5, Thm 3.7], implies

$$n_+(W, s) = \int_W \bar{s}^*\delta_W = \int_W e\big(S^+W, \nabla^{S^+W}\big) - \int_M s^*\psi\big(\nabla^{SM}, g^{SM}\big) \tag{25}$$

by Stokes' theorem. Thus, at least formally, the integral of the Mathai–Quillen form over M is analogous to the η-invariants in (23) and (24). We combine (10) and (12) with (23)–(25) to get an intrinsic formula for the ν-invariant.

Theorem 4.1 *Let $s \in \Gamma(SM)$ define a G_2-structure on a spin 7-manifold M. Then*

$$\nu(s) = 3\,\eta(B_M) - 24\,(\eta + h)(D_M) + 2\int_M s^*\psi\big(\nabla^{SM}, g^{SM}\big) \in \mathbb{Z}/48\,. \square$$

4.2 The Extended ν-Invariant

Let us now assume that (M, g) has holonomy G_2. Then the defining spinor $s \in \Gamma(SM)$ is parallel, and $s^*\psi(g^{SM}, \nabla^{SM})$ vanishes by construction, see [10, Lemma 1.3]. Hence, Theorem 4.1 becomes

$$\nu(s) = 3\,\eta(B_M) - 24\,(\eta + h)(D_M) \in \mathbb{Z}/48\,. \tag{26}$$

We recall that the η-invariants $\eta(B_M)$, $\eta(D_M)$ depend on the spectrum of B_M and D_M, and hence on the Riemannian geometry of (M, g). If one varies the metric g, the corresponding variation formula for η-invariants typically contains two terms. The first term is an integral of a Chern–Simons class over M, which varies continuously in g. Since $\nu(s)$ is always an integer, the variation terms for the two η-invariants involved must cancel for families of metrics with holonomy in G_2.

The second term is a $\mathbb{Z}$-valued spectral flow, so the η-invariant, or more precisely the expression $\frac{\eta+h}{2}$, can jump by integers. However, spectral flow can only occur if eigenvalues of the relevant operator change sign. In this case, the dimension h of the kernel must change. The kernel of B_M describes de Rham cohomology, so $h(B_M)$ is constant and $\eta(B_M)$ never jumps. For the spin Dirac operator this is false in general; this gives an alternative explanation why $\nu(s)$ takes values in $\mathbb{Z}/48$ and not in $\mathbb{Z}$.

However, if the holonomy group of (M, g) is a subgroup of G_2, then (M, g) is Ricci flat. The Lichnerowicz formula becomes $D_M^2 = (\nabla^{SM})^*\nabla^{SM}$. Because M is closed, this implies that every harmonic spinor is parallel. If the holonomy group of M is the full group G_2, then the space of parallel spinors is spanned by the defining spinor s, so we have $h(D_M) = 1$. Otherwise, by Ricci flatness, the entire first de Rham cohomology can be represented by parallel 1-forms, and Clifford multiplication $c\,.\,s$ gives an isomorphism from $H^1(M;\mathbb{R})$ to the subspace of parallel spinors perpendicular to s. Hence $h(D_M) = 1 + b_1(M)$ is constant on the moduli space of G_2-holonomy metrics, and the spin Dirac operator has no spectral flow. Therefore, the right hand side of (26) is locally constant on the G_2-moduli space.

Definition 4.2 ([10, Definition 1.4]) For a closed Riemannian 7-manifold (M, g) with holonomy contained in G_2, put

$$\bar{\nu}(M, g) = 3\eta(B_M) - 24\eta(D_M)\,.$$

Corollary 4.3 *For a closed Riemannian 7-manifold (M, g) with holonomy contained in G_2 and with defining parallel spinor s, we have*

$$\nu(s) = \bar{\nu}(M, g) - 24\,(1 + b_1(M)) \quad \mod 48\,.$$

One could argue that we should have changed either (12) or Definition 4.2 in order to avoid the correction term $24(1 + b_1(M))$. But both definitions are the most natural in their respective realm. In particular, $\bar{\nu}(M, g)$ changes sign under reversing the orientation of M, and so vanishes if (M, g) admits an orientation reversing isometry.

4.3 Extra-Twisted Connected Sums

We return to extra-twisted connected sums and sketch a proof of Theorem 1.6. Let $M_\ell = M_{+,\ell} \cup M_{-,\ell}$ as in Sect. 3 be such that $Y = M_{+,\ell} \cap M_{-,\ell} = \partial M_{+,\ell} = \partial M_{-,\ell}$, and

$$M_{\pm,\ell} = \left(S^1 \times (V_\pm \setminus ((\ell,\infty) \times S^1 \times \Sigma_\pm))\right) / \,\Gamma_\pm$$

with $\Gamma_\pm \cong \mathbb{Z}/k_\pm$. The parameter ℓ stands for the length of the cylindrical neck. There is a closed G_2-structure φ_ℓ on M_ℓ, and a torsion free G_2-structure $\bar{\varphi}_\ell$ nearby.

We apply the $\mathbb{R}$-valued gluing formula for η-invariants by Bunke [6] and Kirk–Lesch [21]. In [10], we construct operators $D_{M,\ell}$ and $B_{M,\ell}$ that are of product type on a neighbourhood of Y, and have the same kernels as the corresponding operators on the G_2-manifold $(M_\ell, \bar{\varphi}_\ell)$. The harmonic spinors on Y that extend to harmonic spinors of the restrictions $D_{M_\pm,\ell}$, $B_{M_\pm,\ell}$ to $M_{\pm,\ell}$ form Lagrangian subspaces $L_{D_\pm} \subset \ker(D_Y)$ independent of ℓ. Similarly, harmonic forms representing $\operatorname{Im}\big(H^\bullet(M_\pm;\mathbb{R}) \to H^\bullet(Y;\mathbb{R})\big)$ form Lagrangians $L_{B_\pm} \subset \ker(B_Y)$. We modify the APS boundary conditions for the operators $D_{M_\pm}$ and $B_{M_\pm}$ on the two halves $M_\pm$ by these Lagrangian subspaces and define $\eta_{\mathrm{APS}}(D_{M_\pm}; L_{D_\pm})$ and $\eta_{\mathrm{APS}}(D_{M_\pm}; L_{B_\pm})$ with respect to those boundary conditions.

Recall the polarising lattices $N_\pm$ inside the K3 lattice L from Sect. 1.2. Let $A_\pm$ denote the reflections of $L \otimes \mathbb{R} = H^2(\Sigma;\mathbb{R})$ in the subspaces $N_\pm$. Then the *configuration angles* are the arguments $\alpha_1^+, \alpha_2^+, \alpha_3^+$ and $\alpha_1^-, \dots, \alpha_{19}^-$ of the eigenvalues of the restrictions of $A_+ \circ A_-$ to an invariant positive or negative subspace of $H^2(\Sigma;\mathbb{R})$, respectively. We always have $\{\alpha_1^+, \alpha_2^+, \alpha_3^+\} = \{0, \pm 2\vartheta\}$. We define

$$m_\rho(L; N_+, N_-) = \operatorname{sign}\rho\left(\#\{\, j \mid \alpha_j^- \in \{\pi - |\rho|, \pi\}\,\} - 1 + 2\#\{\, j \mid \alpha_j^- \in (\pi - |\rho|, \pi)\,\}\right). \tag{27}$$

By [6] and [21], see [10, Thm 1], we find that

$$\bar{\nu}(M) = \bar{\nu}(M_+) + \bar{\nu}(M_-) - 72\frac{\rho}{\pi} + 3m_\rho(L; N_+, N_-)\,,$$
$$\text{where} \quad \bar{\nu}(M_\pm) = \lim_{\ell\to\infty}\big(3\eta_{\mathrm{APS}}(B_{M_\pm,\ell}; L_{B_\pm}) - 24\eta(D_{M_\pm,\ell}; L_{D_\pm})\big)\,.$$

To describe the remaining ingredients of Theorem 1.6, let $\zeta_\pm$ and $\xi_\pm$ denote the lengths of the "interior" and "exterior" circle factors as in Sect. 3, and define $s_\pm$ as in (22a). We will now set the exterior radius to $\xi_\pm = a\zeta_\pm$ instead and consider $M_{\pm,a} = (S^1_{a\zeta_\pm} \times V_\pm)/\Gamma_\pm$. To compute $\bar{\nu}(M_{\pm,a})$, we will compute its limit as $a \to 0$, and the variation of $\bar{\nu}(M_{\pm,a})$ as a changes.

To describe the limit $a \to 0$, let $\gamma_\pm \in \Gamma_\pm$ be the generator that rotates the exterior circle factor by $\frac{2\pi}{k_\pm}$. Let $V_\pm^{0,j} \subset V_\pm$ be the set of isolated fixed points of $\gamma_\pm^j$, and for $p \in V_\pm^{0,j}$, let $\alpha_{j,1}(p), \alpha_{j,2}(p), \alpha_{j,3}(p)$ denote the angles of the $\gamma_\pm^j$-action on $T_pV_\pm$. Because the $\Gamma_\pm$-action preserves the holomorphic volume form, these angles can be chosen such that their sum is 0. Then the isolated fixed points contribute to $\bar{\nu}(M_\pm)$ by

$$\begin{aligned} D_{\gamma_\pm}(V_\pm) &= \lim_{a\to 0}\bar{\nu}(M_{\pm,a}) \\ &= \frac{3}{k_\pm}\sum_{j=1}^{k_\pm-1}\cot\frac{\pi j}{k_\pm}\sum_{p\in V_\pm^{0,j}}\frac{\cos\frac{\alpha_{j,1}(p)}{2}\cos\frac{\alpha_{j,2}(p)}{2}\cos\frac{\alpha_{j,3}(p)}{2} - 1}{\sin\frac{\alpha_{j,1}(p)}{2}\sin\frac{\alpha_{j,2}(p)}{2}\sin\frac{\alpha_{j,3}(p)}{2}}\,, \end{aligned} \tag{28}$$

see [17]. This is proved using methods from [16].

Another contribution arises as a boundary term in the variational formula for η-invariants on manifolds with boundary by Bismut–Cheeger [4] and Dai–Freed [14]. Assume that the generator $\gamma_\pm$ of $\Gamma_\pm$ rotates the interior circle by an angle $\frac{2\pi\varepsilon_\pm}{k_\pm}$ as above. Let $\sigma_{-1}(n) = \sum_{d|n} d^{-1}$, and let $L(\tau)$ denote the logarithm of the Dedekind η-function, defined for $\tau \in \mathcal{H} \subset \mathbb{C}$ in the upper half plane by

$$L(\tau) = \frac{\pi i \tau}{12} - \sum_{n=1}^{\infty} \sigma_{-1}(n)\, e^{2\pi i n \tau} .$$

Then the last contribution to $\bar{\nu}(M)$ is

$$\begin{aligned} F_\pm &= \int_0^{s_\pm} \frac{d}{da} \bar{\nu}(M_{\pm,a}) = \frac{144}{\pi} F_{k_\pm,\varepsilon_\pm}(s_\pm) , \\ \text{where} \qquad F_{k,\varepsilon}(s) &= iL\left(\frac{si+\varepsilon}{k}\right) - iL\left(\frac{si-\varepsilon}{k}\right) + c_{k,\varepsilon} , \end{aligned} \tag{29}$$

see [17]. The constant $c_{k,\varepsilon}$ takes the special values

$$c_{k,\varepsilon} = \begin{cases} -\varepsilon\pi\, \frac{k^2-3k+1}{6k} & \text{if } \varepsilon = \pm 1, \text{ and} \\ \frac{\pi\varepsilon}{6k} & \text{if } \varepsilon^2 \equiv -1 \text{ modulo } k. \end{cases} \tag{30}$$

We are grateful to Don Zagier for the formulas above for $F_{k,\varepsilon}(s)$ and $c_{k,\varepsilon}$.

The explicit values of L are hard to determine. Instead, one may use the functional equations

$$L(\tau+1) = \frac{\pi i}{12} + L(\tau) \qquad \text{and} \qquad L\left(-\frac{1}{\tau}\right) = \frac{1}{2}\log\left(\frac{\tau}{i}\right) + L(\tau) \tag{31}$$

to compute the sum of all values of L occurring in Theorem 1.6 for a particular extra-twisted connected sum.

Example 4.4 We consider Example 3.6, where $k_+ = 3$, $k_- = 1$. By construction in Example 3.3, the group Γ_+ acts without isolated fixed points on V_+, so we have $D_{\gamma_+}(V_+) = 0$. And because $k_- = 1$, also $D_{\gamma_-}(V_-) = 0$.

From the gluing matrix $G = \left(\begin{smallmatrix} m & p \\ n & q \end{smallmatrix}\right) = \left(\begin{smallmatrix} 1 & 1 \\ 2 & -1 \end{smallmatrix}\right)$ in Fig. 4 we conclude that $\varepsilon_+ = -1$, $s_+ = \sqrt{2} = s_-$. Because $k_- = 1$, we have $F_{k_-,\varepsilon_-}(s_-) = 0$. Using (30) and (31), we compute

$$\begin{aligned}F_{k_+,\varepsilon_+}(s_+) &= iL\left(\frac{\sqrt{2}i-1}{3}\right) - iL\left(\frac{\sqrt{2}i+1}{3}\right) + c_{3,-1}\\ &= \frac{i}{2}\log\frac{\sqrt{2}-i}{\sqrt{2}+i} + iL(\sqrt{2}i+1) - iL(\sqrt{2}i-1) + \frac{\pi}{18}\\ &= \frac{i}{2}\log\frac{1-\sqrt{8}i}{3} - \frac{\pi}{6} + \frac{\pi}{18} = \frac{1}{2}\arccos\frac{1}{3} - \frac{\pi}{9}\,.\end{aligned}$$

Because both N_+ and N_- have rank 1, both lie in $H^{2,+}(\Sigma;\mathbb{R})$. So $A_+ \circ A_-$ acts as the identity on $H^{2,-}(\Sigma;\mathbb{R})$, and hence $\alpha_1^- = \cdots = \alpha_{19}^- = 0$. The angle $\vartheta = \arccos\frac{1}{\sqrt{3}}$ is acute, so $\rho > 0$, hence $m_\rho(L; N_+, N_-) = -1$. Combining all this information, Theorem 1.6 gives

$$\bar{\nu}(M) = \frac{144}{\pi}\left(\frac{1}{2}\arccos\frac{1}{3} - \frac{\pi}{9}\right) - \frac{72}{\pi}\left(\pi - 2\arccos\frac{1}{\sqrt{3}}\right) - 3 = -19\,.$$

We see that $3 \nmid \bar{\nu}(M)$, so (M, g) is indeed not G_2-nullbordant.

4.4 Elementary Hyperbolic Geometry

There is an alternative way to treat the variational term $F_+ + F_-$. We can compute it as the area of a certain ideal hyperbolic polygon, see [17]. To this end, we regard the upper half plane $\mathcal{H}$ as space of conformal structures on a fixed torus. Then $\mathcal{H}$ carries a tautological family of flat tori. Let $\tilde{\eta}(\mathbb{A}) \in \Omega^1(\mathcal{H})$ be the η-form of the spin Dirac operator of this family. Using the variation formula for η-invariants on manifolds with boundary in [4] and [14], we represent $F_\pm$ as

$$F_\pm = \pm 288 \int_{\gamma_\pm} \tilde{\eta}(\mathbb{A})\,. \tag{32}$$

Using local index theory, one expresses the exterior derivative of the η-form in terms of the hyperbolic volume form dA_{hyp} as

$$d\tilde{\eta}(\mathbb{A}) = \frac{1}{4\pi}\,dA_{\mathrm{hyp}}\,. \tag{33}$$

Let $\gamma_\pm\colon (0, s_\pm] \to \mathcal{H}$ represent the families $(S^1_{a\zeta_\pm} \times S^1_{\zeta_\pm})/\Gamma_\pm$. Then $\gamma_\pm$ are hyperbolic rays. As we explain in [17], the ray γ_+ goes from $\frac{\varepsilon_+}{k_+} \in \mathbb{R} \cup \{\infty\} = \partial_\infty\mathcal{H}$ vertically to the point $\frac{\varepsilon_+ + is_+}{k_+}$ representing T^2_+. The ray γ_- goes from $\frac{\varepsilon_+}{k_+} - \frac{n}{k_+m}$ to $\frac{\varepsilon_+ + is_+}{k_+}$ along a hyperbolic geodesic with second endpoint $\frac{\varepsilon_+}{k_+} - \frac{q}{k_+p}$. We can now complete $\gamma_+ \cup \gamma_-$ to an ideal hyperbolic polygon P of finite area using geodesics along which $\tilde{\eta}(\mathbb{A})$ vanishes for symmetry reasons; these are hyperbolic geodesics joining

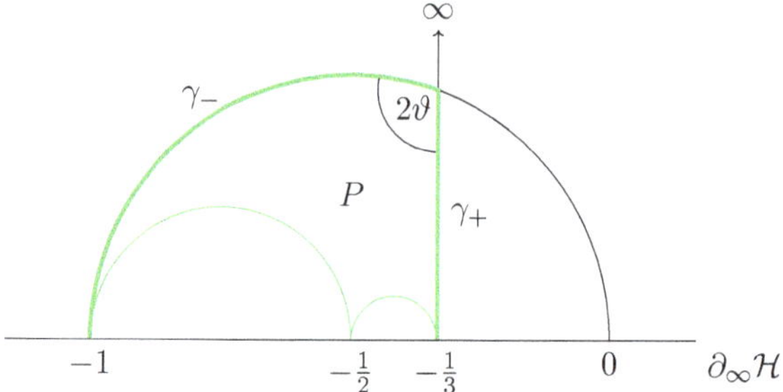

Fig. 5 The hyperbolic polygon for Example 4.5

points $\frac{a}{b}, \frac{c}{d} \in \mathbb{Q}$ with $k = |ad - bc| \in \{1, 2\}$, corresponding to families of rectangular ($k = 1$) and rhombic ($k = 2$) tori, respectively.

By Stokes theorem and (32) and (33), we can express $F_+ + F_-$ as the sum of $\frac{72}{\pi} A_{\mathrm{hyp}}(P)$ and contributions from the cusps of P. Using a strict version of the adiabatic limit formula for families by Bunke and Ma [7], a cusp at $\frac{e}{f}$ between geodesics to x and $y \in \partial_\infty\mathcal{H}$ contributes to $F_+ + F_-$ by $-24\measuredangle_{\frac{e}{f}}(x, y)$, where the *cusp angle* is given as

$$\measuredangle_{\frac{e}{f}}(x, y) = \frac{x - y}{(fx - e)(fy - e)} \in \mathbb{R} \tag{34}$$

if $\frac{e}{f}$ is a reduced fraction. Recall that the hyperbolic area of a polygon can be computed from its angles and the number of corners. Because the rays $\gamma_\pm$ meet at angle 2ϑ, this approach explains in particular why the final value of $\bar{\nu}(M)$ is rational even though the terms $-72\frac{\varrho}{\pi}$ and $F_\pm$ in Theorem 1.6 can be irrational for $k_+ > 2$ or $k_- > 2$.

Example 4.5 We still consider the example above, but compute $F_+ + F_-$ using hyperbolic geometry. Here, γ_+ lies on the vertical line with real part $-\frac{1}{3}$, and the ray γ_- lies on the hyperbolic geodesic from -1 to 0. We complete to a hyperbolic polygon with another cusp at $-\frac{1}{2}$, see Fig. 5. Because P consists of two ideal triangles, we have $A_{\mathrm{hyp}}(P) = 2\pi - 2\vartheta$. By (34), the relevant cusp angles are

$$\measuredangle_{-\frac{1}{1}}\left(0, -\frac{1}{2}\right) = 1\,, \qquad \measuredangle_{-\frac{1}{2}}\left(-1, -\frac{1}{3}\right) = 2\,, \qquad \text{and} \qquad \measuredangle_{-\frac{1}{3}}\left(-\frac{1}{2}, \infty\right) = \frac{2}{3}\,,$$

with sum $\ell(P) = \frac{11}{3}$. Now, we can confirm the computation above because

$$\bar{\nu}(M) = \frac{72}{\pi} A_{\mathrm{hyp}}(P) - 24\,\ell(P) - \frac{72}{\pi}(\pi - 2\vartheta) + 3m_\rho(L; N_+, N_-) = -19\,.$$

Acknowledgements We thank Jean-Michel Bismut, Uli Bunke, Xianzhe Dai, Matthias Lesch for inspiring discussions on adiabatic limits and variational formulas for η-invariants on manifolds with boundary. We thank Alessio Corti, Jesus Martinez Garcia, David Morrison, Emanuel Scheidegger and Katrin Wendland for helpful information about $K3$-surfaces and Fano threefolds. We also thank

Mark Haskins and Arkadi Schelling for talking with us about G_2-manifolds and G_2-bordism. We are particularly indebted to Don Zagier for his formula for $F_{k,\varepsilon}(s)$ in Sect. 4.3. SG and JN would like to thank the Simons foundation for its support of their research under the Simons Collaboration on "Special Holonomy in Geometry, Analysis and Physics" (grants #488617, Sebastian Goette, and #488631, Johannes Nordström).

References

1. Atiyah, M. F., Patodi, V. K., & Singer, I. M. (1975). Spectral asymmetry and Riemannian geometry, I. *Mathematical Proceedings of the Cambridge Philosophical Society*, *77*, 97–118.
2. Atiyah, M. F., & Singer, I. M. (1968). The index of elliptic operators. III. *Annals of Mathematics* (2) *87*, 546–604.
3. Beauville, A. (2004). Fano threefolds and $K3$ surfaces. In *The Fano conference* (pp. 175–184). Turin: University Torino.
4. Bismut, J.-M., & Cheeger, J. (1991). Remarks on the index theorem for families of Dirac operators on manifolds with boundary. *Differential geometry* (Vol. 52, pp. 59–83). Pitman Monographs and Surveys in Pure and Applied Mathematics. Harlow: Longman Scientific and Technical.
5. Bismut, J.-M., & Zhang, W. (1992). An extension of a theorem by Cheeger and Müller. With an appendix by François Laudenbach.
6. Bunke, U. (1995). On the gluing problem for the η-invariant. *Journal of Differential Geometry*, *41*, 397–448.
7. Bunke, U., & Ma, X. (2004). Index and secondary index theory for flat bundles with duality. In *Aspects of boundary problems in analysis and geometry* (Vol. 151, pp. 265–341). Operator theory: Advances and applications. Basel: Birkhäuser.
8. Corti, A., Haskins, M., Nordström, J., & Pacini, T. (2013). Asymptotically cylindrical Calabi-Yau 3-folds from weak Fano 3-folds. *Geometry & Topology*, *17*, 1955–2059.
9. Corti, A., Haskins, M., Nordström, J., & Pacini, T. (2015). G_2-manifolds and associative submanifolds via semi-Fano 3-folds. *Duke Mathematical Journal*, *164*, 1971–2092.
10. Crowley, D., Goette, S., & Nordström J. (2018). An analytic invariant of G_2-manifolds. arXiv:1505.02734v2.
11. Crowley, D., & Nordström, J. (2015). New invariants of G_2-structures. *Geometry & Topology*, *19*, 2949–2992.
12. Crowley, D., & Nordström, J. (2018). Exotic G_2-manifolds. arXiv:1411.0656.
13. Crowley, D., & Nordström, J. (2019). The classification of 2-connected 7-manifolds. *Proceedings of the London Mathematical Society*. https://doi.org/10.1112/plms.12222, arXiv:1406.2226.
14. Dai, X., & Freed, D. (2001). APS boundary conditions, eta invariants and adiabatic limits. *Journal of Mathematical Physics*, *35*, 5155–5194.
15. Eells, J. Jr., & Kuiper, N. (1962). An invariant for certain smooth manifolds. *Annali di Matematica Pura ed Applicata* (4) *60*, 93–110.
16. Goette, S. (2014). Adiabatic limits of Seifert fibrations, Dedekind sums, and the diffeomorphism type of certain 7-manifolds. *Journal of the European Mathematical Society*, 2499–2555.
17. Goette, S., & Nordström, J. (2018). ν-invariants of extra twisted connected sums, with an appendix by D. Zagier, in preparation.
18. Gray, A., & Green, P. S. (1970). Sphere transitive structures and the triality automorphism. *Pacific Journal of Mathematics*, *34*, 83–96.
19. Haskins, M., Hein, H.-J., & Nordström, J. (2015). Asymptotically cylindrical Calabi-Yau manifolds. *Journal of Differential Geometry*, *101*, 213–265.
20. Joyce, D. (2000). *Compact manifolds with special holonomy*. OUP mathematical monographs series. Oxford: Oxford University Press.

21. Kirk, P., & Lesch, M. (2004). The η-invariant, Maslov index, and spectral flow for Dirac-type operators on manifolds with boundary. *Forum Mathematicum*, *16*, 553–629.
22. Kovalev, A. (2003). Twisted connected sums and special Riemannian holonomy. *Journal für die reine und angewandte Mathematik*, *565*, 125–160.
23. Mathai, V., & Quillen, D. (1986). Superconnections, Thom classes, and equivariant differential forms. *Topology*, *25*(1), 85–110.
24. Milnor, J. W. (1956). On manifolds homeomorphic to the 7-sphere. *Annals of Mathematics* (2) *64*(2), 399–405.
25. Milnor, J. W., & Husemöller, D. (1973). *Symmetric bilinear forms* (Vol. 73). Ergebnisse der Mathematik und ihrer Grenzgebiete. New York: Springer.
26. Milnor, J. W., & Stasheff, J. D. (1974). *Characteristic classes* (Vol. 76). Annals of Mathematics Studies. Princeton, N. J.: Princeton University Press.
27. Nikulin, V. (1979). Integer symmetric bilinear forms and some of their applications. *Izvestiya Rossiiskoi Akademii Nauk. Seriya Matematicheskaya*, *43*, 111–177, 238. (English translation: *Mathematics of the USSR Izvestia*, *14*, 103–167 (1980).)
28. Nordström, J. (2018). Extra-twisted connected sum G_2-manifolds. arXiv:1809.09083.
29. Schelling, A. (2014). Die topologische η-Invariante und Mathai-Quillen-Ströme. Diploma thesis, Universität Freiburg. http://www.freidok.uni-freiburg.de/volltexte/9530/.
30. Wallis, D. (2018). Disconnecting the moduli space of G_2-metrics via $U(4)$-coboundary defects. arXiv:1808.09443.
31. Wilkens, D. L. (1971). Closed $(s-1)$–connected $(2s+1)$–manifolds. Ph.D. thesis, University of Liverpool.

Gravitational Instantons and Degenerations of Ricci-flat Metrics on the K3 Surface

Lorenzo Foscolo

Abstract The study of degenerations of metrics with special holonomy is an important theme unifying the study of convergence of Einstein metrics, the study of complete non-compact manifolds with special holonomy and the construction of spaces with special holonomy by singular perturbation methods. We survey three constructions of degenerating sequences of hyperkähler metrics on the (smooth 4-manifold underlying a complex) K3 surface—the classical Kummer construction, Gross–Wilson's work on collapse along the fibres of an elliptic fibration, and the author's construction of sequences collapsing to a 3-dimensional limit—describing how they fit into the general theory and highlighting the role played in each construction by gravitational instantons, i.e. complete non-compact hyperkähler 4-manifolds with decaying curvature at infinity.

1 Hyperkähler Metrics in Dimension 4

Hyperkähler 4-manifolds are the lowest dimensional non-flat examples of manifolds with special holonomy.

Definition 1.1 A Riemannian 4-manifold (M^4, g) is *hyperkähler* if the holonomy $\mathrm{Hol}(g)$ is contained in $\mathrm{SU}(2)$.

Despite its integro-differential definition in terms of parallel transport, the holonomy reduction to $\mathrm{SU}(2)$ can be recast in terms of a PDE for a triple of 2-forms satisfying special algebraic properties at each point [18]. Recall that the space of 2-forms on an oriented 4-dimensional vector space carries a natural non-degenerate bilinear form of signature $(3, 3)$.

Definition 1.2 Let (M^4, μ_0) be an oriented 4-manifold with volume form μ_0. A *definite triple* is a triple $\boldsymbol{\omega} = (\omega_1, \omega_2, \omega_3)$ of 2-forms on M such that $\mathrm{span}(\boldsymbol{\omega}) = \mathrm{span}(\omega_1, \omega_2, \omega_3)$ is a 3-dimensional positive definite subspace of $\Lambda^2 T_x^* M$ at every point $x \in M$.

L. Foscolo (✉)
Department of Mathematics, University College London, London, UK
e-mail: l.foscolo@ucl.ac.uk

S. Karigiannis et al. (eds.), *Lectures and Surveys on G_2-Manifolds and Related Topics*, Fields Institute Communications 84, https://doi.org/10.1007/978-1-0716-0577-6_7

Given a triple $\boldsymbol{\omega}$ of 2-forms on (M, μ_0) we consider the matrix $Q \in \Gamma\big(M, \mathrm{Sym}^2(\mathbb{R}^3)\big)$ defined by

$$\tfrac{1}{2}\,\omega_i \wedge \omega_j = Q_{ij}\,\mu_0. \tag{1}$$

$\boldsymbol{\omega}$ is a definite triple if and only if Q is a positive definite matrix. To every definite triple $\boldsymbol{\omega}$ we associate a volume form μ_ω by

$$\mu_\omega = (\det Q)^{\frac{1}{3}}\,\mu_0 \tag{2}$$

and the new matrix $Q_\omega = (\det Q)^{-\frac{1}{3}}\,Q$ which satisfies (1) with μ_ω in place of μ_0. Note that the volume form μ_ω and the matrix Q_ω are independent of the choice of volume form μ_0.

Now, let (M^4, μ_0) be an oriented 4-dimensional manifold. The choice of a 3-dimensional positive definite subspace of $\Lambda^2 T^*_x M$ for all $x \in M$ is equivalent to the choice of a conformal class on M, see for example [20, Sect. 1.1.5]. Thus every definite triple defines a Riemannian metric g_ω by requiring that $\mathrm{span}(\boldsymbol{\omega})|_x = \Lambda^+ T^*_x M$ for all $x \in M$ and $\mathrm{dv}_{g_\omega} = \mu_\omega$.

Definition 1.5 A definite triple $\boldsymbol{\omega}$ is said to be

(i) *closed* if $d\omega_i = 0$ for $i = 1, 2, 3$;
(ii) an SU(2)-*structure* if $Q_\omega \equiv \mathrm{id}$;
(iii) *hyperkähler* if it is both closed and an SU(2)-structure.

A closed definite triple is also called a *hypersymplectic* triple. The metric g_ω associated to a hyperkähler triple is hyperkähler in the sense of Definition 1.1.

Let $(M, \boldsymbol{\omega})$ be a hyperkähler 4-manifold. We now make a choice of direction in $\mathbb{R}^3$. Up to rotations we can assume that the chosen direction is e_1. We write $\omega = \omega_1$, $\omega_c = \omega_2 + i\omega_3$ and $\overline{\omega}_c = \omega_2 - i\omega_3$. The complex 2-form ω_c defines an almost complex structure $J = J_1$ on M by declaring a complex 1-form α of type $(1, 0)$ if and only if $\alpha \wedge \omega_c = 0$. Since $d\omega_c = 0$ the differential ideal generated by the $(1, 0)$–forms is closed and therefore the almost complex structure J is integrable by the Newlander–Nirenberg Theorem. Moreover, ω_c and ω are, respectively, a holomorphic $(2, 0)$–form and a real $(1, 1)$–form with respect to J. Since ω is closed and non-degenerate (M, ω, J) is a Kähler surface, with g the induced Kähler metric. Moreover, by the expression for the Ricci curvature in Kähler geometry, cf. for example [32, Sect. 4.6], $\omega^2 = \frac{1}{2}\omega_c \wedge \overline{\omega}_c$ implies that g is *Ricci-flat*. Since the choice of direction in $\mathbb{R}^3$ was arbitrary, we see that hyperkähler metrics are Kähler with respect to a 2-sphere of compatible integrable complex structures—this might be the definition of hyperkähler manifolds the reader is already familiar with.

1.1 The K3 Surface

Beside the 4-torus endowed with a flat metric, the only other compact 4-manifold carrying hyperkähler metrics is the *K3 surface*. In this note *the* K3 surface is the smooth 4-manifold M underlying any simply connected complex surface (M, J) with trivial canonical bundle. The fact that all such complex surfaces are diffeomorphic to each other was proved by Kodaira [34, Theorem 13]. We say that (M, J) is a *complex K3 surface* if we make a choice of complex structure. As above, every simply connected hyperkähler 4-manifold is in particular a complex surface (M, J) with trivial canonical bundle (trivialised by ω_c). Conversely, every complex K3 surface is Kähler [47] and therefore admits a Kähler Ricci-flat metric by Yau's Theorem [52]. Since M is simply connected any Kähler Ricci-flat metric has holonomy contained in SU(2) and therefore is hyperkähler. Examples of complex K3 surfaces (M, J) are smooth quartics in $\mathbb{CP}^3$, complete intersections of a cubic and quadric in $\mathbb{CP}^4$ and the double cover of $\mathbb{CP}^2$ branched along a sextic.

Note also that every Einstein metric on the K3 surface must be hyperkähler [30, Theorem 1]. Indeed, given any metric g the Chern–Gauss–Bonnet and Signature Formulas are

$$8\pi^2\chi(M) = \int_M \tfrac{1}{24}\text{Scal}^2 + |W|^2 - \tfrac{1}{2}|\,\mathring{\text{Ric}}\,|^2, \qquad 12\pi^2\tau(M) = \int_M |W_+|^2 - |W_-|^2, \tag{3}$$

where Scal is the scalar curvature, $\mathring{\text{Ric}}$ the traceless Ricci tensor and $W = W_+ + W_-$ is the Weyl tensor of g, decomposed into its self-dual and anti-self-dual parts. We deduce that every Einstein metric g on the K3 surface M must be Ricci-flat and anti-self-dual since

$$\frac{1}{2\pi^2}\int_M \tfrac{1}{48}\text{Scal}^2 + |W_+|^2 = 2\chi(M) + 3\tau(M) = 0.$$

Indeed the Betti numbers of the K3 surface are $b_0 = 1, b_1 = 0, b_+ = 3$ and $b_- = 19$. Furthermore, the Weitzenböck formula on Λ^+ is

$$\triangle_{\Lambda^+} = \nabla^*\nabla - 2W_+ + \tfrac{1}{3}\text{Scal} = \nabla^*\nabla.$$

Since $b_+ = 3$, we deduce that (M, g) carries a 3-dimensional space of parallel self-dual 2-forms and therefore the holonomy of g reduces to SU(2).

Let $\mathcal{M}$ be the moduli space of Ricci-flat metrics of volume 1 on the K3 surface M. The deformation theory of Einstein metrics is governed by an index zero elliptic problem and therefore moduli spaces of Einstein metrics are in general singular. In contrast, metrics with special holonomy often form smooth moduli spaces. This is the case for hyperkähler metrics and thus $\mathcal{M}$ is a smooth manifold. In fact we also know what this manifold is. Let $\text{Gr}^+(3, 19) = \text{SO}(3, 19)/\text{SO}(3) \times \text{SO}(19)$ be the Grassmannian of positive 3-planes in $\mathbb{R}^{3,19} \simeq H^2(M; \mathbb{R})$ and Γ be the automorphism of the lattice $H^2(M; \mathbb{Z})$ endowed with the intersection form (equivalently Γ is the quotient of the group of diffeomorphisms of M by the subgroup of diffeomorphisms acting trivially on cohomology). The period map

$$\mathcal{P}\colon \mathcal{M} \to \mathrm{Gr}^{+}(3,19)/\Gamma \tag{4}$$

associates to each metric the positive definite subspace $\mathrm{span}[\boldsymbol{\omega}] = \mathrm{span}([\omega_1], [\omega_2], [\omega_3]) \subset H^2(M,\mathbb{R})$. The Local Torelli Theorem [34, Theorem 17] implies that $\mathcal{P}$ is a local diffeomorphism.

The period map $\mathcal{P}$ in (4) is *not* surjective: smooth hyperkähler metrics correspond to triples $[\boldsymbol{\omega}] \in H^2(M,\mathbb{R})$ such that

$$[\boldsymbol{\omega}](\Sigma) \neq \mathbf{0} \in \mathbb{R}^3 \text{ for all } \Sigma \in H_2(M,\mathbb{Z}) \text{ such that } \Sigma\cdot\Sigma = -2, \tag{5}$$

cf. [32, Theorem 7.3.16]. Thus the image of $\mathcal{P}$ is the complement of codimension-3 "holes" in $\mathrm{Gr}^{+}(3,19)/\Gamma$. In the next section we describe hyperkähler metrics approaching this excluded codimension-3 locus and explain the significance of (5).

2 Non-collapsed Limits

2.1 *The Kummer Construction*

We begin with a prototypical example. Soon after Yau's proof of the Calabi Conjecture [52] implied that the K3 surface carries hyperkähler metrics, physicists and mathematicians alike have been interested in finding a more explicit description of these Ricci-flat metrics. Gibbons and Pope [23] suggested the construction of explicit approximately Ricci-flat metrics on Kummer surfaces.

Let $\Lambda \simeq \mathbb{Z}^4$ be a lattice in $\mathbb{R}^4$ and consider the flat 4-torus $T^4 = \mathbb{R}^4/\Lambda$. Consider the $\mathbb{Z}_2$–action on T^4 induced by the involution $x \mapsto -x$ of $\mathbb{R}^4$. Then $T^4/\mathbb{Z}_2$ is a flat 4-orbifold which is singular at the 16 points of the half-lattice $\frac{1}{2}\Lambda$. Each singular point is modelled on $\mathbb{R}^4/\mathbb{Z}_2$. If we identify $\mathbb{R}^4$ with $\mathbb{C}^2$ then T^4 becomes a complex manifold and by blow-up we can resolve $T^4/\mathbb{Z}_2$ to a complex surface (M,J) which is simply connected and satisfies $c_1(M,J) = 0$ and therefore is a complex K3 surface. The blow-up replaces each singularity with a holomorphic $\mathbb{CP}^1$ with self-intersection -2. Thus a tubular neighbourhood of each $\mathbb{CP}^1 \simeq S^2$ in M is identified with a disc bundle in the $\mathbb{R}^2$–bundle T^*S^2 over S^2 with Euler class -2.

Gibbons and Pope suggested that Ricci-flat metrics can be brought into this resolution picture. The missing ingredient is a model Ricci-flat metric on T^*S^2 that is asymptotic at infinity to the flat metric on $\mathbb{R}^4/\mathbb{Z}_2$. Such a metric is explicit and is called the *Eguchi–Hanson metric* [21].

Note that T^*S^2 can be identified with the total space of the holomorphic line bundle $\mathcal{O}(-2)$ over $\mathbb{CP}^1$. This identification endows T^*S^2 with a complex structure J. In fact, the blow-down of the zero-section $\pi\colon \mathcal{O}(-2) \to \mathbb{C}^2/\mathbb{Z}_2$ exhibits $\mathcal{O}(-2)$

as a *crepant resolution* of $\mathbb{C}^2/\mathbb{Z}_2$: the standard holomorphic $(2,0)$–form $dz_1 \wedge dz_2$ on $\mathbb{C}^2$ descends to $\mathbb{C}^2/\mathbb{Z}_2$ by $\mathbb{Z}_2$–invariance and its pull-back to $\mathcal{O}(-2)$ extends to a nowhere-vanishing holomorphic $(2,0)$–form ω_c^{eh} on $\mathcal{O}(-2)$. We now define a hyperkähler triple $\boldsymbol{\omega}^{\mathrm{eh}}$ on T^*S^2 by $\omega_2^{\mathrm{eh}} = \operatorname{Re}\omega_c^{\mathrm{eh}}$, $\omega_3^{\mathrm{eh}} = \operatorname{Im}\omega_c^{\mathrm{eh}}$ and ω_1^{eh} the Kähler form defined outside the zero-section by

$$\omega_1^{\mathrm{eh}} = \tfrac{i}{2}\partial\overline{\partial}\varphi^{\mathrm{eh}}, \qquad \varphi^{\mathrm{eh}} = \sqrt{1+r^4} + 2\log r - \log\left(1+\sqrt{1+r^4}\right). \tag{6}$$

Here we identify the complement of the zero-section in T^*S^2 with the complement of the origin in $\mathbb{C}^2/\mathbb{Z}_2$ via π and set $r = \sqrt{|z_1|^2+|z_2|^2}$. One can check that ω_1^{eh} extends to a smooth Kähler form on the whole of T^*S^2. Note that as $r \to \infty$, ω_1^{eh} approaches the flat metric $\frac{i}{2}\partial\overline{\partial}\varphi_0$, $\varphi_0 = r^2$, up to terms that decay as r^{-4}.

Now, Gibbons and Pope suggested to remove neighbourhoods of the 16 singular points of $T^4/\mathbb{Z}_2$ and replace them with 16 copies of a disc bundle in $T^*S^2 \to S^2$. This cut-and-paste construction of the smooth 4-manifold M can be promoted to the construction of a hypersymplectic triple on M by patching together the flat hyperkähler triple $\hat{\boldsymbol{\omega}}$ on $T^4/\mathbb{Z}_2$ with 16 copies of the rescaled Eguchi–Hanson hyperkähler triple. We now provide more details of this construction.

We first need to "prepare" the Eguchi–Hanson metric to be "grafted" into $T^4/\mathbb{Z}_2$. Following [8, Sect. 1.1], fix $t>0$ and consider a cut-off function $\chi = \chi_t$ such that $\chi(r) = 1$ for $r \leq \frac{1}{\sqrt{t}}$ and $\chi(r) = 0$ for $r \geq \frac{2}{\sqrt{t}}$. Define a new triple $\boldsymbol{\omega}^{\mathrm{eh},t}$ by $\omega_i^{\mathrm{eh},t} = \omega_i^{\mathrm{eh}}$ for $i = 2,3$ and $\omega_1^{\mathrm{eh},t} = \frac{i}{2}\partial\overline{\partial}\tilde{\varphi}_t^{\mathrm{eh}}$, where

$$\tilde{\varphi}_t^{\mathrm{eh}}(r) = t^2\tilde{\varphi}^{\mathrm{eh}}\left(t^{-1}r\right), \qquad \tilde{\varphi}^{\mathrm{eh}} = \chi\,\varphi^{\mathrm{eh}} + (1-\chi)\,\varphi_0.$$

The triple $\boldsymbol{\omega}^{\mathrm{eh},t}$ coincides with $t^2\boldsymbol{\omega}^{\mathrm{eh}}$ for $r \leq \sqrt{t}$ and with the flat hyperkähler triple $\boldsymbol{\omega}_0$ on $\mathbb{C}^2/\mathbb{Z}_2$ for $r \geq 2\sqrt{t}$. In the annulus $\sqrt{t} \leq r \leq 2\sqrt{t}$, $\boldsymbol{\omega}^{\mathrm{eh},t}$ differs from $\boldsymbol{\omega}_0$ by terms of order $O(t^2)$. If t is sufficiently small $\boldsymbol{\omega}^{\mathrm{eh},t}$ is a closed definite triple which is approximately hyperkähler in the sense that $Q_{\boldsymbol{\omega}^{\mathrm{eh},t}} - \mathrm{id} = O(t^2)$.

Let $p_1, \ldots, p_{16}$ denote the singular points of $T^4/\mathbb{Z}_2$. We construct a smooth 4-manifold M by replacing (disjoint) balls $B_{3\sqrt{t}}(p_i)$ in $T^4/\mathbb{Z}_2$ with copies of the region $\{r \leq 3\sqrt{t}\} \subset T^*S^2$. Since $\boldsymbol{\omega}^{\mathrm{eh},t}$ coincides with the flat triple $\boldsymbol{\omega}_0$ for $r \geq 2\sqrt{t}$, M comes equipped with a natural hypersymplectic triple $\boldsymbol{\omega}^t$ obtained by gluing $\boldsymbol{\omega}^{\mathrm{eh},t}$ with the flat hyperkähler triple $\hat{\boldsymbol{\omega}}$ of $T^4/\mathbb{Z}_2$. Then $\boldsymbol{\omega}^t$ is an approximate hyperkähler triple in the sense that $Q_{\boldsymbol{\omega}^t} - \mathrm{id} = O(t^2)$.

The question now is to deform the approximate hyperkähler triple $\boldsymbol{\omega}^t$ into an exact solution. A first rigorous proof of such a perturbation was given by LeBrun–Singer [38] (following an earlier attempt by Topiwala [51]); it uses twistor theory and we will not say anything about it. A different approach exploits the fact that a complex structure J on M with $c_1(M,J) = 0$ can be readily constructed by blow-up $\pi\colon M \to T^4/\mathbb{Z}_2$ of the complex orbifold $T^4/\mathbb{Z}_2$: $\pi^*\hat{\omega}_c$, where $\hat{\omega}_c = \hat{\omega}_2 + i\hat{\omega}_3$ is the holomorphic $(2,0)$–form on $T^4/\mathbb{Z}_2$, extends to a nowhere vanishing holomorphic $(2,0)$–form on M. Indeed, we can arrange our gluing so that $\omega_c^t = \omega_2^t + i\omega_3^t$ is closed and satisfies $\omega_c^t \wedge \omega_c^t = 0$ and $\omega_c^t \wedge \overline{\omega}_c^t \neq 0$. Then the problem of perturbing $\boldsymbol{\omega}^t$ to an exact hyperkähler triple reduces to solving the complex Monge–Ampère equation

$$\left(\omega_1^t + i\partial\overline{\partial}u\right)^2 = \tfrac{1}{2}\omega_c^t \wedge \overline{\omega}_c^t. \tag{7}$$

Since $(t, u) = (0, 0)$ is a solution one can hope to solve this equation for small $t > 0$ by the Implicit Function Theorem. The main issue is that $(0, 0)$ correspond to a singular solution to the equation and therefore care is needed in applying the Implicit Function Theorem. This was done by Donaldson [19] exploiting the conformal equivalence between the cone metric $dr^2 + r^2 g_{\mathbb{RP}^3}$ (the model for the singularities of $T^4/\mathbb{Z}_2$ and for the geometry at infinity of the Eguchi–Hanson metric) and the cylindrical metric $dt^2 + g_{\mathbb{RP}^3}$. This conformal rescaling allows one to control constants in the application of the Implicit Function Theorem since the cylindrical metric has bounded geometry. Alternatively, one could work with weighted Banach spaces as in analogous constructions of complete non-compact hyperkähler 4-manifolds by Biquard–Minerbe [8].

The result is a family of Kähler Ricci-flat metrics on the K3 surface that develop 16 orbifold singularities modelled on $\mathbb{R}^4/\mathbb{Z}_2$ in the limit $t \to 0$. Each singularity is associated with a 2-sphere of self-intersection -2 which shrinks to zero size as $t \to 0$. Furthermore, appropriate rescalings of the family close to each singular point converge to the Eguchi–Hanson metric.

We can also introduce further parameters in the construction to recover a full 58-dimensional family of hyperkähler metrics on the K3 surface close to the singular limit $T^4/\mathbb{Z}_2$. Indeed, when gluing the scaled Eguchi–Hanson metric to the flat metric in a neighbourhood of the point p_i we have the choice of an isometric identification between the tangent cone at a singularity of $T^4/\mathbb{Z}_2$ and $\mathbb{R}^4/\mathbb{Z}_2$. Since the Eguchi–Hanson metric is U(2)–invariant, this choice lives in $\mathrm{SO}(4)/\mathrm{U}(2) \simeq S^2$. In other words, at each singular point we can choose a direction in $\mathrm{span}(\hat{\omega}_1, \hat{\omega}_2, \hat{\omega}_3)$ to be identified with the direction of $\omega_1^{\mathrm{eh},t}$ in $\mathrm{span}(\omega_1^{\mathrm{eh},t}, \omega_2^{\mathrm{eh},t}, \omega_3^{\mathrm{eh},t})$. In the previous situation, where we define a complex structure J on M by blow-up, we make the same choice of direction at each singular point $p_1, \ldots, p_{16}$. If different choices are made at different points then M does not come equipped with an integrable complex structure and instead of solving a complex Monge–Ampère equation we need to glue hyperkähler triples directly. This can be done as follows.

Let $\boldsymbol{\omega}^t$ be the closed definite triple on M obtained by gluing 16 copies of $\omega^{\mathrm{eh},t}$ with the flat orbifold triple $\hat{\boldsymbol{\omega}}$. We know that $\| Q_{\omega^t} - \mathrm{id}\|_{C^0} = O(t^2)$. We look for a triple of closed 2-forms $\boldsymbol{\eta} = (\eta_1, \eta_2, \eta_3)$ on M such that

$$\tfrac{1}{2}\left(\omega_i^t + \eta_i\right) \wedge \left(\omega_j^t + \eta_j\right) = \delta_{ij}\, \mu_{\omega^t}. \tag{8}$$

Decompose η into self-dual and anti self dual parts $\eta = \eta^+ + \eta^-$ with respect to g_{ω^t}. The self-dual part can be written in terms of a $M_{3\times 3}(\mathbb{R})$–valued function A by

$$\eta_i^+ = \sum_{j=1}^{3} A_{ij}\,\omega_j.$$

Denote by $\boldsymbol{\eta}^- * \boldsymbol{\eta}^-$ the symmetric (3×3)–matrix with entries $(\frac{1}{2}\,\eta_i^- \wedge \eta_j^-)/\mu_{\omega^t}$. Then we can rewrite (8) as

$$Q_{\omega^t} + Q_{\omega^t}\,A^T + A\,Q_{\omega^t} + A\,Q_{\omega^t}\,A^T + \boldsymbol{\eta}^- * \boldsymbol{\eta}^- = \mathrm{id}. \tag{9}$$

Now, consider the map

$$M_{3\times 3}(\mathbb{R}) \longrightarrow Sym^2(\mathbb{R}^3); \qquad A \longmapsto Q_{\omega^t}\,A^T + A\,Q_{\omega^t} + A\,Q_{\omega^t}\,A^T$$

and its differential $A \mapsto Q_{\omega^t}\,A^T + A\,Q_{\omega^t}$. Since Q_{ω^t} is arbitrarily close to the identity as $t\to 0$, this linear map induces an isomorphism $Sym^2(\mathbb{R}^3)\to Sym^2(\mathbb{R}^3)$ for t sufficiently small. We can therefore define a smooth function $\mathcal{F}\colon Sym^2(\mathbb{R}^3)\to Sym^2(\mathbb{R}^3)$ such that $Q_{\omega^t}\,A^T + A\,Q_{\omega^t} + A\,Q_{\omega^t}\,A^T = S$ if and only if $A=\mathcal{F}(S)$. Hence we reformulate (9) as

$$\boldsymbol{\eta}^+ = \mathcal{F}\left((\mathrm{id} - Q_{\omega^t}) - \boldsymbol{\eta}^- * \boldsymbol{\eta}^-\right). \tag{10}$$

Now, let $\mathcal{H}^+_{\omega^t}$ be the 3-dimensional space of self-dual harmonic 2-forms with respect to g_{ω^t}. Since $\omega_1^t, \omega_2^t, \omega_3^t$ are closed and self-dual (therefore harmonic) and linearly independent (since $\boldsymbol{\omega}^t$ is a definite triple) we deduce that $\mathcal{H}^+_{\omega^t}$ consist of constant linear combinations of $\omega_1,\omega_2,\omega_3$. By Hodge theory with respect to g_{ω^t} we can finally rewrite (10) as the *elliptic* equation

$$d^+\boldsymbol{a} + \boldsymbol{\zeta} = \mathcal{F}\left((\mathrm{id} - Q_{\omega^t}) - \boldsymbol{\eta}^- * \boldsymbol{\eta}^-\right), \qquad d^*\boldsymbol{a} = 0, \tag{11}$$

for a triple $\boldsymbol{a}$ of 1-forms on M and a triple $\boldsymbol{\zeta}\in\mathcal{H}^+_{\omega^t}\otimes\mathbb{R}^3$. Here $2\,d^+a = da + *da$ is the self-dual part of da.

Instead of the Monge–Ampère equation (7), one must now solve (11) applying the Implicit Function Theorem close to the singular limit $t\to 0$ to deform $\boldsymbol{\omega}^t$ into an exact hyperkähler triple. Assuming this can be done, if we now count parameters in the construction we find

(i) 10 moduli of the flat metric on T^4;
(ii) the choice of scale t of the Eguchi–Hanson metric and gauge $\psi\in \mathrm{SO}(4)/\mathrm{U}(2)\simeq S^2$ for each singular point.

Thus we have $10 + 3\times 16 = 58$ parameters in total, exactly the dimension of the moduli space of Ricci-flat metrics (without any normalisation for the volume) on the K3 surface.

2.2 Orbifold Singularities

From a broader perspective the Kummer construction furnishes the prototypical example of the appearance of orbifold singularities in non-collapsing sequences of Einstein 4-manifolds. By work of Anderson [1, Theorem C], Nakajima [43, Theorem 1.3] and Bando–Kasue–Nakajima [7, Theorem 5.1], we know that a sequence of Einstein 4-manifolds (M_i, g_i) with a uniform lower bound on volume and upper bounds on diameter and Euler characteristic converges (after passing to a subsequence) to an Einstein 4-orbifold M_∞ with finitely many singular points. The formation of orbifold singularities is modelled on complete Ricci-flat ALE spaces which appear as rescaled limits, or "bubbles", of the sequence (M_i, g_i) around points that approach one of the singularities of the orbifold M_∞. We now provide a more detailed description of these results.

Theorem 2.7 *Fix $\Lambda, C, V, D > 0$ and let (M_i^4, g_i) be a sequence of Einstein 4-manifolds satisfying*

(i) $|Ric(g_i)| \leq \Lambda$;
(ii) $\chi(M_i) \leq C$;
(iii) $Vol(M_i, g_i) \geq V$;
(iv) $diam(M_i, g_i) \leq D$.

Then a subsequence converges to an Einstein orbifold (M_∞, g_∞) with finitely many isolated singular points $\{x_1, \dots, x_n\}$ with $n \leq n(\Lambda, C, V, D)$. More precisely, (M_i, g_i) converges to (M_∞, g_∞) in the Gromov–Hausdorff sense and there are smooth embeddings $f_i\colon M_\infty \setminus \{x_1, \dots, x_n\} \to M_i$ such that $f_i^* g_i$ converges to the smooth Einstein metric $g_\infty|_{M_\infty \setminus \{x_1,\dots,x_n\}}$ in C^∞ over compact sets of $M_\infty \setminus \{x_1, \dots, x_n\}$.

Here are some ingredients in the proof of the theorem. First of all, there exists a subsequence that converges to a compact metric space (M_∞, d_∞) in Gromov–Hausdorff topology and one has to understand the structure of M_∞. The Bishop–Gromov volume comparison and hypotheses (iii) and (iv) imply the *non-collapsing* condition $\mathrm{Vol}\,(B_1(p)) \geq v$ for all $p \in M_i$ and all i and some uniform $v > 0$. Moreover, the hypotheses of the Theorem guarantee that we have uniform control on the Sobolev constant of (M_i, g_i). Since the Einstein equation implies the differential inequality $\triangle|\mathrm{Rm}_{g_i}| + c\,|\mathrm{Rm}_{g_i}|^2 \geq 0$, Moser iteration now yields the following *ε–regularity* result: there exists $\varepsilon > 0$, $C > 0$, $r_0 > 0$ such that for all $0 < r < r_0$

$$\int_{B_{2r}(p)} |\mathrm{Rm}_{g_i}|^2\, \mathrm{dv}_{g_i} < \varepsilon \Longrightarrow \sup_{B_r(p)} |\mathrm{Rm}_{g_i}| \leq C r^{-2} \left(\int_{B_{2r}(p)} |\mathrm{Rm}_{g_i}|^2\, \mathrm{dv}_{g_i} \right)^{\frac{1}{2}}. \quad (12)$$

Given (i), the bound (ii) is equivalent to a global bound $\|\mathrm{Rm}_{g_i}\|_{L^2} \leq C'$ by the Gauss–Chern–Bonnet Formula (3). Then (12) fails only for a definite number of balls. Together with a bootstrap argument using the Einstein equation, we conclude that (M_∞, g_∞) is a smooth Einstein manifold away from a definite number of points

$x_1, \ldots, x_n$. A first step in analysing the structure of these singular points is to study their *tangent cone*. Fix $a = 1, \ldots, n$. Consider a sequence $r_i \to 0$ and consider the sequence of pointed manifolds $(M_\infty, r_i^{-2} g_\infty, x_a)$. The pointed Gromov–Hausdorff limit (Y_a, o^*) of a subsequence $i_k \to \infty$ is called a tangent cone to M_∞ at x_a. A priori it depends on the sequence of rescaling r_i. Now, since $\|\mathrm{Rm}_{g_\infty}\|_{L^2}$ is bounded by the lower continuity of the energy, we have $\int_{B_{2r}(x_a)\setminus B_r(x_a)} |\mathrm{Rm}_{g_i}|^2\, \mathrm{dv}_{g_i} \to 0$ as $r \to 0$. Then using (12) one can show that the annulus $B_2(o^*) \setminus B_1(o^*)$ in Y_a is flat. In fact Y_a is a flat cone $Y_a = \mathrm{C}(\mathbb{S}^3/\Gamma_a)$ which is smooth outside of its vertex o^*.

Not only does the available theory characterise the singularities of non-collapsed limits of Einstein 4-manifolds; it also explains *how* these singularities arise. The key notion is the one of *ALE (asymptotically locally Euclidean)* manifolds.

Definition 2.9 A complete Riemannian 4-manifold (W^4, h) is ALE of rate $\nu < 0$ if there exists a finite group $\Gamma \subset \mathrm{SO}(4)$ acting freely on $\mathbb{R}^4 \setminus \{0\}$, a compact set $K \subset W$, $R > 0$ and a diffeomorphism $f\colon \left(\mathbb{R}^4 \setminus B_R(0)\right)/\Gamma \to W \setminus K$ such that

$$|\nabla^k (f^* h - h_{\mathbb{R}^4/\Gamma})| = O(r^{\nu - k}).$$

Here the norm and covariant derivative are computed using the flat metric $h_{\mathbb{R}^4/\Gamma}$.

Theorem 2.10 *In the same notation and in addition to the statements of Theorem 2.7, for each $a = 1, \ldots, n$ there exist $x_{a,i} \in M_i$ and $r_i \to \infty$ such that, up to subsequences,*

(i) $B(x_{a,i}, \delta) \subset M_i$ *converges to* $B(x_a, \delta) \subset M_\infty$ *for all* $\delta > 0$ *sufficiently small;*
(ii) $(M_i, r_i^2 g_i, x_{a,i})$ *converges to a Ricci-flat ALE 4-manifold* $(W_a, g_a, x_{a,\infty})$ *of rate* -4 *in the following sense: for each* $R > 0$ *there exists maps* $f_{a,i}\colon B(x_{a,\infty}, R) \to M_i$ *such that* $f_{a,i}^*(r_i^2 g_i)$ *converges in* C^∞ *to* h_a *on* $B(x_{a,\infty}, R) \subset W_a$.

The points $x_{a,i}$ and scales r_i are chosen so that $|\mathrm{Rm}_{g_i}|(x_{a,i}) = r_i^2$ is essentially the maximum of $|\mathrm{Rm}_{g_i}|$ in a small ball that is converging to a neighbourhood of x_a in the Gromov–Hausdorff topology. The existence of a limit (W_a, h_a) which is a complete Ricci-flat manifold with finite energy and maximal volume growth, i.e. $\|\mathrm{Rm}_{h_a}\|_{L^2} < \infty$ and $\lim_{r\to\infty} r^{-4}\mathrm{Vol}(B_r(x_{a,\infty})) > 0$, follows by arguments based on (12) as before. The fact that any such manifold is ALE of rate $\nu = -4$ follows from [7, Theorem 1.5].

In fact, the tangent cone at infinity of W_a might not match the tangent cone at the orbifold singularity $x_a \in M_\infty$ and a series of blow-ups at different scales might be necessary to capture the full picture of the degeneration of M_i to M_∞. Such bubbling-off of a "bubble-tree" of ALE Ricci-flat orbifolds was made precise by Bando [6] and Anderson–Cheeger [3]. We will see later some explicit examples of this phenomenon, cf. Remark 3.3.

2.3 ALE Gravitational Instantons

Theorem 2.10 provides the motivation for the study and ideally the *classification* of all Ricci-flat ALE 4-manifolds. It is here that the hyperkähler case differs dramatically from the more general Ricci-flat case: ALE hyperkähler 4-manifolds were constructed and classified by Kronheimer [35, 36] following earlier work of Eguchi–Hanson, Gibbons–Hawking and Hitchin (the classification was extended to non-simply connected Kähler Ricci-flat ALE 4-manifolds by Suvaina [48]); in contrast, not a single example of an ALE Ricci-flat 4-manifold with generic holonomy SO(4) is currently known and the question of whether Ricci-flat ALE 4-manifolds must have special holonomy is wide open.

A *gravitational instanton* is a complete non-compact hyperkähler 4-manifold with finite energy $\|\mathrm{Rm}\|_{L^2}$. We will see later that often stronger assumptions of curvature decay have to be imposed to obtain better control of the asymptotic geometry at infinity. Note that since every hyperkähler manifold is in particular Ricci-flat, gravitational instantons have constrained volume growth: the volume of a geodesic ball of radius r grows at most as r^4 and at least linearly in r. By the result of Bando–Kasue–Nakajima [7, Theorem 1.5] mentioned above, gravitational instantons of maximal volume growth are ALE hyperkähler 4-manifolds in the sense of Definition 2.9.

We now state Kronheimer's results. Let Γ be a finite subgroup of SU(2) that acts freely on $\mathbb{C}^2 \setminus \{0\}$. Such groups are classified by simply-laced Dynkin diagrams, i.e. the Dynkin diagrams of type ADE. The Kleinian (or Du Val) singularity $\mathbb{C}^2/\Gamma$ admits a (unique) *minimal resolution* $\pi\colon X_\Gamma \to \mathbb{C}^2/\Gamma$: X_Γ is a smooth complex surface, π is an isomorphism outside of $\pi^{-1}(0)$ and X_Γ does not contain any rational curve with self-intersection -1 (which could be blown-down to produce another smooth resolution). The exceptional locus $\pi^{-1}(0)$ is a configuration of rational curves with self-intersection -2 that intersects according to the Dynkin diagram of Γ. Finally, X_Γ has trivial canonical bundle, i.e. it admits a nowhere vanishing holomorphic (2, 0)-form ω_c that outside of $\pi^{-1}(0)$ restricts to the pull-back of the standard complex volume form $dz_1 \wedge dz_2$ on $\mathbb{C}^2/\Gamma$. In the following theorem we forget the complex structure and regard X_Γ as a smooth 4-manifold.

Theorem 2.11 *Let Γ be a finite subgroup of $SU(2)$ that acts freely on $\mathbb{C}^2 \setminus \{0\}$ and X_Γ be the smooth 4-manifold underlying the minimal resolution of $\mathbb{C}^2/\Gamma$.*

(i) Let $\boldsymbol{\alpha} \in H^2(X_\Gamma, \mathbb{R}) \otimes \mathbb{R}^3$ satisfy

$$\boldsymbol{\alpha}(\Sigma) \neq \mathbf{0} \in \mathbb{R}^3 \text{ for all } \Sigma \in H_2(X_\Gamma, \mathbb{Z}) \text{ such that } \Sigma \cdot \Sigma = -2. \quad (13)$$

Then there exists an ALE hyperkähler structure $\boldsymbol{\omega}$ on X_Γ with $[\boldsymbol{\omega}] = \boldsymbol{\alpha}$.

(ii) If $(X, \boldsymbol{\omega})$ is an ALE hyperkähler 4-manifold asymptotic to $\mathbb{C}^2/\Gamma$ then X is diffeomorphic to X_Γ and $[\boldsymbol{\omega}]$ satisfies (13). Moreover, if $(X, \boldsymbol{\omega})$ and $(X', \boldsymbol{\omega}')$ are two such manifolds and there exists a diffeomorphism $f\colon X \to X'$ such that $[f^*\boldsymbol{\omega}'] = [\boldsymbol{\omega}]$ then $(X, \boldsymbol{\omega})$ and $(X', \boldsymbol{\omega}')$ are isomorphic hyperkähler manifolds.

The hyperkähler structures in (i) are obtained by the so-called *hyperkähler quotient* construction. For example, the Eguchi–Hanson metric can be described as the hyperkähler quotient of $\mathbb{H}^2$ by U(1) acting by $e^{i\theta} \cdot (q_1, q_2) = (e^{i\theta} q_1, e^{-i\theta} q_2)$. The hyperkähler moment map for this U(1)–action is $\mu(q_1, q_2) = \overline{q}_1 i q_1 - \overline{q}_2 i q_2 \in \operatorname{Im}\mathbb{H} \simeq \mathbb{R}^3$. Given $\zeta \in \mathbb{R}^3$, the hyperkähler quotient construction guarantees that, when smooth, $\mu^{-1}(\zeta)/\mathrm{U}(1)$ is a hyperkähler manifold. When $\zeta = \mathbf{0}$ we have the flat metric on $\mathbb{C}^2/\mathbb{Z}_2$ and when $\zeta \neq \mathbf{0}$ we have the Eguchi–Hanson hyperkähler structure on T^*S^2 scaled and rotated so that $[\omega](S^2) = \zeta$.

For the classification result in (ii) Kronheimer exploits twistor theory and the natural "1-point" conformal compactification of an ALE gravitational instanton to an anti-self-dual 4-orbifold. More recently, Conlon–Hein [17, Corollary D] have obtained a different proof of this result that does not use twistor theory: with respect to any complex structure, an ALE gravitational instanton asymptotic to $\mathbb{C}^2/\Gamma$ must be the crepant resolution of a member of the versal $\mathbb{C}^*$–deformation of the Kleinian singularity $\mathbb{C}^2/\Gamma$; every such deformation has a unique crepant resolution and the latter admit a unique ALE Kähler Ricci-flat metric in each Kähler class.

3 Codimension One Collapse

If we include hyperkähler orbifolds with finitely many isolated singularities, the period map (4) can be extended as a map from the completion of the moduli space $\mathcal{M}$ of Einstein metrics on the K3 surface with unit volume in the Gromov–Hausdorff topology onto $\mathrm{Gr}^+(3, 19)/\Gamma$ [2, Theorem IV]. However $\mathrm{Gr}^+(3, 19)/\Gamma$ is non-compact so we must still consider sequences of hyperkähler metrics that do not converge in Gromov–Hausdorff topology. This amounts to understanding *collapsing* sequences of hyperkähler metrics on the K3 surface.

Let (M, g_i) be a sequence of unit-volume hyperkähler metrics with $\mathrm{diam}(M, g_i) \to \infty$. Then $\mathrm{Vol}_{g_i}(B_1(p)) \to 0$ as $i \to \infty$ for all $p \in M$, since otherwise we would bound the diameter of (M, g_i) in terms of the total volume [46, Theorem I.4.1]. Under these assumptions, Anderson [2, Theorem II] showed that (M, g_i) collapses in the sense of Cheeger–Gromov outside finitely many points $x_1, \ldots, x_n$, where the number n is controlled by the Euler characteristic $\chi(M)$. This means that for $x \in M \setminus \{x_1, \ldots, x_n\}$ the injectivity radius $\mathrm{inj}_{g_i}(x)$ converges to zero and that we control the curvature after rescaling the metric so that the injectivity radius stays bounded: $\mathrm{inj}_{g_i}(x)^2 |\mathrm{Rm}_{g_i}|_{g_i}(x) \leq \epsilon_0$, for a universal constant $\epsilon_0 > 0$. In fact, Cheeger and Tian [10, Theorems 0.1 and 0.8] have proven the much stronger result that the collapse occurs with *bounded* curvature away from a definite number of points.

Cheeger–Tian's result implies that Cheeger–Fukaya–Gromov's theory of collapse with bounded curvature [9] can be applied outside of finitely many points. The most important feature of this theory in our discussion is that the limiting geometry acquires continuous symmetries. Here we describe these symmetries only at the level of the local geometry around each point in the region that collapses with bounded curvature, referring to [9] for the globalisation of this local picture. Let (M_i^n, g_i) be a sequence of manifolds with sectional curvature bounded by a uniform constant $K > 0$. If $p_i \in M_i$ and $3r \in (0, \frac{1}{\sqrt{K}})$ then we can consider the sequence of Riemannian metrics $\hat{g}_i = \exp_{p_i}^* g_i$ on the ball $B_{3r}(0) \subset \mathbb{R}^n \simeq T_{p_i} M_i$. For each i there exists a pseudo-group Γ_i of local isometries of $(B_r(0), \hat{g}_i)$ whose action induces the equivalence relation $x \sim_{\Gamma_i} y$ if and only if $\exp_{p_i}(x) = \exp_{p_i}(y) \in M_i$. Up to passing to a subsequence, $(B_r(0), \hat{g}_i)$ converges in $C^{1,\alpha}$ to $(B_r(0), \hat{g}_\infty)$ (the limit and the convergence are smooth if we control higher order derivatives of the curvature, as in the Einstein case) and the pseudogroups Γ_i converge to a pseudogroup Γ_∞ of isometries of $(B_r(0), \hat{g}_\infty)$. The Gromov–Hausdorff limit of $(B_r(p_i), g_i)$ is $(B_r(0), \hat{g}_\infty)/\Gamma_\infty$. Since Γ_i acts in an increasingly dense fashion, Γ_∞ contains continuous isometries: in fact, a neighbourhood of the identity in Γ_∞ is isomorphic to a neighbourhood of the identity in a nilpotent Lie group.

Now, this general theory of Riemannian collapse with bounded curvature motivates us to study hyperkähler metrics in dimension 4 with a *triholomorphic* Killing field, i.e. a Killing field that preserves the hyperkähler triple as well as the metric, as models for regions that collapse with bounded curvature. Thought experiments based on the Kummer construction suggest that we should study gravitational instantons with *non-maximal volume growth* as models for regions that collapse with unbounded curvature. Indeed, consider the Kummer construction of Ricci-flat metrics on the K3 surface along a family of split tori $T^4 = T^{4-k} \times T^k_\epsilon$ with a T^k–factor of volume $\epsilon^k \to 0$. We can then think of the 2-spheres arising in the resolution of the 16 singularities of $T^4/\mathbb{Z}_2$ as coming in 2^k–tuples aligned along the collapsing k-torus over each of the 2^{4-k} singular points of $T^{4-k}/\mathbb{Z}_2$. If we now rescale the sequence of Kähler Ricci-flat metrics on the K3 surface by ϵ^{-2} around one of these 2^k–tuples, in the limit $\epsilon \to 0$ we should obtain a complete hyperkähler metric asymptotic to $(\mathbb{R}^{4-k} \times T^k)/\mathbb{Z}_2$. In the case $k = 1$, the appearance of gravitational instantons asymptotic to $(\mathbb{R}^3 \times S^1)/\mathbb{Z}_2$ as rescaled limits was suggested by Page [45]. Hyperkähler metrics asymptotic to $(\mathbb{R}^{4-k} \times T^k)/\mathbb{Z}_2$ for $k = 1, 2, 3$ have been constructed by Biquard–Minerbe [8] using a non-compact version of the Kummer construction (earlier Hitchin [31] used twistor methods in the case $k = 1$).

3.1 The Gibbons–Hawking Ansatz

The Gibbons–Hawking ansatz [24] describes 4-dimensional hyperkähler metrics with a triholomorphic S^1–action (or more generally metrics with a triholomorphic Killing field).

Let U be an open set of $\mathbb{R}^3$ and $\pi\colon\ P \to U$ be a principal U(1)–bundle. Suppose that there exists a positive harmonic function h on U such that $*dh$ is the curvature $d\theta$ of a connection θ on P. Then the metric

$$g^{\mathrm{gh}} = h\,\pi^* g_{\mathbb{R}^3} + h^{-1}\theta^2 \tag{14a}$$

on P is hyperkähler. Indeed, we can exhibit an explicit hyperkähler triple ω^{gh} that induces the metric g^{gh}. Fix coordinates (x_1, x_2, x_3) on $U \subset \mathbb{R}^3$ and define

$$\omega_i^{\mathrm{gh}} = dx_i \wedge \theta + h\, dx_j \wedge dx_k, \tag{14b}$$

where (ijk) is a cyclic permutation of (123). One can check that ω^{gh} is an SU(2)–structure inducing the Riemannian metric g^{gh}. Moreover, the requirement that ω^{gh} is also closed is equivalent to the abelian *monopole equation*

$$* \, dh = d\theta. \tag{15}$$

The fibre-wise circle action on P preserves ω^{gh} and π is nothing but a hyperkähler moment map for this action. Conversely, every 4-dimensional hyperkähler metric with a triholomorphic circle action is locally described by (14).

The basic example of the Gibbons–Hawking construction is given in terms of so-called Dirac monopoles on $\mathbb{R}^3$. Fix a set of distinct points $p_1, \ldots, p_n$ in $\mathbb{R}^3$ and consider the harmonic function

$$h = m + \sum_{j=1}^{n} \frac{k_j}{2|x - p_j|},$$

where $m \geq 0$ and $k_1, \ldots, k_n$ are constants. Since $\mathbb{R}^3 \setminus \{p_1, \ldots, p_n\}$ has non-trivial second homology, we must require $k_j \in \mathbb{Z}$ for all j in order to be able to solve (15). If these integrality constraints are satisfied then $*dh$ defines the curvature $d\theta$ of a connection θ (unique up to gauge transformations) on a principal U(1)–bundle P over $\mathbb{R}^3 \setminus \{p_1, \ldots, p_n\}$ which restricts to the principal U(1)–bundle associated with the line bundle $\mathcal{O}(k_j) \to S^2$ on a small punctured neighbourhood of p_j. The pair (h, θ) is a solution of (15) which we call a *Dirac monopole* with singularities at $p_1, \ldots, p_n$.

The Gibbons–Hawking ansatz (14) associates a hyperkähler metric g^{gh} to every Dirac monopole on the open set where $h > 0$. When $k_j > 0$ then g^{gh} is certainly defined on the restriction of P to a small punctured neighbourhood of p_j. By a change of variables one can check that g^{gh} can be extended to a smooth orbifold metric modelled on $\mathbb{C}^2/\mathbb{Z}_{k_j}$ by adding a single point.

Remark 3.3 By considering clusters of points $p_1, \ldots, p_n$ coalescing together at different rates one can easily construct sequences of (non-compact) hyperkähler metrics developing orbifold singularities modelled on bubble-trees of ALE spaces.

In particular g^{gh} is a complete metric whenever $m \geq 0$ and $k_j = 1$ for all $j = 1, \ldots, n$. When $m = 0$ one can check that g^{gh} is an ALE metric in the sense of Definition 2.9. When $m > 0$ (by scaling we can then assume that $m = 1$) g^{gh} has a drastically different asymptotic geometry called *ALF (asymptotically locally flat)*.

Definition 3.4 A gravitational instanton (M, g) is called ALF if there exists a compact set $K \subset M$ such that the following holds. The (unique) end $M \setminus K$ is the total space of a circle fibration $\pi\colon M \setminus K \to (\mathbb{R}^3 \setminus B_R)/\Gamma$, where $R > 0$ and Γ is a finite subgroup of O(3) acting freely on S^2. Passing to a Γ–cover we can always assume that π is a principal circle bundle. Define a model metric g_∞ on $M \setminus K$ by choosing a connection θ on (the Γ–cover of) π and setting $g_\infty = \pi^* g_{\mathbb{R}^3} + \theta^2$. Then we have

$$|\nabla^k_{g_\infty}(g - g_\infty)|_{g_\infty} = O(r^{\nu-k}) \tag{16}$$

for some $\nu < 0$ and all $k \geq 0$.

There are only two possibilities for Γ: if $\Gamma = \mathrm{id}$ we say that M is an ALF gravitational instanton of *cyclic type*; if $\Gamma = \mathbb{Z}_2$ we say that M is an ALF gravitational instanton of *dihedral type*.

Recall that gravitational instantons have constrained volume growth: $\mathrm{Vol}\left(B_r(p)\right)$ grows at least linearly in r and at most as r^4. Under the assumption of faster than quadratic curvature decay, i.e. $|\mathrm{Rm}| = O(r^{-2-\epsilon})$ for some $\epsilon > 0$ (or a slightly weaker finite weighted energy assumption), Minerbe [40, Theorem 0.1] has shown that if we assume a uniformly submaximal volume growth, $\mathrm{Vol}\left(B_r(p)\right) \leq Cr^a$ for some $3 \leq a < 4$ and all p, say, then the volume growth is at most cubic, $a \leq 3$. Minerbe also described the asymptotic geometry of gravitational instantons of cubic volume growth and faster than quadratic curvature decay: they are all ALF spaces as in Definition 3.4.

3.2 *ALF Gravitational Instantons*

Now we describe the classification of ALF gravitational instantons obtained by Minerbe [41] and Chen–Chen [12] in the cyclic and dihedral case respectively.

Let H^k be the total space of the principal U(1)–bundle associated with the line bundle $\mathcal{O}(k)$ over S^2 radially extended to $\mathbb{R}^3 \setminus B_R$ for any $R > 0$. Let θ_k denote the (unique up to gauge transformation) SO(3)–invariant connection on H^k. The Gibbons–Hawking ansatz (14) yields a hyperkähler metric

$$g_k = \left(1 + \frac{k}{2r}\right)(dr^2 + r^2 g_{\mathbb{S}^2}) + \left(1 + \frac{k}{2r}\right)^{-1} \theta_k^2 \tag{17}$$

on H^k for all $k \in \mathbb{Z}$. Here r is a radial function on $\mathbb{R}^3$. Finally, on H^{2k} we consider the $\mathbb{Z}_2$–action which is generated by the simultaneous involutions on the base $\mathbb{R}^3$ and the fibre: on $\mathbb{R}^3$ we act by the standard involution $x \mapsto -x$ and the involution on the fibre $S^1 = \mathbb{R}/2\pi\mathbb{Z}$ is the one induced by the standard involution on the universal cover $\mathbb{R}$. We refer to this involution of H^{2k} as its standard involution.

Definition 3.7 Let (M^4, g) be an ALF gravitational instanton.

(i) We say that M is of type A_k for some $k \geq -1$ if there exists a compact set $K \subset M$, $R > 0$ and a diffeomorphism $\phi\colon H^{k+1} \to M \setminus K$ such that

$$|\nabla^l_{g_{k+1}}(g_{k+1} - \phi^* g)|_{g_{k+1}} = O(r^{-3-l})$$

for every $l \geq 0$.

(ii) We say that M is of type D_m for some $m \geq 0$ if there exists a compact set $K \subset M$, $R > 0$ and a double cover $\phi \colon H^{2m-4} \to M \setminus K$ such that the group $\mathbb{Z}_2$ of deck transformations is generated by the standard involution on H^{2m-4} and

$$|\nabla^l_{g_{2m-4}}(g_{2m-4} - \phi^* g)|_{g_{2m-4}} = O(r^{-3-l})$$

for every $l \geq 0$.

Chen–Chen [12, Theorem 1.1] have shown that every ALF gravitational instanton is either of type A_k for some $k \geq -1$ or D_m for some $m \geq 0$. The constraints $k \geq -1$ and $m \geq 0$ were derived earlier by Minerbe [39, Theorem 0.1] in the cyclic case and by Biquard–Minerbe [8, Corollary 3.2] in the dihedral case.

3.2.1 ALF Spaces of Cyclic Type

We saw that gravitational instantons of type A_k can be constructed from Dirac monopoles on $\mathbb{R}^3$ with $k+1$ singularities via the Gibbons–Hawking ansatz. These are usually called *multi-Taub–NUT* metrics. The case $k = 0$ is the Taub–NUT metric on $\mathbb{R}^4$ and $k = -1$ is $\mathbb{R}^3 \times \mathbb{S}^1$ with its flat metric. Minerbe [41, Theorem 0.2] has shown that every ALF space of cyclic type must be isometric to a multi-Taub–NUT metric.

3.2.2 ALF Spaces of Dihedral Type

ALF metrics of dihedral type are not globally given by the Gibbons–Hawking construction and in most cases are not explicit. A number of different constructions have appeared over the past decades, but only recently Chen–Chen [12, Theorem 1.2] have shown that all these constructions yield equivalent families of ALF metrics.

$m = 0$: The D_0 ALF manifold is the moduli space of centred charge 2 monopoles on $\mathbb{R}^3$ with its natural L^2–metric, known as the *Atiyah–Hitchin manifold*. The metric admits a cohomogeneity one isometric action of SO(3) and is explicitly given in terms of elliptic integrals [4, Chap. 11]. The D_0 ALF metric is rigid modulo scaling.

$m = 1$: The Atiyah–Hitchin manifold is diffeomorphic to the complement of a Veronese $\mathbb{RP}^2$ in S^4 and therefore it retracts to $\mathbb{RP}^2$ and is not simply connected. The double cover of the Atiyah–Hitchin manifold is a D_1 ALF space. As a smooth manifold it is diffeomorphic to the complement of $\mathbb{RP}^2$ in $\mathbb{CP}^2$, or equivalently to the total space of $\mathcal{O}(-4)$ over S^2. This rotationally invariant D_1 ALF metric admits a 3-dimensional family of D_1 ALF deformations, sometimes referred to as the *Dancer metrics*.

$m = 2$: D_2 ALF metrics were constructed by Hitchin [31, Sect. 7] using twistor methods and by Biquard–Minerbe [8, Theorem 2.4] using a non-compact version of the Kummer construction: one considers the quotient of $\mathbb{R}^3 \times \mathbb{S}^1$ by an involution and resolves the two singularities gluing in copies of the Eguchi–Hanson metric.

$m \geq 3$: D_m ALF metrics (for all $m \geq 1$) appeared in the work of Cherkis–Kapustin [16] on moduli spaces of singular monopoles on $\mathbb{R}^3$ and were rigorously constructed by Cherkis–Hitchin [15] using twistor methods and the generalised Legendre transform. In the case $m \geq 3$ a more transparent construction due to Biquard–Minerbe [8, Theorem 2.5] yields D_m ALF metrics by desingularising the quotient of the Taub–NUT metric by the binary dihedral group $\mathcal{D}_m$ of order $4(m-2)$ using ALE dihedral spaces. Using complex Monge–Ampère methods Auvray [5] has then constructed $3m$–dimensional families of D_m ALF metrics on the smooth 4-manifold underlying the minimal resolution of $\mathbb{C}^2/\mathcal{D}_m$.

3.3 ALF Gravitational Instantons and Collapsing Ricci-flat Metrics on the K3 Surface

Despite this rich theory of ALF gravitational instantons, until recently it has remained unclear how they can appear as models for the formation of singularities in collapsing sequences of hyperkähler metrics on the K3 surface. In [22] the author exploited singular perturbation methods to construct examples of Ricci-flat metrics on the K3 surface collapsing to a 3-dimensional limit and exhibit ALF gravitational instantons as the "bubbles" appearing in the process.

Theorem 3.8 *Let $T^3 = \mathbb{R}^3/\Lambda$ be a 3-torus for some lattice $\Lambda \simeq \mathbb{Z}^3$. Endow T^3 with a flat metric g_{T^3}. Let $\tau\colon T^3 \rightarrow T^3$ be the standard involution $x \mapsto -x$ and denote by $q_1, \ldots, q_8$ its fixed points. Fix a τ–symmetric configuration of further $2n$ distinct points $p_1, \tau(p_1), \ldots, p_n, \tau(p_n)$. Denote by T^* the complement of $\{q_1, \ldots, q_8, p_1, \ldots, \tau(p_n)\}$ in T^3.*

Let $m_1, \ldots, m_8 \in \mathbb{Z}_{\geq 0}$ and $k_1, \ldots, k_n \in \mathbb{Z}_{\geq 1}$ satisfy

$$\sum_{j=1}^{8} m_j + \sum_{i=1}^{n} k_i = 16. \tag{18}$$

For each $j = 1, \ldots, 8$ fix a D_{m_j} ALF space M_j and for each $i = 1, \ldots, n$ an A_{k_i-1} ALF space N_i.

Then there exists a 1-parameter family of hyperkähler metrics $\{g_\epsilon\}_{\epsilon\in(0,\epsilon_0)}$ on the K3 surface with the following properties. We can decompose the K3 surface into the union of open sets $K^\epsilon \cup \bigcup_{j=1}^{8} M_j^\epsilon \cup \bigcup_{i=1}^{n} N_i^\epsilon$ such that as $\epsilon \rightarrow 0$:

(i) *(K^ϵ, g_ϵ) collapses to the flat orbifold $T^*/\mathbb{Z}_2$ with bounded curvature away from the punctures;*
(ii) *for each $j = 1, \ldots, 8$ and $k \geq 0$, $(M_j^\epsilon, \epsilon^{-2} g_\epsilon)$ converges in $C^{k,\alpha}_{loc}$ to the D_{m_j} ALF space M_j;*
(iii) *for each $i = 1, \ldots, n$ and $k \geq 0$, $(N_i^\epsilon, \epsilon^{-2} g_\epsilon)$ converges in $C^{k,\alpha}_{loc}$ to the $A_{k_i - 1}$ ALF space N_i.*

The metric g_ϵ is constructed by gluing methods: we first construct an approximate hyperkähler metric by patching together known models and then perturb it to an exact solution. The construction of the approximate hyperkähler metric proceeds as follows. The ALF gravitational instantons provide models for the collapsing geometry near points of curvature concentration. We aim to construct a model for the collapsing sequence of hyperkähler metrics on regions where the curvature remains bounded using the Gibbons–Hawking ansatz over the punctured 3-torus T^*. We look for a Dirac monopole (h, θ) on T^* with the following singular behaviour: h is a harmonic function on T^* with prescribed singularities at the punctures

$$h \sim \frac{2m_j - 4}{2r_j} \text{ as } r_j \to 0, \qquad h \sim \frac{k_i}{2r_i} \text{ as } r_i \to 0.$$

Here r_j and r_i denote the distance functions from the points q_j and $p_i, \tau(p_i)$ with respect to the flat metric g_{T^3}. The balancing condition (18) guarantees the existence of the harmonic function h. Since the weights m_j and k_i are integers and the configuration of punctures is τ–invariant, one can also show the existence of a connection θ with curvature $*dh$ on a principal circle bundle over T^*.

Fix a (small) positive number $\epsilon > 0$. The Gibbons–Hawking ansatz (14) yields a hyperkähler metric

$$g_\epsilon^{\mathrm{gh}} = (1 + \epsilon h)\, \pi^* g_{T^3} + \epsilon^2 (1 + \epsilon h)^{-1}\, \theta^2$$

over the region where $1 + \epsilon h > 0$. Unless h is constant (which corresponds to Page's suggestion of considering the Kummer construction starting with $T^3 \times S^1_\epsilon$ for a circle factor of length $2\pi\epsilon \to 0$) there must exists some j with $m_j = 0, 1$ and therefore the harmonic function $1 + \epsilon h$ must become negative somewhere. The key observation is that by taking ϵ sufficiently small (which geometrically corresponds to making the circle fibres have small length) it is possible to construct highly collapsed hyperkähler metrics g_ϵ^{gh} outside of an arbitrarily small neighbourhood of the punctures. More precisely, one can prove that there exists $\epsilon_0 > 0$ such that for every $\epsilon < \epsilon_0$ we have $1 + \epsilon h > \frac{1}{2}$ on the complement of $\bigcup_{j=1}^k B_{8\epsilon}(q_j)$.

Now, as we know from Definition 3.7 the asymptotic model of any ALF metric (up to a double cover in the dihedral case) can be written in Gibbons–Hawking coordinates. The configuration of punctures and weights on T^3 was chosen so that, after taking a $\mathbb{Z}_2$–quotient, we are able to glue in copies of ALF spaces to extend the Gibbons–Hawking metric g_ϵ^{gh} to an approximately hyperkähler triple ω_ϵ: close to the fixed point q_j of the $\mathbb{Z}_2$–action on T^3 we glue in the D_{m_j} ALF space M_j (this explains why we need 8 of them in the theorem); close to the image of $p_i, \tau(p_i)$ in $T^3/\mathbb{Z}_2$ we glue in the A_{k_i-1} ALF space N_i. In this way one obtains a closed definite triple $\boldsymbol{\omega}_\epsilon$ which is approximately hyperkähler in the sense that $|Q_{\boldsymbol{\omega}_\epsilon} - \mathrm{id}| \to 0$ as $\epsilon \to 0$. The approximate hyperkähler triple is then deformed into an exact solution by solving an equation like (11) using the Implicit Function Theorem in appropriately chosen weighted Hölder spaces.

4 Collapse and Elliptic Fibrations

In this final section we describe an influential work of Gross–Wilson [27] on the behaviour of hyperkähler metrics on the K3 surface collapsing to a 2-dimensional limit along the fibres of an elliptic fibration. We will also discuss more recent work of Hein [29] and related work by Chen–Chen [11–13] on gravitational instantons with non-maximal volume growth, in which elliptic fibrations also play a key role.

4.1 The Gross–Wilson's Construction

A complex surface (i.e. a complex manifold of complex dimension 2) (M, J) is said to be elliptic if it admits a holomorphic map $\pi\colon M \to C$ onto a smooth complex curve C such that the generic fibre is a smooth curve of genus 1. If $\pi\colon M \to C$ has a holomorphic section σ, then the generic fibre becomes a smooth elliptic curve. We say that M is a minimal elliptic surface if there are no (-1)–curves contained in the fibres.

If (M, J) is an elliptic complex K3 surface not all fibres can be smooth elliptic curves because $\chi(M) = 24$. The possible singular fibres of elliptic surfaces have been classified by Kodaira. They are distinguished by the monodromy. Work locally with a minimal elliptic surface $\pi\colon M \to \Delta$ over a disc with a section σ and assume that all fibres except possibly the one over the origin are smooth elliptic curves. Using σ, one can describe the restriction $M|_{\Delta^*}$ of M to the punctured disc as $\pi\colon (\Delta^* \times \mathbb{C})/\Lambda \to \Delta^*$, for a family of lattices $\Lambda \subset \mathbb{C}$ defined by (possibly multi-valued) holomorphic functions τ_1, τ_2 on Δ^*. The monodromy is the representation of the fundamental group of Δ^* on the mapping class group of the smooth fibre. We can think of it as the conjugacy class of the matrix $A \in \mathrm{SL}(2, \mathbb{Z})$ generating the action of $\pi_1(\Delta^*)$ on the oriented pair (τ_1, τ_2). We refer to [42, Tables I.4.1 and I.4.2] for Kodaira's list and limit ourself to the example of a singular fibre of type I_1. In this case $M|_{\Delta^*}$ is isomorphic to $(\Delta^* \times \mathbb{C})/(\tau_1\mathbb{Z} + \tau_2\mathbb{Z})$ with $\tau_1 = 1, \tau_2 = \frac{1}{2\pi i}\log z$. Since $\tau_2(e^{i\theta}z) = \tau_2(z) + \frac{\theta}{2\pi}$, the monodromy around an I_1 fibre is

$$\begin{pmatrix} 1 & 1 \\ 0 & 1 \end{pmatrix}. \tag{19}$$

The singular fibre $\pi^{-1}(0)$ is a pinched torus, i.e. a 2-sphere with south and north poles identified.

Generically a complex K3 surface that admits an elliptic fibration (necessarily over $\mathbb{CP}^1$) has exactly 24 singular fibres, all of type I_1. Let $\pi\colon M \to \mathbb{CP}^1$ be such such an elliptic K3 surface with 24 singular I_1 fibres. Up to changing the complex structure of M preserving the fibration $\pi\colon M \to \mathbb{CP}^1$ we can always reduce to the case that π has a holomorphic section. Gross–Wilson studied the behaviour of Kähler Ricci-flat metrics on M as we fix this complex structure and deform the Kähler class so that the elliptic fibres of π shrink to zero size. In other words, they considered a sequence of Kähler classes converging to the class $[\pi^*\omega_{\mathrm{FS}}]$ at the boundary of the Kähler cone of M and described the behaviour of the Kähler Ricci-flat metric given by Yau's Theorem along this sequence. Here ω_{FS} is the Fubini–Study metric on $\mathbb{CP}^1$. Gross–Wilson's description of the collapsing Ricci-flat metrics is achieved by a gluing construction.

4.1.1 The Semi-flat Metric

The model for the collapsing Ricci-flat metrics away from the singular fibres is provided by a certain semi-flat metric [27, Sect. 2], i.e. a metric that restricts to a flat metric on each elliptic fibre.

Let $\pi\colon M \to \mathbb{CP}^1$ be an elliptic K3 surface with a section and restrict the fibration to a small disc $\Delta \subset \mathbb{CP}^1$. Fix a holomorphic coordinate z on Δ. We assume that $\pi\colon M|_\Delta \to \Delta$ is a minimal elliptic fibration with a section such that all fibres are smooth. Using the given holomorphic section, we can identify $M|_\Delta$ with $(\Delta \times \mathbb{C}_w)/(\tau_1\mathbb{Z} + \tau_2\mathbb{Z})$ as before. Without loss of generality we assume that $\mathrm{Im}(\overline{\tau_1}\tau_2) > 0$. Fix a holomorphic symplectic form ω_c on M. In coordinates z, w we can assume that $\omega_c = dz \wedge dw$. Given $\varepsilon > 0$, we construct a semi-flat metric $\omega_{sf,\varepsilon}$ using the following ingredients.

(i) For each $z \in \Delta$ define a flat Kähler metric $\omega_{z,\varepsilon}$ on $\pi^{-1}(z)$ by choosing a dual basis $\xi_1(z), \xi_2(z)$ to $\tau_1(z), \tau_2(z)$ and setting $\omega_{z,\varepsilon} = \varepsilon\, \xi_1(z) \wedge \xi_2(z)$. Changing basis to $dw, d\overline{w}$ yields $\omega_{z,\varepsilon} = \frac{i}{2} W dw \wedge d\overline{w}$, with $W = \frac{\varepsilon}{\mathrm{Im}(\overline{\tau_1}\tau_2)}$.

(ii) Define $\omega_{\Delta,\varepsilon}$ as the unique Kähler metric on Δ such that the pairing $T^{1,0}\Delta \times (\Delta \times \mathbb{C}) \to \mathbb{C}$ induced by ω_c is isometric with respect to the Hermitian metrics induced by $\omega_{z,\varepsilon}$ and $\omega_{\Delta,\varepsilon}$. Explicitly, $\omega_{\Delta,\varepsilon} = \frac{i}{2} W^{-1} dz \wedge d\overline{z}$.

(iii) The family of lattices $\tau_1\mathbb{Z} + \tau_2\mathbb{Z}$ defines a flat connection on the trivial bundle $\Delta \times \mathbb{C}$ by declaring τ_1 and τ_2 to be flat sections. The associated connection 1-form is

$$\Gamma\, dz = \frac{1}{\mathrm{Im}(\overline{\tau_1}\tau_2)} \big(\mathrm{Im}(\overline{\tau_1} w) d\tau_2 - \mathrm{Im}(\overline{\tau_2} w) d\tau_1\big).$$

The semi-flat metric is then

$$\omega_{sf,\varepsilon} = \tfrac{i}{2} W^{-1} dz \wedge d\overline{z} + \tfrac{i}{2} W (dw - \Gamma dz) \wedge (d\overline{w} - \overline{\Gamma} d\overline{z}). \tag{20}$$

Note that the triple $(\omega_{sf,\varepsilon}, \operatorname{Re}\omega_c, \operatorname{Im}\omega_c)$ is hyperkähler and that $\omega_{sf,\varepsilon}|_{\pi^{-1}(z)}$ is the flat metric with volume ε.

The construction of the semi-flat metric can be extended to the situation where $M|_{\Delta}$ has a singular fibre over the origin $z = 0$. We simply replace Δ with the punctured disc Δ^* and use Kodaira's normal form for $M|_{\Delta^*}$. For example, if $\pi^{-1}(0)$ is a fibre of type I_1 then $M|_{\Delta^*}$ is isomorphic to $(\Delta^* \times \mathbb{C})/(\tau_1\mathbb{Z} + \tau_2\mathbb{Z})$ with $\tau_1 = 1$, $\tau_2 = \frac{1}{2\pi i}\log z$. Note that the assumption $\operatorname{Im}(\overline{\tau}_1\tau_2) > 0$ forces Δ to be strictly contained in the unit disc in $\mathbb{C}$. The semi-flat metric (20) on the complement of a singular fibre of type I_1 admits a tri-holomorphic S^1–action and, following [27], we can rewrite it in Gibbons–Hawking coordinates.

First of all, since $W \operatorname{Im}(\Gamma dz) = -\operatorname{Im}(w)\, dW$, the imaginary part of $W(dw - \Gamma\, dz)$ is closed. Hence there exists a function $t\colon\ \Delta^* \times \mathbb{C} \to \mathbb{R}$, unique up to the addition of a constant, such that $-W^{-1}dt = \operatorname{Im}(dw - \Gamma\, dz)$. Then $\pi\colon\ (\Delta^* \times \mathbb{C})/\tau_1\mathbb{Z} \to \Delta^* \times \mathbb{R}_t$ is a principal U(1)–bundle. Explicitly, $t = -\frac{2\pi\varepsilon \operatorname{Im}(w)}{\log|z|}$. Taking the quotient by $\tau_2\mathbb{Z}$ we obtain a principal U(1)–bundle $(\Delta^* \times \mathbb{C})/(\tau_1\mathbb{Z} + \tau_2\mathbb{Z}) \to \Delta^* \times \mathbb{R}/\varepsilon\mathbb{Z}$. Its Euler class evaluated on $|z| = \text{const}$ is ± 1, depending on the orientation. Now set $h = W^{-1}, dw - \Gamma\, dz = \theta - ih\, dt$ and use polar coordinates $re^{i\psi} = z$. The semi-flat metric (20) can then be written in Gibbons–Hawking coordinates (14)

$$g_{sf,\varepsilon} = \frac{-\log r}{2\pi\varepsilon}\left(dr^2 + r^2 d\psi^2 + dt^2\right) + \frac{2\pi\varepsilon}{-\log r}\theta^2. \tag{21}$$

4.1.2 The Ooguri–Vafa Metric

The second building block in Gross–Wilson's construction is the Ooguri–Vafa metric, an explicit (incomplete) hyperkähler metric defined in a neighbourhood of a singular fibre of type I_1. This metric was first constructed in [44]. A more thorough analysis is given in [27, Sect. 3]. The Ooguri–Vafa metric is a periodic version of the Taub–NUT metric, in the sense that it can be constructed by the Gibbons–Hawking ansatz on $\mathbb{R}^2 \times S^1$ with a harmonic function h with a Green's function singularity at a point. Since the Green's function of $\mathbb{R}^2 \times S^1$ changes sign (we say that $\mathbb{R}^2 \times S^1$ is *parabolic*), the Ooguri–Vafa metric is only defined on a small enough neighbourhood of the Green's function singularity.

Fix $\varepsilon > 0$ sufficiently small so that $2\varepsilon < 1$. Let Δ be the unit disc in $\mathbb{C}$ with coordinate $z = re^{i\psi}$. Let t be a periodic coordinate of period ε and consider the product $\Delta \times S^1_t$, where $S^1_t = \mathbb{R}/\varepsilon\mathbb{Z}$. By abuse of notation we denote by 0 the point with coordinates $z = 0$ and $t = 0 \pmod{\varepsilon\mathbb{Z}}$. Consider the power series

$$h(z,t) = \frac{1}{2}\sum_{m\in\mathbb{Z}}\left(\frac{1}{\sqrt{r^2 + 4\pi^2(t - m\varepsilon)^2}} - a_{|m|}\right), \tag{22}$$

where

$$a_{|m|} = \tfrac{1}{2|m|\pi\varepsilon} \text{ if } m \neq 0 \text{ and } a_0 = \tfrac{\log 4\pi\varepsilon - 2\gamma}{\pi\varepsilon}.$$

Here γ is the Euler constant, $\gamma = \lim_{n\to\infty} \sum_{k=1}^{n} k^{-1} - \log n$. The series converges uniformly on compact subsets of $(\Delta \times S^1_t) \setminus \{0\}$ to the Green's function of $\mathbb{R}^2 \times S^1_t$ with singularity at 0. Whenever $z \neq 0$, h can be expressed as

$$h(z,t) = -\frac{1}{2\pi\varepsilon}\log r + \frac{1}{2\pi\varepsilon}\sum_{m\in\mathbb{Z}^*} K_0\left(\frac{|m|r}{\varepsilon}\right) e^{\frac{2\pi m i}{\varepsilon}t},$$

where K_0 is the second modified Bessel function. In particular, due to the exponential decay of the Bessel function, for all $k \geq 0$ there exists a constant $C_k > 0$ such that

$$\left|\nabla^k\left(h(z,t) + \frac{1}{2\pi\varepsilon}\log r\right)\right| \leq \frac{C_k}{\varepsilon}e^{-\frac{r}{\varepsilon}} \tag{23}$$

for all $r \geq 2\varepsilon$.

One can now use the harmonic function h defined in (22) in the Gibbons–Hawking ansatz (14) to produce a hyperkähler metric—the Ooguri–Vafa metric—on a circle bundle X over $\Delta \times S^1_t$. As in the case of the multi-Taub–NUT metrics, a change of coordinates shows that the Gibbons–Hawking metric on X extends smoothly over a point corresponding to the singular points 0 of h.

By (23) the Ooguri–Vafa metric approaches the semi-flat metric (21) up to terms that decay exponentially fast as $\varepsilon \to 0$. It remains to check that the Ooguri–Vafa metric is defined on an elliptic fibration over a disc with a singular fibre of type I_1 over the origin. Choose the complex structure such that dz and $\theta - ihdt$ span the space of $(1,0)$–forms. In this complex structure the projection $\pi : X \to \Delta$ is an elliptic fibration and $\pi^{-1}(0)$ is the only singular fibre. One can identify the periods and therefore the monodromy of this elliptic fibration by integrating the $(1,0)$–form $\theta - ihdt$ over a basis $\{\gamma_1, \gamma_2\}$ of the first homology of a fibre $\pi^{-1}(z)$. If one chooses γ_1 to be an orbit for the S^1–action on the circle bundle $X \to \Delta \times S^1_t$ and γ_2 to be the circle parametrised by t in the base then one finds easily that the monodromy coincides with (19). Alternatively, one can identify $\pi^{-1}(0)$ with a pinched torus, since the restriction of the circle fibration X over $\{z = 0\} \times S^1_t$ degenerating at the point 0 is a 2-sphere with the two poles identified.

4.1.3 Behaviour of Ricci-flat Metrics

For $\varepsilon > 0$ sufficiently small, Gross–Wilson now patch together the semi-flat metric (20) with 24 copies of the Ooguri–Vafa metric to obtain an approximate Ricci-flat metric ω_ε on the elliptic K3 surface M. The error (measured in terms of appropriate Hölder norms of the Ricci-potential of ω_ε) is of order $e^{-C/\varepsilon}$. This exponential decay is crucial for the perturbation argument to work. Indeed, by Yau's proof of the Calabi Conjecture there exists a unique function u_ε on M such that

$$(\omega_\varepsilon + i\partial\bar{\partial}u_\varepsilon)^2 = \tfrac{1}{4}\omega_c \wedge \overline{\omega}_c, \qquad \int_M u_\varepsilon\, \omega_\varepsilon^2 = 0. \tag{24}$$

Gross–Wilson run through Yau's proof of the existence of u_ε keeping careful track of all the constants involved (e.g. the Sobolev constant in the Moser iteration argument to prove the C^0–estimate). All these constants do blow-up as $\varepsilon \to 0$, but only polynomially in ε^{-1}. Since the error is exponentially small the Implicit Function Theorem can still be applied to obtain the following theorem [27, Theorems 5.6 and 6.4].

Theorem 4.7 *Let $\pi\colon (M, \omega_c) \to \mathbb{CP}^1$ be an elliptic K3 surface with a holomorphic section and* 24 *singular I_1 fibres. For $\varepsilon > 0$ sufficiently small let ω_ε be the Kähler metric on X constructed by gluing the semi-flat metric* (20) *to* 24 *copies of the Ooguri–Vafa metric. Let u_ε be the unique solution to* (24).

(i) *For every $k \geq 2$, $\alpha \in (0, 1)$ and every simply connected open subset $U \subset \mathbb{CP}^1$ with closure contained in the complement of the* 24 *points $p_1, \ldots, p_{24}$ corresponding to singular fibres there exist constants $C, c > 0$ such that $\|u_\varepsilon\|_{C^{k,\alpha}(U)} \leq Ce^{-c/\varepsilon}$.*

(ii) *(X, ω_ϵ) converges in the Gromov–Hausdorff sense to $\mathbb{CP}^1$ endowed with the distance induced by the (singular) metric ω_0 limit of the semi-flat metric* (20) *away from the* 24 *singular points. Away from $p_1, \ldots, p_{24}$, ω_0 satisfies $Ric(\omega_0) = \omega_{WP}$, where ω_{WP} is the pull-back to $\mathbb{CP}^1 \setminus \{p_1, \ldots, p_{24}\}$ of the Weil–Peterson metric on the moduli space of elliptic curves.*

Similar results—convergence after rescaling to the semi-flat metric on the locus of smooth fibres and global Gromov–Hausdorff convergence to $\mathbb{CP}^1$ as in (ii)—have been obtained more recently for arbitrary elliptic K3 surfaces without a detailed picture of the collapsing hyperkähler metrics in a neighbourhood of the singular fibres, cf. [25, 26]. A complete detailed picture as in [26] for arbitrary configurations of singular fibre was recently given by Chen-Viaclovsky-Zhang [14] exploiting gluing definite triples instead of complex Monge-Ampère methods.

4.2 ALG and ALH Gravitational Instantons

In [29] Hein constructs families of gravitational instantons with quadratic and lower-than-quadratic volume growth. The metrics are constructed by applying Tian–Yau's method to a rational elliptic surface, i.e. a complex surface (X, J) which is birationally equivalent to $\mathbb{CP}^2$ and which admits a minimal elliptic fibration with a section. All rational elliptic surfaces can be constructed in the following way. Let C_1 be a smooth plane cubic and C_2 a second distinct cubic. The pencil $\{\lambda_1 C_1 + \lambda_2 C_2 \,|\, [\lambda_1 : \lambda_2] \in \mathbb{CP}^1\}$ has $C_1 \cdot C_2 = 9$ base points (counted with multiplicities). After blowing them up we obtain a rational elliptic surface $\pi\colon X \to \mathbb{CP}^1$; X is a minimal elliptic surface because we blew-up just enough to resolve all the tangencies of the pencil and X has at least a section given by the (-1)–curve obtained in the last blow-up. As for the K3 surface, if X is a rational elliptic surface not all fibres can be smooth elliptic curves because $\chi(X) = \chi\left(\mathbb{CP}^2\#9\overline{\mathbb{CP}}^2\right) = 12$.

The crucial point now is that the class of an elliptic fibre in a rational elliptic surface is an anti-canonical divisor: there exists a holomorphic symplectic form ω_c on $M = X \setminus \pi^{-1}(\infty)$ with simple poles along $\pi^{-1}(\infty)$. (Here we choose an affine coordinate on the base of the fibration $\mathbb{CP}^1$ so that the chosen elliptic fibre is the fibre over ∞.) Assuming the existence of an appropriate complete background metric ω_0 on M, Tian and Yau's method [49, 50] can be applied to construct a Ricci-flat Kähler metric on M by solving the complex Monge–Ampère equation $\left(\omega_0 + i\partial\overline{\partial}u\right)^2 = \frac{1}{2}\omega_c \wedge \overline{\omega}_c$ on the complement of $\pi^{-1}(\infty)$. In order to be able to solve this Monge–Ampère equation it is necessary to assume that ω_0 is already an approximate solution at infinity, in the sense that the Ricci potential of ω_0 decays with a certain rate. Note that the choice of the background ω_0 is not obvious nor unique: the flat and Taub–NUT metrics on $\mathbb{C}^2 = \mathbb{CP}^2 \setminus \mathbb{CP}^1$ are different complete hyperkähler metrics with the same holomorphic symplectic form [37]. In the case of rational elliptic surfaces, Hein exploits the elliptic fibration to construct a good background Kähler metric ω_0 which is approximately Ricci-flat at infinity. The type of fibre $\pi^{-1}(\infty)$ removed dictates the asymptotics of the metric ω_0 using Kodaira's normal form for a neighbourhood of $\pi^{-1}(\infty)$ and a semi-flat metric as in Gross–Wilson's construction.

The simplest examples of Hein's construction are those obtained by removing a smooth elliptic fibre (a fibre of type I_0 in Kodaira's classification): in this case the metric is ALH.

Definition 4.8 A gravitational instanton (M, g) is called ALH if there exists a compact subset $K \subset M$ and a diffeomorphism $f\colon \mathbb{R}_+ \times T^3 \to M \setminus K$ such that

$$|\nabla^k_{g_{\text{flat}}}(f^*g - g_{\text{flat}})|_{g_{\text{flat}}} = O\left(e^{-\delta t}\right)$$

for all $k \geq 0$ and some $\delta > 0$. Here $g_{\text{flat}} = dt^2 + g_{T^3}$ for a flat metric g_{T^3} on T^3.

Examples of ALH metrics have also been obtained by Biquard–Minerbe [8] by desingularising the flat orbifold $(\mathbb{R} \times T^3)/\mathbb{Z}_2$ by gluing in 8 copies of the Eguchi–Hanson metric. More recently, Chen–Chen [13, Theorem 1.5] have given a complete classification of ALH gravitational instantons.

Theorem 4.9 *Let M be the smooth 4-manifold underlying the minimal resolution of $(\mathbb{R} \times T^3)/\mathbb{Z}_2$, where $T^3 = \mathbb{R}^3/(\mathbb{Z}v_1 + \mathbb{Z}v_2 + \mathbb{Z}v_3)$. For each $i = 1, 2, 3$ let F_i be the element of $H_2(T^3, \mathbb{Z})$ corresponding to span(v_j, v_k), where $\epsilon_{ijk} = 1$. Then $H_2(M, \mathbb{Z})$ is spanned by F_1, F_2, F_3 and the classes of the 8 (-2)–curves introduced by the resolution $M \to (\mathbb{R} \times T^3)/\mathbb{Z}_2$.*

(i) Let $\alpha \in H^2(M, \mathbb{R}) \otimes \mathbb{R}^3$ satisfy

$$\begin{gathered} \alpha(\Sigma) \neq \mathbf{0} \in \mathbb{R}^3 \textit{ for all } \Sigma \in H_2(M, \mathbb{Z}) \textit{ such that } \Sigma \cdot \Sigma = -2, \\ \textit{the matrix with rows } \alpha(F_i),\ i = 1, 2, 3, \textit{ is positive definite.} \end{gathered} \tag{25}$$

Then there exists an ALH hyperkähler structure ω on M with $[\omega] = \alpha$, unique up to triholomorphic isometries acting trivially on $H^2(M, \mathbb{R})$.

(ii) If (X, ω) is an ALH gravitational instanton then X is diffeomorphic to M and $[\omega]$ satisfies (25).

ALH gravitational instantons can be used to produce hyperkähler metrics on the K3 surface that develop a long neck. Indeed, if (M, ω) and (M, ω') are two ALH gravitational instantons asymptotic to the same flat cylinder $dt^2 + g_{T^3}$, then one can cut off their cylindrical ends for $t \gg 1$ and glue the resulting manifolds with boundary to produce a sequence of approximately Ricci-flat metrics on the K3 surface that develop a very long neck. Alternatively, by rescaling the metrics in the sequence so that the diameter stays bounded, one produces in this way a sequence of approximately Ricci-flat metrics that collapse to a closed interval with curvature concentration at the two end points. Chen–Chen [13, Sect. 5] show that these approximate solutions can be perturbed into exact hyperkähler metrics with the same collapsing behaviour.

Hein's examples of gravitational instantons defined on the complement of a singular fibre of type I_0^*, II, III, IV in Kodaira's classification are also easily understood, in particular those examples that arise from isotrivial elliptic fibrations. Let E be a smooth elliptic curve admitting a $\mathbb{Z}_r$–subgroup of automorphisms for $r = 2, 3, 4$ or 6. Thus E is any elliptic curve if $r = 2$; a Weierstrass equation for E is $y^2 = x^3 + x$ if $r = 4$, with $\mathbb{Z}_4$–action generated by $(x, y) \mapsto (-x, iy)$; if $r = 3$ or 6 then $E\colon\ y^2 = x^3 + 1$ and the $\mathbb{Z}_3$ and $\mathbb{Z}_6$–actions are generated by $(x, y) \mapsto (e^{2\pi i/3}x, y)$ and $(x, y) \mapsto (e^{2\pi i/3}x, -y)$ respectively. Now consider the orbifold $(\mathbb{CP}^1 \times E)/\mathbb{Z}_r$, where the cyclic group $\mathbb{Z}_r$ acts diagonally on $\mathbb{CP}^1$ and E. Resolve the singularities and blow down all (-1)–curves in the fibres to obtain a rational elliptic surface with only two singular fibres over 0 and ∞ and such that all smooth fibres are isomorphic. Corresponding to $r = 2, 3, 4, 6$ this construction yields four pairs of singular fibres—(I_0^*, I_0^*), (II, II^*), (III, III^*) and (IV, IV^*) in Kodaira's notation. Unless $r = 2$, the two fibres in each pair are different because the $\mathbb{Z}_r$–action on $\mathbb{CP}^1$ has different weights at 0 and ∞. By removing the fibre of non–$*$–type in each pair, one obtains a crepant resolution of $T^*E/\mathbb{Z}_r$ and the resulting semi-flat metric coincides with the flat metric on $T^*E/\mathbb{Z}_r$. In fact, in this case some of Hein's Ricci-flat metrics can also be obtained from the Kummer-type construction of Biquard–Minerbe [8], gluing rescaled ALE spaces to resolve the singularities of the flat orbifold. When we remove the fibre of $*$–type in each pair, Hein's Ricci-flat metric is asymptotic to the twisted product of a flat metric on E and of a flat 2-dimensional cone which is not a quotient of $\mathbb{C}$ [29, Theorem 1.5 (ii)].

All these examples have faster than quadratic curvature decay and their asymptotic geometry is called ALG. The recent classification result of Chen–Chen [13, Theorem 1.4] states that all ALG gravitational instantons arise from (a slight improvement of) Hein's construction on the complement of a fibre of type $I_0^*, II, II^*, III, III^*, IV$ or IV^*. Furthermore, we note that constructions of sequences of Ricci-flat metrics on the K3 surface obtained by desingularising orbifolds $(E_1 \times E_2)/\mathbb{Z}_r$ for a product of $\mathbb{Z}_r$–invariant elliptic curves with $\mathrm{Vol}(E_2) \to 0$ could provide examples of collapsing sequences of hyperkähler metrics with ALG spaces of type I_0^*, II, III, IV as rescaled limits.

By removing a singular fibre with infinite monodromy, Hein is also able to produce examples with more exotic asymptotic geometry, often referred to as gravitational instantons of type ALG* and ALH*. The examples of type ALG* (ALH*) have quadratic volume growth (volume growth $r^{\frac{4}{3}}$) and are obtained by removing a fibre of Kodaira type I_b^*, $b = 1, \ldots, 4$, (I_b, $b = 1, \ldots, 9$) from a rational elliptic surface. These examples do not have faster than quadratic curvature decay and do not fit into Chen–Chen's classification.

The asymptotic geometry of the ALG* and ALH* examples can be constructed using the Gibbons–Hawking ansatz on (the $\mathbb{Z}_2$–quotient of) $\mathbb{R}^2 \times S^1$ and $\mathbb{R} \times T^2$, respectively, with a finite number of punctures. Since $\mathbb{R}^2 \times S^1$ and $\mathbb{R} \times T^2$ are parabolic, the sum of Green's functions used as the harmonic function in the Gibbons–Hawking construction is only positive at infinity and the construction provides only good asymptotic models. We expect that a gluing construction as in Theorem 3.8 using Atiyah–Hitchin spaces as building blocks together with the Gibbons–Hawking construction on $\mathbb{R}^2 \times S^1$ and $\mathbb{R} \times T^2$ will yield families of ALG* and ALH* gravitational instantons close to a collapsed limit $(\mathbb{R}^2 \times S^1)/\mathbb{Z}_2$ and $(\mathbb{R} \times T^2)/\mathbb{Z}_2$, respectively. We also expect that extensions of Theorem 3.8 where one considers sequences of flat metrics on T^3 collapsing to T^2 and S^1 should provide examples of collapsing Ricci-flat metrics with ALG* and ALH* gravitational instantons as rescaled limits. More generally, it is expected that ALG, ALH, ALG* and ALH* gravitational instantons will play an important role in understanding relations between collapsing sequences of Ricci-flat metrics on the K3 surface and degenerations of a compatible complex structure, cf. for example [33]. Very recently Hein–Sun–Viaclovsky–Zhang [28] gave a general construction of families of Ricci-flat metrics on the K3 surface that collapse to a closed bounded interval with curvature concentrating at a finite number of points (always including the two endpoints). The building blocks for this gluing construction are a pair ALH* metrics bubbling off at the endpoints of the interval and an incomplete "neck" joining the two obtained from the Gibbons–Hawking ansatz on $\mathbb{R} \times T^3$.

Acknowledgements The author is supported by a Royal Society University Research Fellowship. He would also like to thank NSF for the partial support of his work under grant DMS-1608143.

References

1. Anderson, M. T. (1989). Ricci curvature bounds and Einstein metrics on compact manifolds. *Journal of the American Mathematical Society*, *2*(3), 455–490.
2. Anderson, M. T. (1992). The L^2 structure of moduli spaces of Einstein metrics on 4-manifolds. *Geometric and Functional Analysis*, *2*(1), 29–89.
3. Anderson, M. T., & Cheeger, J. (1991). Diffeomorphism finiteness for manifolds with Ricci curvature and $L^{n/2}$- norm of curvature bounded. *Geometric and Functional Analysis*, *1*(3), 231–252.

4. Atiyah, M., & Hitchin, N. (1988). *The geometry and dynamics of magnetic monopoles*. M. B. Porter lectures. Princeton: Princeton University Press.
5. Auvray, H. (2018). From ALE to ALF gravitational instantons. *Compositio Mathematica*, *154*(6), 1159–1221.
6. Bando, S. (1990). Bubbling out of Einstein manifolds. *Tohoku Mathematical Journal (2)*, *42*(2), 205–216.
7. Bando, S., Kasue, A., & Nakajima, H. (1989). On a construction of coordinates at infinity on manifolds with fast curvature decay and maximal volume growth. *Inventiones Mathematicae*, *97*(2), 313–349.
8. Biquard, O., & Minerbe, V. (2011). A Kummer construction for gravitational instantons. *Communications in Mathematical Physics*, *308*(3), 773–794.
9. Cheeger, J., Fukaya, K., & Gromov, M. (1992). Nilpotent structures and invariant metrics on collapsed manifolds. *Journal of the American Mathematical Society*, *5*(2), 327–372.
10. Cheeger, J., & Tian, G. (2006). Curvature and injectivity radius estimates for Einstein 4-manifolds. *Journal of the American Mathematical Society*, *19*(2), 487–525.
11. Chen, G., & Chen, X. (2015). *Gravitational instantons with faster than quadratic curvature decay (I)*
12. Chen, G., & Chen, X. (2019). Gravitational instantons with faster than quadratic curvature decay (II). *Journal Für Die Reine und Angewandte Mathematik, 756*, 259–284. https://doi.org/10.1515/crelle-2017-0026
13. Chen, G., & Chen, X. (2016). *Gravitational instantons with faster than quadratic curvature decay (III)*.
14. Chen, G., Viaclovsky, J., Zhang, R. (2019). Collapsing Ricci-flat metrics on elliptic K3 surfaces. *Math.DG* arXiv:1910.11321
15. Cherkis, S. A., & Hitchin, N. J. (2005). Gravitational instantons of type D_k. *Communications in Mathematical Physics*, *260*(2), 299–317.
16. Cherkis, S. A., & Kapustin, A. (1999). Singular monopoles and gravitational instantons. *Communications in Mathematical Physics*, *203*(3), 713–728.
17. Conlon, R. J., & Hein, H.-J. (2014). *Asymptotically conical Calabi-Yau manifolds, III*. arXiv:1405.7140.
18. Donaldson, S. K. (2006). *Two-forms on four-manifolds and elliptic equations*. Inspired by S. S. Chern, Nankai Tracts in Mathematics (Vol. 11, pp. 153–172). Hackensack: World Scientific Publishing.
19. Donaldson, S. K. (2012). *Calabi-Yau metrics on Kummer surfaces as a model gluing problem*. Advances in geometric analysis. Advanced lectures in mathematics (ALM) (Vol. 21, pp. 109–118). Somerville: International Press.
20. Donaldson, S. K., & Kronheimer, P. B. (1990). *The geometry of four-manifolds*. Oxford Mathematical Monographs. New York: The Clarendon Press; Oxford University Press; Oxford Science Publications.
21. Eguchi, T., & Hanson, A. J. (1979). Self-dual solutions to Euclidean gravity. *Annals of Physics*, *120*(1), 82–106.
22. Foscolo, L. (2019). ALF gravitational instantons and collapsing Ricci-flat metrics on the K3 surface. *Journal of Differential Geometry, 112*(1), 79–120. https://doi.org/10.4310/jdg/1557281007.
23. Gibbons, G. W., & Pope, C. N. (1979). The positive action conjecture and asymptotically Euclidean metrics in quantum gravity. *Communications in Mathematical Physics*, *66*(3), 267–290.
24. Gibbons, G., & Hawking, S. (1978). Gravitational multi-instantons. *Physics Letters B*, *78*(4), 430–432.
25. Gross, M., Tosatti, V., & Zhang, Y. (2013). Collapsing of abelian fibered Calabi-Yau manifolds. *Duke Mathematical Journal*, *162*(3), 517–551.
26. Gross, M., Tosatti, V., & Zhang, Y. (2016). Gromov-Hausdorff collapsing of Calabi-Yau manifolds. *Communications in Analysis and Geometry*, *24*(1), 93–113.

27. Gross, M., & Wilson, P. M. H. (2000). Large complex structure limits of K3 surfaces. *Journal of Differential Geometry*, *55*(3), 475–546.
28. Hein, H.-J., Sun, S., Viaclovsky, J., & Zhang, R. (2018). *Nilpotent structures and collapsing Ricci-flat metrics on K3 surfaces*. arXiv:1807.09367.
29. Hein, H.-J. (2012). Gravitational instantons from rational elliptic surfaces. *Journal of the American Mathematical Society*, *25*(2), 355–393.
30. Hitchin, N. (1974). Compact four-dimensional Einstein manifolds. *Journal of Differential Geometry*, *9*, 435–441.
31. Hitchin, N. (1984). *Twistor construction of Einstein metrics*. Global Riemannian geometry (Durham, 1983). Ellis Horwood series in mathematics and its applications (pp. 115–125). Chichester: Horwood.
32. Joyce, D. D. (2000). *Compact manifolds with special holonomy*. Oxford mathematical monographs. Oxford: Oxford University Press.
33. Kobayashi, R. (1990). *Ricci-flat Kähler metrics on affine algebraic manifolds and degenerations of Kähler-Einstein K3 surfaces*. Kähler metric and moduli spaces. Advanced studies in pure mathematics (Vol. 18, pp. 137–228). Boston: Academic.
34. Kodaira, K. (1964). On the structure of compact complex analytic surfaces. *American Journal of Mathematics I*, *86*, 751–798.
35. Kronheimer, P. B. (1989). The construction of ALE spaces as hyper-Kähler quotients. *Journal of Differential Geometry*, *29*(3), 665–683.
36. Kronheimer, P. B. (1989). A Torelli-type theorem for gravitational instantons. *Journal of Differential Geometry*, *29*(3), 685–697.
37. LeBrun, C. (1991). *Complete Ricci-flat Kähler metrics on C^n need not be flat*. Several complex variables and complex geometry, Part 2 (Santa Cruz, CA, 989). Proceedings of Symposia in Pure Mathematics (Vol. 52, pp. 297–304). Providence: American Mathematical Society.
38. LeBrun, C., & Singer, M. (1994). A Kummer-type construction of self-dual 4-manifolds. *Mathematische Annalen*, *300*(1), 165–180.
39. Minerbe, V. (2009). A mass for ALF manifolds. *Communications in Mathematical Physic*, *289*(3), 925–955.
40. Minerbe, V. (2010). On the asymptotic geometry of gravitational instantons. *Annales Scientifiques de École Normale Supérieure (4)*, *43*(6), 883–924.
41. Minerbe, V. (2011). Rigidity for multi-Taub-NUT metrics. *Journal Für Die Reine und Angewandte*, *656*, 47–58.
42. Miranda, R. (1989). *The basic theory of elliptic surfaces*. Dottorato di Ricerca in Matematica. Pisa: ETS Editrice.
43. Nakajima, H. (1988). Hausdorff convergence of Einstein 4-manifolds. *Journal of the Faculty of Science, University of Tokyo. Section IA Mathematics*, *35*(2), 411–424.
44. Ooguri, H., & Vafa, C. (1996). Summing up Dirichlet instantons. *Physical Review Letters*, *77*(16), 3296–3298.
45. Page, D. N. (1981). A periodic but nonstationary gravitational instanton. *Physics Letters B*, *100*(4), 313–315.
46. Schoen, R., & Yau, S.-T. (1994). Lectures on differential geometry. In *Conference Proceedings and Lecture Notes in Geometry and Topology, I*. Cambridge: International Press.
47. Siu, Y. T. (1983). Every K3 surface is Kähler. *Inventiones Mathematicae*, *73*(1), 139–150.
48. Suvaina, I. (2012). ALE Ricci-flat Kähler metrics and deformations of quotient surface singularities. *Annals of Global Analysis and Geometry*, *41*(1), 109–123.
49. Tian, G., & Yau, S.-T. (1990). Complete Kähler manifolds with zero Ricci curvature. I. *Journal of the American Mathematical Society*, *3*(3), 579–609.
50. Tian, G., & Yau, S.-T. (1991). Complete Kähler manifolds with zero Ricci curvature. II. *Inventiones Mathematicae*, *106*(1), 27–60.
51. Topiwala, P. (1987). A new proof of the existence of Kähler-Einstein metrics on K3. I, II. *Inventiones Mathematicae*, *89*(2), 425–448, 449–454.
52. Yau, S. T. (1977). Calabi's conjecture and some new results in algebraic geometry. *Proceedings of the National Academy of Sciences of the United States of America*, *74*(5), 1798–1799.

Frölicher–Nijenhuis Bracket on Manifolds with Special Holonomy

Kotaro Kawai, Hông Vân Lê, and Lorenz Schwachhöfer

Abstract In this article, we summarize our recent results on the study of manifolds with special holonomy via the Frölicher–Nijenhuis bracket. This bracket enables us to define the Frölicher–Nijenhuis cohomologies which are analogues of the d^c and the Dolbeault cohomologies in Kähler geometry, and assigns an L_∞-algebra to each associative submanifold. We provide several concrete computations of the Frölicher–Nijenhuis cohomology.

2010 Mathematic Subject Classification Primary: 53C25 53C29 53C38, Secondary: 17B56 17B66

1 Introduction

The Frölicher–Nijenhuis bracket, which was introduced in [3, 4], defines a natural structure of a graded Lie algebra on the space of tangent bundle valued differential forms $\Omega^*(M, TM)$ on a smooth manifold M.

On a Riemannian manifold (M, g), if there is a parallel differential form of even degree, we can define canonical cohomologies which are analogues of the d^c and

The first named author is supported by JSPS KAKENHI Grant Numbers JP17K14181, and the research of the second named author was supported by the GAČR project 18-00496S and RVO:67985840. The third named author was supported by the Deutsche Forschungsgemeinschaft by grant SCHW 893/5-1.

K. Kawai (✉)
Gakushuin University, 1-5-1, Mejiro, Toshima, Tokyo 171-8588, Japan
e-mail: kkawai@math.gakushuin.ac.jp

H. Vân Lê
Institute of Mathematics, Czech Academy of Sciences, Zitna 25, 11567 Praha 1, Czech Republic
e-mail: hvle@math.cas.cz

L. Schwachhöfer
Faculty of Mathematics, TU Dortmund University, Vogelpothsweg 87, 44221 Dortmund, Germany
e-mail: lschwach@math.tu-dortmund.de

S. Karigiannis et al. (eds.), *Lectures and Surveys on G_2-Manifolds and Related Topics*, Fields Institute Communications 84, https://doi.org/10.1007/978-1-0716-0577-6_8

the Dolbeault cohomologies in Kähler geometry. See Sect. 3. We compute these cohomologies for G_2-manifolds in Theorem 4.1 and give a sketch of the proof in Sect. 4. A similar statement holds for Spin(7) and Calabi–Yau manifolds. See [7, 8].

In the second part of our note, using the Frölicher–Nijenhuis bracket, we assign to each associative submanifold an L_∞-algebra.

Notation: Let (V, g) be an n-dimensional oriented real vector space with a scalar product g. Define the map ∂ by contraction of a form with the metric g, i.e.

$$\partial = \partial_g : \Lambda^k V^* \longrightarrow \Lambda^{k-1} V^* \otimes V, \qquad \partial_g(\alpha^k) := (\iota_{e_i}\alpha^k) \otimes e^i, \tag{1.1}$$

where (e_i) is an orthonormal basis of V with the dual basis (e^i).

2 Preliminaries

2.1 Graded Lie Algebras and Differentials

We briefly recall some basic notions and properties of graded (Lie) algebras. Let $V := (\bigoplus_{k\in\mathbb{Z}} V_k, \cdot)$ be a graded real vector space with a graded bilinear map $\cdot : V \times V \to V$, called a *product on* V. A *graded derivation of* $(V, \cdot)$ *of degree* l is a linear map $D^l : V \to V$ of degree l (i.e., $D^l(V_k) \subset V_{k+l}$) such that

$$D^l(x \cdot y) = (D^l x) \cdot y + (-1)^{l|x|} x \cdot (D^l y), \tag{2.1}$$

where $|x|$ denotes the degree of an element, i.e. $|x| = k$ for $x \in V_k$. If we denote by $\mathcal{D}^l(V)$ the graded derivations of $(V, \cdot)$ of degree l, then $\mathcal{D}(V) := \bigoplus_{l\in\mathbb{Z}} \mathcal{D}^l(V)$ is a graded Lie algebra with the Lie bracket

$$[D_1, D_2] := D_1 D_2 - (-1)^{|D_1||D_2|} D_2 D_1, \tag{2.2}$$

i.e., the Lie bracket is graded anti-symmetric and satisfies the graded Jacobi identity,

$$[x, y] = -(-1)^{|x||y|}[y, x] \tag{2.3}$$

$$(-1)^{|x||z|}[x, [y, z]] + (-1)^{|y||x|}[y, [z, x]] + (-1)^{|z||y|}[z, [x, y]] = 0. \tag{2.4}$$

In general, if $L = (\bigoplus_{k\in\mathbb{Z}} L_k, [\cdot, \cdot])$ is a graded Lie algebra, then an *action of* L *on* V is a Lie algebra homomorphism $\pi : L \to \mathcal{D}(V)$, which yields a graded bilinear map $L \times V \to V$, $(x, v) \mapsto \pi(x)(v)$ such that the map $\pi(x) : V \to V$ is a graded derivation of degree $|x|$ and such that

$$[\pi(x), \pi(y)] = \pi[x, y].$$

For instance, a graded Lie algebra acts on itself via the adjoint representation $ad : L \to \mathcal{D}(L)$, where $ad_x(y) := [x, y]$.

For a graded Lie algebra L we define the set of *Maurer-Cartan elements of* L *of degree* $2k + 1$ as

$$\mathcal{MC}^{2k+1}(L) := \{\xi \in L_{2k+1} \mid [\xi, \xi] = 0\}.$$

If $\pi : L \to \mathcal{D}(V)$ is an action of L on $(V, \cdot)$, then for $\xi \in \mathcal{MC}^{2k+1}(L)$ we have $0 = [\pi(\xi), \pi(\xi)] = 2\pi(\xi)^2$, so that $\pi(\xi) : V \to V$ is a differential on V. We define the *cohomology of* $(V, \cdot)$ *w.r.t.* ξ as

$$H^i_\xi(V) := \frac{\ker(\pi(\xi) : V_i \to V_{i+2k+1})}{\mathrm{Im}\,(\pi(\xi) : V_{i-(2k+1)} \to V_i)} \qquad \text{for } \xi \in \mathcal{MC}^{2k+1}(L). \tag{2.5}$$

Since $\pi(\xi)$ is a derivation, it follows that $\ker \pi(\xi) \cdot \ker \pi(\xi) \subset \ker \pi(\xi)$, whence there is an induced product on $H^*_\xi(V) := \bigoplus_{i \in \mathbb{Z}} H^i_\xi(V)$.

If $L = \bigoplus_{k \in \mathbb{Z}} L_k$ is a graded Lie algebra, then for $v \in L_0$ and $t \in \mathbb{R}$, we define the formal power series

$$\exp(tv) : L \longrightarrow L[[t]], \qquad \exp(tv)(x) := \sum_{k=0}^{\infty} \frac{t^k}{k!} ad_v^k(x). \tag{2.6}$$

Observe that $ad_\xi(v) = 0$ for some $v \in L_0$ iff $ad_v(\xi) = 0$ iff $\exp(tv)(\xi) = \xi$ for all $t \in \mathbb{R}$. In this case, we call v an *infinitesimal stabilizer of* ξ.

For $\xi \in \mathcal{MC}^{2k+1}(L)$, we say that $x \in L_{2k+1}$ is an *infinitesimal deformation of* ξ *within* $\mathcal{MC}^{2k+1}(L)$ if $[\xi + tx, \xi + tx] = 0 \mod t^2$. Evidently, this is equivalent to $[\xi, x] = 0$ or $x \in \ker ad_\xi$. Such an infinitesimal deformation is called *trivial* if $x = [\xi, v]$ for some $v \in L_0$, since in this case, $\xi + tx = \exp(-tv)(\xi) \mod t^2$, whence up to second order, it coincides with elements in the orbit of ξ under the (formal) action of $\exp(tv)$. Thus, we have the following interpretation of some cohomology groups.

Proposition 2.1 *Let* $(L = \bigoplus_{i \in \mathbb{Z}} L_i, [\cdot, \cdot])$ *be a real graded Lie algebra, acting on itself by the adjoint representation, and let* $\xi \in \mathcal{MC}^{2k+1}(L)$*. Then the following holds.*

(1) If $L_{-(2k+1)} = 0$*, then* $H^0_\xi(L)$ *is the Lie algebra of infinitesimal stabilizers of* ξ*.*
(2) $H^{2k+1}_\xi(L)$ *is the space of infinitesimal deformations of* ξ *within* $\mathcal{MC}^{2k+1}(L)$ *modulo trivial deformations.*

2.2 The Frölicher–Nijenhuis Bracket

We shall apply our discussion from the preceding section to the following example. Let M be a manifold and $(\Omega^*(M), \wedge) = (\bigoplus_{k \geq 0} \Omega^k(M), \wedge)$ be the graded algebra of differential forms. Evidently, the exterior derivative $d : \Omega^k(M) \to \Omega^{k+1}(M)$ is a derivation of $\Omega^*(M)$ of degree 1, whereas insertion $\iota_X : \Omega^k(M) \to \Omega^{k-1}(M)$ of a vector field $X \in \mathfrak{X}(M)$ is a derivation of degree -1.

More generally, for $K \in \Omega^k(M, TM)$ we define $\iota_K \alpha^l$ as the *insertion of K into* $\alpha^l \in \Omega^l(M)$ pointwise by

$$\iota_{\kappa^k \otimes X} \alpha^l := \kappa^k \wedge (\iota_X \alpha^l) \in \Omega^{k+l-1}(M),$$

where $\kappa^k \in \Omega^k(M)$ and $X \in \mathfrak{X}(M)$, and this is a derivation of $\Omega^*(M)$ of degree $k-1$. Thus, the *Nijenhuis-Lie derivative along* $K \in \Omega^k(M, TM)$ defined as

$$\mathcal{L}_K(\alpha^l) := [\iota_K, d](\alpha^l) = \iota_K(d\alpha^l) + (-1)^k d(\iota_K \alpha^l) \in \Omega^{k+l}(M) \tag{2.7}$$

is a derivation of $\Omega^*(M)$ of degree k.

Observe that for $k = 0$ in which case $K \in \Omega^0(M, TM)$ is a vector field, both ι_K and $\mathcal{L}_K$ coincide with the standard notion of insertion of and Lie derivative along a vector field.

In [3, 4], it was shown that $\Omega^*(M, TM)$ carries a unique structure of a graded Lie algebra, defined by the so-called *Frölicher–Nijenhuis bracket*,

$$[\cdot, \cdot]^{FN} : \Omega^k(M, TM) \times \Omega^l(M, TM) \to \Omega^{k+l}(M, TM)$$

such that $\mathcal{L}$ defines an action of $\Omega^*(M, TM)$ on $\Omega^*(M)$, that is,

$$\mathcal{L}_{[K_1, K_2]^{FN}} = [\mathcal{L}_{K_1}, \mathcal{L}_{K_2}] = \mathcal{L}_{K_1} \circ \mathcal{L}_{K_2} - (-1)^{|K_1||K_2|} \mathcal{L}_{K_2} \circ \mathcal{L}_{K_1}. \tag{2.8}$$

It is given by the following formula for $\alpha^k \in \Omega^k(M)$, $\beta^l \in \Omega^l(M)$, $X_1, X_2 \in \mathfrak{X}(M)$ [9, Theorem 8.7 (6), p. 70]:

$$\begin{aligned} [\alpha^k \otimes X_1, \beta^l \otimes X_2]^{FN} &= \alpha^k \wedge \beta^l \otimes [X_1, X_2] \\ &+ \alpha^k \wedge (\mathcal{L}_{X_1} \beta^l) \otimes X_2 - (\mathcal{L}_{X_2} \alpha^k) \wedge \beta^l \otimes X_1 \\ &+ (-1)^k \left(d\alpha^k \wedge (\iota_{X_1} \beta^l) \otimes X_2 + (\iota_{X_2} \alpha^k) \wedge d\beta^l \otimes X_1 \right). \end{aligned} \tag{2.9}$$

In particular, for a vector field $X \in \mathfrak{X}(M)$ and $K \in \Omega^*(M, TM)$ we have [9, Theorem 8.16 (5), p. 75]

$$\mathcal{L}_X(K) = [X, K]^{FN}, \tag{2.10}$$

that is, the Frölicher–Nijenhuis bracket with a vector field coincides with the Lie derivative of the tensor field $K \in \Omega^*(M, TM)$. This means that $\exp(tX) : \Omega^*(M, TM) \to \Omega^*(M, TM)[[t]]$ is the action induced by (local) diffeomorphisms of M. Thus, Proposition 2.1 now immediately implies the following result.

Theorem 2.2 *Let* M *be a manifold and* $K \in \Omega^{2k+1}(M, TM)$ *be such that* $[K, K]^{FN} = 0$, *and define the differential* $d_K(K') := [K, K']^{FN}$. *Then*

(1) $H^0_K(\Omega^*(M, TM))$ *is the Lie algebra of vector fields stabilizing* K.
(2) $H^{2k+1}_K(\Omega^*(M, TM))$ *is the space of infinitesimal deformations of* K *within the differentials of* $\Omega^*(M, TM)$ *of the form* $ad_{\xi^{2k+1}}$, *modulo (local) diffeomorphisms.*

3 Frölicher–Nijenhuis Cohomology

Suppose that (M, g) is an n-dimensional Riemannian manifold with Levi-Civita connection ∇, and $\Psi \in \Omega^{2k}(M)$ is a parallel form of even degree. We now make the following simple but crucial observation. The proof is given by a straightforward computation in geodesic normal coordinates.

Proposition 3.1 *Let* $\hat{\Psi} := \partial_g \Psi \in \Omega^{2k-1}(M, TM)$ *with the contraction map* ∂_g *from (1.1). Then* $\hat{\Psi}$ *is a Maurer-Cartan element, i.e.,* $[\hat{\Psi}, \hat{\Psi}]^{FN} = 0$.

Thus, by the discussion in Sect. 2.1, the Lie derivative $\mathcal{L}_{\hat{\Psi}}$ and the adjoint map $ad_{\hat{\Psi}}$ are differentials on $\Omega^*(M)$ and $\Omega^*(M, TM)$, respectively, and for simplicity, we shall denote these by

$$\mathcal{L}_\Psi : \Omega^*(M) \longrightarrow \Omega^*(M), \qquad ad_\Psi : \Omega^*(M, TM) \longrightarrow \Omega^*(M, TM),$$

or, if we wish to specify the degree,

$$\mathcal{L}_{\Psi;l} : \Omega^{l-2k+1}(M) \longrightarrow \Omega^l(M), \qquad ad_{\Psi;l} : \Omega^{l-2k+1}(M, TM) \longrightarrow \Omega^l(M, TM).$$

The cohomology algebras we denote by $H^*_\Psi(M)$ and $H^*_\Psi(TM)$ instead of $H^*_{\hat{\Psi}}(\Omega^*(M))$ and $H^*_{\hat{\Psi}}(\Omega^*(M, TM))$, respectively. That is, the i-th cohomologies are defined as

$$\begin{aligned} H^i_\Psi(M) &:= \frac{\ker \mathcal{L}_\Psi : \Omega^i(M) \to \Omega^{i+2k-1}(M)}{\operatorname{Im} \mathcal{L}_\Psi : \Omega^{i-2k+1}(M) \to \Omega^i(M)}, \\ H^i_\Psi(M, TM) &:= \frac{\ker ad_\Psi : \Omega^i(M, TM) \to \Omega^{i+2k-1}(M, TM)}{\operatorname{Im} ad_\Psi : \Omega^{i-2k+1}(M, TM) \to \Omega^i(M, TM)}. \end{aligned} \tag{3.1}$$

Example 3.2 In the case of a Kähler manifold, using the Kähler form $\Psi = \omega$, the differential $\mathcal{L}_\omega$ on $\Omega^*(M)$ is the complex differential $d^c := i(\bar{\partial} - \partial)$, whereas on $\Omega^*(M, TM)$, ad_ω coincides with the Dolbeault differential $\bar{\partial} : \Omega^{p,q}(M, TM) \to \Omega^{p,q+1}(M, TM)$ [4]. Thus, these differentials recover well known and natural cohomology theories. In particular, the cohomology algebras $H^*_\omega(M)$ and $H^*_\omega(M, TM)$ are finite dimensional if M is closed.

Now we give some general strategies to compute $H^*_\Psi(M)$. First we summarize formulas of $\mathcal{L}_\Psi$.

Lemma 3.3 ([7, Sect. 2.4])

$$\mathcal{L}_\Psi d\alpha^l = -d\mathcal{L}_\Psi\alpha^l, \quad \mathcal{L}_\Psi d^*\alpha^l = -d^*\mathcal{L}_\Psi\alpha^l \quad \text{and thus} \quad \mathcal{L}_\Psi\Delta\alpha^l = \Delta\mathcal{L}_\Psi\alpha^l.$$

$$\Delta(\alpha \wedge \Psi) = (\Delta\alpha) \wedge \Psi, \qquad \Delta(\alpha \wedge *\Psi) = (\Delta\alpha) \wedge *\Psi,$$

where Δ *is the Hodge Laplacian. As in the case of* d^*, *the formal adjoint* $\mathcal{L}^*_{\Psi;l} : \Omega^l(M) \to \Omega^{l-2k+1}(M)$ *of* $\mathcal{L}_{\Psi;l} : \Omega^{l-2k+1}(M) \to \Omega^l(M)$ *is given by*

$$\mathcal{L}^*_{\Psi;l}\alpha^l = (-1)^{n(n-l)+1} * \mathcal{L}_\Psi * \alpha^l. \tag{3.2}$$

Recall that for a closed oriented Riemannian manifold (M, g) there is the *Hodge decomposition* of differential forms

$$\Omega^l(M) = \mathcal{H}^l(M) \oplus d\Omega^{l-1}(M) \oplus d^*\Omega^{l+1}(M), \tag{3.3}$$

where $\mathcal{H}^l(M) \subset \Omega^l(M)$ denotes the space of harmonic forms.

We define the *space of* $\mathcal{L}_\Psi$*-harmonic forms* as

$$\begin{aligned} \mathcal{H}^l_\Psi(M) :=& \ \{\alpha \in \Omega^l(M) \mid \mathcal{L}_\Psi\alpha = \mathcal{L}^*_\Psi\alpha = 0\} \\ \overset{(3.2)}{=}& \ \{\alpha \in \Omega^l(M) \mid \mathcal{L}_\Psi\alpha = \mathcal{L}_\Psi * \alpha = 0\} \end{aligned} \tag{3.4}$$

Evidently, the Hodge-$*$ yields an isomorphism

$$* : \mathcal{H}^l_\Psi(M) \longrightarrow \mathcal{H}^{n-l}_\Psi(M). \tag{3.5}$$

Since $\mathcal{H}^l_\Psi(M) \subset \ker \mathcal{L}_{\Psi;l+2k-1}$ and $\mathcal{H}^l_\Psi(M) \cap \mathrm{Im}\,(\mathcal{L}_{\Psi;l}) = 0$, there is a canonical injection

$$\iota_l : \mathcal{H}^l_\Psi(M) \hookrightarrow H^l_\Psi(M). \tag{3.6}$$

This is analogous to the inclusion of harmonic forms into the de Rham cohomology of a manifold, which for a closed manifold is an isomorphism due to the Hodge decomposition (3.3). Therefore, one may hope that the maps ι_l are isomorphisms as well. It is not clear if this is always true, but we shall give conditions which assure this to be the case and show that in the applications we have in mind, this condition is satisfied.

Definition 3.4 We say that the differential $\mathcal{L}_\Psi$ is *l-regular* for $l \in \mathbb{N}$ if there is a direct sum decomposition

$$\Omega^l(M) = \ker(\mathcal{L}^*_{\Psi;l}) \oplus \mathrm{Im}\,(\mathcal{L}_{\Psi;l}). \tag{3.7}$$

A standard result from elliptic theory states that $\mathcal{L}_\Psi$ is l-regular if the differential operator $\mathcal{L}_{\Psi;l} : \Omega^{l-2k+1}(M) \to \Omega^l(M)$ is elliptic, overdetermined elliptic or underdetermined elliptic, see e.g. [1, p. 464, 32 Corollary].

The following theorem now relates the cohomology $H^*_\Psi(M)$ to the $\mathcal{L}_\Psi$-harmonic forms $\mathcal{H}^*_\Psi(M)$.

Theorem 3.5 ([7, Theorem 2.7])

(1) If $\mathcal{L}_\Psi$ is l-regular, then the map ι_l from (3.6) is an isomorphism.
(2) There are direct sum decompositions

$$H^l_\Psi(M) = \mathcal{H}^l(M) \oplus H^l_\Psi(M)_d \oplus H^l_\Psi(M)_{d^*} \tag{3.8}$$
$$\mathcal{H}^l_\Psi(M) = \mathcal{H}^l(M) \oplus \mathcal{H}^l_\Psi(M)_d \oplus \mathcal{H}^l_\Psi(M)_{d^*}, \tag{3.9}$$

where $\mathcal{H}^l(M)$ is the space of harmonic l-forms on M, $H^l_\Psi(M)_d$ and $H^l_\Psi(M)_{d^*}$ are the cohomologies of $(d\Omega^*(M), \mathcal{L}_\Psi)$ and $(d^*\Omega^*(M), \mathcal{L}_\Psi)$, respectively, and where $\mathcal{H}^l_\Psi(M)_d := \mathcal{H}^l_\Psi(M) \cap d\Omega^{l-1}(M)$, $\mathcal{H}^l_\Psi(M)_{d^*} := \mathcal{H}^l_\Psi(M) \cap d^*\Omega^{l+1}(M)$. Moreover, the injective map ι_l from (3.6) preserves this decomposition, i.e.,

$$\iota_l : \mathcal{H}^l_\Psi(M)_d \hookrightarrow H^l_\Psi(M)_d \quad \text{and} \quad \iota_l : \mathcal{H}^l_\Psi(M)_{d^*} \hookrightarrow H^l_\Psi(M)_{d^*}.$$

(3) There are isomorphisms

$$d : H^l_\Psi(M)_{d^*} \to H^{l+1}_\Psi(M)_d \quad \text{and} \quad d^* : H^l_\Psi(M)_d \to H^{l-1}_\Psi(M)_{d^*} \tag{3.10}$$
$$d : \mathcal{H}^l_\Psi(M)_{d^*} \to \mathcal{H}^{l+1}_\Psi(M)_d \quad \text{and} \quad d^* : \mathcal{H}^l_\Psi(M)_d \to \mathcal{H}^{l-1}_\Psi(M)_{d^*} \tag{3.11}$$

(4) If $\mathcal{L}_\Psi$ is $(l+1)$-regular and $(l-1)$-regular, then it is also l-regular.

Next, we consider another important case. We call a form $\Psi \in \Omega^k(M)$ *multi-symplectic*, if $d\Psi = 0$ and for all $v \in TM$

$$\iota_v \Psi = 0 \Longleftrightarrow v = 0. \tag{3.12}$$

Lemma 3.6 *If $\Psi \in \Omega^{2k}(M)$ is multi-symplectic, then the differential operator $\mathcal{L}_{\Psi;l} : \Omega^{l-2k+1}(M) \to \Omega^l(M)$ is overdetermined elliptic for $l = 2k-1$ and underdetermined elliptic for $l = n$.*

We also can make some statement for $H^l_\Psi(M)$ for special values of l.

Proposition 3.7 *For a parallel form $\Psi \in \Omega^{2k}(M)$, we have*

$$H^0_\Psi(\Omega^*(M)) \cong \mathcal{H}^0_\Psi(M) = \{f \in C^\infty(M) \mid \iota_{df^\#}\Psi = 0\},$$

If Ψ is multi-symplectic, then $\mathcal{H}^0_\Psi(M) = \mathcal{H}^0(M)$ and $H^n_\Psi(\Omega^*(M)) \cong \mathcal{H}^n_\Psi(M) = \mathcal{H}^n(M)$.

Indeed, it can be shown that $H^0_\Psi(\Omega^*(M))$ is infinite dimensional if Ψ is not multi-symplectic.

Proposition 3.8 *Let $\Psi \in \Omega^{2k}(M)$ be a parallel multi-symplectic form. Then*

$$H_{\Psi}^{2k-1}(\Omega^*(M)) = \mathcal{H}_{\Psi}^{2k-1}(M),$$
$$\ker(\mathcal{L}_{\Psi;2k}) = \{\alpha \in \Omega^1(M) \mid \mathcal{L}_{\alpha^\#}(*\Psi) = 0 \text{ and } d^*\alpha = 0\}.$$

In particular, if $k \geq 2$ then $\ker \mathcal{L}_{\Psi;2k} = \mathcal{H}_{\Psi}^1(M) \cong H_{\Psi}^1(M)$ *and*

$$\mathcal{H}_{\Psi}^{n-1}(M) = \{\alpha \in \Omega^{n-1}(M) \mid \mathcal{L}_{(*\alpha)^\#}(*\Psi) = 0 \text{ and } d\alpha = 0\}.$$

The first statement is an immediate consequence of Theorem 3.5 and Lemma 3.6. The second and the third statements follow from a direct computation and an integration by parts argument.

4 The Frölicher–Nijenhuis Cohomology of Manifolds with Special Holonomy

On a G_2-manifold, there is a canonical parallel 4-form $*\varphi$, the Hodge dual of the G_2-structure φ. We may consider the differentials $\mathcal{L}_{*\varphi}$ and $ad_{*\varphi}$. On closed manifolds, we obtain the following results on their cohomology groups.

Theorem 4.1 *Let (M^7, φ) be a closed G_2-manifold. Then for the cohomologies $H_{*\varphi}^i(M^7)$ and $H_{*\varphi}^i(M^7, TM^7)$ defined above, the following hold.*

(1) There is a Hodge decomposition

$$H_{*\varphi}^i(M^7) = \mathcal{H}^i(M^7) \oplus (H_{*\varphi}^i(M^7) \cap d\Omega^{i-1}(M^7))$$
$$\oplus (H_{*\varphi}^i(M^7) \cap d^*\Omega^{i+1}(M^7)),$$

where $\mathcal{H}^i(M^7)$ denotes the spaces of harmonic forms.

(2) The Hodge-$*$ induces an isomorphism $* : H_{*\varphi}^i(M^7) \to H_{*\varphi}^{7-i}(M^7)$.

(3) $H_{*\varphi}^i(M^7) = \mathcal{H}^i(M^7)$ for $i = 0, 1, 6, 7$. For $i = 2, 3, 4, 5$, $H_{*\varphi}^i(M^7)$ is infinite dimensional.

(4) $\dim H_{*\varphi}^0(M^7, TM^7) = b^1(M^7)$; in particular, $H_{*\varphi}^0(M^7) = 0$ if M^7 has full holonomy G_2.

(5) $\dim H_{*\varphi}^3(M^7, TM^7) \geq b^3(M^7) > 0$, as it contains all torsion free deformations of the G_2-structure modulo deformations by diffeomorphisms.

In [7, Theorem 3.5], we give a precise description of the cohomology ring $H_{*\varphi}^*(M^7)$

Remark 4.2 On a Spin(7)-manifold, there is also a canonical parallel 4-form and we obtain the similar results. For more details, see [7, Theorem 4.2].

Recently, we also computed $H^i_\Phi(M)$ for the real part of a holomorphic volume form Ψ in $4n$-dimensional Calabi–Yau manifolds in [8]. When $n = 1$, it is isomorphic to the de Rham cohomology. When $n \geq 2$, as in the G_2 and Spin(7)-case, it is regular again, and all summands involved other than the harmonic forms are infinite dimensional.

Outline of the proof of Theorem 4.1 We begin by showing the l-regularity of $\mathcal{L}_{*\varphi}$. For $l < 3$ and $l > 7$, this is obvious as then $\mathcal{L}_{*\varphi;l} = 0$. By Lemma 3.6, $\mathcal{L}_{*\varphi,l}$ is overdetermined elliptic for $l = 3$ and underdetermined elliptic for $l = 7$, whence $\mathcal{L}_{*\varphi}$ is also 3- and 7-regular.

By a simple calculation, it follows that $\mathcal{L}_{*\varphi,l}$ is overdetermined elliptic for $l = 4$ and underdetermined elliptic for $l = 6$, whence $\mathcal{L}_{*\varphi}$ is 4-regular and 6-regular. Thus it is also 5-regular by Theorem 3.5(4). Therefore, the l-regularity of $\mathcal{L}_{*\varphi,l}$ for all l is established, whence by Theorem 3.5(1), $H^l_{*\varphi}(M) = \mathcal{H}^l_{*\varphi}(M)$.

For $l = 0, 7$, $\mathcal{H}^l_{*\varphi}(M) \cong \mathcal{H}^l(M)$ by Proposition 3.7.

For $l = 1$, $H^1_{*\varphi}(M) = \ker \mathcal{L}_{*\varphi}|_{\Omega^1(M)}$. Thus, by Proposition 3.8, $\alpha \in H^1_{*\varphi}(M)$ implies that $\mathcal{L}_{\alpha^\#}(\varphi) = 0$, which in turn implies that $\alpha^\#$ is a Killing vector field. Since a G_2-manifold is Ricci flat, it follows by Bochner's theorem that $\alpha^\#$ is parallel, whence so is α. In particular, α is harmonic, showing that $H^1_{*\varphi}(M) = \mathcal{H}^1(M)$. For $l = 6$, we have $H^6_{*\varphi}(M) = *H^1_{*\varphi}(M) = *\mathcal{H}^1(M) = \mathcal{H}^6(M)$. This shows that $\mathcal{H}^l_{*\varphi}(M) \cong \mathcal{H}^l(M)$ for $l = 1, 6$.

Next, for $l = 2$, we have $\mathcal{H}^2_{*\varphi}(M)_d = 0$ by (3.10). Thus, we need to determine

$$\mathcal{H}^2_{*\varphi}(M)_{d^*} = \{\alpha^2 \in d^*\Omega^3(M) \mid d^*(\alpha^2 \wedge *\varphi) = 0\}.$$

We can investigate this space in detail by the irreducible decomposition of $\Omega^*(M)$ under the G_2-action and the Hodge decomposition. Then we can prove that $\mathcal{H}^2_{*\varphi}(M)_{d^*}$ is isomorphic to an infinite dimensional function space. We can prove the case of $l = 3$ similarly.

Again, since $* : \mathcal{H}^l_{*\varphi}(M) \to \mathcal{H}^{7-l}_{*\varphi}(M)$ is an isomorphism, the assertions for $l = 4, 5$ follow.

Next, we consider $H^*_{*\varphi}(M^7, TM^7)$. First, note the following.

Lemma 4.3 *Let V be an oriented 7-dimensional vector space, and let $\Lambda^3_{G_2}V^* \subset \Lambda^3 V^*$ be the set of G_2-structures on V. By definition, the group $GL^+(V)$ of orientation preserving automorphisms of V act transitively on $\Lambda^3_{G_2}V^*$ so that $\Lambda^3_{G_2}V^* = GL^+(V)/G_2$. Then the map*

$$\mathfrak{C} : \Lambda^3_{G_2}V^* \longrightarrow \Lambda^3 V^* \otimes V, \qquad \varphi \longmapsto \partial_{g_\varphi}(*_{g_\varphi}\varphi)$$

is a $GL^+(V)$-equivariant injective immersion. Here, ∂_g is the map from (1.1), and g_φ denotes the metric induced by φ.

Proposition 4.4

$$H^0_{*\varphi}(M^7, TM^7) = \{X \in \mathfrak{X}(M^7) \mid \mathcal{L}_X\varphi = 0\} = \{X \in \mathfrak{X}(M^7) \mid \nabla X = 0\}.$$

This proposition implies the 4th part of Theorem 4.1.

Proof Let $X \in \mathfrak{X}(M^7)$ be a vector field, $p \in M^7$ and denote by F^t_X the local flow along X, defined in a neighborhood of p. Then because of the pointwise equivariance of $\mathfrak{C}$ we have

$$(F^t_X)^*\Big(\partial_g * \varphi\Big)_{F^t_X(p)} = (F^t_X)^*\Big(\mathfrak{C}(\varphi)_{F^t_X(p)}\Big) = \mathfrak{C}\Big((F^t_X)^*(\varphi_{F^t_X(p)})\Big)$$

and taking the derivative at $t = 0$ yields

$$\mathcal{L}_X(\partial_g * \varphi)_p = \mathcal{L}_X(\mathfrak{C}(\varphi))_p = \mathfrak{C}_*(\mathcal{L}_X\varphi)_p. \tag{4.1}$$

Now $\mathcal{L}_X(\partial_g * \varphi) = [X, \partial_g * \varphi]^{FN}$, and since $\mathfrak{C}$ is an immersion by Lemma 4.3, it follows that $X \in H^0_{*\varphi}(M^7, TM^7) = \ker ad_{\partial_g * \varphi}$ iff $\mathcal{L}_X\varphi = 0$.

Since φ uniquely determines the Riemannian metric g_φ on M^7, any vector field satisfying $\mathcal{L}_X\varphi = 0$ must be a Killing vector field. Since M^7 is closed, the Ricci flatness of G_2-manifolds and Bochner's theorem imply that X must be parallel, showing that in this case, $\dim H^0_\Phi(M^7, TM^7) = b^1(M^7)$. □

It was shown in [6, Theorem 1.1] that a G_2-structure φ' is torsion-free if and only if $[\chi_{\varphi'}, \chi_{\varphi'}]^{FN} = 0$, where $\chi_{\varphi'} := \mathfrak{C}(\varphi') = \partial_{g_{\varphi'}} *_{g_{\varphi'}} \varphi' \in \Omega^3(M^7, TM^7)$. Therefore, for a family of torsion-free G_2-structures $\{\varphi_t\}$ with $\varphi_0 = \varphi$, we have

$$0 = \frac{d}{dt}\bigg|_{t=0} [\chi_{\varphi_t}, \chi_{\varphi_t}] = 2\left[\chi_{\varphi_0}, \frac{d}{dt}\bigg|_{t=0} \chi_{\varphi_t}\right] = 2\left[\chi_{\varphi_0}, \mathfrak{C}_*(\dot{\varphi}_0)\right],$$

so that $\dot{\varphi}_0 \in \Omega^3(M^7)$ is a torsion free infinitesimal deformation of φ_0 iff $\mathfrak{C}_*(\dot{\varphi}_0) \in \ker(ad_{\chi_{\varphi_0}} : \Omega^3(M^7, TM^7) \to \Omega^6(M^7, TM^7))$. Since $\mathfrak{C}$ is an immersion and hence $\mathfrak{C}_*$ injective by Lemma 4.3, we have an isomorphism

$$\{\text{torsion free infinitesimal deformations of } \varphi_0\} \overset{\mathfrak{C}_*}{\cong} \ker\Big(ad_{\chi_{\varphi_0}} : \Omega^3(M^7, TM^7) \to \Omega^6(M^7, TM^7)\Big) \cap \mathrm{Im}\,(\mathfrak{C}_*).$$

Observe that by (4.1)

$$\mathfrak{C}_*(\mathcal{L}_X\varphi_0) = \mathcal{L}_X(\mathfrak{C}(\varphi_0)) = [X, \chi_{\varphi_0}]^{FN} = -ad_{\chi_{\varphi_0}}(X),$$

whence there is an induced inclusion

$$\frac{\{\text{torsion free infinitesimal deformations of } \varphi_0\}}{\{\text{trivial deformations of } \varphi_0\}} \xrightarrow{\mathfrak{C}_*} H^3_{\varphi_0}(M^7, TM^7).$$

This implies the 5th part of Theorem 4.1. □

5 Strongly Homotopy Lie Algebra Associated with Associative Submanifolds

In this section we assign to each associative submanifold in a G_2-manifold an L_∞-algebra, using the Frölicher–Nijenhuis bracket and Voronov's derived bracket construction of L_∞-algebras. The main purpose of this section is to explain the motivation that led us to study the Frölicher–Nijenhuis bracket on G_2-manifolds. We refer the reader to [2] for detailed and general treatment of the theory discussed here.

5.1 *Voronov's Construction of L_∞-Algebras*

Strongly homotopy Lie algebras, also called L_∞-algebras, were defined by Lada and Stasheff in [10], see also [14] for a historical account. In [14] Voronov suggested a relatively simple method to construct an L_∞-algebra based on algebraic data, now called V-data. A *V-data* is a quadruple $(L, P, \mathfrak{a}, \Delta)$, where

(1) L is a $\mathbb{Z}_2$-graded Lie algebra $L = L_0 \oplus L_1$ (we denote its bracket by $[.,.]$),
(2) $\mathfrak{a}$ is an abelian Lie subalgebra of L,
(3) $P : L \to \mathfrak{a}$ is a projection whose kernel is a Lie subalgebra of L,
(4) $\Delta \in (\ker P) \cap L_1$ is an element such that $[\Delta, \Delta] = 0$.

When Δ is an arbitrary element of L_1 instead of $\ker(P) \cap L_1$, we refer to $(L, \mathfrak{a}, P, \Delta)$ as a *curved V-data.*

Recall that a (k, l)-shuffle is a permutation of indices $1, 2, \ldots, k + l$ such that $\sigma(1) < \cdots < \sigma(k)$ and $\sigma(k + 1) < \cdots < \sigma(k + l)$.

Definition 5.1 ([14, Definition 1]) A vector space $V = V_0 \oplus V_1$ endowed with a sequence of odd n-linear operations $\mathfrak{m}_n$, $n = 0, 1, 2, 3, \ldots$, is a *strongly homotopy Lie algebra or L_∞-algebra* if: (a) all operations are symmetric in the $\mathbb{Z}_2$-graded sense:

$$\mathfrak{m}_n(a_1, \ldots, a_i, a_{i+1}, \ldots, a_n) = (-1)^{\tilde{a}_i \tilde{a}_{i+1}} \mathfrak{m}_n(a_1, \ldots, a_{i+1}, a_i, \ldots, a_n),$$

and (b) the "generalized Jacobi identities"

$$\sum_{k+l=n} \sum_{(k,l)-\text{shuffles}} (-1)^\alpha \mathfrak{m}_{l+1}(\mathfrak{m}_k(a_{\sigma(1)}, \ldots, a_{\sigma(k)}), a_{\sigma(k+1)}, \ldots, a_{\sigma(k+l)}) = 0$$

hold for all $n = 0, 1, 2, \ldots$. Here $\bar{a}$ is the degree of $a \in V$ and $(-1)^{\alpha}$ is the sign prescribed by the sign rule for a permutation of homogeneous elements $a_1, \ldots, a_n \in V$.

Henceforth symmetric will mean $\mathbb{Z}_2$-graded symmetric.

A 0-ary bracket is just a distinguished element Φ in V. We call the L_∞-algebras with $\Phi = 0$ *strict*. In this case $\mathfrak{m}_1^2 = 0$ and we also write d instead of $\mathfrak{m}_1$. For strict L_∞-algebras, the first three "generalized Jacobi identities" simplify to

$$d^2 a = 0,$$

$$d\mathfrak{m}_2(a, b) + \mathfrak{m}_2(da, b) + (-1)^{\bar{a}\bar{b}}\mathfrak{m}_2(db, a) = 0,$$

$$d\mathfrak{m}_3(a, b, c) + \mathfrak{m}_2(\mathfrak{m}_2(a, b), c) + (-1)^{\bar{b}\bar{c}}\mathfrak{m}_2(\mathfrak{m}_2(a, c), b) + (-1)^{\bar{a}(\bar{b}+\bar{c})}\mathfrak{m}_2(\mathfrak{m}_2(b, c), a)$$

$$+\mathfrak{m}_3(da, b, c) + (-1)^{\bar{a}\bar{b}}\mathfrak{m}_3(db, a, c) + (-1)^{(\bar{a}+\bar{b})\bar{c}}\mathfrak{m}_3(dc, a, b) = 0.$$

Proposition 5.2 ([14, Theorem 1, Corollary 1]) *Let $(L, \mathfrak{a}, P, \triangle)$ be a curved V-data. Then $\mathfrak{a}$ is a curved L_∞-algebra for the multibrackets*

$$\mathfrak{m}_n(a_1, \ldots, a_n) = P[\cdots[[\triangle, a_1], a_2], \ldots, a_n].$$

We obtain a strict L_∞-algebra exactly when $\triangle \in \ker(P)$.

Remark 5.3 Usually in the literature a strict L_∞-algebra is called *an L_∞-algebra,* which we also adopt in this paper.

5.2 L_∞-algebra Associated with an Associative Submanifold $L \subset M^7$

Let L be a closed submanifold in a Riemannian M. There exists an open neighborhood $N_\varepsilon L$ of the zero section L in the normal bundle NL such that the exponential mapping $Exp : N_\varepsilon L \to M$ is a local diffeomorphism.

Given such an open neighborhood $N_\varepsilon L$, we consider the pullback operator $Exp^* : \Omega^*(M, TM) \to \Omega^*(N_\varepsilon L, TNL)$

$$Exp^*(K)_x(X_1, \ldots, X_k) = (Exp_*)_x^{-1}(K_{Exp(x)}((Exp_*)_x(X_1), \ldots, (Exp_*)_x(X_k))$$

for $x \in N_\varepsilon L$, $X_i \in T_x N_\varepsilon L$ and $K \in \Omega^k(M, TM)$ given by [9, 8.16, p. 74]. It is known that [9, Theorem 8.16 (2), p. 74]

$$[Exp^*(K), Exp^*(L)]^{FN} = Exp^*([K, L]^{FN}). \tag{5.1}$$

Let π denote the projection $NL \to L$. For each section $X \in \Gamma(NL)$, define the vector field $\hat{X}$ on $N_\varepsilon L \subset NL$ by the restriction of the vertical lift of X to $N_\varepsilon L$. That is,

$$\hat{X}(y) = \frac{d}{dt}(y + tX(\pi(y)))\Big|_{t=0} \qquad \text{for } y \in N_\varepsilon L.$$

Let $\Omega^*(N_\varepsilon L, TNL)$ be the space of all smooth TNL-valued forms on $N_\varepsilon L$. We define a linear embedding $I : \Omega^*(L, NL) \to \Omega^*(N_\varepsilon L, TNL)$ by

$$I(\phi \otimes X) := \pi^*(\phi) \otimes \hat{X}$$

and extend it linearly on the whole $\Omega^*(L, NL)$.

Let P denote the composition of the restriction $r : \Omega^*(N_\varepsilon L, TNL) \to \Omega^*(L, TNL)$ and the projection $Pr^N : \Omega^*(L, TNL) \to \Omega^*(L, NL)$ defined via the decomposition $TNL_{|L} = NL \oplus TL$. Set

$$\tilde{P} := I \circ P : \Omega^*(N_\varepsilon L, TNL) \to \Omega^*(N_\varepsilon L, TNL).$$

Lemma 5.4 *The image of the map $\tilde{P}$ is an abelian subalgebra of the $\mathbb{Z}_2$-graded Lie algebra $(\Omega^*(N_\varepsilon L, TNL), [.,.]^{FN})$. The space* ker $\tilde{P}$ *is closed under the Frölicher–Nijenhuis bracket.*

Proof Note that the image of $\tilde{P}$ is equal to the image of I. To prove the first assertion of Lemma 5.4 it suffices to prove that

$$[\pi^*(\alpha_1) \otimes \hat{X}_1, \pi^*(\alpha_2) \otimes \hat{X}_2]^{FN} = 0 \tag{5.2}$$

for any $X, Y \in \Gamma(NL)$ and any $\alpha_1, \alpha_2 \in \Omega^*(L)$.

Using (2.9) and taking into account the following identities

$$[\hat{X}_1, \hat{X}_2]^{FN} = 0,$$

$$i_{\hat{X}_1}\pi^*(\alpha_2) = 0 = i_{\hat{X}_2}\pi^*(\alpha_1),$$

$$d\pi^*(\alpha_1) = \pi^*(d\alpha_1),$$

for any $\alpha_1, \alpha_2 \in \Omega^*(L)$ and any $X_1, X_2 \in \mathfrak{X}(L)$ we obtain (5.2) immediately.

Now let us prove the second assertion of Lemma 5.4. Since I is injective, we have $\ker \tilde{P} = \ker P$. Moreover, $\ker P$ is generated by the TNL-valued differential forms $\alpha \otimes X$ such that $X(x) \in T_x L$ for all $x \in L$. Assume that $\alpha_1 \otimes X_1, \alpha_2 \otimes X_2 \in \ker P$. Using (2.9) and the fact that if $X_1, X_2 \in \mathfrak{X}(N_\varepsilon L)$ and $(X_1)_{|L} \in \mathfrak{X}(L)$, $(X_2)_{|L} \in \mathfrak{X}(L)$ then

$$[X_1, X_2]_{|L} \in \mathfrak{X}(L),$$

we obtain immediately

$$[\alpha_1 \otimes X_1, \alpha_2 \otimes X_2] \in \ker P = \ker \tilde{P}.$$

This completes the proof of Lemma 5.4. □

Theorem 5.5 *Assume that L is an associative submanifold of a G_2-manifold (M^7, φ). There is an L_∞-algebra structure on the space $\Omega^*(L, NL)$ given by the following family of graded multi-linear maps*

$$\mathfrak{m}_k : \Omega^*(L, NL)^{\otimes k} \to \Omega^*(L, NL)$$

$$\mathfrak{m}_k(\omega_1, \ldots, \omega_k) = P[\cdots[[Exp^*(\chi), I(\omega_1)]^{FN}, I(\omega_2)]^{FN} \ldots, I(\omega_k)]^{FN},$$

where $\chi = \partial_g * \varphi \in \Omega^3(M, TM)$ with ∂_g from (1.1).

Proof By Proposition 3.1 and using (5.1), we have

$$[Exp^*(\chi), Exp^*(\chi)]^{FN} = 0. \tag{5.3}$$

□

Lemma 5.6 *A submanifold L in a G_2-manifold (M^7, φ) is associative iff*

$$Exp^*(\chi) \in \ker \tilde{P}.$$

Proof It is known that a 3-submanifold L in a G_2-manifold is associative iff $\chi_{|L} = 0$ [5]. Since $\chi(x \wedge y \wedge z)$ is orthogonal to each x, y, z, it follows that L is associative, iff $Pr^N(\chi_{|L}) = 0 \in \Omega^*(L, NL)$, where Pr^N is the orthogonal projection from $TM|_L$ to the normal bundle of L. This implies Lemma 5.6, taking into account the injectivity of I and Exp^*. □

Lemmas 5.4, 5.6 and the Eq. (5.3) imply that $(\Omega^*(N_\varepsilon L, TNL), I(\Omega^*(L, NL)), \tilde{P}, Exp^*(\chi))$ is a V-data. This and Proposition 5.2 yield Theorem 5.5 immediately. □

Lemma 5.7 *Let $V_1, \ldots, V_k \in \Gamma(NL) = \Omega^0(L, NL)$. Then*

$$\mathfrak{m}_k(V_1, \ldots, V_k) = P(\mathcal{L}_{I(-V_1)} \ldots \mathcal{L}_{I(-V_k)}(Exp^*(\chi))).$$

Proof Let $V_i \in \Gamma(NL)$. Then $\{I(V_i)\}$ are mutually commuting vector fields on $N_\varepsilon L$. Using (2.10) and noting that

$$[Exp^*(\chi), I(V_i)]^{FN} = [I(-V_i), Exp^*(\chi)]^{FN}$$

we obtain Lemma 5.7 immediately. □

We shall denote the map $\mathfrak{m}_1 : \Omega^*(L, NL) \to \Omega^{*+3}(L, NL)$ by d_L. Since $\dim L = 3$, d_L is non-trivial only on $\Omega^0(L, NL)$.

Remark 5.8 Using the formal deformation theories developed in [11, 12] it is not hard to see that the L_∞-algebra associated to a closed associative submanifold L encodes the formal and smooth associative deformations of L. This and further generalization have been considered in [2]. A search for an L_∞-algebra associated to each associative submanifold in G_2-manifolds led us to discover the role of the Frölicher–Nijenhuis bracket in geometry of G_2-manifolds.

References

1. Besse, A. (1987). *Einstein manifolds*. Berlin: Springer.
2. Fiorenza, D., Lê, H. V., Schwachhöfer, L., & Vitagliano, L. Strongly homotopy Lie algebras and deformations of calibrated submanifolds. arXiv:1804.05732.
3. Frölicher, A., & Nijenhuis, A. (1956). Theory of vector-valued differential forms. I. Derivations of the graded ring of differential forms. *Indagationes Mathematicae*, *18*, 338–359 (1956).
4. Frölicher, A., & Nijenhuis, A. (1956). Some new cohomology invariants for complex manifolds. I, II. *Indagationes Mathematicae*, *18*, 540–564.
5. Harvey, R., & Lawson, H. B. (1982). Calibrated geometry. *Acta Mathematica*, *148*, 47–157.
6. Kawai, K., Lê, H. V., & Schwachhöfer, L. (2018). The Frölicher-Nijenhuis bracket and the geometry of G_2- and Spin(7)-manifolds. *Annali di Matematica Pura ed Applicata*, *197*, 411–432.
7. Kawai, K., Lê, H. V., & Schwachhöfer, L. (2018). Frölicher-Nijenhuis cohomology on G_2 and Spin(7)-manifolds. *International Journal of Mathematics*, *29*, 1850075.
8. Kawai, K., Lê, H. V., & Schwachhöfer, L. Calculation of Frölicher-Nijenhuis cohomology of Calabi-Yau manifolds, preprint.
9. Kolar, I., Michor, P. W., & Slovak, J. (1993). *Natural operators in differential geometry*. Berlin: Springer.
10. Lada, T., & Stasheff, J. (1993). Introduction to SH Lie algebras for physicists. *International Journal of Theoretical Physics*, *32*, 1087–1103.
11. Lê, H. V., & Oh, Y.-G. (2016). Deformations of coisotropic submanifolds in locally conformal symplectic manifolds. *Asian Journal of Mathematics*, *20*, 555–598.
12. Lê, H. V., & Schwachhöfer, L. (2019). Lagrangian submanifolds in strict nearly Kähler 6-manifolds. *Osaka Journal of Mathematics*, *56*, 601–629.
13. McLean, R. (1998). Deformations of Calibrated submanifolds. *Communications in Analysis and Geometry*, *6*, 705–747.
14. Voronov, T. (2005). Higher derived brackets and homotopy algebras. *Journal of Pure and Applied Algebra*, *202*, 133–153.

Distinguished G_2-Structures on Solvmanifolds

Jorge Lauret

Abstract Among closed G_2-structures there are two very distinguished classes: *Laplacian solitons* and *Extremally Ricci-pinched* G_2-structures. We study the existence problem and explore possible interplays between these concepts in the context of left-invariant G_2-structures on solvable Lie groups. Also, some Ricci pinching properties of G_2-structures on solvmanifolds are obtained, in terms of the extremal values and points of the functional $F = \frac{\mathrm{Scal}^2}{|\mathrm{Ric}|^2}$, $0 < F < 7$. Many natural open problems have been included.

1 Introduction

Our main motivation in this article is the following heuristic though very natural and intriguing question, which we borrowed from the first page of Besse's book [6] and adapted to G_2-geometry:

> Given a 7-dimensional differentiable manifold M, are there any best (or nicest, or most distinguished) G_2-structures on M?

The question remains natural when restricted to special kinds of manifolds or particular classes of G_2-structures, like the set of all G_2-structures with the same associated metric, left-invariant G_2-structures on a given Lie group, etc. The meaning of the adjectives in the question are of course part of the problem, and any good candidate is expected to be weak enough to allow existence results but also sufficiently strong to imply some kind of uniqueness or finiteness results.

As a first reduction, we consider closed G_2-structures, but there are many other reasonable and natural special classes to start with. We have included in an appendix (see Sect. 4.3) the definition of several of them, as well as a diagram describing the

This research was partially supported by grants from CONICET, FONCYT and SeCyT (Universidad Nacional de Córdoba).

J. Lauret (✉)
Universidad Nacional de Córdoba, FaMAF and CIEM, 5000 Córdoba, Argentina
e-mail: lauret@famaf.unc.edu.ar

S. Karigiannis et al. (eds.), *Lectures and Surveys on G_2-Manifolds and Related Topics*, Fields Institute Communications 84, https://doi.org/10.1007/978-1-0716-0577-6_9

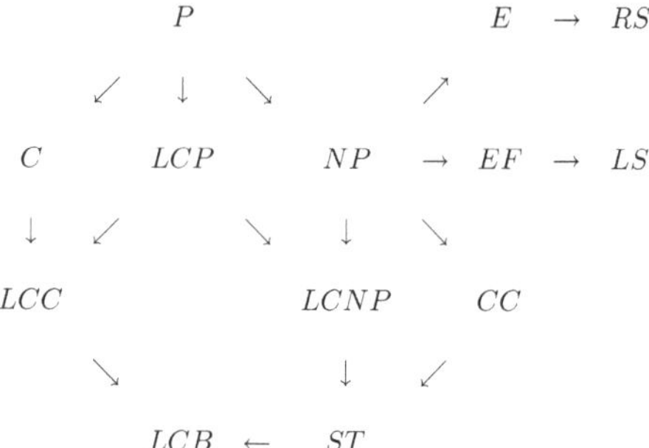

Fig. 1 Special classes of G_2-structures

inclusion relationships between such classes (see Fig. 1). No topological obstruction on M to admit a closed G_2-structure is known, other than the ones for admitting just a G_2-structure, i.e. orientable and spin.

In the case when a G_2-structure φ is closed, the only torsion that survives is contained in a 2-form τ, and the starting situation can be described as follows:

$$d\varphi = 0, \qquad \tau = - * d * \varphi, \qquad d * \varphi = \tau \wedge \varphi, \qquad d\tau = \Delta\varphi,$$

where $*$ and Δ denote the Hodge star and Laplacian operator, respectively, defined by the metric attached to φ.

Among closed G_2-structures, one finds two concepts which are both distinguished but from points of view of a very different taste:

- *Laplacian solitons*: $d\tau = c\varphi + \mathcal{L}_X\varphi$ for some $c \in \mathbb{R}$ and $X \in \mathfrak{X}(M)$.
- *Extremally Ricci-pinched*: $d\tau = \frac{1}{6}|\tau|^2\varphi + \frac{1}{6} * (\tau \wedge \tau)$.

In this paper, we mainly work in the homogeneous setting (see [21] for further information); more specifically, in the context of left-invariant G_2-structures on solvable Lie groups (or *solvmanifolds*). We aim to overview what is known on the existence of the above two special classes of G_2-structures and explore possible interplays. Diverse open problems have been included throughout the paper.

Any G_2-structure or metric on a Lie group is always assumed to be left-invariant.

1.1 Solitons

The space $\mathcal{G}$ of all G_2-structures on a given 7-dimensional manifold M is an open cone in $\Omega^3 M$, whose equivalence classes are $\mathrm{Diff}(M)$-orbits. Assume that at each $\varphi \in \mathcal{G}$, we have a preferred direction $q(\varphi) \in \Omega^3 M$, an optimal 'direction of improvement' in some sense (e.g. the gradient of a natural functional on $\mathcal{G}$). It is therefore reasonable to consider an element $\varphi \in \mathcal{G}$ distinguished when $q(\varphi)$ is tangent to its equivalence class (up to scaling), i.e.

$$q(\varphi) \in T_\varphi \left(\mathbb{R}^* \operatorname{Diff}(M) \cdot \varphi\right). \tag{1}$$

Heuristically, it is like such a φ is nice enough that it does not need to be improved. A G_2-structure for which condition (1) holds will be called a *q-soliton*. It is easy to see that if q is Diff(M)-equivariant, then the following conditions are equivalent:

- φ is a q-soliton.
- $q(\varphi) = c\varphi + \mathcal{L}_X\varphi$ for some $c \in \mathbb{R}$, $X \in \mathfrak{X}(M)$.
- The solution $\varphi(t)$ starting at φ to the corresponding geometric flow

$$\frac{\partial}{\partial t}\varphi(t) = q(\varphi(t)),$$

is self-similar, i.e. $\varphi(t) = c(t) f(t)^*\varphi$ for some $c(t) \in \mathbb{R}$ and $f(t) \in \operatorname{Diff}(M)$.

The q-soliton is said to be *expanding*, *steady* or *shrinking* if $c > 0$, $c=0$ or $c<0$, respectively. The corresponding self-similar solutions are respectively immortal, eternal and ancient if $q(a\varphi) = a^\alpha q(\varphi)$ for any $a \in \mathbb{R}^*$ and some fixed $\alpha < 1$ (see [20, Sect. 4.4]).

We consider in this paper the direction $q(\varphi) = \Delta\varphi$, which determines the so called *Laplacian solitons* and the Laplacian flow introduced by Bryant in [9]. Many other types of q-solitons have also been studied in the literature, see for example [5, 15, 28, 29].

We next list all the results on Laplacian solitons in the literature that we are aware of:

- [24, Corollary 1] There are no compact shrinking Laplacian solitons, and the only compact steady Laplacian solitons are the torsion-free G_2-structures (see also [25, Proposition 9.4] for a shorter proof in the closed case).
- Any nearly parallel G_2-structure φ satisfies $\Delta\varphi = c^2\varphi$ and so it is a coclosed expanding Laplacian soliton. Examples are given by the round and squashed spheres (see [29, Sect. 4.1]).
- [17, Section 6] Examples of non-compact expanding coclosed Laplacian solitons which are not nearly parallel. However, they still are all *eigenforms* (i.e. $\Delta\varphi = c\varphi$ for some $c \in \mathbb{R}$).
- [25, Proposition 9.1] The only compact and closed Laplacian solitons which are eigenforms are the torsion-free G_2-structures.
- [20, Section 7] A closed G_2-structure on a nilpotent Lie group which is an expanding Laplacian soliton and is not an eigenform was found.
- [26] Closed expanding Laplacian solitons were exhibited on seven of the twelve nilpotent Lie groups admitting a closed G_2-structure. There is even a one-parameter family of pairwise non-homothetic closed Laplacian solitons on one of them.
- [21] Homogeneous Laplacian solitons are studied using the algebraic soliton approach. Many continuous families of expanding Laplacian solitons on almost-abelian Lie groups were given (see [21, Sect. 5.2]).

- [22, Section 4] Examples of steady and shrinking closed Laplacian solitons were found on solvmanifolds by using coupled SU(3)-structures.
- [14] Closed expanding Laplacian solitons were found on solvmanifolds from symplectic half-flat SU(3)-structures.

The following are open questions on Laplacian solitons:

- Are there compact and closed expanding Laplacian solitons?
- Are there compact expanding Laplacian solitons other than nearly parallel G_2-structures?

1.2 Extremally Ricci Pinched G_2-Structures

The following nice interplay between the metric and the torsion 2-form of a closed G_2-structure was discovered by R. Bryant. Let Scal and Ric denote the scalar and Ricci curvature of the metric attached to a G_2-structure.

Theorem 1.1 ([9, Corollary 3]) *If φ is a closed G_2-structure on a compact manifold M, then*

$$\int_M \mathrm{Scal}^2 * 1 \leq 3 \int_M |\operatorname{Ric}|^2 * 1,$$

and equality holds if and only if $d\tau = \frac{1}{6}|\tau|^2\varphi + \frac{1}{6} * (\tau \wedge \tau)$.

The special G_2-structures for which equality holds were called *extremally Ricci-pinched* (ERP for short) in [9, Remark 13]. Notice the factor of 3 on the right hand side, much smaller than the factor of 7 provided by the Cauchy-Schwartz inequality, only attained at Einstein metrics.

As far as we know, there are only two examples of ERP G_2-structures in the literature and they are both homogeneous: the first one was given in [9, Example 1] on the homogeneous space $\mathrm{SL}_2(\mathbb{C}) \ltimes \mathbb{C}^2/\mathrm{SU}(2)$ (and on any compact quotient by a lattice), which also has a presentation as a G_2-structure on the solvable Lie group given in [22, Examples 4.13, 4.10], and the one found on a unimodular solvable Lie group in [22, Example 4.7]. Surprisingly (or not), both examples are also steady Laplacian solitons. We do not know if there could be an interplay between the two notions.

Motivated by Theorem 1.1, we consider the invariant (up to isometry and scaling) functional

$$F := \frac{\mathrm{Scal}^2}{|\operatorname{Ric}|^2}, \qquad 0 \leq F \leq 7,$$

on the space of all non-flat homogeneous closed G_2-structures (recall that a Ricci flat homogeneous Riemannian manifold is necessarily flat; see [2]). Note that after integrating on M in the compact case, $F \leq 3$ by Theorem 1.1. On the other hand, we know that $F < 7$ always since no solvable Lie group admits an Einstein closed (non-parallel) G_2-structure (see [11]) and the Alekseevskii Conjecture, asserting that any homogeneous Einstein metric of negative scalar curvature is isometric to a solvmanifold, has recently been proved in dimension 7 (see [4]).

Within the class of closed G_2-structures on almost-abelian Lie groups, $F \leq 1$ and equality holds precisely at non-nilpotent expanding Laplacian solitons (see [21, Sect. 5]). At some point it was not unreasonable to expect that $F \leq 3$ would also hold in the homogeneous case. However, we found in [22] a curve φ_t, $\frac{1}{4} \leq t \leq 1$, of closed G_2-structures on pairwise non-isomorphic solvmanifolds such that $F(\varphi_t)$ is strictly decreasing and

$$F(\varphi_{1/4}) = \tfrac{81}{17} \; (\approx 4.76) \; > \; 3 = F(\varphi_1).$$

Furthermore, φ_t is a shrinking Laplacian soliton for any $\frac{1}{4} \leq t < 1$ and φ_1 is the ERP steady Laplacian soliton given by Bryant. It is therefore natural to wonder about what would be the 'extremally Ricci pinched' G_2-structures in the homogeneous case:

What is the value of sup F and its meaning? Is it a maximum? Are the maximal G_2-structures distinguished in some sense?

We study the behavior of the functional F on solvmanifolds in Sect. 4, after giving a summary on what is known on Ricci pinching of solvmanifolds in the Riemannian case in Sect. 3.

2 The Space of Closed G_2-Structures on Solvmanifolds

We fix a 7-dimensional real vector space $\mathfrak{s}$ endowed with a basis $\{e_1, \ldots, e_7\}$ and the positive 3-form

$$\varphi := e^{127} + e^{347} + e^{567} + e^{135} - e^{146} - e^{236} - e^{245},$$

whose associated inner product $\langle \cdot, \cdot \rangle$ is the one making the basis $\{e_i\}$ orthonormal. Let $\mathcal{S} \subset \Lambda^2\mathfrak{s}^* \otimes \mathfrak{s}$ denote the algebraic subset of all Lie brackets on $\mathfrak{s}$ which are solvable. Each $\mu \in \mathcal{S}$ will be identified with the left-invariant G_2-structure determined by φ on the simply connected solvable Lie group S_μ with Lie algebra $(\mathfrak{s}, \mu)$:

$$\mu \longleftrightarrow (S_\mu, \varphi).$$

In this way, the isomorphism class $\mathrm{GL}_7(\mathbb{R}) \cdot \mu$ stands for the set of all left-invariant G_2-structures on S_μ:

$$(S_{h\cdot\mu}, \varphi) \longleftrightarrow (S_\mu, \varphi(h\cdot, h\cdot, h\cdot)), \qquad \forall h \in \mathrm{GL}_7(\mathbb{R}).$$

Note that h^{-1} is an isomorphism determining an equivalence between these two Lie groups endowed with G_2-structures. Recall that any G_2-structure or metric on a Lie group is assumed to be left-invariant.

Thus any two Lie brackets in the same G_2-orbit are equivalent as G_2-structures, and if they are in the same O(7)-orbit then they are isometric as Riemannian metrics. Both converse assertions hold for completely real Lie brackets. We note that the orbit $O(7) \cdot \mu$ consists of all the G_2-structures on S_μ defining some fixed metric.

By intersecting $\mathcal{S}$ with the linear subspace $\{\mu \in \Lambda^2\mathfrak{s}^* \otimes \mathfrak{s} : d_\mu\varphi = 0\}$, one obtains the G_2-invariant algebraic subset

$$\mathcal{S}_{closed} := \{\mu \in \mathcal{S} : d_\mu\varphi = 0\}.$$

The space $\mathcal{S}_{closed}$ therefore parameterizes the set of all closed G_2-structures on solvmanifolds. Note that a Lie group S_μ admits a closed G_2-structure if and only if the orbit $GL_7(\mathbb{R}) \cdot \mu$ meets $\mathcal{S}_{closed}$ (or equivalently, the above linear subspace). We do not know much about the topology of the cone $GL_7(\mathbb{R}) \cdot \mu \cap \mathcal{S}_{closed}$ of all closed G_2-structures on a given Lie group: is it connected? Is its intersection with a sphere connected?

Recall that for each $\mu \in \mathcal{S}_{closed}$, the only torsion form that survives is the 2-form $\tau_\mu \in \Lambda^2\mathfrak{s}^*$ given by

$$\tau_\mu = - * d_\mu * \varphi, \qquad d_\mu * \varphi = \tau_\mu \wedge \varphi.$$

We also consider the G_2-invariant subset of torsion-free G_2-structures,

$$\mathcal{S}_{tf} := \{\mu \in \mathcal{S} : d_\mu\varphi = 0, \ d_\mu * \varphi = 0\} = \left\{\mu \in \mathcal{S}_{closed} : \tau_\mu = 0\right\},$$

and the O(7)-invariant subset

$$\mathcal{S}_{flat} := \left\{\mu \in \mathcal{S} : (S_\mu, \langle\cdot,\cdot\rangle) \text{ is flat}\right\} = \left\{\mu \in \mathcal{S} : \mathrm{Scal}_\mu = 0\right\}.$$

Since the scalar curvature of μ (i.e. of $(S_\mu, \langle\cdot,\cdot\rangle)$) equals $\mathrm{Scal}_\mu = -\frac{1}{2}|\tau_\mu|^2$, we obtain that

$$\mathcal{S}_{tf} = \mathcal{S}_{closed} \cap \mathcal{S}_{flat}.$$

2.1 Nilpotent Case

There are exactly twelve nilpotent Lie algebras admitting a closed G_2-structure (see [10]). Thus the space $\mathcal{S}_{closed}$ meets twelve nilpotent $GL(\mathfrak{s})$-orbits, say $GL(\mathfrak{s}) \cdot \mu_1, \ldots, GL(\mathfrak{s}) \cdot \mu_{12}$. In [12], the authors classified which of these twelve Lie groups admit a closed G_2-structure which is in addition a Ricci soliton (called *nilsolitons* in the nilpotent case), and in [26], the existence of closed Laplacian solitons was studied. The following information has been extracted from these two articles (we use the same enumeration of the algebras):

- μ_1: This is the abelian Lie algebra and so $S_{\mu_1} = \mathbb{R}^7$ admits a unique G_2-structure up to equivalence which is torsion-free.
- μ_2: The Lie group S_{μ_2} admits a unique closed G_2-structure up to equivalence and scaling, which is a nilsoliton and also a Laplacian soliton.
- μ_3: There exists a curve of closed Laplacian solitons on S_{μ_3} which are not nilsolitons; however, there are no nilsoliton closed G_2-structures on this group.
- μ_4: The Lie group S_{μ_4} admits a pairwise non-equivalent one-parameter family of closed G_2-structures, among which there are a nilsoliton and a (different) Laplacian soliton.
- μ_5: S_{μ_5} does not admit any closed G_2-structure which is a nilsoliton. There is though a Laplacian soliton belonging to a curve of closed G_2-structures.
- μ_6: The Lie group S_{μ_6} admits a curve of closed G_2-structures, one of them being a nilsoliton and another one a Laplacian soliton.
- μ_7: S_{μ_7} does not admit any closed G_2-structure which is a nilsoliton. However, there exists a curve of closed G_2-structures containing a Laplacian soliton.
- μ_8, μ_9, μ_{11}: None of these Lie groups admit a closed G_2-structure which is a nilsoliton.
- μ_{10}: The existence of a nilsoliton closed G_2-structure in $S_{\mu_{10}}$ is still open.
- μ_{12}: The Lie group $S_{\mu_{12}}$ admits a closed G_2-structure which is also a nilsoliton.

The existence of closed Laplacian solitons on the Lie groups $S_{\mu_8}, \ldots, S_{\mu_{12}}$ remains open.

2.2 *Almost-Abelian Case*

Closed G_2-structures in the class of *almost-abelian* Lie algebras (i.e. with a codimension-one abelian ideal) were studied in [21, Sect. 5], we refer the reader there for further information. One attaches to each matrix $A \in \mathfrak{gl}_6(\mathbb{R})$ a Lie bracket $\mu_A \in \mathcal{S}$ as follows: relative to a fixed orthonormal basis $\{e_1, \ldots, e_7\}$, $\mathfrak{n} := \operatorname{span}\{e_1, \ldots, e_6\}$ is an abelian ideal for μ_A and $\operatorname{ad}_{\mu_A} e_7|_{\mathfrak{n}} = A$.

We have that $\mu_A \in \mathcal{S}_{closed}$ if and only if $A \in \mathfrak{sl}_3(\mathbb{C}) \subset \mathfrak{gl}_6(\mathbb{R})$, where the complex structure defining $\mathfrak{sl}_3(\mathbb{C})$ is $Je_i = e_{i+1}$, $i = 1, 3, 5$, and $\mu_A \in \mathcal{S}_{tf}$ if and only if $A \in \mathfrak{su}(3)$. It is easy to see that $\mu_B \in \mathrm{GL}(\mathfrak{s}) \cdot \mu_A$ for $A, B \in \mathfrak{sl}_3(\mathbb{C})$ if and only if $B \in \mathbb{R}^* \mathrm{SL}_3(\mathbb{C}) \cdot A$, where the last action is by conjugation. This implies that every almost-abelian Lie algebra admitting a closed G_2-structure is isomorphic to μ_A for some matrix A in the following list:

$$\begin{bmatrix} \alpha & & \\ & \beta & \\ & & \gamma \end{bmatrix}, \begin{bmatrix} \alpha & 1 & \\ & \alpha & \\ & & -2\alpha \end{bmatrix}, \begin{bmatrix} 0 & 1 & \\ & 0 & \\ & & 0 \end{bmatrix}, \begin{bmatrix} 0 & 1 & \\ & 0 & 1 \\ & & 0 \end{bmatrix}, \begin{bmatrix} \mathrm{i} & & \\ & a\mathrm{i} & \\ & & b\mathrm{i} \end{bmatrix}, \begin{bmatrix} \mathrm{i} & 1 & \\ & \mathrm{i} & \\ & & -2\mathrm{i} \end{bmatrix},$$

where $\alpha, \beta, \gamma \in \mathbb{C}$, $\alpha+\beta+\gamma=0$, $|\alpha|=1$, $\alpha \neq \pm \mathrm{i}$ and $a, b \in \mathbb{R}$, $1+a+b=0$. The two nilpotent matrices in the middle define groups isomorphic to S_{μ_2} and S_{μ_6}, respectively (see Sect. 2.1). Moreover, each Lie group S_{μ_A} admits an $\mathrm{SL}_3(\mathbb{C})$-orbit of closed G_2-structures up to scaling and each SU(3)-orbit consists of pairwise equivalent structures. Thus there are continuous families of closed G_2-structures depending on many parameters on most of these Lie groups (see e.g. [21, Example 5.9]).

It is easy to see that if $A = S + N$ for $A, S, N \in \mathfrak{sl}_3(\mathbb{C})$, where S is semisimple, N nilpotent and $[S, N] = 0$, then $\mu_S, \mu_N \in \overline{\mathbb{R}^*\mathrm{SL}_3(\mathbb{C}) \cdot \mu_A} \subset \mathcal{S}_{closed}$. It is worth observing that any kind of geometric quantity associated to closed G_2-structures depends continuously on the Lie bracket $\mu \in \mathcal{S}_{closed}$, so μ_S and μ_N inherit any property that μ_A may have. This can also be used to study pinching curvature properties (see [20, Sect. 3.3] and the next sections).

It is proved in [21, Proposition 5.22] that μ_A is a Laplacian soliton for any normal matrix $A \in \mathfrak{sl}_3(\mathbb{C})$ (see [21, Propositions 5.22, 5.27] for the Laplacian soliton conditions for a nilpotent A).

3 Ricci Pinching of Solvmanifolds

In this section, we give a short overview on Ricci pinching of solvmanifolds. We refer to [23] for a more detailed treatment with a complete list of references.

We fix an n-dimensional real vector space $\mathfrak{s}$ endowed with an inner product $\langle\cdot,\cdot\rangle$. Let $\mathcal{S} \subset \Lambda^2\mathfrak{s}^* \otimes \mathfrak{s}$ denote the algebraic subset of all Lie brackets on $\mathfrak{s}$ which are solvable. Given $\mu \in \mathcal{S}$, its isomorphism class $\mathrm{GL}(\mathfrak{s}) \cdot \mu$ can be identified with the set of all left-invariant metrics on the corresponding simply connected solvable Lie group S_μ in the following way:

$$(S_{h\cdot\mu}, \langle\cdot,\cdot\rangle) \longleftrightarrow (S_\mu, \langle h\cdot, h\cdot\rangle), \qquad \forall h \in \mathrm{GL}(\mathfrak{s}).$$

Consider the following $\mathrm{GL}(\mathfrak{s})$-invariant subsets of $\mathcal{S}$:

$$\begin{aligned}
\mathcal{S}_{\mathrm{i}\mathbb{R}} &:= \left\{\mu \in \mathcal{S} : \mathrm{Spec}(\mathrm{ad}_\mu X) \subset \mathrm{i}\mathbb{R}, \ \forall X \in \mathfrak{s}\right\}, \\
\mathcal{S}_{\mathbb{R}} &:= \left\{\mu \in \mathcal{S} : \text{either } \mathrm{ad}_\mu X \text{ is nilpotent or } \mathrm{Spec}(\mathrm{ad}_\mu X) \not\subset \mathrm{i}\mathbb{R}, \ \forall X \in \mathfrak{s}\right\}, \\
\mathcal{S}_{c\mathbb{R}} &:= \left\{\mu \in \mathcal{S} : \mathrm{Spec}(\mathrm{ad}_\mu X) \subset \mathbb{R}, \ \forall X \in \mathfrak{s}\right\}, \\
\mathcal{S}_{unim} &:= \left\{\mu \in \mathcal{S} : \mathrm{tr}\,\mathrm{ad}_\mu X = 0, \ \forall X \in \mathfrak{s}\right\}, \\
\mathcal{N} &:= \{\mu \in \mathcal{S} : \mu \text{ is nilpotent}\},
\end{aligned}$$

where $\mathrm{Spec}(\mathrm{ad}_\mu X)$ is the set of eigenvalues of the operator $\mathrm{ad}_\mu X$.

The Lie algebras in $\mathcal{S}_{\mathrm{i}\mathbb{R}}$ and $\mathcal{S}_{\mathbb{R}}$ are called of *imaginary* and *real type*, respectively (see e.g. [8, Sect. 3]). The closed subset $\mathcal{S}_{c\mathbb{R}}$ is known in the literature as the class of *completely real* or *completely solvable* Lie algebras, and $\mathcal{S}_{unim}$ is the subset of *unimodular* solvable Lie algebras. It easily follows that $\mathcal{S}_{i\mathbb{R}}$ is closed, $\mathcal{S}_{\mathbb{R}} \setminus \mathcal{N}$ is open in $\mathcal{S}$ and $\mathcal{S}_{i\mathbb{R}} \cap \mathcal{S}_{\mathbb{R}} = \mathcal{N}$. We also consider the subset

$$\mathcal{S}_{flat} := \left\{\mu \in \mathcal{S} : (S_\mu, \langle\cdot,\cdot\rangle) \text{ is flat}\right\}.$$

The following inclusions hold,

$$\{0\} \subset \mathcal{S}_{flat} \subset \mathrm{GL}(\mathfrak{s}) \cdot \mathcal{S}_{flat} \subset \mathcal{S}_{i\mathbb{R}} \subset \mathcal{S}_{unim}, \qquad \{0\} \subset \mathcal{N} \subset \mathcal{S}_{c\mathbb{R}} \subset \mathcal{S}_{\mathbb{R}}.$$

and the following lemma will be very useful.

Lemma 3.1 ([8, Lemma 3.4]) *If* $\mu \in \mathcal{S}_{\mathbb{R}}$ *then* $\overline{\mathrm{GL}(\mathfrak{s}) \cdot \mu} \cap \mathcal{S}_{flat} = \{0\}$.

From our point of view, the Ricci pinching is captured by the extremal values of the functional

$$F : \mathcal{S} \setminus \mathcal{S}_{flat} \longrightarrow \mathbb{R}, \qquad F(\mu) := \frac{\mathrm{Scal}_\mu^2}{|\,\mathrm{Ric}_\mu\,|^2},$$

where Scal_μ and Ric_μ are respectively the scalar curvature and Ricci operator of μ (recall that $\mu \leftrightarrow (S_\mu, \langle\cdot,\cdot\rangle)$). Note that F is invariant up to isometry and scaling; in particular, F is $\mathrm{O}(\mathfrak{s})$-invariant. Since $\mathrm{Scal}_\mu = 0$ if and only if $\mu \in \mathcal{S}_{flat}$, one obtains from the Cauchy-Schwartz inequality that

$$0 < F(\mu) \leq n, \qquad \forall \mu \in \mathcal{S} \setminus \mathcal{S}_{flat},$$

with $F(\mu) = n$ if and only if μ is Einstein. For each $\mu \in \mathcal{S}$ we define,

$$m_\mu := \inf F(\mathrm{GL}(\mathfrak{s}) \cdot \mu), \qquad M_\mu := \sup F(\mathrm{GL}(\mathfrak{s}) \cdot \mu),$$

that is, the infimum and supremum of F among all left-invariant metrics on the Lie group S_μ. It follows that

$$(m_\mu, M_\mu) \subset F(\mathrm{GL}(\mathfrak{s}) \cdot \mu) \subset F\left(\overline{\mathrm{GL}(\mathfrak{s}) \cdot \mu}\right) \subset [m_\mu, M_\mu].$$

Recall that F is not defined on $\mathcal{S}_{flat}$, so when we write $F(\mathcal{C})$ for some subset $\mathcal{C} \subset \mathcal{S}$ we always mean $F(\mathcal{C} \setminus \mathcal{S}_{flat})$.

An element $\mu \in \mathcal{S}$ is called a *solvsoliton* when $\mathrm{Ric}_\mu = cI + D$ for some $c \in \mathbb{R}$ and $D \in \mathrm{Der}(\mu)$. Any solvsoliton belongs to $\mathcal{S}_{\mathbb{R}}$ (see [18, Theorem 4.8]) and any Ricci soliton metric on a solvable Lie group is isometric to a solvsoliton (see [16]).

The only non-abelian Lie groups with $m_\mu = M_\mu$ are precisely those admitting a unique metric up to isometry and scaling, i.e.

$$\mu_{heis}(e_1, e_2) = e_3, \qquad \mu_{hyp}(e_n, e_i) = e_i, \quad i = 1, \ldots, n-1,$$

and zero otherwise. Note that $S_{\mu_{hyp}}$ is isometric to the real hyperbolic space $\mathbb{R}H^n$ and so $m_{\mu_{hyp}} = M_{\mu_{hyp}} = n$. On the other hand, $m_{\mu_{heis}} = M_{\mu_{heis}} = \frac{1}{3}$, and since $\mu_{heis} \in \overline{\mathrm{GL}(\mathfrak{s}) \cdot \mu}$ for any $\mu \notin \mathrm{GL}(\mathfrak{s}) \cdot \mu_{hyp}$, we have that

$$m_\mu \leq \tfrac{1}{3} \text{ for any } \mu \in \mathcal{S} \text{ such that } \mu \notin \mathrm{GL}(\mathfrak{s}) \cdot \mu_{hyp}.$$

The maximum of F among all left-invariant metrics on a nilpotent Lie group is attained at a nilsoliton, which is known to be unique up to isometry and scaling, if one exists. For any nonzero $\mu \in \mathcal{N}$,

$$F(\mathrm{GL}(\mathfrak{s}) \cdot \mu) = \begin{cases} \left(\frac{1}{3}, M_\mu\right], & S_\mu \text{ admits a nilsoliton and } \mu \notin \mathrm{GL}(\mathfrak{s}) \cdot \mu_{heis}, \\ \left(\frac{1}{3}, M_\mu\right), & S_\mu \text{ does not admit any nilsoliton}, \\ \left\{\frac{1}{3}\right\}, & \mu \in \mathrm{GL}(\mathfrak{s}) \cdot \mu_{heis}. \end{cases}$$

We also have that $F(\mathcal{N}) = [\frac{1}{3}, C_n]$ for some constant $C_n < n$ depending only on n, which is necessarily the value of F at some nilsoliton. The nilsolitons with $\mathrm{Ric} = \mathrm{Dg}(1, 2, \ldots, n)$ have $F = \frac{n(n-1)}{2(2n+1)}$, showing that $\frac{1}{5}n \leq C_n$ for all $7 \leq n$. In the nilpotent case, the functional F is strictly increasing along any Ricci flow solution $g(t)$, unless $g(0)$ is a nilsoliton (see [19]).

As in Sect. 2.2, in order to study the almost-abelian case, one fixes an orthogonal decomposition $\mathfrak{s} = \mathfrak{n} \oplus \mathbb{R}e_n$ and attaches to each matrix $A \in \mathfrak{gl}_{n-1}(\mathbb{R})$ (identified with $\mathfrak{gl}(\mathfrak{n})$ via any fixed orthonormal basis) the Lie bracket μ_A defined by $\mu_A(\mathfrak{n}, \mathfrak{n}) = 0$ and $\mathrm{ad}_{\mu_A} e_n|_{\mathfrak{n}} = A$. The construction covers, up to isometry, all left-invariant metrics on almost abelian Lie groups. Note that the class of almost-abelian Lie brackets is contained in $\mathcal{S}_{\mathrm{i}\mathbb{R}} \cup \mathcal{S}_{\mathbb{R}}$.

Using well-known formulas for the Ricci curvature of solvmanifolds, one obtains that

$$F(A) \leq 1 + \frac{(\mathrm{tr}\, A)^2}{\mathrm{tr}\, S(A)^2} \leq n,$$

where $S(A) := \frac{1}{2}(A + A^t)$. Moreover, $F(A) = n$ if and only if $S(A) = aI$, $a \neq 0$, if and only if μ_A is isometric to the real hyperbolic space $\mathbb{R}H^n$. It was proved in [3, Proposition 3.3] that A is a solvsoliton if and only if either A is normal or A is nilpotent and $[A, [A, A^t]] = cA$ for some $c \in \mathbb{R}$.

In the case when $\mathrm{tr}\, A = 0$, i.e. μ_A unimodular, it follows that

$$F(A) = \frac{\left(\mathrm{tr}\, S(A)^2\right)^2}{\left(\mathrm{tr}\, S(A)^2\right)^2 + \frac{1}{4}|[A, A^t]|^2},$$

hence $F(A) \leq 1$ and equality holds if and only if $[A, A^t] = 0$. Thus $M_A = 1$ for any $\mu_A \in \mathcal{S}_{\mathbb{R}}$ and it is a maximum if and only if A is semisimple. Note that the maxima of F on one of these Lie groups are precisely solvsolitons, as in the nilpotent case.

More generally, it is proved in [7, (2)] that for any $\mu \in \mathcal{S}_{unim}$, $F(\lambda) = M_\mu$ for some $\lambda \in \mathrm{GL}(\mathfrak{s}) \cdot \mu$ if and only if λ is a solvsoliton.

Some general results on Ricci pinching of solvmanifolds follow.

Theorem 3.2 ([23])

(i) $0 < m_\mu$ *and* $F\left(\overline{\mathrm{GL}(\mathfrak{s}) \cdot \mu}\right) = [m_\mu, M_\mu]$ *for any* $\mu \in \mathcal{S}_{\mathbb{R}}$.
(ii) *For* $n \geq 4$, $\inf\{m_\mu : \mu \in \mathcal{S}_{\mathbb{R}}\} = 0$; *in particular,* $F(\mathcal{S}_{\mathbb{R}}) = (0, n]$.
(iii) $F(\mathrm{GL}(\mathfrak{s}) \cdot \mu) = (m_\mu, M_\mu]$ *for every* $\mu \in \mathcal{S}_{\mathbb{R}} \cap \mathcal{S}_{unim}$.
(iv) $m_\mu = 0$ *for any* $\mu \in \mathcal{S} \setminus \mathcal{S}_{\mathbb{R}}$.

4 Ricci Pinching of G_2-Structures on Solvmanifolds

With Sects. 1.2 and 3 as our motivation, we now study the extremal points and values of the functional

$$F : \mathcal{S}_{closed} \setminus \mathcal{S}_{tf} \longrightarrow \mathbb{R}, \qquad F(\mu) := \frac{\mathrm{Scal}_\mu^2}{|\,\mathrm{Ric}_\mu\,|^2},$$

where Scal_μ and Ric_μ are respectively the scalar curvature and Ricci operator of $(S_\mu, \langle\cdot,\cdot\rangle)$. Since no solvable Lie group admits an Einstein (non-flat) and closed G_2-structure (see [11]),

$$0 < F(\mu) < 7, \qquad \forall \mu \in \mathcal{S}_{closed} \setminus \mathcal{S}_{tf}.$$

For each $\mu \in \mathcal{S}_{closed}$ we define,

$$n_\mu := \inf F(\mathrm{GL}(\mathfrak{s}) \cdot \mu \cap \mathcal{S}_{closed}), \qquad N_\mu := \sup F(\mathrm{GL}(\mathfrak{s}) \cdot \mu \cap \mathcal{S}_{closed}),$$

that is, the infimum and supremum of F among all closed G_2-structures on the Lie group S_μ. Recall that F is not defined on $\mathcal{S}_{tf}$, so when we write $F(\mathcal{C})$ for some subset $\mathcal{C} \subset \mathcal{S}_{closed}$ we always mean $F(\mathcal{C} \setminus \mathcal{S}_{tf})$.

It follows from Theorem 1.1 that if $\mu \in \mathcal{S}_{closed}$ and S_μ admits a lattice (i.e. a cocompact discrete subgroup), then $F(\mu) \leq 3$, and equality holds if and only if μ is ERP. On the other hand, if for a unimodular $\mu \in \mathcal{S}_{closed}$ there is a solvsoliton $\lambda \in \mathrm{GL}_7(\mathbb{R}) \cdot \mu \cap \mathcal{S}_{closed}$, then $F(\lambda) = N_\mu = M_\mu$.

For any class of G_2-structures defined by a closed cone $\mathcal{C} \subset \mathcal{S}$ such that $\mathcal{C} \cap \mathcal{S}_{tf} = \{0\}$, one has that

$$\mathcal{C} \cap \mathcal{S}_{closed} \cap \{\mu : |\mu| = 1\}$$

is a compact subset of $\mathcal{S}_{closed} \setminus \mathcal{S}_{tf}$. This implies that the infimum and supremum of $F(\mathcal{C} \cap \mathcal{S}_{closed})$ are actually minimum and maximum, respectively, and

$$0 < \min F(\mathcal{C} \cap \mathcal{S}_{closed}) \leq \max F(\mathcal{C} \cap \mathcal{S}_{closed}) < 7.$$

Examples of classes $\mathcal{C}$ for which the above holds include

- $\mathcal{N}$ or any closed cone contained in $\mathcal{N}$.
- $\overline{\mathrm{GL}_7(\mathbb{R}) \cdot \mu}$ for any $\mu \in \mathcal{S}_{\mathbb{R}}$ (see Lemma 3.1).

It would be really interesting to know the number

$$N_{closed} := \sup F(\mathcal{S}_{closed}).$$

If N_{closed} turns out to be a maximum, then the closed G_2-structures with $F = N_{closed}$ should be special in some sense. At the moment, the largest known value for F on $\mathcal{S}_{closed}$ is $\frac{81}{17} \sim 4.76$ and was found in [22, Example 4.11] at a shrinking Laplacian soliton.

Proposition 4.1

(i) $0 < n_\mu$ *for any* $\mu \in \mathcal{S}_{closed} \cap \mathcal{S}_{\mathbb{R}}$.
(ii) $\inf\{n_\mu : \mu \in \mathcal{S}_{closed} \cap \mathcal{S}_{\mathbb{R}}\} = 0$.

Proof Given $\mu \in \mathcal{S}_{closed} \cap \mathcal{S}_{\mathbb{R}}$, it follows from [8, Lemma 3.4] that

$$\overline{\mathrm{GL}_7(\mathbb{R}) \cdot \mu} \cap \mathcal{S}_{closed} \cap \{\mu : |\mu| = 1\}$$

is a compact subset of $\mathcal{S}_{closed} \setminus \mathcal{S}_{tf}$, so part (i) follows. Part (ii) was proved in Example 4.2 by using the family C_t.

We note that part (i) also follows from Theorem 3.2, (i) and the fact that $m_\mu \leq n_\mu$. $\square$

Corollary 4.2 *For any non-abelian solvable Lie group S of real type there exists a constant* $C(S) > 0$ *depending only on S such that*

$$|\operatorname{Ric}(\psi)| \leq C(S) |\operatorname{Scal}(\psi)|,$$

for any left-invariant G_2*-structure* ψ *on S.*

This estimate may have some applications in the study of convergence of geometric flows for G_2-structures (see [8]).

4.1 Nilpotent Case

Since μ_{hel3} does not appear in the list $\mu_1, \dots, \mu_{12}$ given in Sect. 2.1, we obtain that $\frac{1}{3} < \min F(\mathcal{S}_{closed} \cap \mathcal{N})$ and so $\frac{1}{3} < n_\mu$ for any $\mu \in \mathcal{S}_{closed} \cap \mathcal{N}$. In what follows, we describe what we know about the behavior of F on each of the nilpotent Lie groups admitting a closed G_2-structure (see [26]):

- μ_1: F is not defined.
- μ_2: $F \equiv \frac{1}{2}$; in particular, $n_{\mu_2} = N_{\mu_2} = \frac{1}{2}$.
- μ_3: $F \equiv \frac{1}{2}$ on the curve of closed Laplacian solitons.
- μ_4: $F = \frac{4}{5} = N_{\mu_4}$ at the nilsoliton and $F = \frac{3}{4}$ at the Laplacian soliton.
- μ_5: $F = \frac{3}{4}$ at the Laplacian soliton, but $F > \frac{3}{4}$ on a certain curve of closed G_2-structures.
- μ_6: At the nilsoliton, $F = \frac{4}{5} = N_{\mu_6}$, and at the Laplacian soliton, $F = \frac{3}{4}$.
- μ_7: At the Laplacian soliton, $F = \frac{3}{4}$, though $F > \frac{3}{4}$ on a curve of closed G_2-structures.
- μ_{12}: $F = 1 = N_{\mu_{12}}$ at the nilsoliton.

We note that Laplacian solitons in general fail to provide the maximum value of F on a given nilpotent Lie group.

4.2 *Almost-Abelian Case*

We work in this section on the class of almost-abelian solvable Lie groups (see Sect. 2.2). For each $A \in \mathfrak{sl}_3(\mathbb{C})$ one has that

$$F(A) = \frac{|H(A)|^4}{|H(A)|^4 + \frac{1}{8}|[A, A^*]|^2},$$

where $H(A) := \frac{1}{2}(A + A^*)$ is the hermitian part of A and $|B|^2 := \operatorname{tr} BB^*$ for any $B \in \mathfrak{sl}_3(\mathbb{C})$. It follows that $F(A) \leq 1$ and equality holds if and only if $[A, A^*] = 0$. Thus the maximum of F on a given non-nilpotent S_{μ_A} is only attained if A is semisimple and it is both a Laplacian and a Ricci soliton. The following example explicitly shows that the maximum value of F is not always attained at a Laplacian soliton in the nilpotent case.

Example 4.3 Consider the set of closed G_2-structures on the 3-step nilpotent Lie group S_{μ_6} parameterized by

$$\begin{bmatrix} 0 & a & 0 \\ & 0 & 1 \\ & & 0 \end{bmatrix}, \qquad a > 0.$$

The Ricci soliton and the Laplacian soliton correspond to $a = 1$ and $a = \sqrt{2}$, respectively. We have that

$$F(a) = \frac{a^4 + 2a^2 + 1}{2a^4 + a^2 + 2}, \qquad 0 < a,$$

a function with only one critical point, a global maximum with $F = \frac{4}{5}$ at the nilsoliton $a = 1$. Note that at the Laplacian soliton, $F(\sqrt{2}) = \frac{3}{4}$.

Concerning the behavior of F close to $\mathcal{S}_{tf}$, we have that

$$A_t := \begin{bmatrix} t & -1 & \\ 1 & -t & \\ & & 0 \end{bmatrix}, \quad F(A_t) = \frac{t^4}{t^4+t^2} \underset{t\to 0}{\longrightarrow} 0; \qquad B_t := \begin{bmatrix} t & -1 & \\ 1 & t & \\ & & -2t \end{bmatrix}, \quad F(B_t) \equiv 1.$$

This implies that F diverges at the torsion-free G_2-structure $A_0 = B_0$. Since $\operatorname{Spec}(A_t) = \{\pm \mathrm{i}\sqrt{1-t^2}\}$ for any $t < 1$, we deduce that $n_{A_0} = 0$. On the other hand, $\operatorname{Spec}(B_t) = \{\pm \mathrm{i} + t, -2t\}$, so the family of Laplacian solitons μ_{B_t} is pairwise non-isomorphic.

More generally, $n_A = 0$ for every $\mu_A \in \mathcal{S}_{\mathrm{i}\mathbb{R}} \smallsetminus \mathcal{N}$. Indeed, if $a \neq b$, then

$$D_t := \begin{bmatrix} a\mathrm{i} & t & \\ & b\mathrm{i} & \\ & & c\mathrm{i} \end{bmatrix}, \qquad F(D_t) = \frac{t^4}{t^4 + (a-b)^2 t^2} \underset{t\to 0}{\longrightarrow} 0.$$

Recall that the class of almost-abelian Lie brackets is contained in the disjoint union of $\mathcal{S}_{\mathbb{R}} \smallsetminus \mathcal{N}$, $\mathcal{N}$ and $\mathcal{S}_{\mathrm{i}\mathbb{R}} \smallsetminus \mathcal{N}$. In the list of matrices given in Sect. 2.2, the first two belong to $\mathcal{S}_{\mathbb{R}} \smallsetminus \mathcal{N}$, the second two to $\mathcal{N}$ and the last two to $\mathcal{S}_{\mathrm{i}\mathbb{R}} \smallsetminus \mathcal{N}$.

The following family C_t, $0 < t$, in $\mathcal{S}_{\mathbb{R}}$ given by

$$C_t := \begin{bmatrix} t & -1 & \\ 1 & 0 & \\ & & -t \end{bmatrix}, \qquad F(C_t) = \frac{4t^4}{4t^4+t^2} \underset{t\to 0}{\longrightarrow} 0,$$

shows that $\inf\{n_A : \mu_A \in \mathcal{S}_{closed} \cap \mathcal{S}_{\mathbb{R}}\} = 0$.

In [21, Example 5.20], the Laplacian flow on the family

$$\begin{bmatrix} 0 & a & \\ b & 0 & \\ & & 0 \end{bmatrix}, \qquad F(a,b) = \frac{(a+b)^4}{(a+b)^4 + (a^2-b^2)^2},$$

was studied. Using the ODE obtained there for $a(t), b(t)$, it is easy to prove that F is strictly decreasing along the Laplacian flow solutions starting at closed G_2-structures with $ab < 0$, $a \neq -b$. This shows that the Laplacian flow does not always improve the Ricci pinching of closed G_2-structures. On the other hand, the functional F was found to be increasing in some other Laplacian flow solutions like in the above example with $ab > 0$ and in the evolution studied in [22, Example 4.9].

4.3 Open Questions

It would be interesting to know the answers to the following natural questions:

- Given $A_0 \in \mathfrak{sl}_3(\mathbb{C})$ such that $\mathrm{Spec}(A_0) \subset i\mathbb{R}$, i.e. $\mu_{A_0} \in \mathcal{S}_{i\mathbb{R}}$, does $\lim F$ as A goes to A_0 exist on the isomorphism class $\mathbb{R}^*\mathrm{SL}_3(\mathbb{C}) \cdot A_0$? Examples A_t and B_t above show that such a limit does not exist on the set of all closed almost-abelian Lie brackets.
- Is $n_\mu = 0$ for any $\mu \notin \mathcal{S}_{closed} \cap \mathcal{S}_{\mathbb{R}}$? This holds in the Riemannian case (see Theorem 3.2, (iv)) and in the G_2 case for almost-abelian Lie groups (see Sect. 4.2).
- Is $F(\mathrm{GL}_7(\mathbb{R}) \cdot \mu \cap \mathcal{S}_{closed}) = (n_\mu, N_\mu]$ for any $\mu \in \mathcal{S}_{\mathbb{R}}$?
- Does $F(\lambda) = N_\mu$ hold for any solvsoliton $\lambda \in \mathrm{GL}_7(\mathbb{R}) \cdot \mu \cap \mathcal{S}_{closed}$? This is known to be true in the unimodular case and it is open in the non-unimodular Riemannian case.
- Is $F(\mathrm{GL}(\mathfrak{s}) \cdot \mu) = (0, M_\mu)$ for any $\mu \in \mathcal{S}_{closed} \setminus \mathcal{S}_{\mathbb{R}}$? What about for $\mu \in \mathcal{S}_{i\mathbb{R}} \cap \mathcal{S}_{closed}$?
- What is the value of $\sup\{M_\mu : \mu \in \mathcal{S}_{i\mathbb{R}} \cap \mathcal{S}_{closed}\}$?

Acknowledgements The author is very grateful to the organizers of the 'Workshop on G_2 Manifolds and Related Topics', August 21–25, 2017 and to The Fields Institute for the great hospitality. The author would also like to thank Ramiro Lafuente for very helpful comments.

Appendix: Special Classes of G_2-structures

The *torsion forms* of a G_2-structure φ on M are the components of the *intrinsic torsion* $\nabla\varphi$, where ∇ is the Levi-Civita connection of the metric g attached to φ. They can be defined as the unique differential forms $\tau_i \in \Omega^i M$, $i = 0, 1, 2, 3$, such that

$$d\varphi = \tau_0 * \varphi + 3\tau_1 \wedge \varphi + *\tau_3, \qquad d * \varphi = 4\tau_1 \wedge *\varphi + \tau_2 \wedge \varphi. \tag{2}$$

Some special classes of G_2-structures are defined or characterized as follows, we refer to [13] for further information:

- *parallel* (P) or *torsion-free*: $d\varphi = 0$ and $d * \varphi = 0$, or equivalently, $\nabla\varphi = 0$ (for M compact, this is equivalent to φ *harmonic* (H), i.e. $\Delta\varphi = 0$);
- *closed* (C) or *calibrated*: $d\varphi = 0$;
- *coclosed* (CC) or *cocalibrated*: $d * \varphi = 0$;
- *locally conformal parallel* (LCP): $d\varphi = 3\tau_1 \wedge \varphi$ and $d * \varphi = 4\tau_1 \wedge *\varphi$;
- *locally conformal closed* (LCC): $d\varphi = 3\tau_1 \wedge \varphi$ (in particular, $d\tau_1 = 0$);
- *nearly parallel* (NP): $d\varphi = \tau_0 * \varphi$ (which implies that $\Delta\varphi = \tau_0^2\varphi$ and $\mathrm{Ric} = \frac{3}{8}\tau_0^2 g$);
- *locally conformal nearly parallel* (LCNP): $d\varphi = \tau_0 * \varphi + 3\tau_1 \wedge \varphi$ and $d * \varphi = 4\tau_1 \wedge *\varphi$;
- *skew-torsion* (ST) or *G_2T-structures*: $d * \varphi = 4\tau_1 \wedge *\varphi$ (in particular, $d\tau_1 = 0$);
- *locally conformal balanced* (LCB): $d\tau_1 = 0$;
- *eigenform* (EF): $\Delta\varphi = c\varphi$ for some $c \in \mathbb{R}$;
- *Einstein* (E): $\mathrm{Rc} = cg$ for some $c \in \mathbb{R}$;

- *Laplacian soliton* (LS): $\Delta\varphi = c\varphi + \mathcal{L}_X\varphi$ for some $c \in \mathbb{R}$ and $X \in \mathfrak{X}(M)$ (called *expanding*, *steady* or *shrinking* if $c > 0$, $c = 0$ or $c < 0$, respectively);
- *Ricci soliton* (RS): $\mathrm{Rc} = cg + \mathcal{L}_X g$ for some $c \in \mathbb{R}$ and $X \in \mathfrak{X}(M)$.

We also refer to [27] for a more detailed study of most of these classes of G_2-structures and their possible intersections. Figure 1 describes the obvious inclusions among them. For a given class $\mathcal{C}$, a G_2-structure φ is said to be *locally conformal* $\mathcal{C}$ if for each $p \in M$, there exist an open neighborhood U and a conformal change $\psi := e^f\varphi$, $f \in C^\infty(U)$ such that (U, ψ) is a G_2-structure of class $\mathcal{C}$.

References

1. Alekseevskii, D. (1971). Conjugacy of polar factorizations of Lie groups. *Matematicheskii Sbornik 84*, 14–26. *English translation: Mathematics of the USSR-Sbornik, 13*, 12–24 (1971).
2. Alekseevskii, D., & Kimel'fel'd, B. (1975). Structure of homogeneous Riemannian spaces with zero Ricci curvature. *Funktional Anal. i Prilozen, 9*, 5–11. *Functional Analysis and its Applications 9*, 97–102 (1975).
3. Arroyo, R. (2013). The Ricci flow in a class of solvmanifolds. *Differential Geometry and its Applications, 31*, 472–485.
4. Arroyo, R., & Lafuente, R. (2017). The Alekseevskii conjecture in low dimensions. *Mathematische Annalen, 367*, 283–309.
5. Bagaglini, L., & Fino, A. (2018). The Laplacian coflow on almost-abelian Lie groups. *Annali di Matematica Pura ed Applicata, 197*, 1855–1873.
6. Besse, A. (1987). *Ergebnisse der Mathematik und ihrer Grenzgebiete* (Vol. 10). Einstein manifolds. Berlin: Springer.
7. Böhm, C., & Lafuente, R. (2018). Immortal homogeneous Ricci flows. *Inventiones Mathematicae, 212*, 461–529.
8. Böhm, C., Lafuente, R. (2017). The Ricci flow on solvmanifolds of real type, preprint.
9. Bryant, R. (2005). Some remarks on G_2-structures. In *Proceedings of Gökova Geometry-Topology Conference*, 75–109.
10. Conti, D., & Fernández, M. (2011). Nilmanifolds with a calibrated G2-structure. *Differential Geometry and its Applications, 29*, 493–506.
11. Fernández, M., Fino, A., & Manero, V. (2015). G_2-structures on Einstein solvmanifolds. *Asian Journal of Mathematics, 19*, 321–342.
12. Fernández, M., Fino, A., & Manero, V. (2016). Laplacian flow of closed G_2-structures inducing nilsolitons. *Journal of Geometric Analysis, 26*, 1808–1837.
13. Fernández, M., & Gray, A. (1982). Riemannian manifolds with structural group G_2. *Annali di Matematica Pura ed Applicata (IV), 32*, 19–45.
14. Fino, A., Raffero, A. (In Press) Closed warped G_2-structures evolving under the Laplacian flow. *The Scuola Superiore Sant'Anna of Pisa*.
15. Grigorian, S. (2013). Short-time behaviour of a modified Laplacian coflow of G2-structures. *Advances in Mathematics, 248*, 378–415.
16. Jablonski, M. (2014). Homogeneous Ricci solitons are algebraic. *Geometry & Topology, 18*, 2477–2486.
17. Karigiannis, S., McKay, B., & Tsui, M.-P. (2012). Soliton solutions for the Laplacian co-flow of some G_2-structures with symmetry. *Differential Geometry and its Applications, 30*, 318–333.
18. Lauret, J. (2011). Ricci soliton solvmanifolds. *The Journal für die reine und angewandte Mathematik, 650*, 1–21.
19. Lauret, J. (2011). The Ricci flow for simply connected nilmanifolds. *Communications in Analysis and Geometry, 19*, 831–854.

20. Lauret, J. (2016). Geometric flows and their solitons on homogeneous spaces (Workshop for Sergio Console). *Rendiconti del Seminario Matematico Torino*, *74*, 55–93.
21. Lauret, J. (2017). Laplacian flow of homogeneous G_2-structures and its solitons. *Proceedings of the London Mathematical Society*, *114*, 527–560.
22. Lauret, J. (2017). Laplacian solitons: Questions and homogeneous examples. *Differential Geometry and its Applications*, *54*, 345–360.
23. Lauret, J., & Will, C. E. (2018). The Ricci pinching functional on solvmanifolds.
24. Lin, C. (2013). Laplacian solitons and symmetry in G_2-geometry. *Journal of Geometry and Physics*, *64*, 111–119.
25. Lotay, J., & Wei, Y. (2017). Laplacian flow for closed G_2 structures: Shi-type estimates, uniqueness and compactness. *Geometric and Functional Analysis*, *27*, 165–233.
26. Nicolini, M. (2018). Laplacian solitons on nilpotent Lie groups. *The Bulletin of the Belgian Mathematical Society*, *25*, 183–196.
27. Raffero, A. (2016). Non-integrable special geometric structures in dimensions six and seven, Ph.D. thesis, Universita degli Studi di Torino.
28. Weiss, H., & Witt, F. (2012). A heat flow for special metrics. *Advances in Mathematics*, *231*(6), 3288–3322.
29. Weiss, H., & Witt, F. (2012). Energy functionals and solitons equations for G_2-forms. *The Annals of Global Analysis and Geometry*, *42*, 585–610.

On G_2-Structures, Special Metrics and Related Flows

Marisa Fernández, Anna Fino, and Alberto Raffero

Abstract We review results about G_2-structures in relation to the existence of special metrics, such as Einstein metrics and Ricci solitons, and the evolution under the Laplacian flow on non-compact homogeneous spaces. We also discuss some examples in detail.

2010 Mathematics Subject Classification 53C10 · 53C25 · 53C44 · 53C30

1 Introduction

A G_2-structure on a seven-dimensional manifold M is characterized by the existence of a globally defined 3-form φ which can be pointwise written as

$$\varphi = e^{127} + e^{347} + e^{567} + e^{135} - e^{146} - e^{236} - e^{245},$$

with respect to a suitable basis $\{e^1, \ldots, e^7\}$ of the cotangent space. Here, the shorthand e^{ijk} stands for $e^i \wedge e^j \wedge e^k$. Such a 3-form φ gives rise to a Riemannian metric g_φ and an orientation dV_φ on M.

The intrinsic torsion of a G_2-structure φ can be identified with the covariant derivative $\nabla^\varphi \varphi$, ∇^φ being the Levi Civita connection of g_φ. By [24], it vanishes identically if and only if both $d\varphi = 0$ and $d *_\varphi \varphi = 0$, where $*_\varphi$ denotes the Hodge operator of g_φ. When this happens, the G_2-structure is said to be *torsion-free*, its

M. Fernández (✉)
Facultad de Ciencia y Tecnología, Departamento de Matemáticas, Universidad del País Vasco (UPV/EHU), Apartado 644, 48080 Bilbao, Spain
e-mail: marisa.fernandez@ehu.es

A. Fino · A. Raffero
Dipartimento di Matematica "G. Peano", Università degli Studi di Torino, Via Carlo Alberto 10, 10123 Torino, Italy
e-mail: annamaria.fino@unito.it

A. Raffero
e-mail: alberto.raffero@unito.it

S. Karigiannis et al. (eds.), *Lectures and Surveys on G_2-Manifolds and Related Topics*, Fields Institute Communications 84, https://doi.org/10.1007/978-1-0716-0577-6_10

associated Riemannian metric g_φ is Ricci-flat and the corresponding Riemannian holonomy group is a subgroup of the exceptional Lie group G_2.

G_2-structures can be divided into classes, which are characterized by the expression of the exterior derivatives $d\varphi$ and $d *_\varphi \varphi$ [11, 24]. A G_2-structure φ is called *closed* (or *calibrated* according to [32]) if $d\varphi = 0$, while it is called *coclosed* (or *cocalibrated*) if $d *_\varphi \varphi = 0$.

Since the Ricci tensor and the scalar curvature of the metric induced by a G_2-structure can be expressed in terms of the intrinsic torsion [11], it may happen that certain restrictions on the curvature give rise to some constraints on the intrinsic torsion. For instance, a calibrated G_2-structure on a compact manifold induces an Einstein metric if and only if it is also cocalibrated, i.e., if and only if it is torsion-free [11, 16]. A natural problem consists then in investigating whether this happens also in the non-compact case, and whether similar results also hold when the metric is a Ricci soliton. These problems were studied for calibrated G_2-structures on homogeneous spaces in [21, 22], and for the wider class of locally conformal calibrated G_2-structures in [26]. We shall review the results in Sect. 3.

A useful tool to study geometric structures on manifolds is represented by geometric flows. Let M be a 7-manifold endowed with a calibrated G_2-structure φ_0. The *Laplacian flow* starting from φ_0 is the initial value problem

$$\begin{cases} \frac{\partial}{\partial t}\varphi(t) = \Delta_{\varphi(t)}\varphi(t), \\ d\varphi(t) = 0, \\ \varphi(0) = \varphi_0, \end{cases}$$

where $\Delta_\varphi \varphi = dd^*\varphi + d^*d\varphi$ is the Hodge Laplacian of φ with respect to the metric g_φ. This flow was introduced by Bryant in [11] to study 7-manifolds admitting calibrated G_2-structures. Short-time existence and uniqueness of the solution when M is compact were proved in the unpublished paper [13]. Recently, the analytic and geometric properties of the Laplacian flow have been deeply investigated in the series of papers [46–48]. In particular, the authors obtained a long-time existence result, and they proved that the solution exists for all positive times and it converges to a torsion-free G_2-structure modulo diffeomorphism provided that the initial datum φ_0 is sufficiently close to a given torsion-free G_2-structure.

The first noncompact examples with long-time existence of the solution were obtained on seven-dimensional nilpotent Lie groups in [22], while further solutions on solvable Lie groups were described in [27, 44, 45, 50]. Moreover, a cohomogeneity one solution converging to a torsion-free G_2-structure on the 7-torus was worked out in [36]. In Sect. 4, we shall discuss the results on nilpotent Lie groups obtained in [22].

2 Preliminaries

Let M be a seven-dimensional manifold endowed with a G$_2$-structure φ. The Riemannian metric g_φ and the volume form dV_φ are determined by φ via the equation

$$g_\varphi(X, Y)\, dV_\varphi = \frac{1}{6}\, \iota_X\varphi \wedge \iota_Y\varphi \wedge \varphi,$$

for all vector fields X, Y on M.

The vanishing of the intrinsic torsion T_φ of a G$_2$-structure φ can be stated in the following equivalent ways.

Theorem 2.1 ([24]) *Let φ be a G$_2$-structure on a seven-dimensional manifold M. Then, the following conditions are equivalent:*

(a) the intrinsic torsion of φ vanishes identically;
(b) $\nabla^\varphi \varphi = 0$, where ∇^φ denotes the Levi Civita connection of g_φ;
(c) $d\varphi = 0$ and $d *_\varphi \varphi = 0$;
(d) $\mathrm{Hol}(g_\varphi)$ is isomorphic to a subgroup of G$_2$.

A G$_2$-structure satisfying any of the above conditions is said to be *torsion-free* or *parallel*. By [9], the Riemannian metric induced by a torsion-free G$_2$-structure φ is Ricci-flat, i.e., $\mathrm{Ric}(g_\varphi) = 0$.

More generally, as the intrinsic torsion T_φ is a section of a vector bundle over M with fibre $\mathbb{R}^{7^*} \otimes \mathfrak{g}_2^\perp$, G$_2$-structures can be divided into classes according to the vanishing of the components of T_φ with respect to the G$_2$-irreducible decomposition

$$\mathbb{R}^{7^*} \otimes \mathfrak{g}_2^\perp \cong \mathcal{X}_1 \oplus \mathcal{X}_2 \oplus \mathcal{X}_3 \oplus \mathcal{X}_4 = \mathbb{R} \oplus \mathfrak{g}_2 \oplus S^2_0(\mathbb{R}^7) \oplus \mathbb{R}^7,$$

where $S^2_0(\mathbb{R}^7)$ denotes the space of traceless symmetric 2-tensors and $\mathfrak{g}_2 = \mathrm{Lie}(\mathrm{G}_2)$. This gives rise to sixteen classes of G$_2$-structures, which were first described in [24].

By [11], it is also possible to characterize each class in terms of the exterior derivatives $d\varphi$ and $d *_\varphi \varphi$. In detail, the spaces $\Lambda^k(\mathbb{R}^{7^*})$, $k = 2, 3$, admit the following G$_2$-irreducible decompositions (cf. [10])

$$\Lambda^2(\mathbb{R}^{7^*}) = \Lambda^2_7(\mathbb{R}^{7^*}) \oplus \Lambda^2_{14}(\mathbb{R}^{7^*}),$$
$$\Lambda^3(\mathbb{R}^{7^*}) = \Lambda^3_1(\mathbb{R}^{7^*}) \oplus \Lambda^3_7(\mathbb{R}^{7^*}) \oplus \Lambda^3_{27}(\mathbb{R}^{7^*}),$$

where the subscript in $\Lambda^k_r(\mathbb{R}^{7^*})$ denotes the dimension of the summand as an irreducible G$_2$-module, and $\Lambda^2_{14}(\mathbb{R}^{7^*}) \cong \mathfrak{g}_2$, $\Lambda^3_{27}(\mathbb{R}^{7^*}) \cong S^2_0(\mathbb{R}^7)$. Consequently, on M there exist a unique function $\tau_0 \in \mathcal{C}^\infty(M)$ and unique differential forms $\tau_1 \in \Omega^1(M)$, $\tau_2 \in \Omega^2_{14}(M) := \{\alpha \in \Omega^2(M) \mid \alpha \wedge *_\varphi\varphi = 0\}$, $\tau_3 \in \Omega^3_{27}(M) := \{\beta \in \Omega^3(M) \mid \beta \wedge \varphi = 0, \beta \wedge *_\varphi\varphi = 0\}$ such that

$$\begin{aligned} d\varphi &= \tau_0 *_\varphi \varphi + 3\tau_1 \wedge \varphi + *_\varphi\tau_3, \\ d *_\varphi \varphi &= 4\tau_1 \wedge *_\varphi\varphi + \tau_2 \wedge \varphi. \end{aligned} \tag{2.1}$$

Table 1 Some classes of G_2-structures

Class	Type	Conditions
$\mathcal{X}_1$	Nearly parallel	$\tau_1, \tau_2, \tau_3 = 0$
$\mathcal{X}_2$	Closed, calibrated	$\tau_0, \tau_1, \tau_3 = 0$
$\mathcal{X}_4$	Locally conformal parallel	$\tau_0, \tau_2, \tau_3 = 0$
$\mathcal{X}_1 \oplus \mathcal{X}_3$	Coclosed, cocalibrated	$\tau_1, \tau_2 = 0$
$\mathcal{X}_2 \oplus \mathcal{X}_4$	Locally conformal calibrated	$\tau_0, \tau_3 = 0$

The differential forms $\tau_0, \tau_1, \tau_2, \tau_3$ are called *intrinsic torsion forms* of the G_2-structure φ, and they can be identified with the components of the intrinsic torsion T_φ belonging to the G_2-modules $\mathcal{X}_1, \mathcal{X}_4, \mathcal{X}_2, \mathcal{X}_3$, respectively.

Some classes of G_2-structures with the defining conditions are recalled in Table 1.

2.1 Link with SU(3)-Structures

An SU(3)-structure on a six-dimensional manifold N is the data of an almost Hermitian structure (g, J) with fundamental 2-form $\omega := g(J\cdot, \cdot)$ and a complex volume form $\Psi = \psi + i\widehat{\psi} \in \Omega^{3,0}(M)$ of nonzero constant length.

By [35], an SU(3)-structure (g, J, Ψ) is completely determined by the real 2-form ω and the real 3-form ψ.

Since G_2 acts transitively on the 6-sphere with isotropy SU(3), every G_2-structure on a 7-manifold M induces an SU(3)-structure on each oriented hypersurface. In particular, if M is endowed with a torsion-free G_2-structure φ, and $N \subset M$ is an oriented hypersurface, then φ induces an SU(3)-structure (ω, ψ) on N which is *half-flat* according to the definition given in [14]. This means that the differential forms ω and ψ satisfy the conditions

$$d(\omega \wedge \omega) = 0, \quad d\psi = 0.$$

The inverse problem, i.e., establishing whether a half-flat SU(3)-structure on a 6-manifold is induced by an immersion into a 7-manifold with a torsion-free G_2-structure, can be analyzed using the so-called *Hitchin flow equations* (see [12, 35] for details).

We now recall the definition of some special types of half-flat SU(3)-structures.

Definition 2.2 A half-flat SU(3)-structure (ω, ψ) such that $d\omega = c\psi$ for some real number c is said to be *coupled* if $c \neq 0$, while it is called *symplectic half-flat* if $c = 0$, i.e., if the 2-form ω is symplectic. A coupled SU(3)-structure satisfying the additional condition $d\widehat{\psi} = -\frac{2}{3}c\,\omega \wedge \omega$ is called *nearly Kähler*.

If N is a 6-manifold endowed with an SU(3)-structure (ω, ψ), then the product manifold $N \times \mathbb{R}$ admits a G_2-structure defined by the 3-form

$$\varphi := \omega \wedge dt + \psi,$$

where dt is the global 1-form on $\mathbb{R}$. Such φ induces the product metric $g_\varphi = g + dt^2$. Moreover, φ is calibrated (resp. locally conformal calibrated) if the SU(3)-structure (ω, ψ) is symplectic half-flat (resp. coupled), while φ is locally conformal parallel if (ω, ψ) is nearly Kähler.

3 G_2-Structures and Special Metrics

By [11], the Ricci tensor and the scalar curvature of the metric induced by a G_2 structure φ can be expressed in terms of the intrinsic torsion forms τ_i. In particular, the scalar curvature is given by

$$\mathrm{Scal}(g_\varphi) = 12d^*\tau_1 + \frac{21}{8}\tau_0^2 + 30|\tau_1|^2 - \frac{1}{2}|\tau_2|^2 - \frac{1}{2}|\tau_3|^2,$$

where $|\cdot|$ denotes the pointwise norm induced by g_φ. Consequently, it has a definite sign for certain classes of G_2-structures. For instance, when φ is calibrated, then $\mathrm{Scal}(g_\varphi) = -\frac{1}{2}|\tau_2|^2$ is non-positive, while a nearly-parallel G_2-structure always induces an Einstein metric with positive scalar curvature $\mathrm{Scal}(g_\varphi) = \frac{21}{8}\tau_0^2$.

A generalization of Einstein metrics is given by Ricci solitons. We recall the definition here.

Definition 3.1 A (complete) Riemannian metric g on a smooth manifold M is a *Ricci soliton* if its Ricci tensor satisfies the equation

$$\mathrm{Ric}(g) = \lambda g + \mathcal{L}_X g,$$

for some real constant λ and some (complete) vector field X, where $\mathcal{L}$ denotes the Lie derivative. If in addition X is the gradient of a smooth function $f \in \mathcal{C}^\infty(M)$, i.e., $X = \nabla f$, then g is said to be of *gradient type*.

Equivalently, a Riemannian metric g on M is a Ricci soliton if and only if there exists a positive real valued function $h(t)$ and a family of diffeomorphisms η_t such that $g(t) = h(t)\,\eta_t^*(g)$ is a solution of the Ricci flow starting from g (see e.g. [15, Lemma 2.4]).

Depending on the sign of λ, a Ricci solitons is called *expanding* ($\lambda < 0$), *steady* ($\lambda = 0$) or *shrinking* ($\lambda > 0$). Moreover, a Ricci soliton is said to be *trivial* if it is either Einstein or the product of a homogeneous Einstein metric with the Euclidean metric. According to [38], if M is a compact manifold with a Ricci soliton g which is steady or expanding, then g is Einstein.

A special class of Ricci solitons is given by *homogeneous* ones, which are defined as follows

Definition 3.2 A Ricci soliton g on a smooth manifold M is *homogeneous* if its isometry group acts transitively on M.

Properties of non-trivial homogeneous Ricci solitons were given by Lauret in [43]. In particular, he proved the following.

Proposition 3.3 *([43]) Let g be a non-trivial homogeneous Ricci soliton on a smooth manifold M. Then, g is expanding and it cannot be of gradient type. Moreover, M has to be non-compact.*

Currently, all known examples of nontrivial homogeneous Ricci solitons are *solv-solitons*, that is left-invariant Ricci solitons on simply connected solvable Lie groups.

Since requiring that the metric induced by a G_2-structure is Einstein might impose some constraints on the intrinsic torsion, a natural problem is to investigate which types of G_2-structures can induce an Einstein (or, more generally, a Ricci soliton) non-Ricci-flat metric, and to see whether there is any difference between the compact and noncompact cases. For instance, if M is a 7-manifold endowed with a locally conformal nearly parallel G_2-structure φ (torsion class $\mathcal{X}_1 \oplus \mathcal{X}_4$) with g_φ complete and Einstein, then (M, φ) is either nearly parallel or conformally equivalent to the standard 7-sphere ([17]).

In what follows, we consider the cases of calibrated and locally conformal calibrated G_2-structures.

3.1 Calibrated G_2-Structures

A calibrated G_2-structure φ satisfies the equations

$$d\varphi = 0, \quad d *_\varphi \varphi = \tau_2 \wedge \varphi,$$

with $\tau_2 \in \Omega^2_{14}(M)$. We collect some known properties of such type of G_2-structures in the next results.

Proposition 3.4 *([11]) Let φ be a calibrated G_2-structure on M. Then,*

1) $\mathrm{Scal}(g_\varphi) \le 0$, and $\mathrm{Scal}(g_\varphi) = 0$ if and only if g_φ is Ricci-flat;
2) φ defines an Einstein metric on M if and only if $d *_\varphi \varphi = \tau_2 \wedge \varphi$, with $d\tau_2 = \frac{3}{14}|\tau_2|^2\varphi + \frac{1}{2} *_\varphi (\tau_2 \wedge \tau_2)$.

Corollary 3.5 *([11, 16]) Let M be a compact 7-manifold with a calibrated G_2-structure φ. If the underlying metric g_φ is Einstein, then $d *_\varphi \varphi = 0$ or, equivalenty, the holonomy group of g_φ is a subgroup of G_2.*

The proof of the corollary follows from the identity $d\left(\frac{1}{3}\tau_2^3\right) = \frac{2}{7}|\tau_2|^4 *_\varphi 1$ and Stokes' theorem. In detail, τ_2 must vanish identically since

$$0 = \int_M d\left(\frac{1}{3}\tau_2^3\right) = \int_M \frac{2}{7}|\tau_2|^4 *_\varphi 1.$$

In the non-compact case, there is a known non-existence result involving the $*$-*Ricci tensor* and the $*$-*scalar curvature*, where

$$\mathrm{Ric}^*(g_\varphi)_{sm} = R_{ijkl}\varphi_{ijs}\varphi_{klm}, \quad \mathrm{Scal}^*(g_\varphi) = \mathrm{tr}_{g_\varphi}(\mathrm{Ric}^*(g_\varphi)).$$

Such a result can be stated as follows.

Theorem 3.6 ([16]) *Let φ be a calibrated G_2-structure on a 7-manifold M. If g_φ is Einstein and $*$-Einstein, i.e., $\mathrm{Ric}^*(g_\varphi) = \frac{\mathrm{Scal}^*(g_\varphi)}{7} g_\varphi$, then g_φ is Ricci-flat.*

In light of the previous results, one might investigate the existence of calibrated G_2-structures that are Einstein but non-Ricci-flat on non-compact manifolds. This problem can be viewed as a G_2-analogue of the Goldberg conjecture [30], which states that a compact Einstein almost-Kähler manifold has to be Kähler. Recall that a non-compact homogeneous example of Einstein strictly almost Kähler 6-manifold was constructed in [2].

In the homogeneous setting, an answer to the above problem for calibrated G_2-structures was given in [21].

All known examples of non-compact homogeneous Einstein manifolds are solvmanifolds, that is, simply connected solvable Lie groups endowed with a left-invariant Einstein metric. The long-standing *Alekseevskii conjecture* [7, Question 7.5] states that a connected homogeneous Einstein space G/K of negative scalar curvature must be diffeomorphic to the Euclidean space. Thus, Einstein solvmanifolds might exhaust the class of non-compact homogeneous Einstein manifolds. The conjecture is known to be true in dimensions five and lower by [39, 51], and in dimension seven by [3]. So, seven-dimensional non-compact homogeneous Einstein manifolds are necessarily solvmanifolds.

We now review some general results about Einstein metrics on solvmanifolds of arbitrary dimension.

Theorem 3.7 ([42]) *Every Einstein solvmanifold (S, g) is standard, i.e., the corresponding solvable metric Lie algebra $(\mathfrak{s}, \langle\cdot,\cdot\rangle)$ admits the orthogonal decomposition $\mathfrak{s} = \mathfrak{n} \oplus \mathfrak{a}$, with $\mathfrak{n} = [\mathfrak{s}, \mathfrak{s}]$ and $\mathfrak{a}$ abelian.*

Recall that the dimension of the abelian summand $\mathfrak{a}$ in the decomposition

$$\mathfrak{s} = \mathfrak{n} \oplus \mathfrak{a}$$

is called the *rank* of the standard solvable metric Lie algebra $(\mathfrak{s}, \langle\cdot,\cdot\rangle)$.

In contrast to the compact homogeneous case (see e.g. [33, §5] and the references therein), standard Einstein metrics are essentially unique.

Theorem 3.8 ([33]) *A standard Einstein metric is unique up to isometry and scaling among invariant metrics.*

Remark 3.9

(1) The study of standard Einstein solvmanifolds reduces to those with $\dim \mathfrak{a} = 1$ (cf. [33, Theorem 4.18]).
(2) The Lie algebra of any standard Einstein solvmanifold resembles an Iwasawa subalgebra of a semisimple Lie algebra, since ad_A is symmetric and non-zero for any $A \neq 0 \in \mathfrak{a}$, and there exists some $A^0 \in \mathfrak{a}$ such that $\mathrm{ad}_{A^0}|_{\mathfrak{n}}$ is positive definite (see [33, Theorem 4.10]).

Using the rank of a standard solvable metric Lie algebra, it is possible to get a classification of seven-dimensional Einstein solvmanifolds (see e.g. [21, Theorem 4.4]). Then, using the obstructions to the existence of calibrated G_2-structures on Lie algebras given in [18], we have the following result.

Theorem 3.10 ([21]) *Let g_φ be the metric determined by a left-invariant calibrated G_2-structure φ on a solvmanifold. Then, g_φ is Einstein if and only if g_φ is flat.*

Remark 3.11 Note that a similar theorem can be proved also for cocalibrated G_2-structures [21]. Moreover, Theorem 3.10 shows that left-invariant calibrated G_2-structures behave differently from almost Kähler structures [2].

The situation is different if we require that g_φ is a non-trivial Ricci soliton. Indeed, non-compact examples of manifolds admitting a calibrated G_2-structure inducing a non-trivial Ricci soliton were constructed in [21], and they are all *nilsolitons* (see Theorem 3.17 and Example 3.19 below).

Definition 3.12 Let N be a simply connected nilpotent Lie group endowed with a left-invariant Riemannian metric g, and denote by $(\mathfrak{n}, \langle \cdot, \cdot \rangle)$ the corresponding metric nilpotent Lie algebra. The metric g is called *nilsoliton* if its Ricci endomorphism $\mathrm{Ric}(g)$ on $\mathfrak{n}$ differs from a derivation D of $\mathfrak{n}$ by a scalar multiple of the identity map I, i.e.,

$$\mathrm{Ric}(g) = \lambda I + D,$$

for some real number λ.

By [41, Proposition 1.1], a left-invariant Riemannian metric on a simply connected nilpotent Lie group is a nilsoliton if and only if it is a Ricci soliton according to Definition 3.1. It is worth recalling here that non-abelian nilpotent Lie groups cannot admit any left-invariant Einstein metric unless it is flat [49].

Remark 3.13 As the existence of a nilsoliton on a simply connected nilpotent Lie group N implies the existence of a non-zero symmetric derivation on the corresponding nilpotent Lie algebra $\mathfrak{n}$, nilsolitons might not exist. This is the case, for instance, of Lie algebras having nilpotent derivation algebra. Such Lie algebras are nilpotent by Engel's Theorem, and they are known as *characteristically nilpotent* in literature.

Before reviewing some properties of nilsolitons, we recall the following.

Definition 3.14 Let $(\mathfrak{n}, [\cdot,\cdot]_{\mathfrak{n}}, \langle\cdot,\cdot\rangle_{\mathfrak{n}})$ be a metric nilpotent Lie algebra. A metric Lie algebra $(\mathfrak{s} = \mathfrak{n} \oplus \mathfrak{a}, [\cdot,\cdot], \langle\cdot,\cdot\rangle)$ is a *metric solvable extension* of $(\mathfrak{n}, [\cdot,\cdot]_{\mathfrak{n}}, \langle\cdot,\cdot\rangle_{\mathfrak{n}})$ if the restriction to $\mathfrak{n}$ of the Lie bracket $[\cdot,\cdot]$ of $\mathfrak{s}$ coincides with $[\cdot,\cdot]_{\mathfrak{n}}$ and $\langle\cdot,\cdot\rangle|_{\mathfrak{n}\times\mathfrak{n}} = \langle\cdot,\cdot\rangle_{\mathfrak{n}}$.

Theorem 3.15 ([41]) *Let* N *be a simply connected nilpotent Lie group with Lie algebra* $\mathfrak{n}$. *Then,*

1) *A nilsoliton metric on* N *is unique up to isometry and scaling;*
2) N *has a nilsoliton metric* g *if and only if the corresponding metric Lie algebra* $(\mathfrak{n}, \langle\cdot,\cdot\rangle)$ *is an Einstein nilradical, i.e., it has a metric solvable extension* $\mathfrak{s} = \mathfrak{n} \oplus \mathfrak{a}$, *with* $\mathfrak{a}$ *abelian, whose corresponding solvmanifold is Einstein.*

From now on, we will use the following notation to define a Lie algebra. Suppose that $\mathfrak{g}$ is a seven-dimensional Lie algebra, whose dual space $\mathfrak{g}^*$ is spanned by $\{e^1, \dots, e^7\}$ satisfying

$$de^i = 0, \quad 1 \le i \le 4, \qquad de^5 = e^{12}, \qquad de^6 = e^{13}, \qquad de^7 = 0,$$

where d is the Chevalley-Eilenberg differential of $\mathfrak{g}$. Then, we will write

$$\mathfrak{g} = (0, 0, 0, 0, e^{12}, e^{13}, 0)$$

with the same meaning.

In order to show the existence of nilpotent Lie algebras with a calibrated G_2-structure inducing a nilsoliton, we need to recall the classification of the nilpotent Lie algebras admitting a calibrated G_2-structure given in [18].

Theorem 3.16 ([18]) *Up to isomorphism, there are exactly twelve nilpotent Lie algebras admitting a calibrated* G_2*-structure. They are:*

$$\begin{aligned}
\mathfrak{n}_1 &= (0, 0, 0, 0, 0, 0, 0),\\
\mathfrak{n}_2 &= (0, 0, 0, 0, e^{12}, e^{13}, 0),\\
\mathfrak{n}_3 &= (0, 0, 0, e^{12}, e^{13}, e^{23}, 0),\\
\mathfrak{n}_4 &= (0, 0, e^{12}, 0, 0, e^{13} + e^{24}, e^{15}),\\
\mathfrak{n}_5 &= (0, 0, e^{12}, 0, 0, e^{13}, e^{14} + e^{25}),\\
\mathfrak{n}_6 &= (0, 0, 0, e^{12}, e^{13}, e^{14}, e^{15}),\\
\mathfrak{n}_7 &= (0, 0, 0, e^{12}, e^{13}, e^{14} + e^{23}, e^{15}),\\
\mathfrak{n}_8 &= (0, 0, e^{12}, e^{13}, e^{23}, e^{15} + e^{24}, e^{16} + e^{34}),\\
\mathfrak{n}_9 &= (0, 0, e^{12}, e^{13}, e^{23}, e^{15} + e^{24}, e^{16} + e^{34} + e^{25}),\\
\mathfrak{n}_{10} &= (0, 0, e^{12}, 0, e^{13} + e^{24}, e^{14}, e^{46} + e^{34} + e^{15} + e^{23}),\\
\mathfrak{n}_{11} &= (0, 0, e^{12}, 0, e^{13}, e^{24} + e^{23}, e^{25} + e^{34} + e^{15} + e^{16} - 3e^{26}),\\
\mathfrak{n}_{12} &= (0, 0, 0, e^{12}, e^{23}, -e^{13}, 2e^{26} - 2e^{34} - 2e^{16} + 2e^{25}).
\end{aligned}$$

Comparing the previous classification with the results in [20], it turns out that, up to isomorphism, $\mathfrak{n}_9$ is the unique nilpotent Lie algebra with a calibrated G_2-structure but not admitting any nilsoliton. Moreover, the existence of a nilsoliton on the Lie algebra $\mathfrak{n}_{10}$ was shown in [19, Example 2], but its explicit expression is not known. Therefore, it is still an open problem to determine whether the Lie algebra $\mathfrak{n}_{10}$ admits a calibrated G_2-structure inducing the nilsoliton. For the remaining Lie algebras, we have the following.

Theorem 3.17 ([22]) *Up to isomorphism, $\mathfrak{n}_2, \mathfrak{n}_4, \mathfrak{n}_6$ and $\mathfrak{n}_{12}$ are the unique s-step nilpotent Lie algebras ($s = 2, 3$) with a nilsoliton inner product determined by a calibrated G_2-structure.*

Remark 3.18 Note that the Lie algebra $\mathfrak{n}_i$, $i = 3, 5, 7, 8, 11$, has a nilsoliton inner product but no calibrated G_2-structure defining the nilsoliton [22].

In the next example, we write the expression of a calibrated G_2-structure inducing the nilsoliton inner product on $\mathfrak{n}_i$, for $i = 2, 4, 6, 12$. Moreover, in each case we also specify the negative real number λ and the derivation D of $\mathfrak{n}_i$ for which $\mathrm{Ric} = \lambda I + D$.

Example 3.19 ([22]) Consider the nilpotent Lie algebras $\mathfrak{n}_2, \mathfrak{n}_4, \mathfrak{n}_6$ with the structure equations given in Theorem 3.16. Then,

$$\begin{aligned}
\mathfrak{n}_2 : \varphi_2 &= e^{147} + e^{267} + e^{357} + e^{123} + e^{156} + e^{245} - e^{346}, \\
\lambda &= -2, \quad D = \mathrm{diag}\left(1, \tfrac{3}{2}, \tfrac{3}{2}, 2, \tfrac{5}{2}, \tfrac{5}{2}, 2\right); \\
\mathfrak{n}_4 : \varphi_4 &= -e^{124} - e^{456} + e^{347} + e^{135} + e^{167} + e^{257} - e^{236}, \\
\lambda &= -\tfrac{5}{2}, \quad D = \mathrm{diag}\left(1, \tfrac{3}{2}, \tfrac{5}{2}, 2, 2, \tfrac{7}{2}, 3\right); \\
\mathfrak{n}_6 : \varphi_6 &= e^{123} + e^{145} + e^{167} + e^{257} - e^{246} + e^{347} + e^{356}, \\
\lambda &= -\tfrac{5}{2}, \quad D = \mathrm{diag}\left(\tfrac{1}{2}, 2, 2, \tfrac{5}{2}, \tfrac{5}{2}, 3, 3\right).
\end{aligned}$$

For $\mathfrak{n}_{12}$, we firstly consider a basis $\{e^1, \ldots, e^7\}$ of its dual space $\mathfrak{n}_{12}{}^*$ for which the structure equations are

$$\left(0, 0, 0, \frac{\sqrt{3}}{6}e^{12}, \frac{\sqrt{3}}{12}e^{13} - \frac{1}{4}e^{23}, -\frac{\sqrt{3}}{12}e^{23} - \frac{1}{4}e^{13}, \frac{\sqrt{3}}{12}e^{16} - \frac{\sqrt{3}}{6}e^{34} + \frac{\sqrt{3}}{12}e^{25} + \frac{1}{4}e^{26} - \frac{1}{4}e^{15}\right).$$

Then, a calibrated G_2-structure satisfying the required properties is

$$\varphi_{12} = -e^{124} + e^{167} + e^{257} + e^{347} - e^{456} + e^{135} - e^{236},$$

with $\lambda = -\frac{1}{4}$ and $D = \frac{1}{8}\,\mathrm{diag}(1, 1, 1, 2, 2, 2, 3)$.

Remark 3.20 The nilsoliton condition is less restrictive for cocalibrated G_2-structures. Indeed, on each 2-step nilpotent Lie algebra admitting cocalibrated

G_2-structures there exists a cocalibrated G_2-structure inducing the nilsoliton inner product (see [4]).

3.2 Locally Conformal Calibrated G_2-Structures

A G_2-structure φ is said to be *locally conformal calibrated* if the intrinsic torsion forms τ_0 and τ_3 vanish identically (cf. Table 1). In this case, Eq. (2.1) reduce to

$$d\varphi = 3\tau_1 \wedge \varphi, \quad d *_\varphi \varphi = 4\tau_1 \wedge *_\varphi\varphi + \tau_2 \wedge \varphi.$$

Let $\theta := 3\tau_1 = -\frac{1}{4} *_\varphi (*_\varphi d\varphi \wedge \varphi)$ denote the *Lee form* of the G_2-structure. Taking the exterior derivative of both sides of the equation $d\varphi = \theta \wedge \varphi$, we get $d\theta \wedge \varphi = 0$. This implies $d\theta = 0$. Consequently, each point of the manifold has an open neighborhood $\mathcal{U}$ where $\theta = df$ for some $f \in \mathcal{C}^\infty(\mathcal{U})$, and the 3-form $e^{-f}\varphi$ defines a calibrated G_2-structure on $\mathcal{U}$. Hence, locally conformal calibrated G_2-structures are locally conformal equivalent to calibrated G_2-structures.

Motivated by Corollary 3.5 and Theorem 3.10, it is natural to investigate the existence of locally conformal calibrated G_2-structures whose associated metric is Einstein and non-Ricci-flat. In what follows, we refer to a locally conformal calibrated G_2-structure φ with g_φ Einstein as an *Einstein locally conformal calibrated* G_2-*structure*.

On compact manifolds, the following constraint on the scalar curvature holds.

Theorem 3.21 ([26]) *An Einstein locally conformal calibrated* G_2-*structure on a compact seven-dimensional manifold has non-positive scalar curvature.*

Since homogeneous Einstein manifolds with negative scalar curvature are non-compact (cf. [7, Theorem 7.4]) and since every homogeneous Ricci-flat metric is flat (see [1]), an immediate consequence of the previous result is the following.

Corollary 3.22 *([26]) A compact homogeneous 7-manifold cannot admit an invariant Einstein locally conformal calibrated* G_2-*structure* φ *unless the underlying metric* g_φ *is flat.*

Remark 3.23 By [8, Theorem 3.1], the result of Corollary 3.22 is valid more generally on every compact locally homogeneous space.

In the non-compact setting, there is an example of a simply connected solvable Lie group endowed with a left-invariant locally conformal calibrated G_2-structure φ such that g_φ is Einstein non-Ricci-flat. Thus, the result of [21] recalled in Theorem 3.10 is not true anymore for locally conformal calibrated G_2-structures. Before describing the example, we recall some useful results.

Consider a 6-manifold N endowed with a coupled SU(3)-structure (ω, ψ), with $d\omega = c\psi$ (cf. Definition 2.2). As we mentioned in §2.1, the 3-form $\varphi := \omega \wedge dt + \psi$

defines a locally conformal calibrated G_2-structure on the product manifold $N \times \mathbb{R}$. It is not difficult to check that the corresponding Lee form is $\theta = -c\,dt$.

In [26], the classification of six-dimensional nilpotent Lie algebras admitting a coupled SU(3)-structure inducing a nilsoliton was achieved (see also [25, §4.1]). We recall it in the next theorem.

Theorem 3.24 ([26]) *A non-abelian six-dimensional nilpotent Lie algebra admitting a coupled* SU(3)*-structure is isomorphic to one of the following*

$$\mathfrak{h}_1 = (0, 0, 0, e^{12}, e^{14} - e^{23}, e^{15} + e^{34}), \qquad \mathfrak{h}_2 = (0, 0, 0, 0, e^{13} - e^{24}, e^{14} + e^{23}).$$

Moreover, the only one admitting a coupled SU(3)*-structure inducing a nilsoliton is* $\mathfrak{h}_2$.

Remark 3.25 Notice that $\mathfrak{h}_2$ is the Lie algebra of the three-dimensional complex Heisenberg group.

We are now ready to describe the example.

Example 3.26 ([26]) Consider the coupled SU(3)-structure on $\mathfrak{h}_2$ defined by the pair

$$\omega = e^{12} + e^{34} - e^{56}, \quad \psi = e^{136} - e^{145} - e^{235} - e^{246}.$$

It satisfies the equation $d\omega = -\psi$, and it induces the nilsoliton inner product $g = \sum_{k=1}^{6} (e^k)^2$ with Ricci operator

$$\mathrm{Ric} = -3I + 4\,\mathrm{diag}\left(\frac{1}{2}, \frac{1}{2}, \frac{1}{2}, \frac{1}{2}, 1, 1\right),$$

where $D = \mathrm{diag}\left(\frac{1}{2}, \frac{1}{2}, \frac{1}{2}, \frac{1}{2}, 1, 1\right)$ is a symmetric derivation of $\mathfrak{h}_2$. Consequently, the metric rank-one solvable extension $\mathfrak{s} = \mathfrak{h}_2 \oplus \langle e_7 \rangle$ of $\mathfrak{h}_2$ with structure equations

$$\left(\frac{1}{2}e^{17}, \frac{1}{2}e^{27}, \frac{1}{2}e^{37}, \frac{1}{2}e^{47}, e^{13} - e^{24} + e^{57}, e^{14} + e^{23} + e^{67}, 0\right)$$

is endowed with the Einstein (non-Ricci-flat) inner product $g + (e^7)^2$. This is precisely the inner product g_φ induced by the 3-form

$$\varphi = \omega \wedge e^7 + \psi,$$

which defines a locally conformal calibrated G_2-structure on $\mathfrak{s}$. A simple computation shows that the non-vanishing intrinsic torsion forms of φ are

$$\tau_1 = -\frac{1}{3}e^7, \quad \tau_2 = -\left(\frac{5}{3}e^{12} + \frac{5}{3}e^{34} + \frac{10}{3}e^{56}\right).$$

Clearly, left multiplication allows to extend φ to a left-invariant Einstein locally conformal calibrated G_2-structure on the simply connected solvable Lie group corresponding to $\mathfrak{s}$.

We conclude this section recalling a general structure result for compact 7-manifolds endowed with a locally conformal calibrated G_2-structure with nowhere vanishing Lee form.

Theorem 3.27 ([23]) *Let M be a compact, connected seven-dimensional manifold endowed with a locally conformal calibrated G_2-structure φ, with nowhere vanishing Lee form θ. Suppose that $\mathcal{L}_X\varphi = 0$, where X is the g_φ-dual vector field of θ. Then,*

1) *M is the total space of a fibre bundle over $\mathbb{S}^1$, and each fibre is endowed with a coupled $SU(3)$-structure;*
2) *M has a locally conformal calibrated G_2-structure $\hat{\varphi}$ such that $d\hat{\varphi} = \hat{\theta} \wedge \hat{\varphi}$, where $\hat{\theta}$ is a 1-form with integral periods.*

The previous theorem implies in particular that M is the mapping torus of a diffeomorphism ν of a certain 6-manifold N, i.e., M is diffeomorphic to the quotient of $N \times \mathbb{R}$ by the infinite cyclic group of diffeomorphisms generated by $(p, t) \mapsto (\nu(p), t + 1)$.

Remark 3.28 It is worth recalling here that compact locally conformal parallel G_2-manifolds can be characterized as fibre bundles over $\mathbb{S}^1$ with compact nearly Kähler fibre (see [37, 52]).

4 The Laplacian Flow on Lie Groups

Consider a 7-manifold M endowed with a calibrated G_2-structure φ_0. The *Laplacian flow* starting from φ_0 is the initial value problem

$$\begin{cases} \frac{\partial}{\partial t}\varphi(t) = \Delta_{\varphi(t)}\varphi(t), \\ d\varphi(t) = 0, \\ \varphi(0) = \varphi_0, \end{cases} \tag{4.1}$$

where Δ_φ denotes the Hodge Laplacian of the Riemannian metric g_φ induced by φ. This flow was introduced by Bryant in [11] to study seven-dimensional manifolds admitting calibrated G_2-structures. Notice that the stationary points of the flow equation in (4.1) are harmonic G_2-structures, which coincide with torsion-free G_2-structures on compact manifolds.

Short-time existence and uniqueness of the solution of (4.1) when M is compact were proved in [13].

Theorem 4.1 ([13]) *Assume that M is compact. Then, the Laplacian flow* (4.1) *has a unique solution defined for a short time $t \in [0, \varepsilon)$, with ε depending on φ_0.*

As a consequence of the condition $d\varphi(t) = 0$, the solution $\varphi(t)$ must belong to the open set

$$[\varphi_0]_+ := [\varphi_0] \cap \Omega^3_+(M)$$

in the cohomology class $[\varphi_0]$ as long as it exists.

Remark 4.2 By [11, 34], the evolution equation in (4.1) is the gradient flow of Hitchin's volume functional

$$[\varphi_0]_+ \ni \varphi \mapsto \int_M dV_\varphi,$$

with respect to a suitable L^2-metric on $[\varphi_0]_+$.

4.1 Solutions to the Laplacian Flow on Nilpotent Lie Groups

Lie groups admitting left-invariant calibrated G_2-structures constitute a convenient setting where it is possible to investigate the behaviour of the Laplacian flow in the non-compact case. In literature, results in this direction have been obtained on nilpotent and solvable Lie groups in various works [22, 27, 44, 45, 50]. In the non-solvable case, the first examples of calibrated G_2-structures have been exhibited only recently in [28], and the study of the Laplacian flow starting from some of them was done in [29].

The main peculiarity of the known non-compact examples is that the solution of (4.1) exists on an infinite time interval. We recall here a result of [22], while we refer the reader to [44, 45] for further examples.

Example 4.3 ([22])

1) On the nilpotent Lie algebra $\mathfrak{n}_2$, the solution of the Laplacian flow starting from the calibrated G_2-structure φ_2 given in Example 3.19 is

$$\varphi(t) = e^{147} + e^{267} + e^{357} + \left(\frac{10}{3}t + 1\right)^{3/5} e^{123} + e^{156} + e^{245} - e^{346},$$

where $t \in \left(-\frac{3}{10}, +\infty\right)$.

2) On the nilpotent Lie algebra $\mathfrak{n}_{12}$, the solution of the Laplacian flow starting from the calibrated G_2-structure φ_{12} given in Example 3.19 is

$$\varphi(t) = -e^{124} + e^{167} + e^{257} + e^{347} - e^{456} + \left(\frac{1}{3}t + 1\right)^{3/4} (e^{135} - e^{236}),$$

with $t \in (-3, +\infty)$.

In the previous example, both the calibrated G_2-structures considered as initial value for the Laplacian flow induce the nilsoliton inner product on the corresponding nilpotent Lie algebra (cf. Theorem 3.17 and Example 3.19). Furthermore, using suitable analytic techniques it is possible to show that the solution $\varphi(t)$ of (4.1) with $\varphi(0) = \varphi_4$ on $\mathfrak{n}_4$ and $\varphi(0) = \varphi_6$ on $\mathfrak{n}_6$ exists for $t \in (T, +\infty)$ with $T < 0$, see [22, Theorems 4.7, 4.8]. In all cases, it is then possible to analyze the behaviour of the solution $\varphi(t)$ when $t \to +\infty$.

Theorem 4.4 ([22]) *On the simply-connected nilpotent Lie groups* N_i, $i = 2, 4, 6, 12$, *the Laplacian flow starting from the left-invariant calibrated* G_2*-structure* φ_i *has a global solution defined for* $t \in (T, +\infty)$, *with* $T < 0$. *Moreover, all solutions converge to a flat* G_2*-structure when* $t \to +\infty$.

Remark 4.5 The nilpotent Lie algebra $\mathfrak{n}_2$ may be seen as a product algebra $\mathfrak{n}_2 = \mathfrak{n}' \oplus \mathbb{R}$ with $\dim(\mathfrak{n}') = 6$ and $\mathbb{R} = \langle e_7 \rangle$, and the calibrated G_2-structure φ_2 on it can be written as

$$\varphi_2 = \omega \wedge e^7 + \psi,$$

where (ω, ψ) is a symplectic half-flat SU(3)-structure on $\mathfrak{n}'$ (cf. Definition 2.2). Moreover, the solution of the Laplacian flow starting from φ_2 at $t = 0$ is of the form

$$\varphi(t) = f(t)\,\omega(t) \wedge e^7 + \psi(t),$$

where $(\omega(t), \psi(t))$ is a family of symplectic half-flat SU(3)-structures on $\mathfrak{n}'$, and the function $f : \left(-\frac{3}{10}, +\infty\right) \to \mathbb{R}^+$ is given by $f(t) = \left(\frac{10}{3}t + 1\right)^{-1/10}$.

This fact is a consequence of a more general result which holds for a suitable class of symplectic half-flat SU(3)-structures and allows to construct new examples of solutions of (4.1) on solvable Lie groups. For more details, we refer the reader to [27].

Remark 4.6 The investigation reviewed in this section can be carried out also for the *Laplacian coflow* for cocalibrated G_2-structures [40] and its modified version introduced in [31]. It turns out that the behaviour of these flows on solvable Lie groups is slightly different from the behaviour of the Laplacian flow. We refer the reader to [5, 6] for a detailed treatment.

Acknowledgements The first author was partially supported by MINECO-FEDER Grant MTM2014-54804-P and Gobierno Vasco Grant IT1094-16, Spain. The second and third authors were partially supported by GNSAGA of INdAM. The second author was also supported by PRIN 2015 "Real and complex manifolds: geometry, topology and harmonic analysis" of MIUR. All authors would like to thank the organizers of the workshop "G_2-manifolds and related topics" (Fields Institute, Toronto, August 2017).

References

1. Alekseevskiĭ, D. V., & Kimel'fel'd, B. N. (1975). Structure of homogeneous Riemannian spaces with zero Ricci curvature. *Funkcional. Anal. i Priložen.*, *9*(2), 5–11.
2. Apostolov, V., Drăghici, T., & Moroianu, A. (2001). A splitting theorem for Kähler manifolds whose Ricci tensors have constant eigenvalues. *International Journal of Mathematics*, *12*(7), 769–789.
3. Arroyo, R. M., & Lafuente, R. A. (2017). The Alekseevskii conjecture in low dimensions. *Mathematische Annalen*, *367*(1–2), 283–309.
4. Bagaglini, L., Fernández, M., & Fino, A. (2018). Coclosed G_2-structures inducing nilsolitons. *Forum of Mathematics*, *30*(1), 109–128.
5. Bagaglini, L., Fernández, M., & Fino, A. Laplacian coflow on the 7-dimensional Heisenberg group. *Asian Journal of Mathematics* arXiv:1704.00295 [math.DG]. (To appear)
6. Bagaglini, L., & Fino, A. (2018). The Laplacian coflow on almost-abelian Lie groups. *Annali di Matematica Pura ed Applicata*, *197*(6), 1855–1873.
7. Besse, A. L. (1987). *Einstein manifolds* (Vol. 10). Ergebnisse der Mathematik und ihrer Grenzgebiete. Berlin: Springer.
8. Böhm, C. (2015). On the long time behavior of homogeneous Ricci flows. *Commentarii Mathematici Helvetici*, *90*(3), 543–571.
9. E. Bonan. Sur des variétés riemanniennes à groupe d'holonomie G_2 ou Spin(7). *C. R. Acad. Sci. Paris Sér. A-B*, **262**, A127–A129, 1966.
10. Bryant, R. L. (1987). Metrics with exceptional holonomy. *Annals of Mathematics*, *126*(3), 525–576.
11. Bryant,R. L. (2006). Some remarks on G_2-structures. In *Proceedings of Gökova Geometry-Topology Conference Gökova Geometry/Topology Conference (GGT), 2005* (pp. 75–109). Gökova
12. Bryant, R . L. (2010). Non-embedding and non-extension results in special holonomy. *The many facets of geometry* (pp. 346–367). Oxford: Oxford University Press.
13. R. L. Bryant and F. Xu. Laplacian flow for closed G_2-structures: Short time behavior. arXiv:1101.2004 [math.DG].
14. Chiossi, S., & Salamon, S. (2002). *Differential geometry Valencia 2001* (pp. 115–133). The intrinsic torsion of SU(3) and G_2 structures River Edge: World Sci. Publ.
15. Chow, B., & Knopf, D. (2004). *The Ricci flow: an introduction* (Vol. 110). Mathematical surveys and monographs. Providence: American Mathematical Society.
16. Cleyton, R., & Ivanov, S. (2007). On the geometry of closed G_2-structures. *Communications in Mathematical Physics*, *270*(1), 53–67.
17. Cleyton, R., & Ivanov, S. (2008). Conformal equivalence between certain geometries in dimension 6 and 7. *Mathematical Research Letters*, *15*(4), 631–640.
18. Conti, D., & Fernández, M. (2011). Nilmanifolds with a calibrated G_2-structure. *Differential Geometry and its Applications*, *29*(4), 493–506.
19. Fernández-Culma, E. A. (2012). Classification of 7-dimensional Einstein Nilradicals. *Transformation Groups*, *17*(3), 639–656.
20. Fernández-Culma, E. A. (2014). Classification of Nilsoliton metrics in dimension seven. *Journal of Geometry and Physics*, *86*, 164–179.
21. Fernández, M., Fino, A., & Manero, V. (2015). G_2-structures on Einstein solvmanifolds. *Asian Journal of Mathematics*, *19*(2), 321–342.
22. Fernández, M., Fino, A., & Manero, V. (2016). Laplacian flow of closed G_2-structures inducing nilsolitons. *Journal of Geometric Analysis*, *26*(3), 1808–1837.
23. Fernández, M., Fino, A., & Raffero, A. (2016). Locally conformal calibrated G_2-manifolds. *Annali di Matematica Pura ed Applicata*, *195*(5), 1721–1736.
24. Fernández, M., & Gray, A. (1982). Riemannian manifolds with structure group G_2. *Annali di Matematica Pura ed Applicata*, *132*, 19–45.
25. Fino, A., & Raffero, A. (2015). Coupled SU(3)-structures and supersymmetry. *Symmetry*, *7*(2), 625–650.

26. Fino, A., & Raffero, A. (2015). Einstein locally conformal calibrated G_2-structures. *Mathematische Zeitschrift*, *280*(3–4), 1093–1106.
27. Fino, A., & Raffero, A. (2020). Closed warped G_2-structures evolving under the Laplacian flow. Annali Della Scuola Normale Superiore Di Pisa - Classe di Scienze. *20*(1), 315–348.
28. Fino, A., & Raffero, A. (2019). Closed G_2-structures on non-solvable Lie groups. *Rev. Mat. Complut.*, *32*(3), 837–851.
29. Fino, A. Raffero A. Remarks on homogeneous solutions of the G_2-Laplacian, arXiv:1905.13078, to appear in C. R. Math. Acad. Sci. Paris.
30. Goldberg, S. I. (1969). Integrability of almost Kähler manifolds. *Proceedings of the American Mathematical Society*, *21*, 96–100.
31. Grigorian, S. (2013). Short-time behavior of a modified Laplacian coflow of G_2-structures. *Advances in Mathematics*, *248*, 378–415.
32. Harvey, R., & Lawson, H. B, Jr. (1982). Calibrated geometries. *Acta Mathematica*, *148*, 47–157.
33. Heber, J. (1998). Noncompact homogeneous Einstein spaces. *Inventiones Mathematicae*, *133*(2), 279–352.
34. Hitchin, N. (2000). The geometry of three-forms in six and seven dimensions. *Journal of Differential Geometry*, *55*(3), 547–576.
35. Hitchin, N. (2001). Stable forms and special metrics. *Global differential geometry: the mathematical legacy of Alfred Gray (Bilbao, 2000)* (Vol. 288, pp. 70–89)., Contemporary mathematics Providence: American Mathematical Society.
36. Huang, H., Wang, Y., & Yao, C. (2018). Cohomogeneity-one G_2-Laplacian flow on 7-torus. *Journal of the London Mathematical Society*, *98*(2), 349–368.
37. Ivanov, S., Parton, M., & Piccinni, P. (2006). Locally conformal parallel G_2 and Spin(7) manifolds. *Mathematical Research Letters*, *13*(2–3), 167–177.
38. Ivey, T. (1993). Ricci solitons on compact three-manifolds. *Differential Geometry and its Applications*, *3*(4), 301–307.
39. Jensen, G. R. (1969). Homogeneous Einstein spaces of dimension four. *Journal of Differential Geometry*, *3*, 309–349.
40. Karigiannis, S., McKay, B., & Tsui, M.-P. (2012). Soliton solutions for the Laplacian coflow of some G_2-structures with symmetry. *Differential Geometry and its Applications*, *30*(4), 318–333.
41. Lauret, J. (2001). Ricci soliton homogeneous nilmanifolds. *Mathematische Annalen*, *319*(4), 715–733.
42. Lauret, J. (2010). Einstein solvmanifolds are standard. *Annals of Mathematics*, *172*(3), 1859–1877.
43. Lauret, J. (2011). Ricci soliton solvmanifolds. *Journal für die reine und angewandte Mathematik*, *650*, 1–21.
44. Lauret, J. (2017). Laplacian flow of homogeneous G_2-structures and its solitons. *Proceedings of the London Mathematical Society*, *114*(3), 527–560.
45. Lauret, J. (2017). Laplacian solitons: Questions and homogeneous examples. *Differential Geometry and its Applications*, *54*(B), 345–360.
46. Lotay, J. D., & Wei, Y. (2017). Laplacian flow for closed G_2 structures: Shi-type estimates, uniqueness and compactness. *Geometric and Functional Analysis*, *27*(1), 165–233.
47. Lotay, J.D., & Wei, Y. (2019). Stability of torsion-free G_2 structures along the Laplacian flow. *Journal of Differential Geometry*. *111*(3), 495–526.
48. Lotay, J.D., & Wei, Y. (2019). Laplacian flow for closed G_2 structures: Real analyticity. *Communications in Analysis and Geometry*. *27*(1), 73–109.
49. Milnor, J. (1976). Curvatures of left invariant metrics on Lie groups. *Advances in Mathematics*, *21*(3), 293–329.
50. Nicolini, M. (2018). Laplacian solitons on nilpotent Lie groups. *Bulletin of the Belgian Mathematical Society*. *25*(2), 183 196.
51. Nikonorov, Y. G. (2000). On the Ricci curvature of homogeneous metrics on noncompact homogeneous spaces. *Sibirskii Matematicheskii Zhurnal*, *41*(2), 421–429.
52. Verbitsky, M. (2008). An intrinsic volume functional on almost complex 6-manifolds and nearly Kähler geometry. *Pacific Journal of Mathematics*, *235*(2), 323–344.

Laplacian Flow for Closed G_2 Structures

Yong Wei

Abstract This is an expository article based on the author's talk in *Workshop on* G_2 *Manifolds and Related Topics* held in August 2017 at The Fields Institute. The aim is to explain the results obtained recently by the author and Jason D. Lotay on the Laplacian flow for closed G_2 structures and some related progress.

2010 Mathematics Subject Classification 53C44 · 53C25 · 53C10

1 G_2 Structures on 7-Manifolds

The group G_2 is one of the exceptional holonomy groups and is defined as the stabilizer of the following 3-form on the 7-dimensional Euclidean space $\mathbb{R}^7$:

$$\phi = e^{123} + e^{145} + e^{167} + e^{246} - e^{257} - e^{347} - e^{356},$$

where $e^{ijk} = e^i \wedge e^j \wedge e^k$ with respect to the basis $\{e^1, e^2, \ldots, e^7\}$ of $\mathbb{R}^7$. The group G_2 is a compact, connected, simply-connected, simple Lie subgroup of SO(7) of dimension 14. The group G_2 acts irreducibly on $\mathbb{R}^7$ and preserves the Euclidean metric and orientation on $\mathbb{R}^7$. If $*_\phi$ denotes the Hodge star determined by the metric and orientation, then G_2 also preserves the 4-form $*_\phi\phi$.

Let M be a 7-manifold. We say a 3-form φ on M is definite if for $x \in M$ there exists an homomorphism $u \in \mathrm{Hom}_{\mathbb{R}}(T_xM, \mathbb{R}^7)$ such that $u^*\phi = \varphi_x$. The space of definite 3-forms on M will be denoted by $\Omega^3_+(M)$. Since ϕ is invariant under the action of the group G_2, each definite 3-form will define a G_2 structure on M. The existence of G_2 structures is equivalent to the property that the manifold M is oriented

Y. Wei (✉)
School of Mathematical Sciences, University of Science and Technology of China, Hefei 230026, P.R. China
e-mail: yongwei@ustc.edu.cn

S. Karigiannis et al. (eds.), *Lectures and Surveys on G_2-Manifolds and Related Topics*, Fields Institute Communications 84, https://doi.org/10.1007/978-1-0716-0577-6_11

and spin. Note that as G_2 is a subgroup of SO(7), a G_2 structure φ defines a unique Riemannian metric $g = g_\varphi$ on M and an orientation such that

$$g_\varphi(u, v)\mathrm{vol}_{g_\varphi} = \frac{1}{6}(u \lrcorner \varphi) \wedge (v \lrcorner \varphi) \wedge \varphi, \quad \forall\, u, v \in C^\infty(TM).$$

The metric and orientation determine the Hodge star operator $*_\varphi$, and we define $\psi = *_\varphi \varphi$, which is sometimes called a positive 4-form. Notice that the relationship between g_φ and φ, and hence between ψ and φ, is nonlinear.

1.1 Type Decomposition of k-Forms

The group G_2 acts irreducibly on $\mathbb{R}^7$ (and hence on $\Lambda^1(\mathbb{R}^7)^*$ and $\Lambda^6(\mathbb{R}^7)^*$), but it acts reducibly on $\Lambda^k(\mathbb{R}^7)^*$ for $2 \le k \le 5$. Hence a G_2 structure φ induces splittings of the bundles $\Lambda^k T^*M$ $(2 \le k \le 5)$ into direct summands, which we denote by $\Lambda^k_l(T^*M, \varphi)$ so that l indicates the rank of the bundle. We let the space of sections of $\Lambda^k_l(T^*M, \varphi)$ be $\Omega^k_l(M)$. We have that

$$\Omega^2(M) = \Omega^2_7(M) \oplus \Omega^2_{14}(M), \qquad \Omega^3(M) = \Omega^3_1(M) \oplus \Omega^3_7(M) \oplus \Omega^3_{27}(M),$$

where

$$\begin{aligned} \Omega^2_7(M) &= \{\beta \in \Omega^2(M) | \beta \wedge \varphi = 2 *_\varphi \beta\} = \{X \lrcorner \varphi | X \in C^\infty(TM)\}, \\ \Omega^2_{14}(M) &= \{\beta \in \Omega^2(M) | \beta \wedge \varphi = - *_\varphi \beta\} = \{\beta \in \Omega^2(M) | \beta \wedge \psi = 0\}, \end{aligned}$$

and

$$\begin{aligned} \Omega^3_1(M) &= \{f\varphi | f \in C^\infty(M)\}, \quad \Omega^3_7(M) = \{X \lrcorner \psi | X \in C^\infty(TM)\}, \\ \Omega^3_{27}(M) &= \{\gamma \in \Omega^3(M) | \gamma \wedge \varphi = 0 = \gamma \wedge \psi\}. \end{aligned}$$

Hodge duality gives corresponding decompositions of $\Omega^4(M)$ and $\Omega^5(M)$.

The space $\Omega^3_{27}(M)$ deserves more attention. As in [3] we define a map i_φ : $\mathrm{Sym}^2(T^*M) \to \Omega^3(M)$ from the space of symmetric 2-tensors to the space of 3-forms, given locally by

$$\mathrm{i}_\varphi(h) = \frac{1}{2} h^l_i \varphi_{ljk} dx^i \wedge dx^j \wedge dx^k \tag{1.1}$$

where $h = h_{ij} dx^i dx^j \in \mathrm{Sym}^2(T^*M)$. Then $C^\infty(M) \otimes g_\varphi$ is mapped isomorphically onto $\Omega^3_1(M)$ under the map i_φ with $\mathrm{i}_\varphi(g_\varphi) = 3\varphi$, and the space of trace-free symmetric 2-tensors $\mathrm{Sym}^2_0(T^*M)$ is mapped isomorphically onto the space $\Omega^3_{27}(M)$.

1.2 Torsion of G_2 Structures

Given a G_2 structure $\varphi \in \Omega^3_+(M)$, if ∇ denotes the Levi-Civita connection with respect to g_φ, we can interpret $\nabla\varphi$ as the torsion of the G_2 structure φ. Following [25], we see that $\nabla\varphi$ lies in the space $\Omega^1_7(M) \otimes \Omega^3_7(M)$. Thus we can define a 2-tensor T which we shall call the full torsion tensor such that

$$\nabla_i \varphi_{jkl} = T_{im} g^{mn} \psi_{njkl}. \tag{1.2}$$

Using the decomposition of the spaces of forms on M determined φ, we can also decompose $d\varphi$ and $d\psi$ into types. Bryant [3] showed that there exist unique differential forms $\tau_0 \in \Omega^0(M)$, $\tau_1 \in \Omega^1(M)$, $\tau_2 \in \Omega^2_{14}(M)$ and $\tau_3 \in \Omega^3_{27}(M)$ such that

$$d\varphi = \tau_0\psi + 3\tau_1 \wedge \varphi + *_\varphi \tau_3, \tag{1.3}$$

$$d\psi = 4\tau_1 \wedge \psi + \tau_2 \wedge \varphi. \tag{1.4}$$

We call $\{\tau_0, \tau_1, \tau_2, \tau_3\}$ the intrinsic torsion forms of the G_2 structure φ. The full torsion tensor T_{ij} is related to the intrinsic torsion forms by the following (see [25]):

$$T_{ij} = \frac{\tau_0}{4} g_{ij} - (\tau_1^\# \lrcorner \varphi)_{ij} - (\bar{\tau}_3)_{ij} - \frac{1}{2}(\tau_2)_{ij}, \tag{1.5}$$

where $\bar{\tau}_3$ is the trace-free symmetric 2-tensor such that $\tau_3 = \mathrm{i}_\varphi(\bar{\tau}_3)$.

If $\nabla\varphi = 0$, we say the G_2 structure φ is torsion-free on M. The torsion-free condition clearly implies that $d\varphi = 0 = d^*_\varphi\varphi$ on M. Fernández and Gray [12] showed that $d\varphi = 0 = d^*_\varphi\varphi$ also implies $\nabla\varphi = 0$ on M, which also follows from the Eq. (1.5). The key property of a torsion-free G_2 structure φ is that the holonomy group $\mathrm{Hol}(g_\varphi) \subseteq G_2$, and thus the manifold (M, g_φ) is Ricci-flat. Moreover, one can characterise the compact G_2 manifolds (i.e., compact manifolds with torsion-free G_2 structures) with $\mathrm{Hol}(g_\varphi) = G_2$ as those with finite fundamental group. Thus understanding torsion-free G_2 structures is crucial for constructing Riemannian manifolds with holonomy G_2.

While there are some explicit examples of manifolds which admit torsion-free G_2 structures for which the holonomy of the induced metric is properly contained in G_2, for example the product of circle S^1 with a Calabi-Yau 3-fold and the product of 3-torus $\mathbb{T}^3$ with a Calabi-Yau 2-fold, the construction of manifolds which admit torsion-free G_2 structures with holonomy equal to G_2 is a hard and important problem. The first local existence result of metrics with holonomy G_2 was obtained by Bryant [2] using the theory of exterior differential systems. Then Bryant–Salamon [4] constructed the first complete non-compact manifolds with holonomy G_2, which are the spinor bundle of S^3 and the bundles of anti-self-dual 2-forms on S^4 and $\mathbb{CP}^2$. In [22], Joyce constructed the first examples of compact 7-manifolds with holonomy G_2 and many further compact examples have now been constructed [7, 24, 29].

1.3 Closed G_2 Structures

If φ is closed, i.e. $d\varphi = 0$, then (1.3) implies that τ_0, τ_1 and τ_3 are all zero, so the only non-zero torsion form is $\tau_2 \in \Omega^2_{14}(M)$. In this case, we write $\tau = \tau_2$ for simplicity. Then from (1.5) we have that the full torsion tensor satisfies $T_{ij} = -\frac{1}{2}\tau_{ij}$ and is a skew-symmetric 2-tensor. By (1.4) and $\tau \in \Omega^2_{14}(M)$, we have $d\psi = \tau \wedge \varphi = -*_\varphi \tau$, which implies that

$$d^*\tau = *_\varphi d *_\varphi \tau = - *_\varphi d^2\psi = 0 \tag{1.6}$$

and the Hodge Laplacian of φ is equal to $\Delta_\varphi \varphi = -d *_\varphi d\psi = d\tau$. We computed in [34] (see also [3]) that

$$\Delta_\varphi \varphi = \mathrm{i}_\varphi(h) \in \Omega^3_1(M) \oplus \Omega^3_{27}(M) \tag{1.7}$$

where h is the symmetric 2-tensor given as follows:

$$h_{ij} = -\nabla_m T_{ni} \varphi_j^{\ mn} - \frac{1}{3}|T|^2 g_{ij} - T_{ik} g^{kl} T_{lj}. \tag{1.8}$$

Since φ determines a unique metric $g = g_\varphi$ on M, we then have the Riemann curvature tensor $\mathrm{Rm} = \{R_{ijkl}\}$, the Ricci tensor $R_{ij} = g^{kl} R_{ijkl}$ and the scalar curvature $\mathrm{R} = g^{ij} R_{ij}$ of (M, g_φ). For closed G_2 structure φ, we computed in [34] that the Ricci curvature is equal to

$$R_{ij} = \nabla_m T_{ni} \varphi_j^{\ mn} - T_{ik} g^{kl} T_{lj}, \tag{1.9}$$

and then the scalar curvature $R = -|T|^2$. With (1.9) we can write the symmetric tensor h in (1.8) as

$$h_{ij} = -R_{ij} - \frac{1}{3}|T|^2 g_{ij} - 2T_{ik} g^{kl} T_{lj}. \tag{1.10}$$

2 Laplacian Flow for Closed G_2 Structures

Since Hamilton [16] introduced the Ricci flow in 1982, geometric flows have been an important tool in studying geometric structures on manifolds. For example, Ricci flow was instrumental in proving the Poincaré conjecture and the $\frac{1}{4}$-pinched differentiable sphere theorem, and Kähler–Ricci flow has proved to be a useful tool in Kähler geometry, particularly in low dimensions. In 1992, Bryant (see [3]) proposed the Laplacian flow for closed G_2 structures

$$\begin{cases} \dfrac{\partial}{\partial t}\varphi(t) &= \Delta_{\varphi(t)}\varphi(t), \\ d\varphi(t) &= 0, \\ \varphi(0) &= \varphi_0, \end{cases} \tag{2.1}$$

where $\Delta_\varphi = dd_\varphi^* + d_\varphi^* d$ is the Hodge Laplacian with respect to g_φ and φ_0 is an initial closed G_2 structure. The stationary points of the flow are harmonic φ, which on a compact manifold are precisely the torsion-free G_2 structures, so the Laplacian flow provides a tool for studying the existence of torsion-free G_2 structures on a manifold admitting closed G_2 structures. The goal is to understand the long-time behavior of the Laplacian flow on compact manifolds M; specifically, to understand conditions under which the flow will converge to a torsion-free G_2 structure. We remark that there are other proposed flows which also have torsion-free G_2 structures as stationary points (e.g. [15, 26, 42]).

2.1 Gradient Flow of Volume Functional

Another motivation for studying the Laplacian flow comes from work of Hitchin [19] (see also [5]), which demonstrates its relationship to a natural volume functional. Let $\bar{\varphi}$ be a closed G_2 structure on a compact 7-manifold M and let $[\bar{\varphi}]_+$ be the open subset of the cohomology class $[\bar{\varphi}]$ consisting of G_2 structures. Define the volume functional on M by

$$\mathrm{Vol}(M, \varphi) = \frac{3}{7}\int_M \varphi \wedge *_\varphi \varphi, \quad \varphi \in [\bar{\varphi}]_+. \tag{2.2}$$

In the arXiv version of [19], Hitchin showed that $\varphi \in [\bar{\varphi}]_+$ is a critical point of $\mathrm{Vol}(M, \varphi)$ if and only if $d *_\varphi \varphi = 0$, i.e. φ is torsion-free.

Moreover, the Laplacian flow (2.1) can be viewed as the gradient flow of the volume functional (2.2). Since $\varphi(t)$ evolves in the same cohomology class with the initial data φ_0, we can write $\varphi(t) = \varphi_0 + d\eta(t)$ for some time dependent 2-form $\eta(t)$. To calculate the variation of the volume functional, we need to compute the variation of $*_{\varphi(t)}\varphi(t)$. This has already been computed in [3, 23]:

$$\frac{\partial}{\partial t}(*_{\varphi(t)}\varphi(t)) = \frac{4}{3} *_{\varphi(t)} \pi_1\left(\frac{\partial\varphi(t)}{\partial t}\right) + *_{\varphi(t)}\pi_7\left(\frac{\partial\varphi(t)}{\partial t}\right) - *_{\varphi(t)}\pi_{27}\left(\frac{\partial\varphi(t)}{\partial t}\right), \tag{2.3}$$

where π_k's are the respective projections to the invariant subspaces of $\Omega^3(M)$ and are determined by $\varphi(t)$. Then

$$\begin{aligned}
\frac{d}{dt}\mathrm{Vol}(M, \varphi(t)) &= \frac{3}{7}\int_M \left(\frac{\partial\varphi(t)}{\partial t} \wedge *_{\varphi(t)}\varphi(t) + \varphi(t) \wedge \frac{\partial}{\partial t}(*_{\varphi(t)}\varphi(t))\right) \\
&= \int_M \frac{\partial\varphi(t)}{\partial t} \wedge *_{\varphi(t)}\varphi(t) \\
&= \int_M \langle \frac{\partial\eta(t)}{\partial t}, d_{\varphi(t)}^*\varphi(t)\rangle *_{\varphi(t)} 1.
\end{aligned}$$

Thus gradient flow of the volume functional within the same cohomology class is given by

$$\frac{\partial \varphi(t)}{\partial t} = d\frac{\partial \eta(t)}{\partial t} = dd^*_{\varphi(t)}\varphi(t) = \Delta_{\varphi(t)}\varphi(t),$$

which is exactly the Laplacian flow. Then along the Laplacian flow, the volume will increase unless $\varphi(t)$ is torsion-free. By examining the second variation of the volume functional, Bryant [3] showed that if $\bar{\varphi}$ is torsion-free, then $\mathrm{Diff}^0(M)\cdot\bar{\varphi}$ is a local maximum of the volume functional on the moduli space $\mathrm{Diff}^0(M)\setminus[\bar{\varphi}]_+$. This gives rise to the following natural question:

Question 2.1 ([3]) *Starting from a initial data $\varphi_0 \in [\bar{\varphi}]_+$ which is sufficiently close to $\bar{\varphi}$ in an appropriate norm, does the Laplacian flow converge to a point on $\mathrm{Diff}^0(M)\cdot\bar{\varphi}$?*

In the statement of Question 2.1, we assumed the existence of a torsion-free G_2 structure $\bar{\varphi}$ on M. In 1996, Joyce [22] proved a criterion for the existence of torsion-free G_2 structures, which says that if one can find a G_2 structure φ with $d\varphi = 0$ on a compact 7-manifold M, whose torsion is sufficiently small in a certain sense, then there exists a torsion-free G_2-structure $\bar{\varphi} \in [\varphi]$ on M which is close to φ. This result has been used to construct compact examples of manifolds with G_2 holonomy. It would be interesting to give a new proof of Joyce's result [22] using the Laplacian flow.

Generally, one cannot expect that the Laplacian flow will converge to a torsion-free G_2 structure, even if it has long-time existence. There are compact 7-manifolds with closed G_2 structures that cannot admit holonomy G_2 metrics for topological reasons (c.f. [9, 10]), and Bryant [3] showed that the Laplacian flow starting with a particular one of these examples will exist for all time but it does not converge; for instance, the volume of the associated metrics will increase without bound. Some explicit examples of the solution to the Laplacian flow which exist for all time and converge can be found in [11, 13, 21].

2.2 *Short Time Existence*

Recall that the Hodge Laplacian Δ_φ is related to the analyst's Laplacian $\Delta = g^{ij}\nabla_i\nabla_j$ by the Weitzenbock formula:

$$\Delta_\varphi \omega = -\Delta\omega + \mathcal{R}(\omega) \tag{2.4}$$

for any $(0, k)$-tensor ω, where $\mathcal{R}$ is the Weitzenbock curvature operator. Since the Laplacian flow (2.1) is defined by the Hodge Laplacian, it appears at first sight to have the wrong sign for the parabolicity. However, if $d\varphi = 0$, using definition (1.2) of the torsion tensor and the divergence-free property (1.6) of τ, we see that $\Delta\varphi$ involves only up to first order derivatives of φ and thus the second order part of the

Hodge Laplacian $\Delta_\varphi\varphi$ lies in the part $\mathcal{R}(\varphi)$ of (2.4). Using DeTurck's trick in the Ricci flow, Bryant–Xu [5] modified the Laplacian flow by an operator of the form $\mathcal{L}_{V(\varphi)}\varphi = d(V \lrcorner \varphi) + V \lrcorner d\varphi = d(V \lrcorner \varphi)$ for some vector field $V(\varphi)$ and showed that the Laplacian–DeTurck flow

$$\frac{\partial \varphi(t)}{\partial t} = \Delta_{\varphi(t)}\varphi(t) + \mathcal{L}_{V(\varphi)}\varphi(t) \tag{2.5}$$

is strictly parabolic in the direction of closed forms by choosing a special vector field $V(\varphi)$. In fact, if $d\theta = 0$, they calculated that the linearization of RHS of (2.5) is

$$\frac{d}{d\epsilon}\bigg|_{\epsilon=0} \left(\Delta_{\varphi+\epsilon\theta}(\varphi+\epsilon\theta) + \mathcal{L}_{V(\varphi+\epsilon\theta)}(\varphi+\epsilon\theta)\right) = -\Delta_\varphi\theta + d\Phi(\theta) \tag{2.6}$$

where $d\Phi(\theta)$ is algebraic linear in θ and $d\Phi(\theta) = 0$ if φ is torsion-free. However, no existing theory of parabolic equations can be used directly since the parabolicity of (2.5) is only true in the direction of closed forms. Fortunately, by using the Nash Moser inverse function theorem [17] for tame Féchet spaces, Bryant and Xu proved the following short time existence theorem.

Theorem 2.2 (Bryant–Xu [5]) *Assume that M is compact and φ_0 is a closed G_2 structure on M. Then the Laplacian flow has a unique solution for a short time $t \in [0, \varepsilon)$ with ε depending on φ_0.*

As in the Ricci flow, we can also write the Laplacian–DeTurck flow (2.5) explicitly in local coordinates. Let $\tilde{g}$ be a fixed Riemannian metric on M and $\tilde{\nabla}$, $\tilde{\Gamma}^k_{ij}$ be the corresponding Levi-Civita connection and Christoffel symbols. We know that the difference $\Gamma^j_{kl} - \tilde{\Gamma}^j_{kl}$ of the Levi-Civita connections of the metrics g and $\tilde{g}$ is a well-defined tensor on M. This gives us a vector field V on M with

$$V_i = g_{ij}g^{kl}(\Gamma^j_{kl} - \tilde{\Gamma}^j_{kl}), \tag{2.7}$$

which is just the vector field chosen in Ricci-DeTurck flow [41]. By a direct but lengthy computation, we can show that if $d\varphi = 0$, the Laplacian–DeTurck flow Eq. (2.5) with V given by (2.7) has the following expression in local coordinates:

$$\frac{\partial}{\partial t}\varphi_{ijk} = g^{pq}\tilde{\nabla}_p\tilde{\nabla}_q\varphi_{ijk} + l.o.t \tag{2.8}$$

and the associated metric g_{ij} evolves by

$$\frac{\partial}{\partial t}g_{ij} = g^{pq}\tilde{\nabla}_p\tilde{\nabla}_q g_{ij} + l.o.t \tag{2.9}$$

where the lower order terms only involve the φ, g, $\tilde{\nabla}g$ and $\tilde{\nabla}\varphi$ and can be written down explicitly. The readers may find that the vector field V is different at first sight with the one chosen by Bryant–Xu [5]. However, we can see that they are essentially the same by considering the linearization of V in the direction of closed forms (see also [15, pp. 400–401]).

2.3 Evolution Equations

Since each G_2 structure induces a unique Riemannian metric on the manifold, the Laplacian flow (2.1) induces a flow for the associated Riemannian metric $g(t) = g_{\varphi(t)}$. Recall that under a general flow for G_2 structures

$$\frac{\partial}{\partial t}\varphi(t) = \mathrm{i}_{\varphi(t)}(h(t)) + X \lrcorner \psi(t), \tag{2.10}$$

where $h(t) \in \mathrm{Sym}^2(T^*M)$ and $X(t) \in C^\infty(TM)$, it is well known that (see [3, 23] and explicitly [25]) the associated metric tensor $g(t)$ evolves by

$$\frac{\partial}{\partial t}g(t) = 2h(t). \tag{2.11}$$

By (1.7) and (1.10), we deduce that the associated metric $g(t)$ of the solution $\varphi(t)$ of the Laplacian flow evolves by

$$\frac{\partial}{\partial t}g_{ij} = -2R_{ij} - \frac{2}{3}|T|^2 g_{ij} - 4T_{ik}g^{kl}T_{lj}, \tag{2.12}$$

which corresponds to the Ricci flow plus some lower order terms involving the torsion tensor, as already observed in [3]. Then it's easy to see that the volume form $\mathrm{vol}_{g(t)}$ evolves by

$$\frac{\partial}{\partial t}\mathrm{vol}_{g(t)} = \frac{1}{2}\mathrm{tr}_g\left(\frac{\partial}{\partial t}g(t)\right)\mathrm{vol}_{g(t)} = \frac{2}{3}|T|^2\mathrm{vol}_{g(t)}, \tag{2.13}$$

where we used the fact that the scalar curvature $R = -|T|^2$. Hence, along the Laplacian flow, the volume of M with respect to the associated metric $g(t)$ will nondecrease (as already noted in Sect. 2.1). Since the torsion tensor T is defined by the first covariant derivative of φ and the Riemannian curvature tensor Rm involves up to second order derivatives of the metric, we calculated in [34] that the evolution equations of the torsion tensor and Riemannian curvature tensor along the Laplacian flow are of the form

$$\frac{\partial}{\partial t}T = \Delta T + Rm * T + Rm * T * \psi + \nabla T * T * \varphi + T * T * T, \tag{2.14}$$

$$\frac{\partial}{\partial t}Rm = \Delta Rm + Rm * Rm + Rm * T * T + \nabla^2 T * T + \nabla T * \nabla T, \quad (2.15)$$

where we use $*$ to mean some contraction using the metric $g(t)$ associated with $\varphi(t)$.

3 Foundational Results of Laplacian Flow

In this section, we discuss several foundational results on the Laplacian flow, which are important for further studies.

3.1 Shi-type Estimates

The first result is the derivative estimates of the solution to the Laplacian flow. For a solution $\varphi(t)$ of the Laplacian flow (2.1), we define the quantity

$$\Lambda(x,t) = \left(|\nabla T(x,t)|^2_{g(t)} + |Rm(x,t)|^2_{g(t)}\right)^{\frac{1}{2}}. \quad (3.1)$$

Notice that the torsion tensor T is determined by the first order derivative of φ and the curvature tensor Rm is second order in the metric g_φ, so both Rm and ∇T are second order in φ. We show that a bound on $\Lambda(x,t)$ will induce a priori bounds on all derivatives of Rm and ∇T for positive time. More precisely, we have the following.

Theorem 3.1 ([34]) *Suppose that $K > 0$ and $\varphi(t)$ is a solution of the Laplacian flow* (2.1) *for closed* G_2 *structures on a compact manifold M^7 for $t \in [0, \frac{1}{K}]$. For all $k \in \mathbb{N}$, there exists a constant C_k such that if $\Lambda(x,t) \le K$ on $M^7 \times [0, \frac{1}{K}]$, then*

$$|\nabla^k Rm(x,t)|_{g(t)} + |\nabla^{k+1} T(x,t)|_{g(t)} \le C_k t^{-\frac{k}{2}} K, \quad t \in (0, \frac{1}{K}]. \quad (3.2)$$

We call the estimates (3.2) Shi-type estimates for the Laplacian flow, because they are analogues of the well-known Shi derivative estimates in the Ricci flow. In Ricci flow, a Riemann curvature bound will imply bounds on all the derivatives of the Riemann curvature: this was proved by Bando [1] and comprehensively by Shi [41] independently. The techniques used in [1, 41] were introduced by Bernstein (in the early twentieth century) for proving gradient estimates via the maximum principle, and was also the key in [34] to prove Theorem 3.1. A key motivation for defining $\Lambda(x,t)$ as in (3.1) is that the evolution equations of $|\nabla T(x,t)|^2$ and $|Rm(x,t)|^2$ both have some bad terms, but the chosen combination kills these terms and yields an effective evolution equation for $\Lambda(x,t)$ which looks like

$$\frac{\partial}{\partial t}\Lambda(x,t)^2 \leq \Delta\Lambda(x,t)^2 + C\Lambda(x,t)^3$$

for some positive constant C. This shows that the quantity Λ has similar properties to Riemann curvature under Ricci flow. Moreover, it implies that the assumption $\Lambda(x,t) \leq K$ in Theorem 3.1 is reasonable as $\Lambda(x,t)$ cannot blow up quickly. We remark that the constant C_k depends on the order of differentiation. In a joint work with Lotay [36], we showed that C_k are of sufficiently slow growth in the order k and then we deduced that the G_2 structure $\varphi(t)$ and associated metric $g_{\varphi(t)}$ are real analytic at each fixed time $t > 0$.

The Shi-type estimates could be used to study finite-time singularities of the Laplacian flow. Given an initial closed G_2 structure φ_0 on a compact 7-manifold, Theorem 2.2 tells us there exists a solution $\varphi(t)$ of the Laplacian flow on a maximal time interval $[0, T_0)$. If T_0 is finite, we call T_0 the singular time. Using our global derivative estimates (3.2), we have the following long time existence result on the Laplacian flow.

Theorem 3.2 ([34]) *If $\varphi(t)$ is a solution of the Laplacian flow* (2.1) *on a compact manifold M^7 in a maximal time interval* $[0, T_0)$ *with $T_0 < \infty$, then*

$$\lim_{t\nearrow T_0}\sup_{x\in M}\Lambda(x,t) = \infty.$$

Moreover, there exists a positive constant C such that the blow-up rate satisfies

$$\sup_{x\in M}\Lambda(x,t) \geq \frac{C}{T_0 - t}.$$

In other words, Theorem 3.2 shows that the solution $\varphi(t)$ of the Laplacian flow for closed G_2 structures will exist as long as the quantity $\Lambda(x,t)$ in (3.1) remains bounded.

3.2 Uniqueness

Given a closed G_2 structure φ_0 on a compact 7-manifold, Theorem 2.2 says that there exists a unique solution to the Laplacian flow for a short time interval $t \in [0, \varepsilon)$. The proof in [5] relies on the Nash–Moser inverse function theorem [16] and the DeTurck's trick. In [34], we gave a new proof the forward uniqueness by adapting an energy approach used previously by Kotschwar [28] for Ricci flow. The idea is to define an energy quantity $\mathcal{E}(t)$ in terms of the differences of the G_2 structures, metrics, connections, torsion tensors and Riemann curvatures of two Laplacian flows, which vanishes if and only if the flows coincide. By deriving a differential inequality for $\mathcal{E}(t)$, it can be shown that $\mathcal{E}(t) = 0$ if $\mathcal{E}(0) = 0$, which gives the forward uniqueness. We also proved in [34] a backward uniqueness result for the solution of Laplacian flow by applying a general backward uniqueness theorem in [27] for time-dependent sections of vector bundles satisfying certain differential inequalities.

Theorem 3.3 ([34]) *Suppose $\varphi(t)$, $\tilde{\varphi}(t)$ are two solutions to the Laplacian flow* (2.1) *on a compact manifold M^7 for $t \in [0, \epsilon]$, $\epsilon > 0$. If $\varphi(s) = \tilde{\varphi}(s)$ for some $s \in [0, \epsilon]$, then $\varphi(t) = \tilde{\varphi}(t)$ for all $t \in [0, \epsilon]$.*

An application of Theorem 3.3 is that on a compact manifold M^7, the subgroup $I_{\varphi(t)}$ of diffeomorphisms of M isotopic to the identity and fixing $\varphi(t)$ is unchanged along the Laplacian flow. Since I_φ is strongly constrained for a torsion-free G_2 structure φ on M, this gives a test for when the Laplacian flow with a given initial condition could converge.

3.3 Compactness and κ-Non-collapsing

In the study of Ricci flow, Hamilton's compactness theorem [18] and Perelman's κ-non-collapsing estimate [38] are two essential tools to study the behavior of the flow near a singularity. We also have the analogous results for the Laplacian flow, which were proved by the author and Lotay [34] and Chen [6] respectively.

Theorem 3.4 ([34]) *Let M_i be a sequence of compact 7-manifolds and let $p_i \in M_i$ for each i. Suppose that, for each i, $\varphi_i(t)$ is a solution to the Laplacian flow* (2.1) *on M_i for $t \in (a, b)$, where $-\infty \leq a < 0 < b \leq \infty$. Suppose that*

$$\sup_i \sup_{x \in M_i, t \in (a,b)} \Lambda_{\varphi_i}(x, t) < \infty \tag{3.3}$$

and

$$\inf_i inj(M_i, g_i(0), p_i) > 0. \tag{3.4}$$

Then there exists a 7-manifold M, a point $p \in M$ and a solution $\varphi(t)$ of the Laplacian flow on M for $t \in (a, b)$ such that, after passing to a subsequence, $(M_i, \varphi_i(t), p_i)$ converge to $(M, \varphi(t), p)$ as $i \to \infty$.

To prove Theorem 3.4, we first proved in [34] a Cheeger–Gromov-type compactness theorem for the space of G_2 structures, which states that the space of G_2 structures with bounded $|\nabla^{k+1}T| + |\nabla^k Rm|$, $k \geq 0$, and bounded injectivity radius is compact. Given this, Theorem 3.4 follows from a similar argument for the analogous compactness theorem in Ricci flow as in [18], with the help of the Shi-type estimate in Theorem 3.1.

The κ-non-collapsing estimate is an estimate on the volume ratio which only involves the Riemannian metric. A Riemannian metric g on a manifold M is κ-non-collapsed relative to an upper bound on the scalar curvature of the metric on the scale ρ if for any geodesic ball $B_g(p, r)$ with $r < \rho$ such that $\sup_{B_g(p,r)} R_g \leq r^{-2}$, there holds $\mathrm{Vol}(B_g(p,r)) \geq \kappa r^n$. By using the same $\mathcal{W}$ functional, Chen [6] generalized Perelman's κ-non-collapsing theorem [38] for Ricci flow to any flow

$$\frac{\partial}{\partial t}g(t) = -2\mathrm{Ric}(g(t)) + E(t) \tag{3.5}$$

for the Riemannian metric $g(t)$, where $E(t)$ is a symmetric 2-tensor.

Theorem 3.5 ([6]) *If $|E(t)|_{g(t)}$ is bounded along the flow* (3.5) *for $t \in [0,s)$ with $s < \infty$, then there exists $\kappa > 0$ such that for all $t \in [0,s)$, $g(t)$ is κ-non-collapsed relative to the upper bound on the scalar curvature on the scale $\rho = \sqrt{s}$.*

Theorem 3.5 applies effectively to our Laplacian flow since the induced metric flow is just a perturbation of the Ricci flow, see (2.12). The κ-non-collapsing estimate is useful to estimate the lower bound on the injectivity radius, which together with the Shi-type estimate in Theorem 3.1 guarantees the condition of the compactness theorem for the purpose of the blow up analysis.

3.4 Solitons

Given a 7-manifold M, a Laplacian soliton on M is a triple (φ, X, λ) satisfying

$$\Delta_\varphi \varphi = \lambda\varphi + \mathcal{L}_X\varphi, \tag{3.6}$$

where $d\varphi = 0$, $\lambda \in \mathbb{R}$, X is a vector field on M and $\mathcal{L}_X\varphi$ is the Lie derivative of φ in the direction of X. Laplacian solitons give self-similar solutions to the Laplacian flow. Specifically, suppose (φ_0, X, λ) satisfies (3.6). Define $\rho(t) = (1 + \frac{2}{3}\lambda t)^{\frac{3}{2}}$, $X(t) = \rho(t)^{-\frac{2}{3}}X$, and let ϕ_t be the family of diffeomorphisms generated by the vector fields $X(t)$ such that ϕ_0 is the identity. Then $\varphi(t)$ defined by $\varphi(t) = \rho(t)\phi_t^*\varphi_0$ is a solution of the Laplacian flow (2.1), which only differs by a scaling factor $\rho(t)$ and pull-back by a diffeomorphism ϕ_t for different times t. We say a Laplacian soliton (φ, X, λ) is expanding if $\lambda > 0$; steady if $\lambda = 0$; and shrinking if $\lambda < 0$.

The soliton solutions of the Laplacian flow are expected to play a role in understanding the behavior of the flow near singularities. Thus the classification is an important problem. In this direction, Lin [30] proved that there are no compact shrinking solitons, and the only compact steady solitons are given by torsion-free G_2 structures. In [34], we show that any Laplacian soliton that is an eigenform (i.e., $X = 0$ in (3.6)) must be an expander or torsion-free. Hence, stationary points of the Laplacian flow on 7-manifold (not necessarily compact) are given by torsion-free G_2 structures. Moreover, we show that there are no compact Laplacian solitons that are eigenforms unless φ is torsion-free. Combining this with Lin's result, any nontrivial Laplacian soliton on a compact manifold M (if it exists) must satisfy (3.6) for $\lambda > 0$ and $X \neq 0$. This phenomenon is somewhat surprising, since it is very different from

Ricci solitons $\mathrm{Ric} + \mathcal{L}_X g = \lambda g$: when $X = 0$, the Ricci soliton equation is just the Einstein equation and there are many examples of compact Einstein metrics.

Since a G_2 structure φ determines a unique metric g, it is natural to ask what condition the Laplacian soliton Eq. (3.6) on φ will impose on g. By writing $\mathcal{L}_X\varphi$ with respect to the type decomposition of 3-forms, we derived from the Laplacian soliton Eq. (3.6) that the induced metric g_φ satisfies, in local coordinates,

$$-R_{ij} - \frac{1}{3}|T|^2 g_{ij} - 2T_{ik}g^{kl}T_{lj} = \frac{1}{3}\lambda g_{ij} + \frac{1}{2}(\mathcal{L}_X g)_{ij} \tag{3.7}$$

and the vector field X satisfies $d^*(X \lrcorner \varphi) = 0$. In particular, we deduce that any Laplacian soliton (φ, X, λ) must satisfy $7\lambda + 3\mathrm{div}(X) = 2|T|^2 \geq 0$, which leads to a new short proof of Lin's result [30] for the closed case.

Remark 3.6 We remark that there are many new results concerning the soliton solutions of the Laplacian flow. We refer the readers to [11, 31–33, 37] for details.

4 Extension Theorem

As we said in Sect. 3, the compactness theorem and the non-collapsing estimate could be used to study the singularities of the Laplacian flow. Theorem 3.2 already characterized the finite time singularities as the points where the quantity $\Lambda(x, t)$ (defined in (3.1)) blow up. This means that the solution of the Laplacian flow exists as long as $\Lambda(x, t)$ remains bounded. The quantity $\Lambda(x, t)$ consists of the full information of the G_2 structure $\varphi(t)$ up to second derivatives. It's interesting to see whether some weaker quantity can control the behavior of the flow. Using the compactness theorem, we improved Theorem 3.2 to the following desirable result, which states that the Laplacian flow will exist as long as the velocity of the flow remains bounded.

Theorem 4.1 ([34]) *Let M be a compact 7-manifold and $\varphi(t)$, $t \in [0, T_0)$, where $T_0 < \infty$, be a solution to the Laplacian flow* (2.1) *with associated metric $g(t)$ for each t. If the velocity of the flow satisfies $\sup_{M\times[0,T_0)} |\Delta_\varphi\varphi(x, t)|_{g(t)} < \infty$, then the solution $\varphi(t)$ can be extended past time T_0.*

Note that for closed G_2 structures, the velocity $\Delta_\varphi\varphi = d\tau$ is just some components of the first derivative of the torsion tensor. Theorem 4.1 is the G_2 analogue of Sesum's [39] theorem that the Ricci flow exists as long as the Ricci tensor remains bounded. It is an open question whether the scalar curvature (the trace of the Ricci tensor) is enough to control the behavior of the Ricci flow, though it is known for Type-I Ricci flow [8] and Kähler–Ricci flow [44]. For a closed G_2 structure φ, the velocity $\Delta_\varphi\varphi = \mathrm{i}_\varphi(h)$ is equivalent to a symmetric 2-tensor h with trace equal to $\frac{2}{3}|T|^2$. Since the scalar curvature of the metric induced by φ is $-|T|^2$, comparing with Ricci flow one may ask whether the Laplacian flow for closed G_2 structures will exist as long as the torsion tensor remains bounded. This is also the natural question to ask from the point of view of G_2 geometry. However, even though $-|T|^2$ is the scalar curvature, it is only *first order* in φ, rather than second order like $\Delta_\varphi\varphi$, so it would be a major step forward to control the Laplacian flow using just a bound on the torsion tensor.

The Proof of Theorem 4.1 involves a standard blow up analysis using the compactness theorem in Sect. 3. However, the non-collapsing estimate is not required for the proof. In fact, for a closed G_2 structure φ, $\Delta_\varphi\varphi = i_\varphi(h)$ and $|\Delta_\varphi\varphi|_g^2 = (\mathrm{tr}_g(h))^2 + 2|h|^2$ with h given by (1.10). Then the condition $|\Delta_{\varphi(t)}\varphi(t)|_{g(t)} < \infty$ is equivalent to $\sup_{M\times[0,T_0)}|h(t)| < \infty$, which implies the uniform continuity of the metric $g(t)$. A desired injectivity radius estimate then follows and the blow up analysis works.

Remark 4.2 By applying the compactness theorem and the non-collapsing estimate and using the method in [43], Chen [6] improved the result in Theorem 4.1. See [6] for the details. Moreover, Fine and Yao studied in [14] the hypersymplectic flow on a compact 4-manifold X related to the Laplacian flow on the 7-manifold $X \times \mathbb{T}^3$ and proved that the flow extends as long as the scalar curvature of the corresponding G_2 structure remains bounded.

5 Stability of Torsion-Free G_2 Structures

As we stated in Question 2.1, Bryant asked the question whether the Laplacian flow with initial G_2 structure φ_0 which is sufficiently close to a torsion-free G_2 structure $\bar\varphi$ will converge to a point in the diffeomorphism orbit of $\bar\varphi$. Jointly with Lotay, we gave a positive answer in [35].

Theorem 5.1 ([35]) *Let $\bar\varphi$ be a torsion-free* G_2 *structure on a compact 7-manifold M. Then there is a neighborhood $\mathcal{U}$ of $\bar\varphi$ such that for any $\varphi_0 \in [\bar\varphi]_+ \cap \mathcal{U}$, the Laplacian flow* (2.1) *with initial value φ_0 exists for all $t \in [0,\infty)$ and converges to $\varphi_\infty \in \mathrm{Diff}^0(M)\cdot\bar\varphi$ as $t \to \infty$. In other words, torsion-free* G_2 *structures are (weakly) dynamically stable along the Laplacian flow for closed* G_2 *structures.*

The Proof of Theorem 5.1 is inspired by the proof of an analogous result in Ricci flow: Ricci-flat metrics are dynamically stable along the Ricci flow. The idea is to combine arguments for the Ricci flow case [20, 40] with the particulars of the geometry of closed G_2 structures and new higher order estimates for the Laplacian flow derived by the author with Lotay in [34]. We first look at the Laplacian–DeTurck flow (2.5). By linearizing (2.5) at the torsion-free G_2 structure $\bar\varphi$, we have (see (2.6)):

$$\frac{d}{d\epsilon}\Big|_{\epsilon=0}\left(\Delta_{\bar\varphi+\epsilon\theta}(\bar\varphi+\epsilon\theta) + \mathcal{L}_{V(\bar\varphi+\epsilon\theta)}(\bar\varphi+\epsilon\theta)\right) = -\Delta_{\bar\varphi}\theta, \tag{5.1}$$

where θ is an exact 3-form. Note that the operator $-\Delta_{\bar\varphi}$ is strictly negative on the space of exact 3-forms by Hodge decomposition theorem. Let $\tilde\varphi(t)$ be the solution of Laplacian–DeTurck flow and denote $\theta(t) = \tilde{\varphi}(t) - \bar{\varphi}$. By the linearization (5.1), there exists $\epsilon > 0$ such that for all t for which $\|\theta(t)\|_{C^k_{\bar{g}}} < \epsilon$, we have

$$\frac{\partial}{\partial t}\theta(t) = -\Delta_{\bar{\varphi}}\theta + dF(\bar{\varphi}, \tilde{\varphi}(t), \theta(t), \bar{\nabla}\theta(t)),$$

where F is a 2-form which is smooth in the first two arguments and linear in the last two arguments. The idea is that if $\theta(t)$ is sufficiently small, the behavior of the Laplacian–DeTurck flow is dominated by the linear term $-\Delta_{\bar{\varphi}}\theta$. If the initial φ_0 is sufficiently close to $\bar{\varphi}$, i.e., $\theta(0)$ is sufficiently small, by estimating the velocity of the Laplacian–DeTurck flow we can show that the solution exists and remains small at least for time $t \in [0, 1]$. By using the strict negativity of the operator $-\Delta_{\bar{\varphi}}$, we show that $\theta(t)$ has an exponential decay in L^2 norm as long as the solution exists and remains small. By deriving higher order integral estimates, we can in fact show that the solution of the Laplacian–DeTurck flow exists for all time and also converges to $\bar{\varphi}$ exponentially and smoothly as time goes to infinity. The final step is to transform back to Laplacian flow via time-dependent diffeomorphisms $\phi(t)$ determined by the vector field $V(\tilde{\varphi}(t))$. The Shi-type esimate and compactness result apply here to show the smooth convergence of Laplacian flow and completes the proof.

As we mentioned in Sect. 2, Joyce [22] proved an existence result for torsion-free G_2 structures, which states that if we control the C^0 and L^2-norms of γ and the L^{14}-norm of $d^*_{\varphi_0}\gamma = d^*_{\varphi_0}\varphi_0$, we can deform φ_0 in its cohomology class to a unique C^0-close torsion-free G_2 structure $\bar{\varphi}$. By choosing a neighbourhood $\mathcal{U}$ appropriately, controlling derivatives up to at least order 8, we can ensure that we can apply both the theory in [22] and Theorem 5.1, and thus deduce the following corollary.

Corollary 5.2 ([35]) *Let φ_0 be a closed G_2 structure on a compact 7-manifold M. There exists an open neighbourhood $\mathcal{U}$ of 0 in $\Omega^3(M)$ such that if $d^*_{\varphi_0}\varphi_0 = d^*_{\varphi_0}\gamma$ for some $\gamma \in \mathcal{U}$, then the Laplacian flow* (2.1) *with initial value φ_0 exists for all time and converges to a torsion-free G_2 structure.*

The neighbourhood $\mathcal{U}$ given by Corollary 5.2 is not optimal, and one would like to be able to prove this result directly using the Laplacian flow with optimal conditions and without recourse to [22], but nevertheless, Corollary 5.2 gives significant evidence that the Laplacian flow will play an important role in understanding the problem of existence of torsion-free G_2 structures on 7-manifolds admitting closed G_2 structures.

Our results also motivate us to study an approach to the following problem, as pointed out by Thomas Walpuski. The work of Joyce [22] shows that the natural map from the moduli space $\mathcal{M}$ of torsion-free G_2 structures to $H^3(M)$ given by $\mathrm{Diff}^0(M) \cdot \bar{\varphi} \mapsto [\bar{\varphi}]$ is locally injective, but the question of whether this map is globally injective, raised by Joyce (c.f. [23]), is still open. Suppose we have two torsion-free G_2 structures $\bar{\varphi}_0$ and $\bar{\varphi}_1$ which lie in the same cohomology class, so we can write $\bar{\varphi}_1 = \bar{\varphi}_0 + d\eta$ for some 2-form η. We would like to see whether $\bar{\varphi}_1 \in \mathrm{Diff}^0(M) \cdot \bar{\varphi}_0$. By our main theorem (Theorem 5.1) we know that the Laplacian flow starting at $\varphi_0(s) = \bar{\varphi}_0 + sd\eta$ (which is closed) will exist for all time and converge to

$\phi_s^* \bar{\varphi}_0$ for some $\phi_s \in \mathrm{Diff}^0(M)$ when s is sufficiently small. Similarly, the Laplacian flow starting at $\varphi_0(s)$ for s near 1 will also exist for all time and now converge to $\phi_s^* \bar{\varphi}_1$ for some $\phi_s \in \mathrm{Diff}^0(M)$. The aim would be to study long-time existence and convergence of the flow starting at any $\varphi_0(s)$.

Acknowledgements I would like to thank Spiro Karigiannis, Naichung Conan Leung and Jason D. Lotay for the invitation to the event *Workshop on* G_2 *Manifolds and Related Topics*. I would also like to acknowledge the support of the Fields Institute during my stay in Toronto.

References

1. Bando, S. (1987). Real analyticity of solutions of Hamilton's equation. *Mathematische Zeitschrift, 195*, 93–97.
2. Bryant, R. L. (1987). Metrics with exceptional holonomy. *Annals of Mathematics, 126*, 525–576.
3. Bryant, R. L. (2005). Some remarks on G_2 -structures. In *Proceedings of Gökova Geometry-Topology Conference* (pp. 75–109).
4. Bryant, R. L., & Salamon, S. (1989). On the construction of some complete metrics with exceptional holonomy. *Duke Mathematical Journal, 58*, 829–850.
5. Bryant, R. L., & Xu, F. Laplacian flow for closed G_2-structures: Short time behavior. arXiv:1101.2004.
6. Chen, G. (2018). Shi-type estimates and finite time singularities of flows of G_2 structures. *Quarterly Journal of Mathematics, 69*(3), 779–797.
7. Corti, A., Haskins, M., Nordström, J., & Pacini, T. (2015). G_2 manifolds and associative submanifolds via semi-Fano 3-folds. *Duke Mathematical Journal, 164*(10), 1971–2092.
8. Enders, J., Müller, R., & Topping, P. M. (2011). On type-I singularities in Ricci flow. *Communications in Analysis and Geometry, 19*, 905–922.
9. Fernández, M. (1987). A family of compact solvable G_2 -calibrated manifolds. *Tohoku Mathematical Journal, 39*, 287–289.
10. Fernández, M. (1987). An example of a compact calibrated manifold associated with the exceptional Lie group G_2. *Journal of Differential Geometry, 26*, 367–370.
11. Fernández, M., Fino, A., & Manero, V. (2016). Laplacian flow of closed G_2 -structures inducing nilsolitons. *Journal of Geometric Analysis, 26*, 1808–1837.
12. Fernández, M., & Gray, A. (1982). Riemannian manifolds with structure group G_2. *Annali di Matematica Pura ed Applicata (IV), 32*, 19–45.
13. Fino, A., & Raffero, A. Closed warped G_2-structures evolving under the Laplacian flow. *Annali della Scuola Normale Superiore di Pisa, Classe di Scienze*. arXiv:1708.00222.
14. Fine, J., & Yao, C. (2018). Hypersymplectic 4-manifolds, the G_2-Laplacian flow and extension assuming bounded scalar curvature. *Duke Mathematical Journal, 167*(18), 3533–3589.
15. Grigorian, S. (2013). Short-time behaviour of a modified Laplacian coflow of G_2-structures. *Advances in Mathematics, 248*, 378–415.
16. Hamilton, R. S. (1982). Three-manifolds with positive Ricci curvature. *Journal of Differential Geometry, 17*, 255–306.
17. Hamilton, R. (1982). The inverse function theorem of Nash and Moser. *Bulletin of the American Mathematical Society, 7*(1), 65–222.
18. Hamilton, R. S. (1995). A compactness property for solutions of the Ricci flow. *American Journal of Mathematics, 117*, 545–572.
19. Hitchin, N. (2000). The geometry of three-forms in six dimensions. *Journal of Differential Geometry, 55*, 547–576.

20. Haslhofer, R. (2012). Perelman's lambda-functional and the stability of Ricci-flat metrics. *Calculus of Variations and Partial Differential Equations*, *45*, 481–504.
21. Huang, H., Wang, Y., & Yao, C. (2018). Cohomogeneity-one G_2-Laplacian flow on 7-torus. *Journal of the London Mathematical Society (2)*, *98*, 349–368.
22. Joyce, D. D. (1996). Compact Riemannian 7-manifolds with holonomy G_2. I, II. *Journal of Differential Geometry*, *43*, 291–328, 329–375.
23. Joyce, D. D. (2000). *Compact manifolds with special holonomy*. Oxford: OUP.
24. Joyce, D., & Karigiannis, S. A new construction of compact G_2-manifolds by gluing families of Eguchi-Hanson spaces. *Journal of Differential Geometry*. arXiv:1707.09325.
25. Karigiannis, S. (2009). Flows of G_2 structures. *I. Q. J. Math.*, *60*, 487–522.
26. Karigiannis, S., McKay, B., & Tsui, M.-P. (2012). Soliton solutions for the Laplacian co-flow of some G_2 -structures with symmetry. *Differential Geometry and its Applications*, *30*, 318–333.
27. Kotschwar, B. (2010). Backwards uniqueness of the Ricci flow. *International Mathematics Research Notices*, *2010*(21), 4064–4097.
28. Kotschwar, B. (2014). An energy approach to the problem of uniqueness for the Ricci flow. *Communications in Analysis and Geometry*, *22*, 149–176.
29. Kovalev, A. G. (2003). Twisted connected sums and special Riemannian holonomy. *Journal fur die reine und angewandte Mathematik*, *565*, 125–160.
30. Lin, C. (2013). Laplacian solitons and symmetry in G_2 -geometry. *Journal of Geometry and Physics*, *64*, 111–119.
31. Lauret, J. (2016). Geometric flows and their solitons on homogeneous spaces. *Rendiconti Seminario Matematico Univ. Pol. Torino*, *74*(1), 55 – 93. (Workshop for Sergio Console).
32. Lauret, J. (2017). Laplacian flow of homogeneous G_2-structures and its solitons. *Proceedings of the London Mathematical Society (3)*, *114*, 527–560.
33. Lauret, J. (2017). Laplacian solitons: Questions and homogeneous examples. *Differential Geometry and Its Applications*, *54*, 345–360.
34. Lotay, J. D., & Wei, Y. (2017). Laplacian flow for closed G_2 structures: Shi-type estimates, uniqueness and compactness. *Geometric and Functional Analysis*, *27*(1), 165–233.
35. Lotay, J. D., & Wei, Y. (2019). Stability of torsion-free G_2 structures along the Laplacian flow. *Journal of Differential Geometry*, *111*(3), 495–526.
36. Lotay, J. D., & Wei, Y. (2019). Laplacian flow for closed G_2 structures: Real Analyticity. *Communications in Analysis and Geometry*, *27*(1), 73–109.
37. Nicolini, M. (2018). Laplacian solitons on nilpotent Lie groups. *Bulletin of the Belgian Mathematical Society-Simon Stevin*, *25*(2), 183–196.
38. Perelman, G. The entropy formula for the Ricci flow and its geometric applications. arXiv:math/0211159.
39. Sesum, N. (2005). Curvature tensor under the Ricci flow. *American Journal of Mathematics*, *127*, 1315–1324.
40. Sesum, N. (2006). Linear and dynamical stability of Ricci-flat metrics. *Duke Mathematical Journal*, *133*, 1–26.
41. Shi, W.-X. (1989). Deforming the metric on complete Riemannian manifolds. *Journal of Differential Geometry*, *30*, 223–301.
42. Weiss, H., & Witt, F. (2012). A heat flow for special metrics. *Advances in Mathematics*, *231*, 3288–3322.
43. Wang, B. (2012). On the conditions to extend Ricci flow (II). *International Mathematics Research Notices*, *14*, 3192–3223.
44. Zhang, Z. (2010). Scalar curvature behaviour for finite-time singularity of Kähler-Ricci flow. *Michigan Mathematical Journal*, *59*, 419–433.

Flows of Co-closed G_2-Structures

Sergey Grigorian

Abstract We survey recent progress in the study of G_2 structure Laplacian coflows, that is, heat flows of co closed G_2-structures. We introduce the properties of the original Laplacian coflow of G_2-structures as well as the modified coflow, reviewing short-time existence and uniqueness results for the modified coflow and well as recent Shi-type estimates that apply to a more general class of G_2-structure flows.

1 Introduction

One of the most successful techniques in geometric analysis has been the application of geometric flows to various problems in geometry and topology, most notably the use of the Ricci flow [20, 30] to solve the Poincaré Conjecture [31]. The Ricci flow is a non-linear weakly parabolic partial differential equation for the Riemannian metric g

$$\frac{\partial g}{\partial t} = -2\mathrm{Ric}_g \tag{1.1}$$

so that the evolution of the metric is given by the Ricci curvature defined by the metric. This can further be interpreted as a heat equation for the metric. In G_2-geometry, there have been a number of proposals for geometric flows of G_2-structures. The general idea is that given an initial G_2-structure with weaker assumptions than vanishing torsion, the flow should eventually seek out a torsion-free G_2-structure, if one exists on the given manifold. A G_2-structure is defined by a positive 3-form φ, which in turn defines the metric g, and the corresponding Hodge dual 4-form $*\varphi =: \psi$. Therefore, a natural equation to consider is the analog of the heat equation for the 3-form φ

$$\frac{\partial \varphi}{\partial t} = \Delta_\varphi \varphi. \tag{1.2}$$

S. Grigorian (✉)
School of Mathematical and Statistical Sciences, University of Texas Rio Grande Valley, Edinburg, TX 78539, USA
e-mail: sergey.grigorian@utrgv.edu

S. Karigiannis et al. (eds.), *Lectures and Surveys on G_2-Manifolds and Related Topics*, Fields Institute Communications 84, https://doi.org/10.1007/978-1-0716-0577-6_12

This Laplacian flow of the 3-form φ is now nonlinear in φ, because the metric and hence the Laplacian depend on φ itself. A particular case of this flow has been first studied by Bryant [5], where he restricted it to closed G_2-structures, that is ones where $d\varphi = 0$. For a closed G_2-structure, $\Delta\varphi = dd^*\varphi$, so in this case, the 3-form φ stays closed under the flow (1.2), and in fact remains within the same cohomology class since $\Delta\varphi$ is exact. Short-time existence and uniqueness of solutions to (1.2) was proved in [6]. Moreover, on a compact manifold M, this flow can be interpreted as the gradient flow of the Hitchin functional V given by

$$V(\varphi) = \frac{1}{7}\int_M \varphi \wedge *_\varphi \varphi. \tag{1.3}$$

The functional V is then the volume of the manifold M. It was shown by Hitchin in [21] that if φ is closed, then the critical points of the functional V within the cohomology class $[\varphi]$ correspond precisely to torsion-free G_2-structures, and in particular, these critical points are maxima in the directions transverse to diffeomorphisms. Under the flow (1.2), V increases monotonically, so if the growth of V is bounded, then $\varphi(t)$ would be expected to approach a torsion-free G_2-structure as $t \longrightarrow \infty$. The stability and analyticity of this flow has recently been proved by Lotay and Wei [26–28].

Alternatively, a G_2-structure and the corresponding metric may also be defined by the 4-form ψ (up to a choice of orientation). Therefore, instead of deforming φ, we may deform ψ. Using this idea, Karigiannis, McKay, and Tsui, introduced the *Laplacian coflow* for the 4-form ψ in [25]. Instead of considering the heat flow equation for φ, they instead considered the flow:

$$\frac{\partial\psi}{\partial t} = \Delta_\psi \psi. \tag{1.4}$$

If restricted to *co-closed* G_2-structures (that is, ones with $d\psi = 0$ and equivalently, those with a symmetric torsion tensor T) this flow preserves the co-closed condition and in fact preserves the cohomology class of ψ. In [14], it was shown that this flow has similar characteristics to the original Laplacian flow for closed G_2-structures. In fact, (1.4) can also be regarded as a gradient flow of the Hitchin functional (but now reformulated via 4-forms). However, a major difference compared with the Laplacian flow of closed G_2-structures (1.2) is that (1.4) is not even a weakly parabolic equation. In fact, the symbol of the linearized equation is indefinite. In order to have any hope of proving the existence of solutions, a *modified Laplacian coflow* of co-closed G_2-structures was introduced in [14]:

$$\frac{d\psi}{dt} = \Delta_\psi \psi + 2d\left(\left(A - \operatorname{Tr} T\right)\varphi\right) \tag{1.5}$$

where $\operatorname{Tr} T$ is the trace of the full torsion tensor T of the G_2-structure defined by ψ, and A is a positive constant. This flow is now weakly parabolic in the direction of closed forms and hence it is possible to relate it to a strictly parabolic flow using an application of DeTurck's trick. Recently, the methods of Lotay and Wei for Shi-type estimates for the flow (1.2) have been extended by Chen [7] to cover a more general class of G_2-structure flows that includes (1.5) as well. We will first survey the properties of G_2-structures and the Laplacian $\Delta_\varphi \varphi$ in Sects. 2 and 3. Then, in Sect. 4 we will focus on Laplacian coflows.

Despite the apparent similarity between closed and co-closed G_2-structures, there are also important differences. As shown in [10], co-closed G_2-structures always satisfy the h-principle (on both open and closed manifolds) and hence always exist whenever a manifold admits G_2-structures. This is in contrast to closed G_2-structures for which the h-principle only holds on open manifolds. Therefore, co-closed G_2-structures are in some sense more generic than closed ones. This is both good and bad—it's good because they always exist, but bad because one cannot expect their flows to always behave nicely. This is also in part shown by the non-parabolicity of the original coflow (1.4).

In this survey we will focus on analytic properties of flows on general 7-manifolds, however another approach to understand the specific behavior of geometric flows and obtain explicit solutions has been to consider manifolds with some symmetry, in which case the number of degrees of freedom in the PDE will be reduced. Both the original Laplacian coflow (1.4) and the modified Laplacian coflow (1.5) have been studied on a variety of such manifolds with symmetry. Note that while in these situations mostly the original coflow (1.4) with the negative sign has been studied, results for the coflow with the positive sign (1.4) would be similar because equations reduce to ODEs. In [16, 25], the coflow and the modified coflow, respectively, have been studied on warped product manifolds of the form $N^6 \times L$ where N^6 is a 6-dimensional manifold with $SU(3)$-structure such as a Calabi–Yau or nearly Kähler manifold and L is either $\mathbb{R}$ or S^1. In particular, soliton solutions in both cases have been obtained. In [1], Bagaglini, Fernandez, and Fino, also studied both the coflows on the 7-dimensional Heisenberg group. In particular, they have shown that the long-term existence properties of the flow (1.5) depend on the constant A. Similarly, in [2], Bagaglini and Fino studied the Laplacian coflow on 7-dimensional almost-abelian Lie groups and showed long-term existence properties and constructed soliton solutions. In [29], Manero, Otal, and Villacampa studied both the Laplacian flow (4.1) and the coflow (1.4) on solvmanifolds, but instead of restricting to closed or co-closed G_2-structures, they instead restricted to *locally conformally parallel* G_2-structures, which are the ones where only the 7-dimensional τ_1 component of the torsion may be nonvanishing.

2 Laplacian of a G_2-Structure

Suppose M is a smooth 7-dimensional manifold with a G_2-structure φ. Then we know φ uniquely defines a compatible Riemannian metric g_φ, the volume form vol_φ, Hodge star $*_\varphi$, and the dual 4-form $\psi = *_\varphi \varphi$. There is arbitrary choice of orientation, which affects the relative sign of ψ. We use the same convention as [4, 13–16, 18], which is opposite from the convention used in [23, 24]. For further properties of φ and ψ, as well as different identities that they satisfy, we refer the reader to the above references. We will also use the following notation. The symbol $\lrcorner$ will denote contraction of a vector with the differential form:

$$(u \lrcorner \varphi)_{mn} = u^a \varphi_{amn}. \tag{2.1}$$

Note that we will also use this symbol for contractions of differential forms using the metric, for example $(T \lrcorner \varphi)_a = T^{mn} \varphi_{mna}$. Given a symmetric 2-tensor h on M, we define the map $\mathrm{i}_\varphi : \Gamma\left(\mathrm{Sym}\left(T^*M\right)\right) \longrightarrow \Lambda_1^3 \oplus \Lambda_{27}^3$ as

$$\mathrm{i}_\varphi\left(h\right)_{abc} = h_{[a}^d \varphi_{bc]d}.$$

We will define the operators π_1, π_7, π_{14} and π_{27} to be the projections of differential forms onto the corresponding representations. Sometimes we will also use $\pi_{1\oplus 27}$ to denote the projection of 3-forms or 4-forms into $\Lambda_1^3 \oplus \Lambda_{27}^3$ or $\Lambda_1^4 \oplus \Lambda_{27}^4$ respectively. For convenience, when writing out projections of forms, we will sometimes just give the vector that defines the 7-dimensional component, the function that defines the 1-dimensional component or the symmetric 2-tensor that defines the $1 \oplus 27$ component whenever there is no ambiguity. For instance,

$$\begin{array}{lll} \pi_1\left(f\varphi\right) = f & \pi_1\left(f\psi\right) = f & \\ \pi_7\left(X \lrcorner \varphi\right)^a = X^a & \pi_7\left(X \lrcorner \psi\right)^a = X^a & \pi_7\left(X \wedge \varphi\right)^a = X^a \\ \pi_{1\oplus 27}\left(\mathrm{i}_\varphi\left(h\right)\right)_{ab} = h_{ab} & \pi_{1\oplus 27}\left(*\mathrm{i}_\varphi\left(h\right)\right)_{ab} = h_{ab} & \end{array} \tag{2.2}$$

The above-mentioned references give more information regarding the properties of decomposition of differential forms with respect to G_2 representations.

The *intrinsic torsion* of a G_2-structure is defined by $\nabla\varphi$, where ∇ is the Levi-Civita connection for the metric g that is defined by φ. Following [24], we have

$$\nabla_a \varphi_{bcd} = T_a{}^e \psi_{ebcd} \tag{2.3a}$$

$$\nabla_a \psi_{bcde} = -4T_{a[b}\varphi_{cde]} \tag{2.3b}$$

where T_{ab} is the *full torsion tensor*. In general we can split T_{ab} according to representations of G_2 into *torsion components* :

$$T = \frac{1}{4}\tau_0 g - \tau_1 \lrcorner \varphi + \frac{1}{2}\tau_2 - \frac{1}{3}\tau_3 \tag{2.4}$$

where τ_0 is a function, and gives the **1** component of T. We also have τ_1, which is a 1-form and hence gives the **7** component, and, $\tau_2 \in \Lambda_{14}^2$ gives the **14** component and τ_3 is traceless symmetric, giving the **27** component. As shown by Karigiannis in [24], the torsion components τ_i relate directly to the expression for $d\varphi$ and $d\psi$. In fact, in our notation,

$$d\varphi = \tau_0\psi + 3\tau_1 \wedge \varphi + *\mathrm{i}_\varphi(\tau_3) \tag{2.5a}$$
$$d\psi = 4\tau_1 \wedge \psi + *\tau_2. \tag{2.5b}$$

Note that in [14–16, 18] a different convention is used: τ_1 in that convention corresponds to $\frac{1}{4}\tau_0$ here, τ_7 corresponds to $-\tau_1$ here, $\mathrm{i}_\varphi(\tau_{27})$ corresponds to $-\frac{1}{3}\tau_3$, and τ_{14} corresponds to $\frac{1}{2}\tau_2$. The notation used here is widely used elsewhere in the literature.

An important special case is when the G_2-structure is said to be torsion-free, that is, $T = 0$. This is equivalent to $\nabla\varphi = 0$ and also equivalent, by Fernández and Gray [12], to $d\varphi = d\psi = 0$. Moreover, a G_2-structure is torsion-free if and only if the holonomy of the corresponding metric is contained in G_2 [22]. On a compact manifold, the holonomy group is then precisely equal to G_2 if and only if the fundamental group π_1 is finite. If $d\varphi = 0$, then we say φ defines a *closed* G_2-structure. In that case, $\tau_0 = \tau_1 = \tau_3 = 0$ and only τ_2 is in general non-zero. In this case, $T = -\frac{1}{2}\tau_2$ and is hence skew-symmetric. If instead, $d\psi = 0$, then we say that we have a *co-closed* G_2-structure. In this case, τ_1 and τ_2 vanish in (2.5b) and we are left with τ_0 and τ_3 components. In particular, the torsion tensor T_{ab} is now symmetric.

We will be using the following notation, as in [14]. Given a tensor ω, the rough Laplacian is defined by

$$\Delta\omega = g^{ab}\nabla_a\nabla_b\omega = -\nabla^*\nabla\omega. \tag{2.6}$$

whereas the Hodge Laplacian defined by φ or ψ will be denoted by Δ_φ or Δ_ψ, respectively. For a vector field X, define the *divergence* of X as

$$\operatorname{div} X = \nabla_a X^a. \tag{2.7}$$

This operator can be extended to a 2-tensor β:

$$(\operatorname{div}\beta)_b = \nabla^a\beta_{ab}. \tag{2.8}$$

Also, for a vector X, we can use the G_2-structure 3-form φ to define a "curl" operator, similar to the standard one on $\mathbb{R}^3$:

$$(\operatorname{curl} X)^a = (\nabla_b X_c)\,\varphi^{abc}. \tag{2.9}$$

This curl operator can then also be extended to 2-tensor β:

$$(\operatorname{curl}\beta)_{ab} = \left(\nabla_m\beta_{na}\right)\varphi_b{}^{mn}. \tag{2.10}$$

Note that when β_{ab} is symmetric, curl β is traceless. It is also not difficult to see that schematically,

$$\operatorname{curl}\left((\operatorname{curl}\beta)^t\right) = -\Delta\beta^t + \nabla(\operatorname{div}\beta) + \operatorname{Riem} \circledast \beta + T \circledast \nabla\beta + (\nabla T) \circledast \beta + T \circledast T \circledast \beta \tag{2.11}$$

where ^{t} denotes transpose and $\circledast$ is some multilinear operator involving g, φ, ψ. From the context it will be clear whether the curl operator is applied to a vector or a 2-tensor.

As in [14], we can also use the G_2-structure 3-form to define a commutative product $\alpha \circ \beta$ of two 2-tensors α and β

$$(\alpha \circ \beta)_{ab} = \varphi_{amn}\varphi_{bpq}\alpha^{mp}\beta^{nq} \tag{2.12}$$

Note that $(\alpha \circ \beta)^t = \left(\alpha^t \circ \beta^t\right)$. If α and β are both symmetric or both skew-symmetric, then $\alpha \circ \beta$ is a symmetric 2-tensor. Also, for a 2-tensor we have the standard matrix product $(\alpha\beta)_{ab} = \alpha_a^{\ k}\beta_{kb}$.

From [8, 15, 24] we know that the torsion of a G_2-structure satisfies the following integrability condition:

$$\frac{1}{2}\operatorname{Riem}_{ij}^{\ \ \beta\gamma}\varphi^{\alpha}_{\ \beta\gamma} = \nabla_i T_j^{\ \alpha} - \nabla_j T_i^{\ \alpha} + T_i^{\ \beta} T_j^{\ \gamma}\varphi^{\alpha}_{\ \beta\gamma}. \tag{2.13}$$

Taking projections of (2.13) to different representations of G_2, we obtain the following expressions:

Lemma 2.1 *The torsion tensor T satisfies the following identities*

$$(\nabla T) \lrcorner \psi = -(T \lrcorner \varphi) \lrcorner T + T^2 \lrcorner \varphi + (\operatorname{Tr} T)(T \lrcorner \varphi) \tag{2.14a}$$

$$0 = d(\operatorname{Tr} T) - \operatorname{div}\left(T^t\right) - (T \lrcorner \varphi) \lrcorner T^t \tag{2.14b}$$

$$\operatorname{Ric} = -\operatorname{Sym}\left(\operatorname{curl} T^t - \nabla(T \lrcorner \varphi) + T^2 - \operatorname{Tr}(T)\, T\right) \tag{2.14c}$$

$$\frac{1}{4}\operatorname{Ric}^* = \operatorname{curl} T + \frac{1}{2} T \circ T \tag{2.14d}$$

$$\mathrm{R} = 2\operatorname{Tr}(\operatorname{curl} T) - \psi(T, T) - \operatorname{Tr}\left(T^2\right) + (\operatorname{Tr} T)^2 \tag{2.14e}$$

where $(\operatorname{Ric}^*)_{ab} = \operatorname{Riem}_{mnpq}\varphi^{mn}_{\ \ a}\varphi^{pq}_{\ \ b}$ *and* $\psi(T, T) = \psi_{abcd}T^{ab}T^{cd}$. *Note that from (2.4),* $\operatorname{Tr} T = \frac{7}{4}\tau_0$ *and* $T \lrcorner \varphi = -6\tau_1$.

The symmetric 2-tensor Ric^* has been defined and studied by Cleyton and Ivanov in [8, 9]. Note that $\operatorname{Tr}(\operatorname{Ric}^*) = 2\mathrm{R}$, where R is the scalar curvature. Thus the tensors Ric and Ric^* span the components of Riem that lie in $1 \oplus 27 \oplus 27$ representations of G_2. It is know that Riem has no components in the 7 or 14 dimensional representations of G_2. The identities (2.14a), (2.14b), as well as the projection of (2.14d) to Λ^2_{14} are a consequence of this. In fact, taking the skew-symmetric part of (2.14d) and using the fact that Ric^* is by definition symmetric, gives us

$$\operatorname{Skew}(\operatorname{curl} T) = -\frac{1}{2}\operatorname{Skew}(T \circ T). \tag{2.15}$$

In particular, this shows that curl T is symmetric whenever T is skew-symmetric or symmetric, and in particular, if φ is closed or co-closed.

Let us now look at the properties of $\Delta_\varphi \varphi = dd^*\varphi + d^*d\varphi$.

Proposition 2.2 ([14]) *Suppose φ defines a G_2-structure. Then $\Delta_\varphi \varphi = X \lrcorner \psi + 3\mathrm{i}_\varphi(h)$ with*

$$X = -\operatorname{div} T \tag{2.16a}$$

$$\begin{aligned} h = & -\frac{1}{4}\operatorname{Ric}^* + \frac{1}{6}\left(\mathrm{R} + 2|T|^2\right)g - T^t T - \frac{1}{2}(T\lrcorner\varphi)(T\lrcorner\varphi) \\ & + \frac{1}{4}T\circ T + \frac{1}{4}T^t\circ T^t - \frac{1}{2}T\circ T^t + \operatorname{Sym}\left((T)(T\lrcorner\psi) - \left(T^t\right)(T\lrcorner\psi)\right). \end{aligned} \tag{2.16b}$$

In particular,

$$\operatorname{Tr} h = \frac{2}{3}\mathrm{R} + \frac{4}{3}|T|^2. \tag{2.17}$$

The leading order terms in $\Delta_\varphi\varphi$ are those that contain second derivatives of φ, and hence first derivatives of T . Thus, div T fully defines the Λ^3_7 component of $\Delta_\varphi\varphi$ and the leading order terms in $\Lambda^3_{1\oplus 27}$ are given by

$$-\frac{1}{4}\operatorname{Ric}^* + \frac{1}{6}\mathrm{R}\,g \sim -\operatorname{curl} T + \frac{1}{3}\operatorname{Tr}(\operatorname{curl} T)\,g. \tag{2.18}$$

3 Flows of G_2-Structures

Suppose $\varphi(t)$ is a one-parameter family of G_2-structures on a manifold M that satisfies

$$\frac{\partial\varphi(t)}{\partial t} = X(t)\lrcorner\psi(t) + 3\mathrm{i}_{\varphi(t)}(h(t)). \tag{3.1}$$

As shown by Karigiannis in [24], the associated quantities $g(t)$, vol_t, $\psi(t)$, $T(t)$ satisfy the following evolution equations:

Lemma 3.1 ([24]) *If $\varphi(t)$ satisfies the Eq. (3.1), then we also have the following equations:*

$$\frac{\partial g}{\partial t} = 2h \tag{3.2a}$$

$$\frac{\partial \mathrm{vol}}{\partial t} = \operatorname{Tr}(h)\,\mathrm{vol} \tag{3.2b}$$

$$\frac{\partial\psi}{\partial t} = 4\mathrm{i}_\psi(h) - X\wedge\varphi \tag{3.2c}$$

$$\frac{\partial T}{\partial t} = \nabla X - \operatorname{curl} h + Th - (T)(X\lrcorner\varphi) \tag{3.2d}$$

where $\mathrm{i}_{\psi}(h)_{abcd} = -h^{e}_{[a}\psi_{bcd]e}$ *and equivalently,* $4\mathrm{i}_{\psi}(h) = -3 * \mathrm{i}_{\varphi}(h) + (\mathrm{Tr}\, h)\,\psi$.

Similarly, as in [14], we can consider flows of ψ, given by

$$\frac{\partial \psi(t)}{\partial t} = *\left(X(t) \lrcorner \psi(t)\right) + 3 * \mathrm{i}_{\varphi(t)}(s(t)) \tag{3.3}$$

for some symmetric 2-tensor s. Since $3 * \mathrm{i}_{\varphi}(s) = 4\mathrm{i}_{\psi}\left(\frac{1}{4}(\mathrm{Tr}\, s)\, g - s\right)$, comparing (3.3) with (3.2c) give us corresponding evolution equations for $\varphi(t), g(t)$, vol_t, $T(t)$ from (3.1) and (3.2) by taking $h = \frac{1}{4}(\mathrm{Tr}\, s)\, g - s$.

When constructing geometric flows, there are two main considerations: (1) the flow's stationary points should correspond to geometrically interesting objects; and (2) the flow should be parabolic in some sense. The first property is the main motivation for studying a flow, since we ideally want the flow to deform a geometric structure to one that has nicer or more constrained properties and the second property is a minimal requirement to at least guarantee short-time existence and uniqueness of solutions. In [7], Chen defined a class of *reasonable* flows (3.1) of G_2-structures that satisfy the following 4 general conditions:

1. The metric should evolve by the Ricci flow to leading order, and be no more than quadratic in the torsion, that is

$$\frac{\partial g}{\partial t} = 2h = -2\,\mathrm{Ric} + Cg + L(T) + T \circledast T \tag{3.4}$$

   where C is a constant and L is some linear operator involving g, φ, ψ.
2. The vector field X is at most linear in ∇T and at most quadratic in T:

$$X = L(\nabla T) + L(T) + L(\mathrm{Riem}) + T \circledast T + C. \tag{3.5}$$

3. The torsion tensor should evolve by ΔT to leading order, and be at most linear in Riem and ∇T, and at most cubic in T:

$$\begin{aligned}\frac{\partial T}{\partial t} &= \Delta T + L(\nabla T) + L(\mathrm{Riem}) + \mathrm{Riem} \circledast T + \nabla T \circledast T \\ &\quad + L(T) + T \circledast T + T \circledast T \circledast T.\end{aligned} \tag{3.6}$$

4. The flow (3.1) has short-time existence and uniqueness.

As one of the key properties of *reasonable* flows defined above is that the flow of the metric is the Ricci flow to leading order, we will instead refer to flows that satisfy properties 1–4 as *Ricci-like flows*. This is appropriate because a variety of techniques that originated from the study of the Ricci flow have been applied to these flows. In particular, under the Ricci flow, invariants of the metric Riem, Ric, R, all satisfy heat-like equations. Therefore it is appropriate that for a Ricci-like flow of a G_2-structure, the torsion, which an invariant of the G_2-structure also satisfies a heat-like Eq. (3.6). This is important because then $\nabla^k T$ and $|T|^2$ also satisfy heat-like equations and this is necessary to be able to obtain estimates using the maximum principle.

Using techniques developed by Shi in [32] for the Ricci flow and their adaptation to G_2-structures by Lotay and Wei [26], Chen then showed that a reasonable flow satisfies the following Shi-type estimate.

Theorem 3.2 ([7, Theorem 2.1]) *Suppose (3.1) is a Ricci-like flow of G_2-structures, such that the coefficients in Eqs. (3.1), (3.4), (3.5), and (3.6) are bounded by a constant Λ. Let $B_r(p)$ be a ball of radius r with respect to the initial metric $g(0)$. If*

$$|\mathrm{Riem}(x,t)|_{g(t)} + |T(x,t)|^2_{g(t)} + |\nabla T(x,t)|_{g(t)} < \Lambda \tag{3.7}$$

for any $(x,t) \in B_r(p) \times [0, t_0]$, then

$$\left|\nabla^k \mathrm{Riem}(x,t)\right|_{g(t)} + \left|\nabla^{k+1} T(x,t)\right|_{g(t)} < C(k,r,\Lambda,t) \tag{3.8}$$

for any $(x,t) \in B_{r/2}(p) \times \left[\frac{t_0}{2}, t_0\right]$ for all $k = 1, 2, 3, \ldots$

It should be noted that in [26], the condition analogous to (3.7) does not include a $|T|^2$ term. This is because in the case of a closed G_2-structure, $|T|^2 = -\mathrm{R} \leq C\,|\mathrm{Riem}|$. Therefore, the norm of the torsion can always be bounded in terms of the norm of Riem . For other torsion classes, and in particular, co-closed G_2-structures, this is no longer true, therefore $|T|^2$ needs to be included in (3.7).

Using the estimates from Theorem 3.2, Chen then derived an estimate for the blow-up rate on a compact manifold.

Theorem 3.3 ([7, Theorem 5.1]) *If $\varphi(t)$ is a solution to a Ricci-like flow of G_2-structures on a compact manifold in a finite maximal time interval $[0, t_0)$, then*

$$\sup_M \left(|\mathrm{Riem}(x,t)|^2_{g(t)} + |T(x,t)|^4_{g(t)} + |\nabla T(x,t)|^2_{g(t)}\right)^{\frac{1}{2}} \geq \frac{C}{t_0 - t} \tag{3.9}$$

for some positive constant C.

The estimate (3.9) shows that a solution will exist as long the quantity of the left-hand side of (3.9) remains bounded.

A classic example of a Ricci-like flow of G_2-structures is the Laplacian flow of G_2-structures that was introduced by Bryant in [5]:

$$\frac{\partial \varphi}{\partial t} = \Delta_\varphi \varphi. \tag{3.10}$$

If the initial G_2-structure is closed, then this property is preserved along the flow. It is then natural to think of (3.10) as a flow of closed G_2-structures. In this case, since $T^t = -T$, from (2.14), $\mathrm{Ric}^* = 4\,\mathrm{Ric} + T \circledast T$ and $\mathrm{R} = 2\,\mathrm{Tr}(\mathrm{curl}\,T) - \psi(T,T) -$

$\operatorname{Tr}(T^2) = -|T|^2$; and thus, from (2.16b), $h = -\operatorname{Ric} + T \circledast T$, and so from (3.2a), we do find that (3.4) holds. Moreover, from (2.14b), we see that $\operatorname{div} T = 0$ in this case, and hence $X = 0$. The expression (3.6) comes from (3.2d) and using $h = -\operatorname{curl} T + T \circledast T$

$$\frac{\partial T}{\partial t} = \operatorname{curl}(\operatorname{curl} T) + \nabla T \circledast T + T \circledast T \circledast T. \tag{3.11}$$

Using (2.11) to expand $\operatorname{curl}(\operatorname{curl} T)$ together the facts that $\operatorname{curl} T$ is symmetric, T is skew-symmetric, and $\operatorname{div} T = 0$, allows to express the right-hand side of (3.11) as $\Delta T + \operatorname{Riem} \circledast T + \nabla T \circledast T + T \circledast T \circledast T$. Finally, short-term existence and uniqueness of the flow (3.10) has been first proved by Bryant and Xu in [6]. For more on the properties of this flow, as well as the details of the above calculations, the reader is referred to the series of papers by Lotay and Wei [26–28]. The results in Theorems 3.2 and 3.3 are extensions of similar results for the Laplacian flow of closed G_2-structures in [26].

4 Laplacian Coflow

In [25], Karigiannis, McKay, and Tsui introduced an alternative flow of G_2-structures, called the Laplacian *coflow*:

$$\frac{\partial \psi}{\partial t} = -\Delta_{\psi}\psi. \tag{4.1}$$

If the initial G_2-structure is co-closed, i.e. $d\psi = 0$, then this property is preserved along the flow. Therefore, the coflow may be regarded as a natural flow of co-closed G_2-structures. In order to understand flows of co-closed G_2-structures, we need to understand better the properties of T and the Hodge Laplacian in this case. Rewriting Lemma 2.1 and Proposition 2.2 in the case of symmetric T, we find the following.

Proposition 4.1 *Suppose φ is a co-closed G_2-structure, then the torsion tensor T satisfies the following identities*

$$\operatorname{div} T = d(\operatorname{Tr} T) \tag{4.2a}$$

$$\operatorname{curl} T = (\operatorname{curl} T)^t \tag{4.2b}$$

$$\operatorname{Ric} = \operatorname{curl} T - T^2 + \operatorname{Tr}(T)\, T \tag{4.2c}$$

$$\frac{1}{4}\operatorname{Ric}^* = \operatorname{curl} T + \frac{1}{2} T \circ T = \operatorname{Ric} + \frac{1}{2} T \circ T + T^2 - \operatorname{Tr}(T)\, T \tag{4.2d}$$

$$\mathrm{R} = (\operatorname{Tr} T)^2 - |T|^2. \tag{4.2e}$$

The Hodge Laplacian is given by $\Delta_{\varphi}\varphi = X \lrcorner \psi + 3\mathrm{i}_{\varphi}(s)$ with

$$X = -\operatorname{div} T \tag{4.3a}$$

$$s = -\operatorname{Ric} + \frac{1}{6}\left(\mathrm{R} + 2|T|^2\right) g + \operatorname{Tr}(T)\, T - 2T^2 - \frac{1}{2} T \circ T \tag{4.3b}$$

$$= -\operatorname{curl} T + \frac{1}{6}\left((\operatorname{Tr} T)^2 + |T|^2\right) g - T^2 - \frac{1}{2} T \circ T \tag{4.3c}$$

$$\operatorname{Tr} s = \frac{2}{3}\mathrm{R} + \frac{4}{3}|T|^2 = \frac{2}{3}\left((\operatorname{Tr} T)^2 + |T|^2\right). \tag{4.3d}$$

Comparing (4.1) with (3.3) and using (4.3), we see that to leading order the evolution of the metric is given by 2 Ric, that is the opposite of the Ricci flow. Thus, in order for the flow to be Ricci-like and to have any hope of existence and uniqueness, the sign in (4.1) needs to be reversed. Therefore, let us redefine the Laplacian coflow as

$$\frac{d\psi}{dt} = \Delta_\psi \psi. \tag{4.4}$$

We then find that

$$\frac{\partial g}{\partial t} = -2\operatorname{Ric} + T \circ T + 2(\operatorname{Tr} T)\, T \tag{4.5}$$

which now satisfies (3.4). Also, $X = -\operatorname{div} T$, which satisfies (3.5). To obtain the general form of the evolution of the torsion, note that to leading order, $h = -s = \operatorname{curl} T$, so from (3.2d),

$$\frac{\partial T}{\partial t} = -\nabla(\operatorname{div} T) - \operatorname{curl}(\operatorname{curl} T) + \nabla T \circledast T$$

however, since both T and $\operatorname{curl} T$ are symmetric,

$$\operatorname{curl}(\operatorname{curl} T) = -\Delta T + \nabla(\operatorname{div} T) + \operatorname{Riem} \circledast T + (\nabla T) \circledast T + T \circledast T \circledast T$$

Hence, overall,

$$\frac{\partial T}{\partial t} = \Delta T - 2\nabla(\operatorname{div} T) + \operatorname{Riem} \circledast T + (\nabla T) \circledast T + T \circledast T \circledast T. \tag{4.6}$$

Notice that this does not satisfy (3.6). In fact, we can see that the presence of the $\nabla(\operatorname{div} T)$ term in (4.6) is due to the negative sign of $\operatorname{div} T$ in (4.3a). As it was shown in [14], the sign of $\operatorname{div} T$ also causes problems at a much more fundamental level: it prevents the flow (4.4) from being parabolic even along closed 4-forms. Proposition 4.2 below gives the linearization of Δ_ψ. It is then easy to see that for closed 4-forms, the symbol will be negative in the Λ^4_7 direction, but non-negative in Λ^4_{27}.

Proposition 4.2 ([14, Prop. 4.7]) *The linearization of Δ_ψ at ψ is given by*

$$\pi_7\left(D_\psi\Delta_\psi\right)(\chi) = d\,(\operatorname{div} X)\wedge\varphi + l.o.t. \tag{4.7a}$$

$$\pi_{1\oplus 27}\left(D_\psi\Delta_\psi\right)(\chi) = \frac{3}{2}*\mathrm{i}_\varphi\left(\Delta h + \frac{1}{4}\operatorname{Hess}(\operatorname{Tr} h) - \frac{1}{2}(\Delta \operatorname{Tr} h)\,g \right. \tag{4.7b}$$
$$\left. - \operatorname{Sym}\left(\nabla \operatorname{div} h + \operatorname{curl}(\nabla X)^t\right) + l.o.t.\right)$$

where $\chi = *\left(X\lrcorner\psi + 3\mathrm{i}_\varphi(h)\right)$. *Moreover, if* χ *is closed, we can write* $D_\psi\Delta_\psi$ *as*

$$D_\psi\Delta_\psi(\chi) = -\Delta_\psi\chi - \mathcal{L}_{V(\chi)}\psi + 2d\left((\operatorname{div} X)\,\varphi\right) + dF(\chi) \tag{4.8}$$

where

$$V(\chi) = \frac{3}{4}\nabla \operatorname{Tr} h - 2\operatorname{curl} X \tag{4.9}$$

and $F(\chi)$ *is a 3-form-valued algebraic function of* χ.

Looking closer at the leading terms in the linearization (4.8) evaluated at closed forms, we see that the term $2d\left((\operatorname{div} X)\,\varphi\right)$ appears for exactly the same reason as the term $-2\nabla(\operatorname{div} T)$ in (4.6)—namely the "wrong" sign of the π_7 component of $\Delta_\psi\psi$. To fix this problem, in [14], a *modified Laplacian coflow* has been proposed:

$$\frac{\partial\psi}{\partial t} = \Delta_\psi\psi + 2d\left((A - \operatorname{Tr} T)\,\varphi\right) \tag{4.10}$$

where A is some constant. Since for co-closed G_2-structures, $\operatorname{Tr} T = \operatorname{div} T$, the leading term in the modification precisely reverses the sign of the Λ^4_7 component of the original flow (1.4). However, because we want the right hand side of the flow to be an exact 4-form for co-closed G_2-structures, there are some additional lower order terms. The constant A could be set to zero, however adding it may allow for more flexibility. The linearization of the modified coflow at a closed 4-form is now given by

$$\frac{\partial\chi}{\partial t} = -\Delta_\psi\chi - \mathcal{L}_{V(\chi)}\psi + d\hat{F}(\chi) \tag{4.11}$$

where $V(\chi)$ is as in (4.9) and $\hat{F}(\chi)$ involves no derivatives of χ. Hence, in the direction of closed forms, this flow is now weakly parabolic. Moreover, the undesired term is removed from the evolution equation for T and its evolution is now given by (3.6).

The additional term in (4.10) now also allows to prove short-time existence and uniqueness, hence completing the requirements for (4.10) to be a Ricci-like flow. The proof, as given in [14], follows a procedure similar to the approach taken by Bryant and Xu [6] for the proof of short-time existence and uniqueness for the Laplacian flow (3.10), which is in turn based on DeTurck's [11] and Hamilton's [19] approaches to the proof of short-time existence and uniqueness of the Ricci flow. Let $\psi(t) = \psi_0 + \chi(t)$ where $\chi(t)$ is an exact 4-form with $\chi(0) = 0$. Then, given this initial condition, the flow (4.10) can be rewritten as an initial value problem for

$\chi(t)$. From the linearization (4.11) we see that by adding the term $\mathcal{L}_{V(\chi(t))}\psi(t)$ we obtain a strictly parabolic flow in the direction of closed forms, which is related to the original flow by diffeomorphism:

$$\frac{\partial \chi}{\partial t} = \Delta_\psi \psi + 2d\left(\left(A - \operatorname{Tr} T_\psi\right) *_\psi \psi\right) + \mathcal{L}_{V(\chi)}\psi. \tag{4.12}$$

This is the essence of what is known as "DeTurck's trick"—turning a weakly parabolic flow into a strictly parabolic one. In the case of Ricci flow this is enough to obtain short-time existence and uniqueness, however in this case, the parabolicity is only along closed forms, hence we cannot apply the standard parabolic theory right away, and more steps are needed. Let us also define the spaces of time-dependent and time-independent exact 4-forms $\mathcal{F}$ and $\mathcal{G}$, respectively. Moreover, since we know that $\psi(t)$ always defines a G_2-structure and is thus a positive 4-form, χ will always lie in an open subset $\mathcal{U} \subset \mathcal{F}$ defined by

$$\mathcal{U} = \{\chi \in \mathcal{F} : \psi_0 + \chi \text{ is a positive 4-form}\}. \tag{4.13}$$

Moreover, let us now define a map $F : \mathcal{U} \longrightarrow \mathcal{F} \times \mathcal{G}$ given by

$$\chi \longrightarrow \left(\frac{\partial \chi}{\partial t} - \Delta_\psi \psi - 2d\left(\left(A - \operatorname{Tr} T_\psi\right) *_\psi \psi\right) - \mathcal{L}_{V(\chi)}\psi,\ \chi|_{t=0}\right). \tag{4.14}$$

Adapting the results in [6], it is easy to see $\mathcal{F}$, $\mathcal{G}$, and $\mathcal{H} := \mathcal{F} \times \mathcal{G}$ are *graded tame Fréchet spaces*. Moreover, it was then shown in [14] that F is smooth *tame* map of Fréchet spaces, such that its derivative $DF(\chi) : \mathcal{F} \longrightarrow \mathcal{H}$ is an isomorphism for all $\chi \in \mathcal{U}$ and the inverse $(DF)^{-1} : \mathcal{U} \times \mathcal{H} \longrightarrow \mathcal{F}$ is smooth tame. The significance of these facts are that in the category of Fréchet spaces there exists an inverse function theorem—the Nash–Moser Inverse Function [19], which tells us that the map F is locally invertible. From this it follows that the flow (4.12) has short-time existence and uniqueness.

To prove short-time existence and uniqueness for the flow (4.10) we need to relate (4.10) and (4.12) via diffeomorphisms. Suppose $\bar{\chi}(t)$ is the unique short-time solution to (4.12), and $\bar{\psi} = \psi_0 + \bar{\chi}$. Consider the following ODE for a family of diffeomorphisms ϕ_t:

$$\begin{cases} \frac{\partial \phi_t}{\partial t} = -V(\bar{\chi}(t)) \\ \phi_0 = \mathrm{id} \end{cases} \tag{4.15}$$

This has a unique solution ϕ_t. Now let $\psi(t) = (\phi_t)^* \bar{\psi}(t)$, then $\psi(0) = \psi_0$, and since diffeomorphisms commute with d, $\psi(t)$ is closed for all t. Moreover, as shown in [14, Theorem 6.9], $\psi(t)$ now satisfies (4.10). Uniqueness is obtained similarly using the uniqueness of solutions of (4.15). Hence, overall, we obtain a unique short-time solution for the modified Laplacian coflow (4.10) and can now conclude that it is a Ricci-like flow.

Theorem 4.3 *The Laplacian coflow (4.10) of co-closed G_2-structures is a Ricci-like flow.*

5 Further Directions

There are several important unanswered questions regarding flows of co-closed G_2-structures. An intriguing question is whether it is possible to obtain at least short-time existence and uniqueness of the unmodified Laplacian coflow (1.4). To leading order the only difference with the modified coflow is the sign of the Λ^4_7 component which is given by $\operatorname{div} T$. So in particular, if $\operatorname{div} T$ vanishes, then the two flows agree. It is also known [23] that deformations in the Λ^4_7 directions keep the metric unchanged. Moreover, in [17], the torsion T has been shown to play a role of an octonionic connection on the bundle of G_2-structures that correspond to the same metric, which can be given the structure of an octonion bundle. In this interpretation, on a compact manifold, the condition $\operatorname{div} T = 0$ corresponds to critical points of the functional $\int |T|^2 \operatorname{vol}$, and is hence the analog of a Coulomb gauge. It is therefore tempting to think that to relate the flows (1.4) and (1.5), a gauge-fixing condition such as $\operatorname{div} T = 0$ needs to be introduced.

There are also multiple questions relating to the modified coflow itself. As it is a Ricci-like flow, Shi-type estimates apply to it, so it is likely that in addition to Chen's results in [7], more properties such as real analyticity and stability could be proved using techniques similar to the ones used by Lotay and Wei in [26–28]. Indeed, as this article was being finalized, the author was made aware that Bedulli and Vezzoni [3] have generalized the proof of stability from [28] to a wider class of geometric flows that also includes the modified Laplacian coflow with $A = 0$.

Apart from the Laplacian flow and the coflows, there could be more interesting flows of G_2-structures. For co-closed G_2-structures, it is an open question whether the flow $\frac{\partial \varphi}{\partial t} = d^* d\varphi$ satisfies the co-closed condition. More generally, the conditions for a flow to be Ricci-like is a good set of conditions that flows should satisfy. In particular, one could try to construct flows using the first 3 conditions, but then also making sure that short-time existence and uniqueness is satisfied.

Acknowledgements The author is supported by the National Science Foundation grant DMS-1811754.

References

1. Bagaglini, L., Fernández, M., & Fino, A. *Laplacian coflow on the 7 dimensional Heisenberg group*. arXiv:1704.00295.
2. Bagaglini, L., & Fino, A. *The laplacian coflow on almost-abelian Lie groups*. arXiv:1711.03751.
3. Bedulli, L., & Vezzoni, L. *Stability of geometric flows of closed forms*. arXiv:1811.09416.

4. Bryant, R. L. (1987). Metrics with exceptional holonomy. *Annals of Mathematics (2), 126*(3), 525–576.
5. Bryant, R. L. (2006). Some remarks on G_2-structures. In *Proceedings of Gökova Geometry-Topology Conference 2005* (pp. 75–109). Gökova Geometry/Topology Conference (GGT), Gökova. arXiv:math/0305124.
6. Bryant, R. L., & Xu, F. *Laplacian flow for closed G_2-structures: Short time behavior.* arXiv:1101.2004.
7. Chen, G. (2018). Shi-type estimates and finite time singularities of flows of G_2 structures. *Quarterly Journal of Mathematics*. arXiv:1703.08526.
8. Cleyton, R., & Ivanov, S. (2007). On the geometry of closed G_2-structures. *Communications in Mathematical Physics, 270*(1), 53–67. arXiv:math/0306362.
9. Cleyton, R., & Ivanov, S. (2008). Curvature decomposition of G_2-manifolds. *Journal of Geometry and Physics, 58*(10), 1429–1449.
10. Crowley, D., & Nordström, J. (2015). New invariants of G_2-structures. *Geometry & Topology, 19*(5), 2949–2992. arXiv:1211.0269.
11. DeTurck, D. M. (1983). Deforming metrics in the direction of their Ricci tensors. *Journal of Differential Geometry, 18*(1), 157–162.
12. Fernández, M., & Gray, A. (1982). Riemannian manifolds with structure group G_2. *Annali di Matematica Pura ed Applicata (4), 132*, 19–45.
13. Grigorian, S. (2012). G_2-structure deformations and warped products. In *String-Math 2011, Proceedings of Symposia in Pure Mathematics*, AMS. arXiv:1110.4594.
14. Grigorian, S. (2013). Short-time behaviour of a modified Laplacian coflow of G_2-structures. *Advances in Mathematics, 248*, 378–415. arXiv:1209.4347.
15. Grigorian, S. (2016). Deformations of G_2-structures with torsion. *Asian Journal of Mathematics, 20*(1), 123–155. arXiv:1108.2465.
16. Grigorian, S. (2016). Modified Laplacian coflow of G_2-structures on manifolds with symmetry. *Differential Geometry and its Applications, 46*, 39–78. arXiv:1504.05506.
17. Grigorian, S. (2017). G_2-structures and octonion bundles. *Advances in Mathematics, 308*, 142–207. arXiv:1510.04226.
18. Grigorian, S., & Yau, S.-T. (2009). Local geometry of the G_2 moduli space. *Communications in Mathematical Physics, 287*, 459–488. arXiv:0802.0723.
19. Hamilton, R. S. (1982). The inverse function theorem of Nash and Moser. *Bulletin of the American Mathematical Society Journal (N.S.), 7*(1), 65–222.
20. Hamilton, R. S. (1982). Three-manifolds with positive Ricci curvature. *Journal of Differential Geometry, 17*(2), 255–306.
21. Hitchin, N. J. (2000). The geometry of three-forms in six and seven dimensions. *Journal of Differential Geometry, 55*(3), 547–576. arXiv:math/0010054.
22. Joyce, D. D. (2000). *Compact manifolds with special holonomy*. Oxford mathematical monographs. Oxford: Oxford University Press.
23. Karigiannis, S. (2005). Deformations of G_2 and Spin(7) structures on manifolds. *Canadian Journal of Mathematics, 57*, 1012. arXiv:math/0301218.
24. Karigiannis, S. (2009). Flows of G_2-Structures, I. *Quarterly Journal of Mathematics, 60*(4), 487–522. arXiv:math/0702077.
25. Karigiannis, S., McKay, B., & Tsui, M.-P. (2012). Soliton solutions for the Laplacian coflow of some G_2-structures with symmetry. *Differential Geometry and its Applications, 30*(4), 318–333. arXiv:1108.2192.
26. Lotay, J. D., & Wei, Y. (2017). Laplacian flow for closed G_2 structures: Shi-type estimates, uniqueness and compactness. *Geometric and Functional Analysis, 27*(1), 165–233. arXiv:1504.07367.
27. Lotay, J. D., & Wei, Y. (2018). Laplacian flow for closed G_2 structures: Real analyticity. *Communications in Analysis and Geometry*. In press arXiv:1601.04258.
28. Lotay, J. D., & Wei, Y. (2018). Stability of torsion-free G_2 structures along the Laplacian flow. *Journal of Differential Geometry*. In press arXiv:1504.07771.

29. Manero, V., Otal, A., & Villacampa, R. *Solutions of the Laplacian flow and coflow of a locally conformal parallel* G_2*-structure*. arXiv:1711.08644.
30. Morgan, J., & Tian, G. (2007). *Ricci flow and the Poincaré conjecture* (Vol. 3). Clay mathematics monographs. Providence: American Mathematical Society.
31. Perelman, G. *The entropy formula for the Ricci flow and its geometric applications*. arXiv:math/0211159.
32. Shi, W.-X. (1989). Deforming the metric on complete Riemannian manifolds. *Journal of Differential Geometry*, *30*(1), 223–301.

G_2-Instantons on Noncompact G_2-Manifolds: Results and Open Problems

Jason D. Lotay and Goncalo Oliveira

Abstract We survey the known existence and non existence results for G_2-instantons on non-compact cohomogeneity-1 G_2-manifolds and their consequences, including an explicit example of a family of G_2-instantons where bubbling, removable singularities and conservation of energy phenomena occur. We also describe several open problems for future research.

1 Introduction

A G_2-instanton is a special kind of Yang–Mills connection on a Riemannian 7-manifold with holonomy group contained in G_2 (a so-called G_2-manifold). One can think of G_2-instantons as analogues of anti-self-dual connections in 4 dimensions. This analogy motivates the hope of using G_2-instantons to construct enumerative invariants of G_2-manifolds. In this review article we shall be focusing on G_2-manifolds and G_2-instantons constructed using symmetry techniques. It is important to note that, using the fact that G_2-manifolds are Ricci flat, one sees that holonomy G_2-manifolds[1] admitting continuous symmetries must be noncompact. Symmetry techniques thus have a somewhat limited scope of applicability, but they do lead to simplifications which make hard problems in the field tractable in this special setting, giving in several cases explicit non-trivial examples as well as significant results which may be useful in the general theory. Here, we shall summarize the known existence and non-existence results for G_2-instantons in the symmetric setting, as well as their consequences. For example, we shall see an explicit example of a family of G_2-instantons for which bubbling and removable singularities phenomena happen. We shall also describe several important open problems for future research.

[1] Those G_2-manifolds whose holonomy is exactly G_2 will be referred to as holonomy G_2-manifolds.

J. D. Lotay
University of Oxford, Oxford, UK

G. Oliveira (✉)
Universidade Federal Fluminese, Niterói, Brazil
e-mail: gooliveira@id.uff.br

S. Karigiannis et al. (eds.), *Lectures and Surveys on G_2-Manifolds and Related Topics*, Fields Institute Communications 84, https://doi.org/10.1007/978-1-0716-0577-6_13

1.1 Background

Let (X^7, φ) be a G_2-manifold,[2] which implies that the 7-manifold X^7 is endowed with a 3-form φ which is closed and determines a Riemannian metric g with respect to which φ is also coclosed. We shall denote $*\varphi$ by ψ for convenience. Let $P \to X$ be a principal bundle with structure group G which we suppose to be a compact and semisimple Lie group. A connection A on P is said to be a G_2-instanton if

$$F_A \wedge \psi = 0. \tag{1}$$

Equivalently, G_2-instantons satisfy the following G_2-analogue of the "anti-self-dual" condition:

$$F_A \wedge \varphi = - * F_A. \tag{2}$$

As far as the authors are aware, the first time G_2-instantons appeared in the literature was in [6]. This reference investigates generalizations of the anti-self-dual gauge equations, in dimension greater than 4, and G_2-instantons appear there as an example.

More recently, the study of G_2-instantons has gained a special interest, primarily due to Donaldson–Thomas' suggestion [11] that it may be possible to use G_2-instantons to define invariants for G_2-manifolds, inspired by Donaldson's pioneering work on anti-self-dual connections on 4-manifolds. Later Donaldson–Segal [10], Haydys [16], and Haydys–Walpuski [17] gave further insights regarding this possibility.

On a compact holonomy G_2-manifold (X^7, φ) any harmonic 2-form is "anti-self-dual" as in (2), hence any complex line bundle L on X admits a G_2-instanton, namely that whose curvature is the harmonic representative of $c_1(L)$. However, the construction of non-abelian G_2-instantons on compact G_2-manifolds is much more involved. In the compact case, the first such examples were constructed by Walpuski [30], over Joyce's G_2-manifolds (see [18]). Sá Earp and Walpuski's work [26, 31] gives an abstract construction of G_2-instantons, and currently one example, on the other known class of compact G_2-manifolds, namely "twisted connected sums" (see [8, 19]). More recently, Ménet–Sá Earp–Nordström constructed other examples of G_2-instantons on twisted connected sum G_2-manifolds [21].

On complete, noncompact, holonomy G_2-manifolds, the first examples of G_2-instantons where found by Clarke in [9], and further examples were given by the second author in [23] and by both authors in [20]. We shall describe these examples in this article, and discuss natural open problems which arise from their study.

---

[2]For further background on G_2-manifolds, the reader may wish to consult Joyce's book [18].

2 Preliminaries

In this section we shall be considering manifolds that (in a dense open set) can be written as $X^7 = I_t \times M^6$ with $I_t \subset \mathbb{R}$ an interval with coordinate $t \in \mathbb{R}$. Then, we will write the G_2-instanton conditions as evolution equations in the t coordinate and make some observations about these equations.

2.1 Evolution Equations

Before turning to G_2-instantons, we recall here how to write the equations for a torsion-free G_2-structure on X as evolution equations. This requires the notion of an SU(3)-structure on an almost complex 6-manifold (M, J), which consists of a pair (ω, γ_2) of a real $(1, 1)$-form and a real 3-form respectively, such that

$$\omega \wedge \gamma_2 = 0, \quad \omega^3 = \frac{3}{2}\gamma_1 \wedge \gamma_2,$$

where $\gamma_1 = -J\gamma_2$. Now let $(\omega(t), \gamma_2(t))$ be a 1-parameter family of SU(3)-structures, parametrized by the coordinate $t \in I_t$, and consider the G_2-structure on X given by

$$\varphi = dt \wedge \omega(t) + \gamma_1(t), \quad \psi = \frac{\omega^2(t)}{2} - dt \wedge \gamma_2(t). \tag{3}$$

The equations $d\varphi = 0$ and $d\psi = 0$, for the G_2-structure to be torsion-free, turn into the following evolution equations for the SU(3)-structures $(\omega(t), \gamma_2(t))$:

$$\dot{\gamma}_1 = d\omega, \quad \omega \wedge \dot{\omega} = -d\gamma_2, \tag{4}$$

subject to the constraints $d\gamma_1 = 0 = d\omega^2$ for all t. These evolution equations are the so-called "Hitchin flow"[3] and the constraint $d\gamma_1 = 0 = d\omega^2$ is usually called the half-flat[4] condition. In fact, this constraint is compatible with the Hitchin flow (4), meaning that if one imposes the half-flat condition on the SU(3)-structure at some $t_0 \in I_t$, the evolution equations (4) will preserve it for all $t \in I_t$. See [22] for more on half-flat SU(3)-structures in a case relevant to some of the works reviewed in this article.

[3]The nomenclature "Hitchin flow" may be misleading. Indeed, the system (4) is not parabolic and does not satisfy the usual regularity properties of geometric flows [4].

[4]The name "half-flat" comes from the fact that the condition implies the vanishing of exactly half of the torsion components of (ω, γ_2) as an SU(3)-structure.

The G_2-structure φ on X obtained from solving the Hitchin flow induces the metric $g = dt^2 + g_t$, where g_t is the metric on $\{t\} \times M$ compatible with the SU(3)-structure $(\omega(t), \gamma_2(t))$. For example, if we take (ω, γ_2) to be nearly Kähler on M, i.e.

$$d\omega = 3\gamma_1, \quad d\gamma_2 = -2\omega^2,$$

and g_M is the nearly Kähler metric on M, then the G_2-structure φ given by solving (4) is

$$\varphi = t^2 dt \wedge \omega + t^3\gamma_1, \quad \psi = t^4\omega^2/2 - t^3 dt \wedge \gamma_2, \tag{5}$$

which gives a conical metric $g = dt^2 + t^2 g_M$ on X.

Now let us consider a principal G-bundle P on X pulled back from M. There is no loss of generality in assuming this, as well as in working in temporal gauge, i.e. in setting the connection on P over X to be of the form $A = a(t)$, where $a(t)$ is a 1-parameter family of connections on P, now seen as a vector bundle over M. The curvature of A is given by $F_A = dt \wedge \dot{a} + F_a(t)$, where $F_a(t)$ is the curvature of $a(t)$ as a connection on P over M. Then, the G_2-instanton equation (1) for A, turns into the following evolution equation for $a(t)$:

$$\dot{a} \wedge \frac{\omega^2}{2} - F_a \wedge \gamma_2 = 0, \quad F_a \wedge \frac{\omega^2}{2} = 0. \tag{6}$$

Applying $*_t$, the Hodge-$*$ of the metric g_t, to both sides of (6) we have

$$J_t\dot{a} = - *_t (F_a \wedge \gamma_2), \tag{7}$$

$$\Lambda_t F_a = 0, \tag{8}$$

with Λ_t denoting the metric dual of the operation of wedging with $\omega(t)$. As for the Hitchin flow, the evolution equation (7) is compatible with the constraint (8). The discussion above and this claim can be formally stated as follows.

Lemma 1 *Let $X = I_t \times M$ be equipped with a G_2-structure φ as in (3) satisfying $\omega \wedge d\omega = 0$ and $\omega \wedge \dot{\omega} = -d\gamma_2$, which is equivalent to $d\psi = 0$. Then, G_2-instantons A for φ are in one-to-one correspondence with 1-parameter families of connections $\{a(t)\}_{t \in I_t}$ solving the evolution equation*

$$J_t\dot{a} = - *_t (F_a \wedge \gamma_2), \tag{9}$$

subject to the constraint $\Lambda_t F_a = 0$. Moreover, this constraint is compatible with the evolution: more precisely, if it holds for some $t_0 \in I_t$, then it holds for all $t \in I_t$.

Proof Both the evolution equation and constraint follow immediately from the previous discussion, more precisely equations (7) and (8). The proof that the constraint is preserved by the evolution follows from computing

$$\frac{d}{dt}\left(F_a \wedge \omega^2\right) = d_a \dot{a} \wedge \omega^2 + F_a \wedge \frac{d}{dt}\omega^2 = d_a(\dot{a} \wedge \omega^2) - 2F_a \wedge d\gamma_2$$
$$= 2d_a(F_a \wedge \gamma_2) - 2F_a \wedge d\gamma_2 = 0,$$

where we used (4), (6), (9) and the Bianchi identity $d_a F_a = 0$. □

Proposition 1 *In the setting of Lemma 1, suppose that the family of* SU(3)*-structures* $(\omega(t), \gamma_2(t))$ *depends real analytically on t, and let* $a(0)$ *be a real analytic connection on P such that* $\Lambda_0 F_a(0) = 0$. *Then there is* $\varepsilon > 0$ *and a* G_2*-instanton A on* $(-\varepsilon, \varepsilon) \times M^6$ *with* $A|_{\{0\}\times M^6} = a(0)$.

Proof This is immediate from applying the Cauchy-Kovalevskaya theorem to (9). □

Remark 1 We can similarly derive evolution equations defining G_2-monopoles, i.e. pairs (A, Φ) where A is a connection on P and Φ is a section of the adjoint bundle, $\mathfrak{g}_P$, satisfying

$$*\nabla_A \Phi = F_A \wedge \psi.$$

In this setting we can write $A = a(t)$ in temporal gauge as before and $\Phi = \phi(t) \in \Omega^0(I_t, \Omega^0(M, \mathfrak{g}_P))$ as a 1-parameter family of Higgs fields over M. Then, the family $(a(t), \phi(t))$ of connections and Higgs fields on M gives rise to a G_2-monopole if and only if they satisfy:

$$J_t \dot{a} = -d_a\phi - *_t\left(F_a \wedge \gamma_2\right) \quad \text{and} \quad \dot{\phi} = \Lambda_t F_a.$$

The analysis of these equations for the Bryant–Salamon G_2-manifolds [5] is carried out in [23].

2.2 Hamiltonian Flow

We now turn to a more formal aspect of the theory, which has not yet been used in applications, but which we have decided to point out here in case it may be of use in the future. On each slice $M_t = \{t\} \times M$, we may define a functional $\mathcal{F}_t$ on $\mathcal{A}$, the space of connections a on P, by

$$\mathcal{F}_t(a) = \frac{1}{2}\int_{M_t} \langle F_a \wedge F_a\rangle \wedge \eta(t),$$

where $\eta(t) = \int^t \omega(s)ds$ and the $\langle\cdot,\cdot\rangle$ stands for an Ad-invariant inner product on $\mathfrak{g}_P$, here applied to the $\mathfrak{g}_P$ components of the curvature in both entries. Then, given this 1-parameter family of functionals $\mathcal{F}_t$, which we may also interpret as a single time-dependent functional, we may compute its gradient with respect to the time-dependent L^2-inner product induced by g_t. We see that

$$\frac{d}{ds}\Big|_{s=0}\mathcal{F}_t(a+sb) = \int_{M_t} \langle d_a b \wedge F_a \rangle \wedge \eta(t)$$
$$= \int_{M_t} d(\langle b \wedge F_a \rangle \wedge \eta(t)) + \langle b \wedge F_a \rangle \wedge d\eta(t)$$
$$= \int_{M_t} \langle b \wedge F_a \rangle \wedge d\eta(t), \tag{10}$$

by Stokes' theorem. Moreover, using Hitchin's flow equations (4),

$$d\eta(t) = \int^t d\omega(s)ds = \int^t \frac{\partial \gamma_1}{ds} ds = \gamma_1(t),$$

and so the outcome of the computation (10) is that the gradient of $\mathcal{F}_t$, with respect to the time-dependent L^2-inner product induced by g_t on $\mathcal{A}$, is

$$\nabla \mathcal{F}_t = *_t(F_a \wedge \gamma_1(t)).$$

At this point it is convenient to equip the space of connections on P over each M_t with a time-dependent (almost)-symplectic form given by

$$\omega_t^{\mathcal{A}}(b_1, b_2) = \langle J_t b_1, b_2 \rangle_{L^2(g_t)},$$

for b_1, b_2 two $\mathfrak{g}_P$-valued 1-forms. Then, the Hamiltonian flow of $\mathcal{F}_t$ is $-J_t \nabla \mathcal{F}_t$ and we can regard the flow equation (9) for G_2-instantons as the Hamiltonian flow of the time-dependent Hamiltonian $\mathcal{F}_t$. Thus, define the space of connections whose curvature in orthogonal to ω_t by

$$\mathcal{A}_t = \{a \in \mathcal{A} \mid \Lambda_t F_a = 0\}.$$

We have shown in proposition 1 that the flow Eq. 9 starting at a connection in $\mathcal{A}_0$ always lies in $\mathcal{A}_t$. Putting this together with the discussion above, we have shown the following.

Proposition 2 *On $I_t \times M$, G_2-instantons are the solutions to the time-dependent Hamiltonian flow of $\mathcal{F}_t$ on $(\mathcal{A}, \omega_t^{\mathcal{A}})$ starting at time $t = 0$ in $\mathcal{A}_0$.*

3 Asymptotically Conical (AC) G_2-Manifolds

In this section, we survey the known results on G_2-instantons on G_2-manifolds X which are *asymptotically conical* (AC), i.e. X is complete with one[5] non-compact

[5]A complete non-compact Ricci-flat manifold, which is not an isometric product, can only have one end due to the Cheeger–Gromoll splitting theorem.

end where the G_2-structure is asymptotic to a conical G_2-structure on $\mathbb{R}^+ \times M$, as given in (5), for some nearly Kähler structure (ω, γ_2) on M.

It follows from Proposition 3 in [23] (or easily from (7)–(8)) that on an AC G_2-manifold, a G_2-instanton whose curvature is decaying pointwise at infinity will have as a limit (if it exists) a pseudo-Hermitian–Yang–Mills connection a (or nearly Kähler instanton) on M: i.e. the curvature F_a of a satisfies

$$F_a \wedge \omega^2 = 0 \quad \text{and} \quad F_a \wedge \gamma_2 = 0.$$

The known explicit examples of AC G_2-holonomy metrics (up to scale) are due to Bryant–Salamon [5]. These metrics are either defined on the total space of the bundle of anti-self-dual 2-forms on a self-dual Einstein 4-manifold with positive scalar curvature, or on $\mathbb{R}^4 \times S^3$ (viewed as the spinor bundle of S^3). These examples are cohomogeneity-1, and thus have a lot of symmetry, and so it is natural to look for G_2-instantons with symmetries on these AC G_2-manifolds.

In this section, we describe results from [20, 23] which provide examples of G_2-instantons on the explicitly known AC G_2-manifolds. We also review the results from [20] about the properties of the moduli space of G_2-instantons constructed on $\mathbb{R}^4 \times S^3$. This forms the content of Sects. 3.1–3.2. We conclude the section, in Sect. 3.3, with some open problems we believe are worthy of investigation concerning G_2-instantons in this AC setting.

3.1 On the Bryant–Salamon Manifolds $\Lambda^2_-(N^4)$

Let (N^4, g_N) be a self-dual Einstein 4-manifold with positive scalar curvature. Then N is either S^4 or $\mathbb{CP}^2$ with g_N being respectively either the round or Fubini–Study metric. The AC Bryant–Salamon metric on the total space of the bundle of anti-self-dual 2-forms $X = \Lambda^2_-(N)$ on N is such that the zero section $N \subset \Lambda^2_-(N)$ is the unique compact coassociative submanifold in X (in fact, any compact minimal submanifold in X is contained in N by Theorem 5.5 in [29]). If $\pi : \Lambda^2_-(N) \to N$ denotes the projection (this is the radially extended twistor projection, as the unit sphere bundle in $\Lambda^2_-(N)$ can be identified with the twistor space of N), then the Bryant–Salamon metric can be written as

$$g = f^2(s) g_{\mathbb{R}^3} + f^{-2}(s) \pi^* g_N,$$

where $g_{\mathbb{R}^3}$ is the Euclidean metric along the fibers,

$$f(s) = (1 + s^2)^{-1/4}$$

and s is the Euclidean distance along the fibers to the zero section. The geodesic distance to the zero section in the metric g is $t(s) = \int_0^s f(u)du$ and using it we can write the metric as

$$g = dt^2 + s^2(t) f^2(s(t)) g_{S^2} + f^{-2}(s(t)) \pi^* g_N,$$

where g_{S^2} is the round metric in the unit normal spheres to N (the twistor spheres).

3.1.1 $N = S^4$

There is a cohomogeneity-1 action of Sp(2) on $\Lambda^2_-(S^4)$ whose principal orbits are the distance sphere bundles over S^4, which are diffeomorphic to the twistor space

$$\mathbb{CP}^3 = \mathrm{Sp}(2)/(\mathrm{Sp}(1) \times \mathrm{U}(1)).$$

We shall fix a reductive splitting

$$\mathfrak{sp}(2) = \mathfrak{h} \oplus \mathfrak{m},$$

as follows. Start by writing $\mathfrak{sp}(2) = \mathfrak{m}_1 \oplus \mathfrak{sp}_1(1) \oplus \mathfrak{sp}_2(1)$ and introduce a basis for the dual $\mathfrak{sp}(2)^*$ with

$$\mathfrak{m}_1^* = \langle e^1, e^2, e^3, e^4 \rangle \ , \ \mathfrak{sp}_1^*(1) = \langle \eta^1, \eta^2, \eta^3 \rangle \ , \ \mathfrak{sp}_2^*(1) = \langle \omega^1, \omega^2, \omega^3 \rangle, \tag{11}$$

where the η^i, ω^i form a standard dual basis for $\mathfrak{sp}(1) \cong \mathfrak{su}(2)$. Using the notation $e^{12} = e^1 \wedge e^2$, define the 2-forms:

$$\begin{aligned} \Omega_1 &= e^{12} - e^{34} \ , \quad \Omega_2 = e^{13} - e^{42} \ , \quad \Omega_3 = e^{14} - e^{23} \ ; \\ \overline{\Omega}_1 &= e^{12} + e^{34} \ , \quad \overline{\Omega}_2 = e^{13} + e^{42} \ , \quad \overline{\Omega}_3 = e^{14} + e^{23} \ . \end{aligned} \tag{12}$$

The Maurer–Cartan relations yield

$$d\omega^i = -2\omega^{jk} + \frac{1}{2}\Omega_i \ , \quad d\eta^i = -2\eta^{jk} - \frac{1}{2}\overline{\Omega}_i, \tag{13}$$

for $i = 1, 2, 3$ and (i, j, k) denoting a cyclic permutation of $(1, 2, 3)$. Furthermore, the Maurer–Cartan relations for the de's can be used to compute

$$d\Omega_i = 2\varepsilon_{ijk} \left(\Omega_j \wedge \omega^k - \Omega_k \wedge \omega^j \right), \tag{14}$$

for $i \in \{1, 2, 3\}$. Then, we pick the reductive decomposition $\mathfrak{sp}(2) = \mathfrak{h} \oplus \mathfrak{m}$, such that

$$\mathfrak{m}^* = \mathfrak{m}_1 \oplus \mathfrak{m}_2 = \mathfrak{m}_1 \oplus \mathbb{R}\langle \omega^2, \omega^3 \rangle \tag{15}$$

$$\mathfrak{h}^* = \mathfrak{sp}_1(1) \oplus \mathbb{R}\langle \omega^1 \rangle. \tag{16}$$

Upon fixing the identifications $\mathfrak{m} \cong T_p\mathbb{CP}^3$ and $\mathfrak{m}_1 \cong T_{\pi(p)}S^4$. The 2-forms Ω_i (resp. $\overline{\Omega}_i$) form a basis for the anti-self-dual (resp. self-dual) 2-forms at $\pi(p)$.

In the complement of the zero section $\Lambda^2_-(S^4)\backslash S^4 \cong \mathbb{R}^+ \times \mathbb{CP}^3$, the G_2-holonomy metric can be written as

$$\tilde{g} = dt \otimes dt + a^2(t)\left(\omega^2 \otimes \omega^2 + \omega^3 \otimes \omega^3\right) + b^2(t)\left(\sum_{i=1}^{4} e^i \otimes e^i\right),$$

where $a(s) = 2sf(s^2)$ and $b(s) = \sqrt{2}f^{-1}(s^2)$. A G_2-structure giving rise to this metric can be written as

$$\varphi = dt \wedge \left(a^2\omega^{23} + b^2\Omega_1\right) + ab^2\left(\omega^3 \wedge \Omega_2 - \omega^2 \wedge \Omega_3\right),$$

and

$$\psi = b^4 e^{1234} - a^2b^2\omega^{23} \wedge \Omega_1 - ab^2 dt \wedge \left(\omega^2 \wedge \Omega_2 + \omega^3 \wedge \Omega_3\right). \qquad (17)$$

We now consider the bundle

$$P_\lambda = \mathrm{Sp}(2) \times_{(\lambda, \mathrm{Sp}(1)\times \mathrm{U}(1))} \mathrm{SU}(2),$$

where $\lambda : \mathrm{Sp}(1) \times \mathrm{U}(1) \to \mathrm{SU}(2)$ is given by $\lambda(g, e^{i\theta}) = \mathrm{diag}(e^{il\theta}, e^{-il\theta})$, for some $l \in \mathbb{Z}$ and $(g, e^{i\theta}) \in \mathrm{SU}_1(2) \times \mathrm{U}_2(1)$. There is a canonical invariant connection, which as a 1-form in $\mathrm{Sp}(2)$ with values in $\mathfrak{su}(2)$ can be written as

$$A_c = \omega^1 \otimes T_1,$$

where T_1, T_2, T_3 is a standard basis for $\mathfrak{su}(2)$. Then, one can prove that (up to an invariant gauge transformation) any other connection $A \in \Omega^1(\mathrm{Sp}(2), \mathfrak{su}(2))$ can be written as $A = A_c + (A - A_c)$ with

$$A - A_c = a\left(T_2 \otimes \omega^2 + T_3 \otimes \omega^3\right), \qquad (18)$$

with $a \in \mathbb{R}$.

Now we consider the bundle P pulled back to $\Lambda^2_- S^4\backslash S^4 \cong \mathbb{R}^+ \times \mathbb{CP}^3$ and equip it with an invariant connection $A \in \Omega^1(\mathbb{R}^+_t \times \mathrm{Sp}(2), \mathfrak{su}(2))$ in radial gauge, i.e. $A(\partial_s) = 0$. Thus A must be a 1-parameter family of connections as above. This is determined by a which is now a real-valued function of $t \in \mathbb{R}^+$, as it must be constant along any $\mathrm{Sp}(2)$ orbit. A straightforward computation yields that the curvature F_A of the connection A satisfies the G_2-instanton equation $F_A \wedge \psi = 0$ if and only if

$$s^2 f^4 = 1 - a^2, \quad \frac{da}{ds} = -sf^{-4}a.$$

In terms of $t(s) = \int_0^s f(l^2)dl = \int_0^s \left(1 + l^2\right)^{-\frac{1}{4}} dl$, the second of these is

$$\frac{da}{dt} = -sf^{-3}a. \tag{19}$$

Moreover, solving the first equation, which is algebraic, yields

$$a(t) = \pm f^2(s(t)),$$

which one can check does provide a solution of the ODE (19). This proves the following result.

Theorem 1 *The* SU(2) *connection*

$$A = A_c \pm (1+s^2)^{-\frac{1}{2}} \left(T_2 \otimes \omega^2 + T_3 \otimes \omega^3 \right)$$

on $P \to \Lambda^2_-(S^4)$ *is an irreducible* G_2*-instanton, with curvature given by*

$$F_A = \left(\frac{\Omega_1}{2} - \frac{2s^2}{1+s^2}\omega^{23} \right) \otimes T_1 \pm \frac{1}{2\sqrt{1+s^2}} (\Omega_2 \otimes T_2 + \Omega_3 \otimes T_3)$$
$$\mp \frac{s}{1+s^2} \left(ds \wedge \omega^2 \otimes T_2 + ds \wedge \omega^3 \otimes T_3 \right).$$

Remark 2 These instantons are asymptotic to the canonical invariant connection A_c. This is a t-independent reducible connection which is in fact pseudo-Hermitian–Yang–Mills with respect to the standard nearly Kähler structure on $\mathbb{CP}^3$.

The Levi-Civita connection of the round metric induces a self-dual connection in the Spin bundle over S^4. Lifting this to $\Lambda^2_-(S^4)$ also gives rise to a G_2-instanton. To prove this we must construct the Spin bundle

$$Q = \mathrm{Sp}(2) \times_{(\mu, \mathrm{Sp}(1)\times \mathrm{U}(1))} \mathrm{Sp}(1),$$

where $\mu : \mathrm{Sp}(1) \times \mathrm{U}(1) \to \mathrm{Sp}(1) \cong \mathrm{SU}(2)$ is simply the projection on the first component. The canonical invariant connection in Q is the Spin connection and is given by

$$\theta = \eta^1 \otimes T_1 + \eta^2 \otimes T_2 + \eta^3 \otimes T_3.$$

Using the Maurer–Cartan relations (13) one can compute the curvature to be

$$F_\theta = d\theta + \frac{1}{2}[\theta \wedge \theta]$$
$$= 2\eta^{23} \otimes T_1 + 2\eta^{31} \otimes T_2 + 2\eta^{12} \otimes T_3$$
$$- \left(2\eta^{23} + \frac{1}{2}\Omega_1 \right) \otimes T_1 - \left(2\eta^{31} + \frac{1}{2}\Omega_2 \right) \otimes T_2 - \left(2\eta^{12} + \frac{1}{2}\Omega_3 \right) \otimes T_3.$$

We shall state this as follows.

Proposition 3 *The lift of the Spin connection θ on S^4 to $\Lambda^2_-(S^4)$ is a* G2*-instanton with curvature*

$$F_\theta = -\frac{1}{2}\overline{\Omega}_1 \otimes T_1 - \frac{1}{2}\overline{\Omega}_2 \otimes T_2 - \frac{1}{2}\overline{\Omega}_3 \otimes T_3.$$

Remark 3 Proposition 3 is a consequence of a more general phenomena. Indeed, for N either $\mathbb{CP}^2$ or $\mathbb{S}^4$, the pullback of any self-dual connection on N gives rise to a G_2-instanton on the Bryant-Salamon G_2-manifolds $\Lambda^2_-(N)$. This can be seen immediately from the calibrating 4-form ψ in Eq. 17 as noticed in [23].

3.1.2 $N = \mathbb{CP}^2$

As already remarked above, the sphere bundle in $\Lambda^2_-(\mathbb{CP}^2)$ is the twistor space of $\mathbb{CP}^2$, which is the flag manifold $\mathbb{F}_2$. This is homogeneous and $\mathrm{SU}(3)$ acts transitively with isotropy the maximal torus $\mathrm{U}(1)^2$. The Serre spectral sequence for the fibration $\mathrm{SU}(3) \to \mathbb{F}_2$ gives $H^2(\mathbb{F}_2, \mathbb{Z}) \cong H^1(\mathrm{U}(1)^2, \mathbb{Z})$, which we can further identify with the integral weight lattice in $(\mathfrak{u}(1)^2)^*$. An explicit way to make the identification is as follows. Given an integral weight $\alpha \in (\mathfrak{u}(1)^2)^*$ we construct the line bundle on $\mathbb{F}_2$

$$L_\alpha = \mathrm{SU}(3) \times_{(e^\alpha, \mathrm{U}(1)^2)} \mathbb{C}.$$

Now let $1 \in \mathrm{SU}(3)$ be the identity and $\mathfrak{m} \subset \mathfrak{su}(3)$ be a reductive complement to the Cartan subalgebra generated by the isotropy, i.e. $\mathfrak{su}(3) = \mathfrak{u}(1)^2 \oplus \mathfrak{m}$ with $[\mathfrak{u}(1)^2, \mathfrak{m}] \subset \mathfrak{m}$ (for example, we can let $\mathfrak{m}$ be the real part of the root spaces). Then, we extend α, first to $\mathfrak{su}(3)^*$ by letting it vanish on $\mathfrak{m}$, and secondly to $\Omega^1(\mathrm{SU}(3), i\mathbb{R})$ by left translations. It is now easy to see that α equips L_α with a connection and so its first Chern class $\frac{i}{2\pi}[d\alpha] \in H^2(\mathbb{F}_2, \mathbb{Z})$ gives the corresponding element in the second cohomology induced by α. The connection α is usually called the canonical invariant connection on L_α and is uniquely determined by $\mathfrak{m}$.

We shall now turn to the construction of SO(3)-bundles over $\mathbb{F}_2$, carrying interesting invariant connections. These are constructed by composing the homomorphism $e^\alpha : \mathrm{U}(1)^2 \to \mathrm{U}(1)$ with the embedding of $\mathrm{U}(1) \hookrightarrow \mathrm{SO}(3)$ as the maximal torus, then setting

$$P_\alpha = \mathrm{SU}(3) \times_{(e^\alpha, \mathrm{U}(1)^2)} \mathrm{SO}(3).$$

These SO(3)-bundles are in fact reducible to the circle bundles inducing L_α and can be equipped with the induced connections $\alpha \in \Omega^1(\mathrm{SU}(3), \mathfrak{so}(3))$ viewed as left invariant 1-forms in SU(3) with values in $\mathfrak{so}(3)$ by embedding $i\mathbb{R} \hookrightarrow \mathfrak{so}(3)$. These induced connections are also SU(3)-invariant and it follows from Wang's theorem, [33], that other invariant connections are in 1-to-1 correspondence with morphisms of $\mathrm{U}(1)^2$-representations

$$\Lambda : (\mathfrak{m}, \mathrm{Ad}) \to (\mathfrak{so}(3), \mathrm{Ad} \circ e^\alpha).$$

Decompose these into irreducible components $\mathfrak{m} \cong \mathbb{C}_{\alpha_1} \oplus \mathbb{C}_{\alpha_2} \oplus \mathbb{C}_{\alpha_3}$, where $\alpha_1, \alpha_2, \alpha_3$ are the positive roots of SU(3), while $\mathfrak{so}(3) \cong \mathbb{R}_0 \oplus \mathbb{C}_\alpha$. Hence it follows from Schur's lemma that such morphisms of representations exist if and only if α is one of the roots, in which case Λ restricts to the corresponding root space as an isomorphism onto $\mathbb{C}_\alpha \subset \mathfrak{so}(3)$ and vanishes in all other components. If $\alpha = \alpha_i$ we shall denote these by Λ_i. Then, notice that fixing a basis of $\mathfrak{m}$ and a basis of $\mathfrak{so}(3)$ (i.e. a gauge) each Λ_i is determined up to a constant.

The problem of constructing instantons on the bundles P_α was analysed in [23]. The first point to settle is that the bundle P_α on which one is solving the instanton equations must extend to a bundle over all of $\Lambda^2_-(\mathbb{CP}^2)$, i.e. across the zero section. It turns out that there is only one such α, say $\alpha = \alpha_2$, which can be characterized by being in the image of the map

$$(\pi_2)^* : H^2(\mathbb{CP}^2, \mathbb{Z}^2) \to H^2(\mathbb{F}_2, \mathbb{Z}^2),$$

where $\pi_2 : \mathbb{F}_2 \to \mathbb{CP}^2$ is the twistor projection. Thus, take $\alpha = \alpha_2$, and extend the bundle and the connection to the whole of $\Lambda^2_-(\mathbb{CP}^2)$. Now the connection

$$A = \alpha + \Lambda_2(r)$$

can be seen as an element of $\Omega^1(\mathbb{R}^+ \times \mathrm{SU}(3), \mathfrak{so}(3))$. Then, in [23] the invariant instanton equations for A are computed, very much in the same way as the case of $\Lambda^2_-(S^4)$ above. They appear as an ODE and an algebraic equation for $|\Lambda_2|$, with the ODE being implied by the algebraic equation which is

$$2s^2(r) f^{-2}(r) |\Lambda_2|^2 = 1.$$

In order to explicitly write this connection we fix a standard basis $\{T_1, T_2, T_3\}$ of $\mathfrak{so}(3)$ so that the image of α is parallel to T_1. Then, the complement $\mathbb{C}_\alpha \subset \mathfrak{so}(3)$ is generated by T_2, T_3, and there are left-invariant 1-forms ν_1, ν_2 on SU(3) such that the restriction to the tangent space to the identity of the map

$$\nu_1 \otimes T_2 + \nu_2 \otimes T_3|_{\mathbb{C}_{\alpha_2}} : \mathbb{C}_{\alpha_2} \subset \mathfrak{m} \subset \mathfrak{su}(3) \to \mathbb{C}_\alpha \subset \mathfrak{so}(3),$$

is an isomorphism. Furthermore, as in the case of $\Lambda^2_-(S^4)$ we fix $\Omega_1, \Omega_2, \Omega_3$ a universal basis for the anti-self-dual 2-forms on $\mathbb{CP}^2$. These are chosen so that $\sum_{i=1}^3 \frac{\mathrm{scal}}{24} \Omega_i \otimes T_i$ is the curvature of the Levi-Civita induced connection on Λ^2_-. Then, we can write the G_2-instanton A as in the following result.

Theorem 2 *The connection on P_{α_2} over $\Lambda^2_-(\mathbb{CP}^2)$ given by*

$$A = \alpha \pm (1 + s^2)^{-\frac{1}{2}} (\nu_1 \otimes T_2 + \nu_2 \otimes T_3)$$

is an irreducible G_2-instanton with curvature

$$\begin{aligned} F_A =& \frac{2s^2}{s^2+1}\nu_{12} \otimes T_1 + \Omega_1 \otimes T_1 \pm \frac{1}{\sqrt{s^2+1}} (\Omega_2 \otimes T_2 + \Omega_3 \otimes T_3) \\ & \mp \frac{s}{(1+s^2)^{\frac{3}{2}}} (ds \wedge \nu_1 \otimes T_2 + ds \wedge \nu_2 \otimes T_3) . \end{aligned}$$

Remark 4 This instanton converges (at a polynomial rate) to the canonical invariant connection α, which is the pullback to the cone on $\mathbb{F}_2$ of a reducible pseudo-Hermitian–Yang–Mills connection on $\mathbb{F}_2$ equipped with its standard nearly Kähler structure.

In [23] irreducible G_2-instantons with gauge group $G = \mathrm{SU}(3)$ in this setting are also investigated. For this we consider the bundle

$$Q = \mathrm{SU}(3) \times_{\mathrm{U}(1)^2} \mathrm{SU}(3),$$

where $\mathrm{U}(1)^2$ acts diagonally on both SU(3) factors by fixing a maximal torus. As before we decompose $\mathfrak{su}(3)$ into irreducible $\mathfrak{u}(1)^2$ representations, as

$$\mathfrak{su}(3) = \mathfrak{u}(1)^2 \oplus \mathbb{C}_{\alpha_1} \oplus \mathbb{C}_{\alpha_2} \oplus \mathbb{C}_{\alpha_3}.$$

Then, we fix certain isomorphisms $l : \mathfrak{u}(1)^2 \to \mathfrak{u}(1)^2$ and $\lambda_i : \mathbb{C}_i \to \mathbb{C}_i$, which we interpret as being left-invariant maps from (subspaces of) $T_1\mathrm{SU}(3) \cong \mathfrak{su}(3) \to \mathfrak{su}(3) \cong \mathfrak{g}$, i.e. as left-invariant 1-forms on SU(3) with values in the Lie algebra of the gauge group $\mathrm{G} = \mathrm{SU}(3)$. Then, Theorem 9 in [23] can be written in the following way.

Theorem 3 *There are two real* 1*-parameter families of irreducible* G_2*-instantons on* Q *parametrized by* $c \geq 0$*. These are given by*

$$A = l - \frac{u_c(s)}{\sqrt{1+s^2}}\lambda_2 \mp \frac{\sqrt{u_c^2(s)-1}}{s} (\lambda_3 - \lambda_1)$$

and

$$A = l + \frac{u_c(s)}{\sqrt{1+s^2}}\lambda_2 \mp \frac{\sqrt{u_c^2(s)-1}}{s} (\lambda_3 + \lambda_1) ,$$

where

$$u_c(s) = 1 - 2c\frac{s^2}{s^2(1+c) + 2\left(\sqrt{1+s^2}+1\right)}.$$

In particular, the case $c = -1$ *gives flat connections.*

3.2 On the Bryant–Salamon $\mathbb{R}^4 \times S^3$

The Bryant–Salamon metric on $\mathbb{R}^4 \times S^3$ [5] is $\mathrm{SU}(2)^2 \times \mathrm{U}(1)$-invariant and so we are motivated to study G_2-instantons with the same symmetry: in fact, the metric is $\mathrm{SU}(2)^3$-invariant, but it convenient to take the $\mathrm{SU}(2)^2 \times \mathrm{U}(1)$-invariant point of view for later study.

3.2.1 $\mathrm{SU}(2)^2 \times \mathrm{U}(1)$-Symmetry

We begin with some preparation for studying $\mathrm{SU}(2)^2$-invariant holonomy G_2-metrics and instantons. Split the Lie algebra $\mathfrak{su}(2) \oplus \mathfrak{su}(2)$ as $\mathfrak{su}^+ \oplus \mathfrak{su}^-$, as follows. If $\{T_i\}_{i=1}^3$ is a basis for $\mathfrak{su}(2)$ such that $[T_i, T_j] = 2\varepsilon_{ijk} T_k$, then $T_i^+ = (T_i, T_i)$ and $T_i^- = (T_i, -T_i)$ for $i = 1, 2, 3$ give a basis for $\mathfrak{su}^+$ and $\mathfrak{su}^-$ respectively. (Thus $\mathfrak{su}^+$ and $\mathfrak{su}^-$ are diagonal and anti-diagonal copies of $\mathfrak{su}(2)$ in $\mathfrak{su}(2) \oplus \mathfrak{su}(2)$.) We shall let $\{\eta_i^+\}_{i=1}^3$ and $\{\eta_i^-\}_{i=1}^3$ be dual bases to $\{T_i^+\}_{i=1}^3$ and $\{T_i^-\}_{i=1}^3$ respectively. The Maurer–Cartan relations in this case give

$$d\eta_i^+ = -\varepsilon_{ijk} \left(\eta_j^+ \wedge \eta_k^+ + \eta_j^- \wedge \eta_k^- \right), \tag{20}$$

$$d\eta_i^- = -2\varepsilon_{ijk} \eta_j^- \wedge \eta_k^+ . \tag{21}$$

The complement of the singular orbit can be written as $\mathbb{R}_t^+ \times M$, where M denotes a principal orbit, which is a finite quotient of $S^3 \times S^3$ (for the Bryant–Salamon metric, it will simply be $S^3 \times S^3$). The $\mathrm{SU}(2) \times \mathrm{SU}(2)$-invariant $\mathrm{SU}(3)$-structure on the principal orbit $\{t\} \times M$ is given by ([22])

$$\omega = 4 \sum_{i=1}^{3} A_i B_i \eta_i^- \wedge \eta_i^+ , \tag{22}$$

$$\gamma_1 = 8 B_1 B_2 B_3 \eta_{123}^- - 4 \sum_{i,j,k} \varepsilon_{ijk} A_i A_j B_k \eta_i^+ \wedge \eta_j^+ \wedge \eta_k^- , \tag{23}$$

$$\gamma_2 = -8 A_1 A_2 A_3 \eta_{123}^+ + 4 \sum_{i,j,k} \varepsilon_{ijk} B_i B_j A_k \eta_i^- \wedge \eta_j^- \wedge \eta_k^+ , \tag{24}$$

for real-valued functions A_i, B_i of $t \in \mathbb{R}^+$, where $\eta_{123}^{\pm}$ denotes $\eta_1^{\pm} \wedge \eta_2^{\pm} \wedge \eta_3^{\pm}$. The compatible metric determined by this SU(3) structure on $\{t\} \times M$ is ([22])

$$g_t = \sum_{i=1}^{3} (2A_i)^2 \eta_i^+ \otimes \eta_i^+ + (2B_i)^2 \eta_i^- \otimes \eta_i^- , \tag{25}$$

and the resulting metric on $\mathbb{R}_t \times M$, compatible with the G_2-structure $\varphi = dt \wedge \omega + \gamma_1$, is given by $g = dt^2 + g_t$. Recall also that this metric has holonomy in G_2 if and only if the SU(3)-structure above solves the Hitchin flow equations (4).

All known complete $SU(2)^2$-invariant holonomy G_2 metrics have an extra U(1)-symmetry: this U(1) acts diagonally on $S^3 \times S^3$ with infinitesimal generator T_1^+. As a consequence, we have $A_2 = A_3$ and $B_2 = B_3$ and (4) becomes (as in [1]):

$$\dot{A}_1 = \frac{1}{2}\left(\frac{A_1^2}{A_2^2} - \frac{A_1^2}{B_2^2}\right), \tag{26}$$

$$\dot{A}_2 = \frac{1}{2}\left(\frac{B_1^2 + B_2^2 - A_2^2}{B_1 B_2} - \frac{A_1}{A_2}\right), \tag{27}$$

$$\dot{B}_1 = \frac{A_2^2 + B_2^2 - B_1^2}{A_2 B_2}, \tag{28}$$

$$\dot{B}_2 = \frac{1}{2}\left(\frac{A_2^2 + B_1^2 - B_2^2}{A_2 B_1} + \frac{A_1}{B_2}\right). \tag{29}$$

3.2.2 The Bryant–Salamon Metric

As we stated above, the Bryant–Salamon metric on $\mathbb{R}^4 \times S^3$ is actually $SU(2)^3$-invariant: the principal orbits are $SU(2)^3/SU(2) \cong S^3 \times S^3$ and the (unique) singular orbit is $SU(2)^3/SU(2)^2 \cong S^3$. (Here, the SU(2) in $SU(2)^3$ is the subgroup $SU(2)_3 = 1 \times 1 \times SU(2)$, and $SU(2)^2 \subset SU(2)^3$ is the subgroup $\Delta SU(2) \times SU(2)$, where $\Delta SU(2) \subset SU(2)^2$ is the diagonal.)

In this case the extra symmetry means that $A_1 = A_2 = A_3$ and $B_1 = B_2 = B_3$ and the equations (26)–(29) reduce to:

$$\dot{A}_1 = \frac{1}{2}\left(1 - \frac{A_1^2}{B_1^2}\right) \quad \text{and} \quad \dot{B}_1 = \frac{A_1}{B_1}. \tag{30}$$

Setting $B_1 = s$ and $A_1 = sC(s)$ we see that (30) becomes $\frac{d}{ds}(sC) = \frac{1-C^2}{2C}$ which we can easily solve as $C(s) = \sqrt{\frac{1-c^3 s^{-3}}{3}}$, so that, for $c > 0$ and $s \geq c$,

$$A_1(s) = \frac{s}{\sqrt{3}}\sqrt{1 - c^3 s^{-3}} \quad \text{and} \quad B_1(s) = s. \tag{31}$$

In particular, choosing $c = 1$ and using t, the arc length parameter along the geodesic parametrized by s, we define a coordinate $r \in [1, \infty)$ implicitly by

$$t(r) = \int_1^r \frac{ds}{\sqrt{1 - s^{-3}}}, \tag{32}$$

and solve (30) as follows:

$$A_1 = A_2 = A_3 = \frac{r}{3}\sqrt{1 - r^{-3}} \quad \text{and} \quad B_1 = B_2 = B_3 = \frac{r}{\sqrt{3}}. \tag{33}$$

It is easy to verify that the geometry at infinity is asymptotically conical to the standard holonomy G_2-cone on $S^3 \times S^3$. In fact, we see from (31) that one obtains a one-parameter family[6] of solutions to (30), equivalent up to scaling, whose limit with $c = 0$ is the conical solution. Moreover, the torsion-free G_2-structure has a unique compact associative submanifold which is the singular orbit S^3.

3.2.3 Examples of G_2-Instantons

It is straightforward to write down the evolution equation (9) for $SU(2)^2$-invariant G_2-instantons on a $U(1)$-bundle over the Bryant–Salamon $\mathbb{R}^4 \times S^3$. One can solve this equation explicitly and obtain the following result.

Proposition 4 *Any $SU(2)^2$-invariant G_2-instanton A with gauge group $U(1)$ over the Bryant–Salamon $\mathbb{R}^4 \times S^3$ can be written as*

$$A = \frac{r^3 - 1}{r} \sum_{i=1}^{3} x_i \eta_i^+$$

for some $x_1, x_2, x_3 \in \mathbb{R}$, where $r \in [1, +\infty)$ is determined by (32).

We therefore wish to turn to a non-abelian gauge group, namely $SU(2)$. The only possible homogeneous $SU(2)$-bundle P on the principal orbits $S^3 \times S^3$ is $P = SU(2)^2 \times SU(2)$, i.e. the trivial $SU(2)$-bundle. We therefore consider connections on this bundle with the $SU(2)^3$-symmetry existent in the underlying Bryant–Salamon geometry, and derive the following evolution equations for invariant G_2-instantons in this setting from (9) (after some work).

Proposition 5 *Let A be an $SU(2)^3$-invariant G_2-instanton with gauge group $SU(2)$ on $\mathbb{R}^+ \times SU(2)^2 \cong \mathbb{R}^+ \times SU(2)^3/\Delta SU(2)$. There is a standard basis $\{T_i\}$ of $\mathfrak{su}(2)$, i.e. with $[T_i, T_j] = 2\varepsilon_{ijk}T_k$, such that (up to an invariant gauge transformation) we can write*

$$A = A_1 x \left(\sum_{i=1}^{3} T_i \otimes \eta_i^+ \right) + B_1 y \left(\sum_{i=1}^{3} T_i \otimes \eta_i^- \right), \tag{34}$$

with $x, y : \mathbb{R}^+ \to \mathbb{R}$ satisfying

$$\dot{x} = \frac{\dot{A}_1}{A_1} x + y^2 - x^2 = \frac{1}{2A_1}\left(1 - \frac{A_1^2}{B_1^2}\right) x + y^2 - x^2, \tag{35}$$

$$\dot{y} = \frac{2\dot{A}_1 - 3}{A_1} y + 2xy = -\frac{1}{A_1}\left(2 + \frac{A_1^2}{B_1^2}\right) y + 2xy. \tag{36}$$

---

[6]There are, in fact, distinct $SU(2)^3$-invariant torsion-free G_2-structures on $\mathbb{R}^4 \times S^3$ inducing the same asymptotially conical Bryant–Salamon metric, determined by their image in $H^3(S^3 \times S^3)$.

Next we must determine the initial conditions in order for an $SU(2)^3$-invariant G_2-instanton A, given by a solution to the ODEs in Proposition 5, to extend smoothly over the singular orbit $S^3 = SU(2)^2/\Delta SU(2)$. For that we need to first extend the bundle over the singular orbit. Up to an isomorphism of homogeneous bundles, there are two possibilities: these are

$$P_\lambda = SU(2)^2 \times_{(\Delta SU(2),\lambda)} SU(2), \tag{37}$$

with the homomorphism $\lambda : SU(2) \to SU(2)$ being either the trivial one (which we denote by 1) or the identity id. Depending on the choice of λ, the conditions for the connection A to extend are different, as we show in the following lemma.

Lemma 2 *The connection A in* (34) *extends smoothly over the singular orbit S^3 if $x(t)$ is odd, $y(t)$ is even, and their Taylor expansions around $t = 0$ are*

- *either $x(t) = x_1 t + x_3 t^3 + \ldots$, $y(t) = y_2 t^2 + \ldots$, in which case A extends smoothly as a connection on P_1;*
- *or $x(t) = \frac{2}{t} + x_1 t + \ldots$, $y(t) = y_0 + y_2 t^2 + \ldots$, in which case A extends smoothly as a connection on P_{id}.*

If we set $y = 0$ in the notation of Proposition 5, the ODEs there become the single ODE:

$$\dot{x} = \frac{\dot{A}_1}{A_1} x - x^2. \tag{38}$$

Writing this equation as

$$\frac{d}{dt}\left(\frac{x}{A_1}\right) = -A_1 \left(\frac{x}{A_1}\right)^2 \tag{39}$$

makes it separable. Since $B_1 \dot{B}_1 = A_1$ by (30) and $B_1^2(0) = \frac{1}{3}$, (39) can be readily integrated to show that

$$x(t) = \frac{2x_1 A_1(t)}{1 + x_1(B_1^2(t) - \frac{1}{3})}. \tag{40}$$

We can explicitly see from Lemma 2 that the connection A extends smoothly over S^3 as a connection on P_1. This is precisely the one-parameter family of $SU(2)^3$-invariant G_2-instantons on the Bryant–Salamon $\mathbb{R}^4 \times S^3$ constructed by Clarke [9], and the parameter can be interpreted as how concentrated the instanton is around the associative S^3.

In fact, it is shown in [20] that these are the only irreducible $\mathrm{SU}(2)^2 \times \mathrm{U}(1)$-invariant G_2-instantons on P_1: in particular, this shows that all irreducible $\mathrm{SU}(2)^2 \times \mathrm{U}(1)$-invariant G_2-instantons on P_1 on the Bryant–Salamon $\mathbb{R}^4 \times S^3$ are actually $\mathrm{SU}(2)^3$-invariant.

Theorem 4 *The moduli space $\mathcal{M}_{P_1}^{BS}$ of irreducible $\mathrm{SU}(2)^2 \times \mathrm{U}(1)$-invariant G_2-instantons with gauge group $\mathrm{SU}(2)$ defined on P_1 on the Bryant–Salamon $\mathbb{R}^4 \times S^3$ is parametrized by the open interval $(0, +\infty)$.*

Specifically, let A be an $\mathrm{SU}(2)^2 \times \mathrm{U}(1)$*-invariant* G_2*-instanton with gauge group* $\mathrm{SU}(2)$ *on the Bryant–Salamon* $\mathbb{R}^4 \times S^3$*, which extends smoothly over the singular orbit on* P_1*.*

(a) *If A is irreducible, then it is one of Clarke's examples [9], in which case it is* $\mathrm{SU}(2)^3$*-invariant and there is* $x_1 \in \mathbb{R}$ *such that, in the notation of Proposition 5,*

$$x(r) = \frac{2x_1 r\sqrt{1-r^{-3}}}{3 + x_1(r^2-1)} \quad \textit{and} \quad y(r) = 0,$$

where $r \in [1, +\infty)$ *is determined by* (32)*. That is, A can be written as*

$$A^{x_1} = \frac{2x_1(r^3-1)}{3r\big(3 + x_1(r^2-1)\big)} \left(\sum_{i=1}^{3} T_i \otimes \eta_i^+\right).$$

Observe that A^{x_1} *is defined globally on* $\mathbb{R}^4 \times S^3$ *if and only if* $x_1 \geq 0$ *and that* A^0 *is the trivial flat connection.*

(b) *If A is reducible, it has gauge group* $\mathrm{U}(1)$ *and is given in Proposition 4 with* $x_2 = x_3 = 0$*, i.e.*

$$A = \frac{r^3-1}{r} x_1 \eta_1^+$$

for some $x_1 \in \mathbb{R}$*, where* $r \in [1, +\infty)$ *is as in* (32)*.*

We now turn to $\mathrm{SU}(2)^2 \times \mathrm{U}(1)$-invariant G_2-instantons defined on P_{id}, for which we have a local existence result for a 1-parameter family of such G_2-instantons.

Proposition 6 *Let* S^3 *be the singular orbit in the Bryant–Salamon* $\mathbb{R}^4 \times S^3$*. There is a one-parameter family of* $\mathrm{SU}(2)^2 \times \mathrm{U}(1)$*-invariant* G_2*-instantons, with gauge group* $\mathrm{SU}(2)$*, defined in a neighbourhood of* S^3 *and smoothly extending over* S^3 *on* P_{id}*. The instantons are actually* $\mathrm{SU}(2)^3$*-invariant and parametrized by* $y_0 \in \mathbb{R}$ *satisfying, in the notation of Proposition 5,*

$$x(t) = \frac{2}{t} + \frac{y_0^2 - 1}{4} t + O(t^3), \quad y(t) = y_0 + \frac{y_0}{2}\left(\frac{y_0^2}{2} - 3\right) t^2 + O(t^4).$$

If we set $y = 0$, which corresponds to taking $y_0 = 0$ in Proposition 6, we can again integrate the ODE (38) (or equivalently (39)) and obtain:

$$x(t) = \frac{A_1(t)}{\frac{1}{2}(B_1^2(t) - \frac{1}{3})}.$$

From Proposition 6 we see that the corresponding instanton extends smoothly over S^3 on P_{id}, and hence we find another G_2-instanton on the Bryant–Salamon $\mathbb{R}^4 \times S^3$.

Theorem 5 *The* G_2*-instanton* A^{lim} *arising from the case when* $y_0 = 0$ *in Proposition 6 is given by*

$$A^{\mathrm{lim}} = \frac{2(r^3-1)}{3r(r^2-1)} \left(\sum_{i=1}^{3} T_i \otimes \eta_i^+ \right).$$

Moreover, A^{lim} *extends as a* $\mathrm{SU}(2)^3$*-invariant* G_2*-instanton to the Bryant–Salamon* $\mathbb{R}^4 \times S^3$.

It is straightforward to compute the curvature of A^{x_1} and A^{lim} and see that they decay at infinity but that their curvatures do not lie in L^2.

3.2.4 The Moduli Space

We have seen from Theorem 4 that we have a moduli space $\mathcal{M}_{P_1}^{BS}$ of irreducible $\mathrm{SU}(2)^2 \times \mathrm{U}(1)$-invariant G_2-instantons on P_1 which is parameterized by $x_1 \in (0, +\infty)$. Therefore, this moduli space is clearly non-compact. A natural question is whether it can be compactified and, if so, what the compactification is: it is clear what happens at $x_1 = 0$, since we just take the trivial flat connections, but we need to understand what happens as $x_1 \to +\infty$. In [20] it is shown that $\mathcal{M}_{P_1}^{BS}$ can be compactified to the closed interval: it is demonstrated that A^{lim} is, in a certain precise sense, the limit of the A^{x_1} as $x_1 \to +\infty$. The result, stated below, confirms expectations from [27, 28].

To state the result we now introduce some notation for the re-scaling we wish to perform: for $p \in S^3$ and $\delta > 0$ we define the map s_δ^p from the unit ball $B_1 \subseteq \mathbb{R}^4$ by

$$s_\delta^p : B_1 \subseteq \mathbb{R}^4 \to B_\delta \times \{p\} \subseteq \mathbb{R}^4 \times S^3, \quad x \mapsto (\delta x, p).$$

Recall that if we view $\mathbb{R}^4 \setminus \{0\} = \mathbb{R}_t^+ \times S^3$ then the basic ASD instanton on $\mathbb{R}^4$ with scale $\lambda > 0$ can be written as

$$A_\lambda^{\mathrm{ASD}} = \frac{\lambda t^2}{1+\lambda t^2} \sum_{i=1}^{3} T_i \otimes \eta_i^+. \tag{41}$$

Theorem 6 *Let* $\{A^{x_1}\}$ *be a sequence of Clarke's* G_2*-instantons from Theorem 4 with* $x_1 \to +\infty$.

(a) After a suitable rescaling, the family $\{A^{x_1}\}$ *bubbles off a basic anti-self-dual instanton transversely to the associative* $S^3 = \{0\} \times S^3$.
More precisely, given any $\lambda > 0$*, there is a sequence of positive real numbers* $\delta = \delta(x_1, \lambda) \to 0$ *as* $x_1 \to +\infty$ *such that: for all* $p \in S^3$, $(s_\delta^p)^* A^{x_1}$ *converges uniformly with all derivatives to the basic ASD instanton* A_λ^{ASD} *on* $B_1 \subseteq \mathbb{R}^4$ *as in* (41).

(b) The connections A^{x_1} *converge uniformly with all derivatives to* A^{lim}*, given in Theorem 5, on every compact subset of* $(\mathbb{R}^4 \setminus \{0\}) \times S^3$ *as* $x_1 \to +\infty$.

(c) *The function* $|F_{A^{x_1}}|^2 - |F_{A_{\lim}}|^2$ *is integrable for all* $x_1 > 0$. *Moreover, as* $x_1 \to +\infty$ *it converges to* $8\pi^2\delta_{\{0\}\times S^3}$ *as a current, i.e. for all compactly supported functions* f *we have*

$$\lim_{x_1\to+\infty}\int_{\mathbb{R}^4\times S^3} f(|F_{A^{x_1}}|^2 - |F_{A_{\lim}}|^2)\ \mathrm{dvol}_g = 8\pi^2\int_{\{0\}\times S^3} f\ \mathrm{dvol}_{g|_{\{0\}\times S^3}}.$$

Whilst (a) gives the familiar "bubbling" behaviour of sequences of instantons, with curvature concentrating on an associative S^3 by (c), we can interpret (b) as a "removable singularity" phenomenon since $A^{\lim}$ is a smooth connection on $\mathbb{R}^4 \times S^3$. In proving Theorem 6, we show that as $\{A^{x_1}\}$ bubbles along the associative S^3 one obtains a Fueter section, as in [10, 16, 32]. Here this is just a constant map from S^3 to the moduli space of anti-self dual connections on $\mathbb{R}^4$ (thought of as a fibre of the normal bundle), taking value at the basic instanton on $\mathbb{R}^4$. Since $8\pi^2$ is the Yang–Mills energy of the basic instanton, we can also view (c) as the expected "conservation of energy".

It is also worth observing that all of the G_2-instantons A^{x_1} for $x_1 > 0$ and $A^{\lim}$ are asymptotic to the canonical pseudo-Hermitian–Yang–Mills connection on the standard nearly Kähler $S^3 \times S^3$ given by:

$$a_\infty = \frac{2}{3}\sum_{i=1}^{3} T_i \otimes \eta_i^+. \tag{42}$$

Proposition 7 *Let* a_∞ *be the canonical pseudo-Hermitian–Yang–Mills connection on* $S^3 \times S^3$ *given in* (42).

(a) *If* $A = A^{x_1}$ *for some* $x_1 \in \mathbb{R}^+$, *then for* $t \gg 1$

$$|A^{x_1} - a_\infty| \leq \frac{c}{x_1 t^3},$$

where $c > 0$ *is some constant independent of* x_1;

(b) *If* $A = A^{\lim}$, *then for* $t \gg 1$, $|A^{\lim} - a_\infty| = O(t^{-4})$.

3.3 *Open Problems*

There are several natural open problems which arise for G_2-instantons in the asymptotically conical setting.

(a) Recently, infinitely many new examples of AC G_2-manifolds X have been found [14]. These examples are cohomogeneity-1 for an action of $SU(2) \times SU(2) \times U(1)$ and so the theory of G_2-instantons with these symmetries developed in [20] applies. Therefore, the local existence of G_2-instantons near the singular orbit in X is guaranteed, and the open question is how many of these local solutions extend globally on X. Once one has classified the global solutions and has a non-trivial family, one can then ask about global properties of the moduli space of solutions, such as those discussed above. This would be specially interesting for the family $\mathbb{D}_7$ of [14], as this was not considered in [20].

(b) As we have seen, G_2-instantons on AC G_2-manifolds naturally have limits at infinity which are pseudo-Hermitian–Yang–Mills (also known as nearly Kähler instantons) on the nearly Kähler link of the asymptotic cone at infinity. There are very few examples of such connections on nearly Kähler 6-manifolds, and it is an important open problem to try to construct some examples on the known nearly Kähler 6-manifolds which arise as links of asymptotic cones of AC G_2-manifolds: that is, S^6, $\mathbb{CP}^3$, $S^3 \times S^3$ (and finite quotients thereof) and the flag $\mathbb{F}_2$. Given these examples of nearly Kähler instantons, one can then ask if they arise as limits of G_2-instantons on AC G_2-manifolds. If they do arise, it is then natural to ask how many G_2-instantons have the given nearly Kähler instanton as their limits at infinity.

(c) An obvious problem in this context is to understand the local geometry of the moduli space of G_2-instantons on AC G_2-manifolds; i.e. the deformation theory of such G_2-instantons. This is currently being investigated by Joe Driscoll (a PhD student of Derek Harland) and would potentially help solve several interesting questions. For example, can one prove a uniqueness result for the "basic" G_2-instanton on $\mathbb{R}^7$ [15] (which has gauge group G_2)? Do deformations of G_2-instantons with symmetries on AC G_2-manifolds also have symmetries? A positive answer to the latter question would mean that we could describe (at least a component) of the moduli space of G_2-instantons on AC G_2-manifolds with a cohomogeneity-1 action via the techniques and results described in this survey. There will also be a natural projection map in this context from the moduli space of G_2-instantons to the moduli space of nearly Kähler instantons (studied in [7]), and so it would be interesting to understand the properties of this map, e.g. whether it is surjective.

(d) What is the limit as $c \to +\infty$ of the G_2-instantons in Theorem 3?

(e) We have seen that the local G_2-instanton defined on P_{id} given by Proposition 6 for $y_0 = 0$ extends globally to the Bryant–Salamon $\mathbb{R}^4 \times S^3$ by Theorem 5. A concrete question is whether any of the other local G_2-instantons from Theorem 3. Proposition 6 for $y_0 \neq 0$ extend globally or not. Some numerical investigation suggests that if they do, their curvature is unbounded at infinity.

4 Asymptotically Locally Conical (ALC) G_2-Manifolds

A noncompact G_2-manifold is said to be asymptotically locally conical (ALC), if it is asymptotic (at infinity) to a circle bundle over a 6-dimensional cone. The central part of this section is to summarize the results of [20], where the authors studied G_2-instantons on the so-called BGGG G_2-manifold: this is an ALC holonomy G_2-metric on $\mathbb{R}^4 \times S^3$ constructed in [2], and coming in a 1-parameter family of torsion-free

G_2-structures found a posteriori in [3]. In fact, the authors construction of instantons on the BGGG extends to give instantons for any holonomy G_2-metric in this whole 1-parameter family, see Remark 13 in [20].

This section is organized as follows. In Sect. 4.1 we present some general structure results on ALC G_2-manifolds, for example we describe the induced structure on the asymptotic circle bundle over a cone, since this asymptotic geometry is less familiar. Then, in Sect. 4.2 we characterise the limits of G_2-instantons with pointwise decaying curvature at infinity. Finally, in Sect. 4.3 we summarize the results of [20] and present some open problems in Sect. 4.4.

4.1 The G_2-Structure

A noncompact G_2-manifold (X, φ) is said to be ALC if there is:

- a $U(1)$-bundle $\pi : \Sigma^6 \to \Gamma^5$ and a $U(1)$-invariant G_2-structure φ_∞ on $(1, +\infty) \times \Sigma$, whose associated metric is

$$g_{\varphi_\infty} = dr^2 + m^2\eta_\infty^2 + r^2\pi^* g_5,$$

  where $m \in \mathbb{R}^+$, η_∞ is a connection on Σ and g_5 a metric on Γ;
- a compact set $K \subset X$ and (up to a double cover)[7] a diffeomorphism $p : (1, +\infty)_r \times \Sigma \to X \backslash K$,

such that if ∇ denotes the Levi-Civita connection of g_{φ_∞} then

$$|\nabla^j(\varphi_\infty - p^*\varphi|_{X\backslash K})|_{g_{\varphi_\infty}} = O(r^{\nu-j}) \quad \text{as } r \to +\infty, \tag{43}$$

for some $\nu < 0$ and $j = 0, 1$.

Our next result describes the structure on $(1, +\infty) \times \Sigma$ induced from the torsion-free G_2-structure φ on X and limits the range of rates ν to consider.

Proposition 8 *Let (X, φ) be an ALC G_2-manifold and use the notation above.*

(a) If $\nu < 0$, the metric g_5 is induced by a Sasaki–Einstein $SU(2)$-structure on Γ given by $(\alpha, \omega_1, \omega_2, \omega_3)$ satisfying

$$d\alpha = -2\omega_1, \quad d\omega_2 = 3\alpha \wedge \omega_3, \quad d\omega_3 = -3\alpha \wedge \omega_2. \tag{44}$$

Hence, the cone metric $dr^2 + r^2 g_5$ on $(1, +\infty)_r \times \Gamma$ is Calabi–Yau.

(b) If $\nu < -1$, then $d\eta_\infty = 0$, and thus the connection is flat.

Now we know from Proposition 8 that the asymptotic cone for an ALC G_2-manifold is Calabi–Yau, we can impose a further condition on the connection η_∞ for the definition

[7] The possible need for the double cover is because X may only be asymptotic to an S^1-bundle, but we can get a principal bundle by taking a double cover.

of an ALC G$_2$-manifold: namely, that η_∞ is Hermitian–Yang–Mills, i.e. $d\eta_\infty \wedge \omega^2 = 0$ and $d\eta_\infty \wedge \Omega_2 = 0$.

We now give the example of the standard Sasaki-Einstein structure on $S^2 \times S^3$ in terms of the framework above. We shall see that is the most important for our study.

Let $S^2 \times S^3 = \mathrm{SU}(2)^2/\Delta\mathrm{U}(1)$ and let $\{\eta_i^+, \eta_i^-\}_{i=1}^3$ be as in Sect. 3.2. We can equip $S^3 \times S^3 \to S^2 \times S^3$ with a connection such that $\eta_2^+, \eta_3^+, \eta_1^-, \eta_2^-, \eta_3^-$ is a horizontal coframing. We define:

$$\eta_\infty = 2\eta_1^+, \quad \alpha = -\frac{4}{3}\eta_1^-, \quad \omega_1 = \frac{4}{3}\left(\eta_2^+ \wedge \eta_3^- + \eta_2^- \wedge \eta_3^+\right),$$
$$\omega_2 = \frac{4}{3}\left(\eta_2^+ \wedge \eta_3^+ - \eta_2^- \wedge \eta_3^-\right), \quad \omega_3 = \frac{4}{3}\left(\eta_2^+ \wedge \eta_2^- + \eta_3^+ \wedge \eta_3^-\right).$$

The forms $\alpha, \omega_1, \omega_2, \omega_3$ are basic for the $\Delta\mathrm{U}(1)$-action and equip $S^2 \times S^3$ with an SU(2)-structure. We can check that (44) holds and so this is the standard homogeneous Sasaki–Einstein structure on $S^2 \times S^3$. The conical Calabi–Yau metric arising from this Sasaki–Einstein structure on $S^2 \times S^3$ is known as the conifold or 3-dimensional ordinary double point.

We also see that η_∞ is a connection form on $S^3 \times S^3$ such that

$$d\eta_\infty = -4\left(\eta_2^+ \wedge \eta_3^+ + \eta_2^- \wedge \eta_3^-\right)$$

is basic anti-self-dual: i.e. $d\eta_\infty \wedge \omega_i = 0$ for $i = 1, 2, 3$. This implies that η_∞ is Hermitian–Yang–Mills.

4.2 G2-*Instantons*

We now study the asymptotic behaviour of G$_2$-instantons on ALC G$_2$-manifolds, and begin by examining the G$_2$-instanton condition on the asymptotic U(1)-bundle over a Calabi–Yau cone. We shall use the notation of the previous subsection.

Let $\pi : (1, +\infty)_r \times \Sigma \to (1, +\infty)_r \times \Gamma$ be a U(1)-bundle over a Calabi–Yau cone, equipped with the G$_2$-structure

$$\varphi_\infty = m\eta_\infty \wedge \omega + \Omega_1,$$

as above. Let P be the pullback to $(1, +\infty) \times \Sigma$ of a bundle over M. If A_∞ is a connection on P, then without loss of generality we can write it as

$$A_\infty = a + m\Phi \otimes \eta_\infty, \tag{45}$$

for a connection a pulled back from $(1, +\infty) \times \Gamma$ and $\Phi \in \Omega^0((1, +\infty) \times \Sigma, \mathfrak{g}_P)$.

In our case we will be investigating G$_2$-instantons that are invariant under the U(1)-action on the end of X; that is, we take a lift of the U(1)-action to the total space and the connection is invariant under the lifted action. If we assume η_∞ is Hermitian–Yang–Mills, then the conditions for a U(1)-invariant connection A_∞ as in (45) to be a G_2-instanton are then

$$F_a \wedge \Omega_2 = -\frac{1}{2} d_a \Phi \wedge \omega^2, \quad F_a \wedge \omega^2 = 0. \tag{46}$$

These are the equations for a Calabi–Yau monopole (a, Φ) on $(1, +\infty) \times \Gamma$ equipped with the conical torsion-free $SU(3)$-structure (ω, Ω_2).

These observations lead to the following.

Proposition 9 *Let A be a* G_2*-instanton on an ALC* G_2*-manifold* (X, φ) *and use the notation from the start of Sect. 4.1. Suppose there exists a* U(1)*-invariant connection* $A_\infty = a + m\Phi \otimes \eta_\infty$ *as in* (45) *such that* $p^* F_A|_{X \setminus K}$ *is asymptotic at infinity to* F_{A_∞}. *Then* (a, Φ) *is a Calabi–Yau monopole on the Calabi–Yau cone* $(1, +\infty) \times \Gamma$.

4.3 *On the BGGG-Bogoyavlenskaya* $\mathbb{R}^4 \times S^3$

On $\mathbb{R}^4 \times S^3$, as well as the Bryant–Salamon metric, there is another explicit complete G_2-holonomy metric constructed by Brandhuber and collaborators in [2], which we will abbreviate to BGGG. The BGGG metric is a member of a family of complete $SU(2)^2 \times U(1)$-invariant, cohomogeneity-1, G_2-holonomy metrics on $\mathbb{R}^4 \times S^3$ found in [3].

4.3.1 The BGGG Metric

To derive the BGGG example, we return to the setting of $SU(2)^2 \times U(1)$-symmetry in Sect. 3.2.1: in particular we recall the functions A_1, A_2, B_1, B_2 defining the metric, satisfying (26)–(29). One can choose $c > 0$, set $B_1 = s$ and

$$A_1 = c\frac{ds}{dt} = c\frac{A_2^2 + B_2^2 - s^2}{A_2 B_2}$$

from (28). Letting $C_\pm = A_2^2 \pm B_2^2$ the equations (27) and (29) yield

$$\frac{d}{ds}C_+ = \frac{s^2 C_+ - C_-^2}{s(C_+ - s^2)} \quad \text{and} \quad \frac{d}{ds}C_- = -\frac{C_-}{s} - 2c.$$

The second equation is easily integrated and so we are able to find solutions

$$C_+(s) = \frac{3s^2 - c^2}{2} \quad \text{and} \quad C_-(s) = -cs.$$

We thus obtain a one-parameter family of solutions to (26)–(29):

$$A_1(s) = 2c\sqrt{\frac{s^2-c^2}{9s^2-c^2}}, \quad A_2(s) = \frac{1}{2}\sqrt{(3s+c)(s-c)}, \tag{47}$$

$$B_1(s) = s, \quad B_2(s) = \frac{1}{2}\sqrt{(3s-c)(s+c)}, \tag{48}$$

defined for $s \geq c > 0$. These solutions give holonomy G_2 metrics on $\mathbb{R}^4 \times S^3$. We can further scale the metric from g to $\lambda^2 g$ and the resulting fields scale as $A_i^\lambda(s) = \lambda A_i(s/\lambda)$, $B_i^\lambda(s) = \lambda B_i(s/\lambda)$. These give the following family of solution to the ODEs (26)–(29) above:

$$A_1^\lambda(s) = 2c\lambda\sqrt{\frac{s^2-c^2\lambda^2}{9s^2-c^2\lambda^2}}, \quad A_2^\lambda(s) = \frac{1}{2}\sqrt{(3s+c\lambda)(s-c\lambda)},$$

$$B_1^\lambda(s) = s, \quad B_2^\lambda(s) = \frac{1}{2}\sqrt{(3s-c\lambda)(s+c\lambda)}.$$

We see that under the scaling we have $c \mapsto c\lambda$, so we can always scale so that $c = 1$. In particular, one can set $\lambda = 3/2$, $c = 1$ and as in [2] define the coordinate $r \in [9/4, +\infty)$ implicitly by

$$t(r) = \int_{9/4}^{r} \frac{\sqrt{(s-3/4)(s+3/4)}}{\sqrt{(s-9/4)(s+9/4)}}\,ds \tag{49}$$

and find that

$$A_1 = \frac{\sqrt{(r-9/4)(r+9/4)}}{\sqrt{(r-3/4)(r+3/4)}}, \quad A_2 = A_3 = \sqrt{\frac{(r-9/4)(r+3/4)}{3}},$$

$$B_1 = \frac{2r}{3}, \quad B_2 = B_3 = \sqrt{\frac{(r-3/4)(r+9/4)}{3}}$$

solve (26)–(29). We see in this BGGG case that the principal orbits are again $S^3 \times S^3$ and the singular orbit $\{0\} \times S^3$ is associative.

It is straightforward to see that the BGGG is ALC with rate $\nu = -1$ and $m = 1$: in fact, the metric is asymptotic to

$$h = dt^2 + 4(\eta_1^+)^2 + \frac{4t^2}{3}\left((\eta_2^+)^2 + (\eta_3^+)^2\right) + \frac{16t^2}{9}(\eta_1^-)^2 + \frac{4t^2}{3}\left((\eta_2^-)^2 + (\eta_3^-)^2\right),$$

which is a circle bundle over the Calabi–Yau cone over the standard homogeneous Sasaki–Einstein structure on $S^2 \times S^3$ described in Sect. 4.1. This is in particular shows that Proposition 8(b) is sharp.

In [3], Bogoyavlenskaya constructed a 1-parameter family (up to scaling) of $\mathrm{SU}(2)^2 \times \mathrm{U}(1)$-invariant, cohomogeneity-1, G_2-holonomy metrics on $\mathbb{R}^4 \times S^3$, obtained by continuously deforming the BGGG metric. With these metrics, one can independently vary the size of the circle at infinity and the associative S^3, and thus, in particular, obtain the BS metric as a limit of the family. The BGGG metric is the only one from [3] which is explicitly known.

4.3.2 Examples of G_2-Instantons

It is again straightforward to write down the evolution equation (9) for $\mathrm{SU}(2)^2$-invariant G_2-instantons on a $\mathrm{U}(1)$-bundle over the Bogoyavlenskaya metrics on $\mathbb{R}^4 \times S^3$. One can solve this equation explicitly in the BGGG case and obtain the following result.

Proposition 10 *Any* $\mathrm{SU}(2)^2$*-invariant* G_2*-instanton* A *with gauge group* $\mathrm{U}(1)$ *over the BGGG* $\mathbb{R}^4 \times S^3$ *can be written as*

$$A = \frac{(r-9/4)(r+9/4)}{(r-3/4)(r+3/4)} x_1 \eta_1^+ + \frac{(r-9/4)e^r}{\sqrt{r}(r+9/4)^2}(x_2\eta_2^+ + x_3\eta_3^+)$$

for some $x_1, x_2, x_3 \in \mathbb{R}$*, where* $r \in [9/4, +\infty)$ *is given by* (49). *When* $x_2 = x_3 = 0$*, A is a multiple of the harmonic 1-form dual to the Killing field generating the* $\mathrm{U}(1)$*-action.*

We already observe a marked difference in the behaviour of G_2-instantons on the BS and BGGG $\mathbb{R}^4 \times S^3$ in this simple abelian setting. In particular, the instantons in the BS case all have bounded curvature, whereas those in the BGGG case have bounded curvature only when $x_2 = x_3 = 0$, in which case the curvature also decays to 0 as $r \to \infty$.

We now turn to gauge group $\mathrm{SU}(2)$ and begin by simplifying the ODEs (9) in the $\mathrm{SU}(2)^2 \times \mathrm{U}(1)$-invariant setting.

Proposition 11 *Let* A *be an* $\mathrm{SU}(2)^2 \times \mathrm{U}(1)$*-invariant* G_2*-instanton on* $\mathbb{R}^+ \times \mathrm{SU}(2)^2 \cong \mathbb{R}^+ \times (\mathrm{SU}(2)^2 \times \mathrm{U}(1)/\Delta\mathrm{U}(1))$ *with gauge group* $\mathrm{SU}(2)$*. There is a standard basis* $\{T_i\}_{i=1}^3$ *of* $\mathfrak{su}(2)$*, i.e. with* $[T_i, T_j] = 2\varepsilon_{ijk}T_k$*, such that (up to an invariant gauge transformation) we can write*

$$\begin{aligned} A =& A_1 f^+ T_1 \otimes \eta_1^+ + A_2 g^+ (T_2 \otimes \eta_2^+ + T_3 \otimes \eta_3^+) \\ &+ B_1 f^- T_1 \otimes \eta_1^- + B_2 g^- (T_2 \otimes \eta_2^- + T_3 \otimes \eta_3^-), \end{aligned} \tag{50}$$

with $f^\pm, g^\pm : \mathbb{R}^+ \to \mathbb{R}$ *satisfying*

$$\dot{f}^+ + \frac{1}{2}\left(\frac{A_1}{B_2^2} - \frac{A_1}{A_2^2}\right) f^+ = (g^-)^2 - (g^+)^2, \tag{51}$$

$$\dot{g}^+ + \frac{1}{2}\left(\frac{A_2^2 + B_1^2 + B_2^2}{A_2 B_1 B_2} - \frac{A_1^2 + 2A_2^2}{A_1 A_2^2}\right) g^+ = f^- g^- - f^+ g^+, \tag{52}$$

$$\dot{f}^- + \left(\frac{A_2^2 + B_1^2 + B_2^2}{A_2 B_1 B_2}\right) f^- = 2g^- g^+, \tag{53}$$

$$\dot{g}^- + \frac{1}{2}\left(\frac{A_2^2 + B_1^2 + B_2^2}{A_2 B_1 B_2} + \frac{A_1^2 + 2B_2^2}{A_1 B_2^2}\right) g^- = g^- f^+ + g^+ f^-. \tag{54}$$

We can then determine the local conditions for these connections to extend over the singular orbit.

Lemma 3 *The connection A in* (50) *extends smoothly over the singular orbit S^3 if and only if f^+ and g^+ are odd, f^- and g^- are even, and their Taylor expansions around $t = 0$ are:*

- *either*

$$f^- = f_2^- t^2 + O(t^4), \qquad g^- = g_2^- t^2 + O(t^4),$$
$$f^+ = f_1^+ t + O(t^3), \qquad g^+ = g_1^+ t + O(t^3),$$

  in which case A extends smoothly as a connection on P_1;
- *or*

$$f^- = b_0^- + b_2^- t^2 + O(t^4), \qquad g^- = b_0^- + b_2^- t^2 + O(t^4),$$
$$f^+ = \frac{2}{t} + (b_2^+ - \frac{2}{3}\dddot{A}_1(0))t + O(t^3), \quad g^+ = \frac{2}{t} + (b_2^+ - \frac{2}{3}\dddot{A}_2(0))t + O(t^3),$$

  in which case A extends smoothly as a connection on P_{id}.

We can now answer the question of how many G_2-instantons there are defined near the singular orbit on any $SU(2)^2 \times U(1)$-invariant G_2-manifold, which extend smoothly on P_1.

Proposition 12 *Let $X \subset \mathbb{R}^4 \times S^3$ contain the singular orbit $\{0\} \times S^3$ of the* $SU(2)^2 \times U(1)$ *action and be equipped with an* $SU(2)^2 \times U(1)$*-invariant holonomy* G_2*-metric. There is a 2-parameter family of* $SU(2)^2 \times U(1)$*-invariant* G_2*-instantons A with gauge group* $SU(2)$ *in a neighbourhood of the singular orbit in X smoothly extending on P_1.*

The BS, BGGG and Bogoyavlenskaya G_2-metrics all have $SU(2)^2 \times U(1)$-symmetry and so Proposition 12 yields a 2-parameter family of G_2-instantons in these cases. In the BS case, we already stated in Theorem 4 that only a 1-parameter family extends globally. In contrast, we see in the BGGG case that there is a 2-parameter family of local G_2-instantons which extends globally with bounded curvature and another 2-parameter family which cannot be extended so as to have bounded curvature.

Theorem 7 *The moduli space $\mathcal{M}_{P_1}^{BGGG}$ of irreducible* $\mathrm{SU}(2)^2 \times \mathrm{U}(1)$*-invariant* G_2*-instantons with gauge group* $\mathrm{SU}(2)$ *on the BGGG metric, smoothly extending on* P_1*, contains a nonempty (and unbounded) open set which is parametrised by* $U \subset \mathbb{R}^2$.

Specifically, let A be a $\mathrm{SU}(2)^2 \times \mathrm{U}(1)$*-invariant* G_2*-instanton with gauge group* $\mathrm{SU}(2)$ *defined in a neighbourhood of* $\{0\} \times S^3$ *on the BGGG* $\mathbb{R}^4 \times S^3$ *smoothly extending over* P_1 *as given by Proposition 12.*

(a) *If* $f_1^+ \leq \frac{1}{2}$, *or* $g_1^+ \geq 0$ *with* $g_1^+ \geq f_1^+$, *then A extends globally to* $\mathbb{R}^4 \times S^3$ *with bounded curvature if and only if A has gauge group* $\mathrm{U}(1)$ *and is given in Proposition 10 with* $x_2 = x_3 = 0$; *i.e. we must have* $g_1^+ = 0$ *and for some* $x_1 \in \mathbb{R}$,

$$A = \frac{(r - 9/4)(r + 9/4)}{(r - 3/4)(r + 3/4)} x_1 \eta_1^+.$$

(b) *If* $f_1^+ \geq \frac{1}{2} + g_1^+ > \frac{1}{2}$, *then A is irreducible and extends globally to* $\mathbb{R}^4 \times S^3$ *with bounded curvature.*

We also have the following interesting observations.

Theorem 8 *In the setting of Theorem 7 the following holds.*

(a) *The instantons parametrised by U have quadratically decaying curvature.*
(b) *The map* $Hol_\infty : U \to \mathrm{U}(1) \subset \mathrm{SU}(2)$, *which evaluates the holonomy of the* G_2*-instanton along the finite size circle at* $+\infty$, *is surjective.*

We can now also ask about P_{id} and we see even the local existence theory is different to the P_1 case.

Proposition 13 *Let* $X \subset \mathbb{R}^4 \times S^3$ *contain the singular orbit* $\{0\} \times S^3$ *of the* $SU(2)^2 \times U(1)$ *action and be equipped with an* $SU(2)^2 \times U(1)$*-invariant holonomy* G_2*-metric. There is a* 1*-parameter family of* $SU(2)^2 \times U(1)$*-invariant* G_2*-instantons A with gauge group* $SU(2)$ *in a neighbourhood of the singular orbit in X smoothly extending over* P_{id}.

In this setting, unfortunately, we cannot yet find any global solutions on the BGGG manifold which extend smoothly on P_{id}.

4.4 Open Problems

There are several natural open problems which present themselves for G_2-instantons on ALC G_2-manifolds.

(a) Proposition 9 shows that the natural limits of G_2-instantons on ALC G_2-manifolds (if they exist) are Calabi–Yau monopoles on Calabi–Yau cones. These observations further motivate the study of Calabi–Yau monopoles on cones or AC Calabi–Yau 3-folds. See [24] and [25] for some examples and results on Calabi–Yau monopoles in the AC and conical settings.

(b) It has been shown that [13] there are many ALC G_2-manifolds which are close to the degenerate Calabi–Yau cone limit, and the typically example will only be U(1)-invariant. It is therefore interesting to attempt to construct G_2-instantons on these ALC G_2-manifolds which are, in a sense, close to Calabi–Yau monopoles on the cone. The authors of this article are actively pursuing this problem.

(c) As in the AC case, it would be good to have a deformation theory for G_2-instantons on ALC G_2-manifolds. This would have obvious relations to the deformation theory of Calabi–Yau monopoles on AC Calabi–Yau 3-folds, which also needs to be developed. In particular, one can ask about the image of the projection map from the moduli space of G_2-instantons on an ALC G_2-manifold to the space of Calabi–Yau monopoles on the Calabi–Yau cone which appears at the end of the ALC G_2-manifold.

(d) On the BGGG G_2-manifold $\mathbb{R}^4 \times S^3$, we have shown non-existence for irreducible $SU(2)^2 \times U(1)$-invariant G_2-instantons with gauge group SU(2) and bounded curvature for $g_1^+ > 0$ and $f_1^+ \leq \frac{1}{2}$ or $g_1^+ \geq f_1^+$, and existence for $f_1^+ \geq \frac{1}{2} + g_1^+ > \frac{1}{2}$. This currently leaves open the region where $0 < f_1^+ - \frac{1}{2} < g_1^+ < f_1^+$, which should be investigated so as to describe the full moduli space $\mathcal{M}_{P_1}^{BGGG}$. Some numerical investigation indicates that some of these initial conditions may lead to globally defined instantons with bounded curvature and some may not. Some of the existence and non-existence results for instantons for the BGGG metric extend, with suitable modifications, to all of the Bogoyavlenskaya metrics, but some do not, so it would be good to address this gap.

(e) An interesting problem is to investigate the behaviour of G_2-instantons as the underlying metric is deformed. For instance, we have G_2-instantons on all of the Bogoyavlenskaya G_2-manifolds, and we would want to analyse these instantons as the size of the circle at infinity gets very large or small. When it gets very large we expect them to resemble G_2-instantons for the Bryant–Salamon metric in $\mathbb{R}^4 \times S^3$. When it gets very small, there may be a relation with Calabi–Yau monopoles on the deformed conifold (as in [25] or problem (b) above).

(f) Even if we can describe the moduli space $\mathcal{M}_{P_1}^{BGGG}$, it will be non-compact so, just as in the Bryant–Salamon case, we will want to compactify it. It is therefore certainly an interesting problem to investigate the behaviour of the family of instantons from Theorem 7 when one or both of f_1^+ and g_1^+ go to infinity. We would expect bubbling phenomena as in the Bryant–Salamon case in Theorem 6, with possible relationship to the ASD instantons on Taub–NUT found in [12]. The lack of an explicit formula for our instantons makes the bubbling analysis more difficult, but it should clearly be explored.

Acknowledgements We thank Spiro Karigiannis and Ragini Singhal for their comments and suggestions on a previous version of this manuscript.

References

1. Bazaikin, Y. V., Bogoyavlenskaya, O. A. (2013). Complete Riemannian metrics with holonomy group G_2 on deformations of cones over $S^3 \times S^3$. *Mathematical Notes, 93*(5–6), 643–653. Translation of Mat. Zametki *93*(5), 645–657. MR3206014 15.
2. Brandhuber, A., Gomis, J., Gubser, S. S., & Gukov, S. (2001). Gauge theory at large N and new G_2 holonomy metrics. *Nuclear Physics B, 611*(1–3), 179–204. MR1857379 23, 25, 26.
3. Bogoyavlenskaya, O. A. (2013). On a new family of complete G_2-holonomy Riemannian metrics on $S^3 \times \mathbb{R}^4$. *Siberian Mathematical Journal, 54*(3), 431–440. Translation of Sibirsk. Mat. Zh. *54*(3), 551–562. MR3112613 23, 25, 27.
4. Bryant, R. L. (2010) *Non-embedding and non-extension results in special holonomy*. The many facets of geometry (pp. 346–367). Oxford: Oxford University Press. MR2681703 4.
5. Bryant, R. L., & Salamon, S. M. (1989). On the construction of some complete metrics with exceptional holonomy. *Duke Mathematics Journal, 58*(3), 829–850. MR1016448 (90i:53055) 6, 7, 14.
6. Corrigan, E., Devchand, C., Fairlie, D. B., & Nuyts, J. (1983). First-order equations for gauge fields in spaces of dimension greater than four. *Nuclear Physics B, 214*(3), 452–464. MR698892 (84i:81058) 2.
7. Charbonneau, B., & Harland, D. (2016). Deformations of nearly kähler instantons. *Communications in Mathematical Physics, 348*(3), 959–990. 22.
8. Corti, A., Haskins, M., Nordström, J., & Pacini, T. (2015). G_2-manifolds and associative submanifolds via semi-Fano 3-folds. *Duke Mathematics Journal, 164*(10), 1971–2092. MR3369307 3.
9. Clarke, A. (2014). Instantons on the exceptional holonomy manifolds of Bryant and Salamon. *Journal of Geometry and Physics, 82*, 84–97. MR3206642 3, 18, 19.
10. Donaldson, S. K., & Segal, E. P. (2011). *Gauge theory in higher dimensions, II*, Surveys in differential geometry, vol. XVI. Geometry of special holonomy and related topics (pp. 1–41). MR2893675 2, 21.
11. Donaldson, S. K., & Thomas, R. P. (1996). *Gauge theory in higher dimensions* (pp. 31–47). Oxford: Oxford University Press. 1998. MR1634503 (2000a:57085) 2.
12. Etesi, G., & Hausel, T. (2001). Geometric construction of new Yang-Mills instantons over Taub-NUT space. *Physics Letter B, 514*(1–20), 189–199. MR1850138 31.
13. Foscolo, L., Haskins, M., & Nordström, J. (2017). Complete non-compact G_2-manifolds from asymptotically conical Calabi–Yau 3-folds. arXiv:1709.04904. 30.
14. Foscolo, L., Haskins, M., & Nordström, J. (2018). Infinitely many new families of complete cohomogeneity one G_2-manifolds: G_2 analogues of the Taub-NUT and Eguchi–Hanson spaces. arXiv:1805.02612. 22.
15. Günaydin, M., & Nicolai, H. (1995). Seven-dimensional octonionic yang-mills instanton and its extension to an heterotic string soliton. *Physics Letters B, 351*(1), 169–172. 22.
16. Haydys, A. (2012). Gauge theory, calibrated geometry and harmonic spinors. *Journal of the London Mathematical Society (2), 86*(2), 482–498. MR2980921 3, 21.
17. Haydys, A., & Walpuski, T. (2015). A compactness theorem for the Seiberg-Witten equation with multiple spinors in dimension three. *Geometric and Functional Analysis*, *25*(6), 1799–1821. 3.
18. Joyce, D. D. (2000). *Compact manifolds with special holonomy*. Oxford Mathematical Monographs. Oxford: Oxford University Press. MR1787733 (2001k:53093) 2, 3.
19. Kovalev, A. (2003). Twisted connected sums and special Riemannian holonomy. *The Journal für die reine und angewandte Mathematik*, 565, 125–160. MR2024648 (2004m:53088) 3.
20. Lotay, J. D., & Oliveira, G. (2018). $SU(2)^2$-invariant G_2-instantons. *Mathematische Annalen, 371*, 961–1011. 3, 7, 18, 20, 22, 23.
21. Menet, G., Nordström, J., & Sá Earp, H. N. (2015). *Construction of G_2-instantons via twisted connected sums*. arXiv:1510.03836. 3.

22. Madsen, T. B., & Salamon, S. (2013). Half-flat structures on $S^3 \times S^3$. *Annals of Global Analysis and Geometry, 44*(4), 369–390. 4, 15.
23. Oliveira, G. (2014). Monopoles on the Bryant–Salamon G_2-manifolds. *Journal of Geometry and Physics, 86*, 599–632. MR3282350 3, 6, 7, 11, 12 ,13 ,14.
24. Oliveira, G. (2014). *Monopoles in higher dimensions*. 30.
25. Oliveira, G. (2016). Calabi–Yau monopoles for the Stenzel metric. *Communications in Mathematical Physics, 341*(2), 699–728. MR3440200 30, 31.
26. Sá Earp, H. N., & Walpuski, T. (2015). G_2-instantons over twisted connected sums. *Geometry & Topology, 19*(3), 1263–1285. MR3352236 3.
27. Tian, G. (2000). Gauge theory and calibrated geometry. I. *Annals of Mathematics (2), 151*(1), 193–268. MR1745014 (2000m:53074) 20.
28. Tao, T., & Tian, G. (2004). A singularity removal theorem for Yang–Mills fields in higher dimensions. *Journal of the American Mathematical Society, 17*(3), 557–593 (electronic). MR2053951 (2005f:58013) 20.
29. Tsai, C-J., & Wang, M-T. (2018). Mean curvature flows in manifolds of special holonomy. *Journal of Differential Geometry, 108*(3), 531–569. MR3770850 8.
30. Walpuski, T. (2013). G_2-instantons on generalised Kummer constructions. *Geometry & Topology, 17*(4), 2345–2388. MR3110581 3.
31. Walpuski, T. (2016). G_2-instantons over twisted connected sums: An example. *Mathematical Research Letters, 23*(2), 529–544. MR3512897 3.
32. Walpuski, T. (2017). G_2-instantons, associative submanifolds and Fueter sections. *Communications in Analysis and Geometry, 25*(4), 847–893. 21.
33. Wang, H.-C. (1958). On invariant connections over a principal fibre bundle. *The Nagoya Mathematical Journal, 13*, 1–19. MR0107276 21.

Current Progress on G_2-Instantons over Twisted Connected Sums

Henrique Sá Earp

Abstract We review a method to construct G_2-instantons over compact G_2-manifolds arising as the twisted connected sum of a matching pair of Calabi-Yau 3-folds with cylindrical end, based on the series of articles [16, 24, 32, 33] by the author and others. The construction is based on gluing G_2-instantons obtained from holomorphic bundles over such building blocks, subject to natural compatibility and transversality conditions. Explicit examples are obtained from matching pairs of semi-Fano 3-folds by an algorithmic procedure based on the Hartshorne-Serre correspondence.

1 Introduction

This text addresses the existence problem of G_2-instantons over twisted connected sums, as formulated by the author and Walpuski in [32], and the production of the first examples to date of solutions obtained by a nontrivially *transversal* gluing process [24]. It is aimed at graduate students and researchers in nearby areas who might be interested in a condensed exposition of the main results spread over my articles [16, 24, 32, 33] with Walpuski, Menet et al. and Menet-Nordström. By no means should this survey convey the impression that the subject is somehow closed or even in its best notational setup; indeed there is much ongoing work on this topic. A number of important questions remain open and the most impressive expected results in this theory are surely still ahead of us.

Recall that a G_2-*manifold* (X, g_ϕ) is a Riemannian 7-manifold together with a torsion-free G_2-*structure*, that is, a non-degenerate closed 3-form ϕ satisfying a certain non-linear partial differential equation; in particular, ϕ induces a Riemannian metric g_ϕ with $\mathrm{Hol}(g_\phi) \subset G_2$ [18, Part I]. A G_2-*instanton* is a connection A on some G-bundle $E \to X$ such that $F_A \wedge *\phi = 0$. Such solutions have a well-understood elliptic deformation theory of index 0 [30], and some form of 'instanton count' of their moduli space is expected to yield new invariants of 7-manifolds, much in the same vein as the Casson invariant and instanton Floer homology from flat connections

H. Sá Earp (✉)
University of Campinas (Unicamp), Campinas, Brazil
e-mail: henrique.saearp@ime.unicamp.br

S. Karigiannis et al. (eds.), *Lectures and Surveys on G_2-Manifolds and Related Topics*, Fields Institute Communications 84, https://doi.org/10.1007/978-1-0716-0577-6_14

on 3-manifolds [10, 12]. While some important analytical groundwork has been established towards that goal [35], major compactification issues remain and this suggests that a thorough understanding of the general theory might currently have to be postponed in favour of exploring a good number of functioning examples. This article proposes a method to construct such examples.

Readers interested in a more detailed account of instanton theory on G_2-manifolds are kindly referred to the introductory sections of [32, 33] and works cited therein.

An important method to produce examples of compact G_2-manifolds with $\mathrm{Hol}(g) = G_2$ is the *twisted connected sum construction*, suggested by Donaldson, pioneered by Kovalev [21] and later extended and improved by Kovalev–Lee [20] and Corti–Haskins–Pacini-Nordström [6]. Here is a brief summary of this construction: A *building block* consists of a projective 3-fold Z and a smooth anti-canonical $K3$ surface $\Sigma \subset Z$ with trivial normal bundle (cf. Definition 2.10). Given a choice of hyperkähler structure $(\omega_I, \omega_J, \omega_K)$ on Σ such that $[\omega_I]$ is the restriction of a Kähler class on Z, one can make $V := Z \setminus \Sigma$ into an asymptotically cylindrical (ACyl) Calabi–Yau 3-fold, that is, a non-compact Calabi–Yau 3-fold with a tubular end modelled on $\mathbb{R}_+ \times \mathbb{S}^1 \times \Sigma$, see Haskins–Hein–Nordström [13]. Then $Y := \mathbb{S}^1 \times V$ is an ACylG_2-manifold with a tubular end modelled on $\mathbb{R}_+ \times T^2 \times \Sigma$.

When a pair $(Z_\pm, \Sigma_\pm)$ of building blocks *matches* 'at infinity', in a suitable sense, one can glue $Y_\pm$ by interchanging the $\mathbb{S}^1$-factors. This yields a simply-connected compact 7-manifold Y together with a family of torsion-free G_2-structures $(\phi_T)_{T \geq T_0}$, see Kovalev [21, Sect. 4]. From the Riemannian viewpoint (Y, ϕ_T) contains a "long neck" modelled on $[-T, T] \times T^2 \times \Sigma_+$; one can think of the twisted connected sum as reversing the degeneration of the family of G_2-manifolds that occurs as the neck becomes infinitely long. In [5, 6, 21], building blocks Z are produced by blowing up Fano or semi-Fano 3-folds along the base curve $\mathscr{C}$ of an anticanonical pencil (cf. Proposition 4.6). By understanding the deformation theory of pairs (X, Σ) of semi-Fanos X and anticanonical K3 divisors $\Sigma \subset X$, one can produce hundreds of thousands of pairs with the required matching (see Sect. 4.3).

This construction raises a natural programme in gauge theory, aimed at constructing G_2-instantons over compact manifolds obtained as a TCS, originally outlined in [30]. If (Z, Σ) is a building block and $\mathcal{E} \to Z$ holomorphic bundle such that $\mathcal{E}|_\Sigma$ is stable, then $\mathcal{E}|_\Sigma$ carries a unique ASD instanton compatible with the holomorphic structure [9]. In this situation $\mathcal{E}|_V$ can be given a Hermitian–Yang–Mills (HYM) connection asymptotic to the ASD instanton on $\mathcal{E}|_\Sigma$ [33, Theorem 58], whose pullback over V to $\mathbb{S}^1 \times V$ is a G_2-*instanton*, i.e., a connection A on a G-bundle over a G_2-manifold such that $F_A \wedge \psi = 0$ with $\psi := *\phi$. It is possible to glue a hypothetical pair of such solutions into a G_2-instanton over the *compact* twisted connected sum, provided a number of technical conditions are met (cf. Theorem 3.1).

However, the hypotheses of our G_2-instanton gluing theorem are rather restrictive and it is not immediate to obtain suitable holomorphic bundles $\mathcal{E}_\pm \to Z_\pm$ over the matching blocks. In particular, a transversality condition over the $K3$ surface $\Sigma_\pm$ 'at infinity' requires some more thorough understanding of the deformation theory of data $(Z_\pm, \Sigma_\pm, \mathcal{E}_\pm)$. Assuming the so-called *rigid* case in which the instantons that are glued are isolated points in their moduli spaces, Walpuski [39] was able to exhibit one such example, by a different systematic approach.

Finally, in [24], we use the Hartshorne-Serre construction (cf. Theorem 4.1) to obtain families of bundles over the building blocks. Our method allows one to generate a large number of examples for which the gluing is *nontrivially transversal* (see Sect. 4.4.1). These are particularly relevant, because they open the possibility of obtaining a conjectural instanton number on the G$_2$-manifold X as a genuine Lagrangian intersection within the moduli space $\mathcal{M}_{S_+}$ over the $K3$ cross-section along the neck, which can be addressed by enumerative methods in the future.

2 Background on G$_2$-Geometry

Let us recall some G$_2$-trivia, following the exposition in [31, 33]; of course the immortal introductory references for the topic are [3, 17, 29]. Recall that a G$_2$-*structure* on an oriented smooth 7-manifold Y is a smooth 3-form $\phi \in \Omega^3(Y)$ such that, at every point $p \in Y$, one has $\phi_p = r_p^*(\phi_0)$ for some frame $r_p : T_pY \to \mathbb{R}^7$ and (with the sign conventions of [29])

$$\phi_0 = e^{567} + \omega_1 \wedge e^5 + \omega_2 \wedge e^6 + \omega_3 \wedge e^7 \tag{2.1}$$

with

$$\omega_1 = e^{12} - e^{34}, \quad \omega_2 = e^{13} - e^{42}, \quad \text{and} \quad \omega_3 = e^{14} - e^{23}.$$

Moreover, ϕ determines a Riemannian metric $g(\phi)$ induced by the pointwise inner-product

$$\langle u, v\rangle\, e^{1\ldots7} := -\frac{1}{6}\,(u \lrcorner \phi_0) \wedge (v \lrcorner \phi_0) \wedge \phi_0. \tag{2.2}$$

under which $*_\phi\phi$ is given pointwise by

$$*\,\phi_0 = e^{1234} - \omega_1 \wedge e^{67} - \omega_2 \wedge e^{75} - \omega_3 \wedge e^{56}. \tag{2.3}$$

Such a pair (Y, ϕ) is a G$_2$-*manifold* if $\mathrm{d}\phi = 0$ and $\mathrm{d} *_\phi \phi = 0$. Notice that the co-closed condition is nonlinear in ϕ, since the Hodge star depends on the metric and hence on ϕ itself.

2.1 Gauge Theory on G$_2$-Manifolds

The G$_2$-structure allows for a 7-dimensional analogue of conventional Yang-Mills theory, yielding a notion analogous to (anti-)self-duality for 2-forms. Working in $\mathbb{R}^7$ under the usual identification between 2-forms and matrices, we have $\mathfrak{g}_2 \subset \mathfrak{so}(7) \simeq \Lambda^2$, so we define $\Lambda^2_{14} := \mathfrak{g}_2$ and Λ^2_7 its orthogonal complement in Λ^2:

$$\Lambda^2 = \Lambda^2_7 \oplus \Lambda^2_{14}. \tag{2.4}$$

It is easy to check that $\Lambda^2_7 = \langle e_1 \lrcorner \phi_0, \ldots, e_7 \lrcorner \phi_0 \rangle$, hence the orthogonal projection onto Λ^2_7 in (2.4) is given by

$$\begin{aligned} L_{*\phi_0} : \Lambda^2 &\to \Lambda^6 \\ \eta &\mapsto \eta \wedge *\phi_0 \end{aligned}$$

in the sense that [3, p. 541]

$$L_{*\phi_0}|_{\Lambda^2_7} : \Lambda^2_7 \tilde{\to} \Lambda^6 \quad \text{and} \quad L_{*\phi_0}|_{\Lambda^2_{14}} = 0. \tag{2.5}$$

Furthermore, since (2.4) splits Λ^2 into irreducible representations of G_2, a little inspection on generators reveals that $\left(\Lambda^2\right)_{\substack{7\\14}}$ is respectively the $\substack{-2\\+1}$–eigenspace of the G_2-equivariant linear map

$$\begin{aligned} T_{\phi_0} \; : \; \Lambda^2 &\to \Lambda^2 \\ \eta &\mapsto T_{\phi_0}\eta := *\left(\eta \wedge \phi_0\right). \end{aligned}$$

2.1.1 Yang-Mills Formalism on G₂-Manifolds

Consider now a G-bundle $E \to Y$ over a compact G_2-manifold (Y, ϕ); the curvature $F := F_A$ of some connection A decomposes according to the splitting (2.4):

$$F_A = F_7 \oplus F_{14}, \qquad F_i \in \Omega^2_i(Y, \mathfrak{g}_E), \; i = 7, 14,$$

where $\mathfrak{g}_E$ denotes the adjoint bundle associated to E. The L^2-norm of F_A is the *Yang-Mills functional*, which therefore has two corresponding components:

$$\mathscr{Y}(A) := \|F_A\|^2 = \|F_7\|^2 + \|F_{14}\|^2. \tag{2.6}$$

It is well-known that the values of $\mathscr{Y}(A)$ can be related to a certain characteristic class of the bundle E, given (up to choice of orientation) by

$$\kappa(E) := -\int_Y \operatorname{tr}\left(F_A^2\right) \wedge \phi.$$

Using the property $d\phi = 0$, a standard argument of Chern–Weil theory [26] shows that the de Rham class $\left[\operatorname{tr}\left(F_A^2\right) \wedge \phi\right]$ is independent of A, thus the integral is indeed a topological invariant. The eigenspace decomposition of T_ψ implies (up to a sign)

$$\kappa(E) = -2\|F_7\|^2 + \|F_{14}\|^2,$$

and combining with (2.6) we get

$$\mathscr{Y}(A) = -\frac{1}{2}\kappa(E) + \frac{3}{2}\|F_{14}\|^2 = \kappa(E) + 3\|F_7\|^2.$$

Hence $\mathscr{Y}(A)$ attains its absolute minimum at a connection whose curvature lies either in $\Omega^2_7(Y, \mathfrak{g}_E)$ or in $\Omega^2_{14}(Y, \mathfrak{g}_E)$. Moreover, since $\mathscr{Y} \geq 0$, the sign of $\kappa(E)$ obstructs the existence of one type or the other, so we fix $\kappa(E) \geq 0$ and define G$_2$-*instantons* as connections with $F \in \Omega^2_{14}(Y, \mathfrak{g}_E)$, i.e., such that $\mathscr{Y}(A) = \kappa(E)$. These are precisely the solutions of the G$_2$-*instanton equation*:

$$F_A \wedge *\phi = 0 \tag{2.7a}$$

or, equivalently,

$$F_A - *(F_A \wedge \phi) = 0. \tag{2.7b}$$

If instead $\kappa(E) \leq 0$, we may still reverse orientation and consider $F \in \Omega^2_{14}(Y, \mathfrak{g}_E)$, but then the above eigenvalues and energy bounds must be adjusted accordingly, which amounts to a change of the $(-)$ sign in (2.7b).

2.1.2 The Chern–Simons Functional ϑ

It was pointed out by Simon Donaldson and Richard Thomas in their seminal article on gauge theory in higher dimensions [12] that, formally, G$_2$-instantons are rather similar to flat connections over 3-manifolds; in particular, they are critical points of a Chern–Simons functional and there is hope that counting them could lead to a enumerative invariant for G$_2$-manifolds not unlike the Casson invariant for 3-manifolds, see [11, Sect. 6] and [38, Chap. 6]. Although this interpretation has no immediate bearing on the remainder of this material, let us briefly review the basic formalism, from a purely motivational perspective.

Given a bundle over a compact 3-manifold, with space of connections $\mathscr{A}$ and gauge group $\mathscr{G}$, the *Chern–Simons functional* is a multi-valued real function on the quotient $\mathscr{B} = \mathscr{A}/\mathscr{G}$, with integer periods, whose critical points are precisely the flat connections [8, Sect. 2.5]. Similar theories can be formulated in higher dimensions in the presence of a suitable closed differential form [12, 34]; e.g. on a G$_2$-manifold (Y, ϕ), the coassociative 4–form $\psi := *\phi$ allows for the definition of a functional of Chern–Simons type.[1] Its 'gradient', the Chern–Simons 1-form, vanishes precisely at the G$_2$-instantons, hence it detects the solutions to the Yang-Mills equation [8]. The explicit case of G$_2$-manifolds, which we now describe, was examined in some detail in [30, 31].

[1]In fact only the condition $d\psi = 0$ is required, so the discussion extends to cases in which the G$_2$-structure is not necessarily torsion-free.

The space $\mathscr{A}$ of connections on $E \to Y$ is an affine space modelled on $\Omega^1(\mathfrak{g}_E)$ so, fixing a reference connection $A_0 \in \mathscr{A}$,

$$\mathscr{A} = A_0 + \Omega^1(Y, \mathfrak{g}_E)$$

and, accordingly, vectors at $A \in \mathscr{A}$ are 1-forms $a, b, \cdots \in T_A\mathscr{A} \simeq \Omega^1(Y, \mathfrak{g}_E)$ and vector fields are maps $\alpha, \beta, \cdots : \mathscr{A} \to \Omega^1(Y, \mathfrak{g}_E)$. In this notation we define the *Chern–Simons functional* by

$$\vartheta(A) := \tfrac{1}{2}\int_Y \operatorname{tr}\left(d_{A_0}a \wedge a + \frac{2}{3}a \wedge a \wedge a\right) \wedge *\phi,$$

fixing $\vartheta(A_0) = 0$. This function is obtained by integration of the *Chern–Simons* $1-$*form*

$$\rho(\beta)_A := \int_Y \operatorname{tr}(F_A \wedge \beta_A) \wedge *\phi. \tag{2.8}$$

It is straightforward to check that the co-closedness condition $\mathrm{d} * \phi = 0$ implies that the $1-$form (2.8) is closed, so the procedure doesn't depend on the path $A(t)$. Since $\mathscr{A}$ is contractible, by the Poincaré Lemma ρ is the derivative of some function ϑ, and by Stokes' theorem ρ vanishes along $\mathcal{G}-$orbits $\operatorname{im} \mathrm{d}_A \simeq T_A\{\mathcal{G}.A\}$. Thus ρ descends to the quotient $\mathscr{B}$ and so does ϑ, at least locally. Since $*\phi$ is not, in general, an integral class, the set of periods of ϑ is actually *dense*; however, as long as our interest remains in the study of the moduli space $\mathcal{M} = \operatorname{Crit}(\rho)$ of G_2-instantons, there is no worry, for the gradient $\rho = \mathrm{d}\vartheta$ is unambiguously defined on $\mathscr{B}$.

2.2 Analysis on Manifolds with Tubular Ends

In order to get some more depth into the instanton gluing process of Theorem 3.1, we will need some general results from linear analysis on asymptotically cylindrical manifolds (cf. Definition 2.3).

Definition 2.1 A *manifold with tubular end* (M, X, π) is given by a smooth manifold M with a distinguished compact submanifold-with-boundary $M_0 \subset M$, a Riemannian manifold X, and a diffeomorphism

$$\pi : M_\infty := M \setminus M_0 \to \mathbb{R}_+ \times X.$$

The complement $M_\infty := M \setminus M_0$ is called the tubular end, π is the *tubular model* and X is the *asymptotic cross section*.[2]

[2]The reader interested in analysis on tubular manifolds will find a thorough and very useful toolbox in [27].

Of course one could in principle consider, analogously, manifolds with any number of tubular ends but, in the context of G_2-manifolds, the Ricci-flat geometry constrains that number to one:

Theorem 2.2 ([28, Theorem 1]) *If a connected and orientable manifold M with k tubular ends admits a Ricci-flat metric, then $k \leq 2$. Moreover $k = 2$ if, and only if, M is a cylinder.*

2.2.1 Geometric Structures on Manifolds with Cylindrical End

On a manifold with tubular end (M, X, π), we have the following natural maps on differential forms (which clearly extend to any tensor fields):

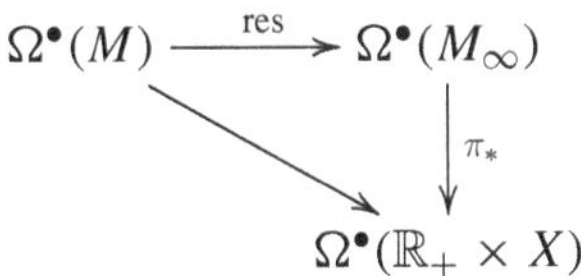

By slight abuse of notation, given $\sigma_\infty \in \Omega^\bullet(X)$, we will also denote by σ_∞ its pullback to the product under $\mathbb{R}_+ \times X \xrightarrow{p_2} X$. Denoting by t the coordinate function on $\mathbb{R}$, we adopt the following notation for asymptotic behaviour:

- $\sigma \overset{\delta}{\rightsquigarrow} \sigma_\infty$, if $|\nabla^k(\pi_*\sigma - \sigma_\infty)| \leq O(e^{-\delta t})$, $t \in \mathbb{R}_+$, $\forall k \geq 0$, for a given $\delta > 0$.
- $\sigma \rightsquigarrow \sigma_\infty$, if $\exists \delta > 0$ such that $\sigma \overset{\delta}{\rightsquigarrow} \sigma_\infty$.

Whenever $\sigma \rightsquigarrow \sigma_\infty$, σ is said to be *asymptotically translation-invariant* and σ_∞ is its *asymptotic limit*.

Definition 2.3 A manifold with tubular end (M, X, π) is said to be *asymptotically cylindrical* (ACyl) if M is also a Riemannian manifold and its metric g_M is asymptotic to the natural cylindrical metric on the tubular model: $g_M \rightsquigarrow g_X + \mathrm{d}t^2$. In this case, we will call the map $\pi\colon M_\infty \to \mathbb{R}_+ \times X$ the *cylindrical model.*

Let $E_\infty \to X$ be a Riemannian vector bundle. By slight abuse of notation we also denote by E_∞ its pullback to $\mathbb{R}_+ \times X$. For $k \in \mathbb{N}_0$, $\alpha \in (0, 1)$ and $\delta \in \mathbb{R}$ we define

$$\|\cdot\|_{C_\delta^{k,\alpha}} := \|e^{-\delta t} \cdot \|_{C^{k,\alpha}},$$

denoting by $C_\delta^{k,\alpha}(X, E_\infty)$ the respective closure of $C_0^\infty(X, E_\infty)$. We set $C_\delta^\infty := \bigcap_k C_\delta^{k,\alpha}$.

Similarly, a Riemannian vector bundle $E \to M$ over an ACyl manifold (M, X, π) is said to be *asymptotic* to $E_\infty \to X$ if there is a bundle isomorphism $\bar{\pi}\colon E|_{M_\infty} \to E_\infty$ covering π such that the push-forward of the metric on E is asymptotic to the metric on E_∞ in the C_δ^∞ tubular norm above (for some $\delta > 0$). Denote by

$t\colon M \to [1,\infty)$ a smooth positive function which agrees with $t \circ \pi$ on $\pi^{-1}([1,\infty)\times X)$, and define

$$\|\cdot\|_{C^{k,\alpha}_\delta} := \|e^{-\delta t}\cdot\|_{C^{k,\alpha}}, \quad \delta \in \mathbb{R},$$

denoting by $C^{k,\alpha}_\delta(M,E)$ the respective closure of $C^\infty_0(M,E)$.

Finally, a connection $A \in \mathscr{A}(E)$ is said to be *asymptotic* to $A_\infty \in \mathscr{A}(E_\infty)$ if $(A - \bar{\pi}^* A_\infty) \rightsquigarrow 0$ (the difference of two connections being a 1-form). We also denote by A_∞ its pullback to $E_\infty \to \mathbb{R}_+ \times X$.

2.2.2 Asymptotically Translation-Invariant Operators on ACyl Manifolds

Let us briefly review some spectral theory for elliptic operators on sections of vector bundles over an ACyl manifold M with asymptotic cross-section X. The primary references for the material in this section are Maz'ya–Plamenevskiĭ [25] and Lockhart–McOwen [22].

Let $F \to X$ be a Riemannian vector bundle, and let $D\colon C^\infty(X,F) \to C^\infty(X,F)$ be a linear self-adjoint elliptic operator of first order. The operator

$$L_\infty := \partial_t - D$$

extends to a bounded linear operator $L_{\infty,\delta}\colon C^{k+1,\alpha}_\delta(X,F) \to C^{k,\alpha}_\delta(X,F)$.

Theorem 2.4 ([25, Theorem 5.1]) *$L_{\infty,\delta}$ is invertible if and only if $\delta \notin$ spec(D).*

Indeed, elements $a \in \ker L_\infty$ can be expanded in terms of the δ-eigensections of D, see [8, Sect. 3.1]:

$$a = \sum_{\delta \in \operatorname{spec} D} e^{\delta t} a_\delta. \tag{2.9}$$

Now let $E \to M$ be a (Riemannian) vector bundle asymptotic to F and consider an elliptic operator

$$L\colon C^\infty_0(M,E) \to C^\infty_0(M,E)$$

asymptotic to L_∞, that is, such that the coefficients of L are asymptotic to the coefficients of L_∞. The operator L extends to a bounded linear operator $L_\delta\colon C^{k+1,\alpha}_\delta(M,E) \to C^{k,\alpha}_\delta(M,E)$.

Proposition 2.5 ([13, Proposition 2.4]) *If $\delta \notin$ spec(D), then L_δ is Fredholm.*

Elements in the kernel of L still have an asymptotic expansion analogous to (2.9). We need the following result which extracts the constant term of this expansion.

Proposition 2.6 ([32, Prop. 3.5]) *There is a constant* $\delta_0 > 0$ *such that, for all* $\delta \in [0, \delta_0]$, *one has* $\ker L_\delta = \ker L_0$ *and there is a linear map* $\iota\colon \ker L_0 \to \ker D$ *such that*

$$a \overset{\delta_0}{\rightsquigarrow} \iota(a).$$

In particular,

$$\ker \iota = \ker L_{-\delta_0}.$$

2.3 Twisted Connected Sums

An important method to produce examples of compact 7-manifolds with holonomy exactly G_2 is the *twisted connected sum* (TCS) construction [5, 6, 21]. It consists of gluing a pair of asymptotically cylindrical (ACyl) Calabi–Yau 3-folds obtained from certain smooth projective 3-folds called *building blocks* (see Definition 2.7). Combining results of Kovalev and Haskins–Hein–Nordström, each matching pair of building blocks yields a one-parameter family of closed G_2-manifolds.

A building block (Z, Σ) is given by a projective morphism $\zeta : Z \to \mathbb{P}^1$ such that $\Sigma := \zeta^{-1}(\infty)$ is a smooth anticanonical $K3$ surface, under some mild topological assumptions (see Definition 2.10); in particular, Σ has trivial normal bundle. Choosing a convenient Kähler structure on Z, one can make $V := Z \setminus \Sigma$ into an ACyl Calabi–Yau 3-fold (cf. Definition 2.9), that is, a non-compact Calabi–Yau manifold with a tubular end modelled on $\mathbb{R}_+ \times \mathbb{S}^1 \times \Sigma$ [6, Theorem 3.4]. Then $\mathbb{S}^1 \times V$ is an ACyl G_2-manifold (cf. Definition 2.15) with a tubular end modelled on $\mathbb{R}_+ \times \mathbb{T}^2 \times \Sigma$.

Definition 2.7 (*cf.* [6, Definition 3.9]) Let $Z_\pm$ be complex 3-folds, $\Sigma_\pm \subset Z_\pm$ smooth anticanonical $K3$ divisors and $\mathrm{k}_\pm \in H^2(Z_\pm)$ Kähler classes. We call a *matching* of $(Z_+, \Sigma_+, \mathrm{k}_+)$ and $(Z_-, \Sigma_-, \mathrm{k}_-)$ a diffeomorphism $\mathfrak{r}\colon \Sigma_+ \to \Sigma_-$ such that $\mathfrak{r}^*\mathrm{k}_- \in H^2(\Sigma_+)$ and $(\mathfrak{r}^{-1})^*\mathrm{k}_+ \in H^2(\Sigma_-)$ have type $(2,0) + (0,2)$.

Given a pair of building blocks $(Z_\pm, \Sigma_\pm)$, a set of *matching data* is a collection $\mathbf{m} = \{(\omega_{I,\pm}, \omega_{J,\pm}, \omega_{K,\pm}), \mathfrak{r}\}$ consisting of a choice of hyper-Kähler structures on $\Sigma_\pm$ such that $[\omega_{I,\pm}] = \mathrm{k}_\pm|_{\Sigma_\pm}$ is the restriction of a Kähler class on $Z_\pm$ and a matching $\mathfrak{r}\colon \Sigma_+ \to \Sigma_-$ such that

$$\mathfrak{r}^*\omega_{I,-} = \omega_{J,+}, \quad \mathfrak{r}^*\omega_{J,-} = \omega_{I,+} \quad \text{and} \quad \mathfrak{r}^*\omega_{K,-} = -\omega_{K,+}.$$

In this case $(Z_\pm, \Sigma_\pm)$ are said to *match* via $\mathbf{m}$ and $\mathfrak{r}$ is called a hyper-Kähler *rotation* (see Remark 2.12 below).

Identifying a matching pair $(Z_\pm, \Sigma_\pm)$ of building blocks by the hyper-Kähler rotation between the $K3$ surfaces 'at infinity', the corresponding pair $\mathbb{S}^1 \times V_\pm$ of ACyl G_2-manifolds is truncated at a large 'neck length' T and, intertwining the circle components in the tori $\mathbb{T}^2_\pm$ along the tubular end, glued to form a compact 7-manifold (Fig. 1)

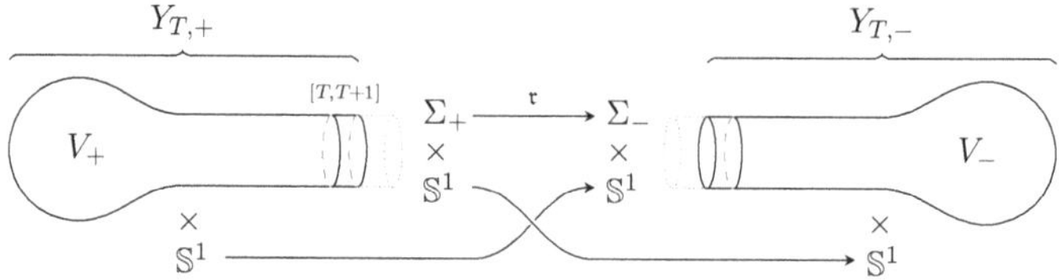

Fig. 1 The twisted connected sum of a matching pair of building blocks

$$Y = Z_+ \#_{\mathfrak{r}} Z_- := \left(\mathbb{S}^1 \times V_+\right) \cup_{\mathfrak{r}} \left(\mathbb{S}^1 \times V_-\right).$$

For large enough T_0, this twisted connected sum Y carries a family of G_2-structures $\{\phi_T\}_{T \geq T_0}$ with $\mathrm{Hol}(\phi_T) = G_2$ [6, Theorem 3.12]. The construction is summarised in the following statement.

Theorem 2.8 ([6, Corollary 6.4]) *Given a matching pair of building blocks* $(Z_\pm, \Sigma_\pm)$ *with Kähler classes* $\mathrm{k}_\pm \in H^{1,1}(Z_\pm)$ *such that* $(\mathrm{k}_+|_{\Sigma_+})^2 = (\mathrm{k}_-|_{\Sigma_-})^2$*, there exists a family of torsion-free* G_2*-structures* $\{\phi_T : T \gg 1\}$ *on the closed 7-manifold* $Y = Z_+ \#_{\mathfrak{r}} Z_-$.

2.3.1 ACyl Calabi–Yau 3-folds from Building Blocks

The twisted connected sum in Theorem 2.8 is based on gluing ACyl G_2-manifolds, which arise as the product of an ACyl Calabi-Yau 3-fold with $\mathbb{S}^1$. Let us review how to produce these from building blocks.

Definition 2.9 Let (V, ω, Ω) be a Calabi–Yau 3-fold with tubular end and asymptotic cross-section $\Sigma \times \mathbb{S}^1$ given by a hyper-Kähler surface $(\Sigma, \omega_I, \omega_J, \omega_K)$. Then V is called an *asymptotically cylindrical Calabi-Yau 3-fold* (ACylCY^3) if

$$\omega \rightsquigarrow \mathrm{d}t \wedge \mathrm{d}\alpha + \omega_I,$$
$$\Omega \rightsquigarrow (\mathrm{d}\alpha - i\mathrm{d}t) \wedge (\omega_J + i\omega_K),$$

where t and α denote the respective coordinates on $\mathbb{R}_+$ and $\mathbb{S}^1$.

Numerous examples of ACylCY^3 can be obtained from the following ingredients:

Definition 2.10 (*Corti–Haskins–Nordström–Pacini* [5, Definition 5.1]) A *building block* is a smooth projective 3-fold Z together with a projective morphism $\zeta\colon Z \to \mathbb{P}^1$ such that the following hold:

- The anticanonical class $-K_Z \in H^2(Z)$ is primitive.
- $\Sigma := \zeta^{-1}(\infty)$ is a smooth $K3$ surface and $\Sigma \sim -K_Z$.

- Identifying $H^2(\Sigma, \mathbb{Z})$ with the $K3$ lattice (i.e. choosing a marking for Σ), the following embedding is primitive:

$$N := \operatorname{im}(H^2(Z, \mathbb{Z}) \to H^2(\Sigma, \mathbb{Z})) \hookrightarrow H^2(\Sigma)$$

- The groups $H^3(Z, \mathbb{Z})$ and $H^4(Z, \mathbb{Z})$ are torsion-free.

In particular, building blocks are simply-connected [5, Sect. 5.1].

Remark 2.11 The existence of the fibration $\zeta\colon Z \to \mathbb{P}^1$ is equivalent to Σ having trivial normal bundle. This is crucial because it means that $Z \setminus \Sigma$ has a cylindrical end, given by an exponential radial coordinate in a tubular neighbourhood of Σ. The last two conditions in the definition of a building block are not essential; they are meant to facilitate the computation of certain topological invariants.

Remark 2.12 Given a matching $\mathfrak{r}$ between a pair of building blocks $(Z_\pm, \Sigma_\pm, \mathrm{k}_\pm)$, one can make the choices in the definition of the ACyl Calabi-Yau structure so that $\mathfrak{r}$ becomes a hyper-Kähler rotation (cf. Definition 2.7) of the induced hyper-Kähler structures [6, Theorem 3.4 & Proposition 6.2].

In his original work, Kovalev [21] used building blocks arising from Fano 3-folds by blowing-up the base-locus of a generic anti-canonical pencil. This method was extended to the much larger class of semi Fano 3-folds (a class of weak Fano 3-folds) by Corti–Haskins–Nordström–Pacini (see Proposition 4.6 below). Kovalev–Lee [20] construct building blocks starting from $K3$ surfaces with non-symplectic involutions, by taking the product with $\mathbb{P}^1$, dividing by $\mathbb{Z}_2$ and blowing up the resulting singularities. In every instance, one obtains an ACylCY^3 by the following theorem:

Theorem 2.13 ([13, Theorem D]) *Let (Z, Σ) be a building block and let $(\omega_I, \omega_J, \omega_K)$ be a hyper-Kähler structure on Σ. If $[\omega_I] \in H^{1,1}(\Sigma)$ is the restriction of a Kähler class on Z, then there is an asymptotically cylindrical Calabi–Yau structure (ω, Ω) on $V := Z \setminus \Sigma$ with asymptotic cross section $(\Sigma, \omega_I, \omega_J, \omega_K)$.*

Remark 2.14 This result was first claimed by Kovalev in [21, Theorem 2.4]; see the discussion in [13, Sect. 4.1].

2.3.2 Gluing ACyl G$_2$-Manifolds

We may now describe the gluing of matching pairs of ACylG$_2$-manifolds, obtained from ACylCY^3 given by Theorem 2.13.

Definition 2.15 Let (Y, ϕ) be a G$_2$-manifold with tubular end and asymptotic cross-section given by a compact Calabi–Yau 3-fold (W, ω, Ω). Then Y is called *asymptotically cylindrical* (ACyl) if

$$\phi \rightsquigarrow \mathrm{d}t \wedge \omega + \operatorname{Re}\Omega,$$

where t denotes the coordinate on $\mathbb{R}_+$.

Taking the product of an ACylCY^3 (V, ω, Ω) with $\mathbb{S}^1$, with coordinate β, yields an ACylG$_2$-manifold

$$(Y := \mathbb{S}^1 \times V, \phi := \mathrm{d}\beta \wedge \omega + \mathrm{Re}\,\Omega)$$

with asymptotic cross section

$$(W := \mathbb{T}^2 \times \Sigma, \omega := \mathrm{d}\alpha \wedge \mathrm{d}\beta + \omega_K, \Omega := (\mathrm{d}\alpha - i\mathrm{d}\beta) \wedge (\omega_J + i\omega_I)).$$

Let $V_\pm$ be a matching pair of ACylCY^3 with asymptotic cross section $\Sigma_\pm$ and suppose that $\mathfrak{r}\colon\ \Sigma_+ \to \Sigma_-$ is a hyper-Kähler rotation. A pair of ACylG$_2$-manifolds $(Y_\pm, \phi_\pm)$ with asymptotic cross sections $(W_\pm, \omega_\pm, \Omega_\pm)$ as above is said to *match* if there exists a diffeomorphism

$$\begin{aligned} q:\quad \mathbb{T}^2 \times \Sigma_+ &\longrightarrow \mathbb{T}^2 \times \Sigma_- \\ f(\alpha, \beta, x) &:= (\beta, \alpha, \mathfrak{r}(x)). \end{aligned}$$

such that

$$q^*\omega_- = -\omega_+ \quad \text{and} \quad q^*\,\mathrm{Re}\,\Omega_- = \mathrm{Re}\,\Omega_+.$$

Remark 2.16 If q did not interchange the $\mathbb{S}^1$-factors, then Y would have infinite fundamental group and, hence, could not carry a metric with holonomy equal to G$_2$ [17, Proposition 10.2.2].

Let $(Y_\pm, \phi_\pm)$ be a matching pair of ACylG$_2$-manifolds. For fixed $T \geq 1$, define

$$\begin{aligned} Q\ :\ [T, T+1] \times Z_+ &\longrightarrow [T, T+1] \times Z_- \\ Q(t, z) &:= (2T + 1 - t, q(z)) \end{aligned}$$

and denote by Y_T the compact 7-manifold obtained by gluing $Y_\pm$ together at neck length T via Q:

$$Y_{T,\pm} := (Y_0)_\pm \cup_Q \pi_\pm^{-1}\left((0, T+1] \times Z_\pm\right).$$

Fix a non-decreasing smooth cut-off function $\chi\colon\ \mathbb{R} \to [0, 1]$ with $\chi(t) = 0$ for $t \leq 0$ and $\chi(t) = 1$ for $t \geq 1$. Define a 3-form $\tilde{\phi}_T$ on Y_T by

$$\tilde{\phi}_T = \phi_\pm - \mathrm{d}[(\chi(t - T + 1)(\omega_\pm - \pi_\pm^*\omega_{\infty,\pm}))]$$

on $Y_{T,\pm}$. If $T \gg 1$, then $\tilde{\phi}_T$ defines a closed G$_2$-structure on Y_T. Clearly, all the Y_T for different values of T are diffeomorphic; hence, we often drop the T from the notation. The G$_2$-structure $\tilde{\phi}_T$ is not torsion-free yet, but can be made so by a small perturbation:

Theorem 2.17 ([21, Theorem 5.34]) *In the above situation there exist a constant $T_0 \geq 1$ and, for each $T \geq T_0$, a 2-form η_T on Y_T such that $\phi_T := \tilde{\phi}_T + \mathrm{d}\eta_T$ defines a torsion-free* G$_2$*-structure and for some $\delta > 0$*

$$\|d\eta_T\|_{C^{0,\alpha}} = O(e^{-\delta T}). \tag{2.10}$$

In summary, the TCS Theorem 2.8 is established by the following procedure. For any building block (Z, Σ), the noncompact 3-fold $V := Z \setminus \Sigma$ admits ACyl Ricci-flat Kähler metrics (Theorem 2.13) hence an ACylCY^3 structure whose asymptotic limit defines a hyper-Kähler structure on Σ. Given a matching pair of such Calabi-Yau manifolds $V_\pm$, one can apply Theorem 2.17 to glue $\mathbb{S}^1 \times V_\pm$ into a closed manifold Y with a 1-parameter family of torsion-free G_2-structures [6, Theorem 3.12].

3 The G_2-Instanton Gluing Theorem

Let A be an ASD instanton on a $\mathbf{P}U(n)$-bundle F over a Kähler surface Σ. The linearisation of the instanton moduli space $\mathcal{M}_\Sigma$ near A is modelled on the kernel of the deformation operator

$$\mathbb{D}_{\mathbb{A}} := d_A^* \oplus d_A^+ : \Omega^1(\Sigma, \mathfrak{g}_F) \to (\Omega^0 \oplus \Omega^+)(\Sigma, \mathfrak{g}_F).$$

Let $\mathcal{F}$ be the corresponding holomorphic vector bundle (*cf.* Donaldson–Kronheimer [10]), and denote by f the Hitchin–Kobayashi isomorphism:

$$f : H^1(\Sigma, \mathcal{E}nd_0(\mathcal{F})) \xrightarrow{\sim} H_A^1 := \ker \mathbb{D}_{\mathbb{A}}. \tag{3.1}$$

Theorem 3.1 ([32, Theorem 1.2]) *Let $Z_\pm$, $\Sigma_\pm$, $\mathrm{k}_\pm$, $\mathfrak{r}$, X and ϕ_T be as in Theorem 2.8. Let $\mathcal{E}_\pm \to Z_\pm$ be a pair of holomorphic vector bundles such that the following hold:*

Asymptotic stability *$\mathcal{E}_\pm|_{\Sigma_\pm}$ is μ-stable with respect to $\mathrm{k}_\pm|_{\Sigma_\pm}$. Denote the corresponding ASD instanton by $A_{\infty,\pm}$.*

Compatibility *There exists a bundle isomorphism $\bar{\mathfrak{r}} : \mathcal{E}_+|_{\Sigma_+} \to \mathcal{E}_-|_{\Sigma_-}$ covering the hyper-Kähler rotation $\mathfrak{r}$ such that $\bar{\mathfrak{r}}^* A_{\infty,-} = A_{\infty,+}$.*

Inelasticity *There are no infinitesimal deformations of $\mathcal{E}_\pm$ fixing the restriction to $\Sigma_\pm$:*

$$H^1(Z_\pm, \mathcal{E}nd_0(\mathcal{E}_\pm)(-\Sigma_\pm)) = 0. \tag{3.2}$$

Transversality *If $\lambda_\pm := f_\pm \circ \mathrm{res}_\pm : H^1(Z_\pm, \mathcal{E}nd_0(\mathcal{E}_\pm)) \to H^1_{A_{\infty,\pm}}$ denotes the composition of restrictions to $\Sigma_\pm$ with the isomorphism (3.1), then the image of λ_+ and $\bar{\mathfrak{r}}^* \circ \lambda_-$ intersect trivially in the linear space $H^1_{A_{\infty,+}}$:*

$$\mathrm{im}(\lambda_+) \cap \mathrm{im}(\bar{\mathfrak{r}}^* \circ \lambda_-) = \{0\}. \tag{3.3}$$

Then there exists a $U(r)$-bundle E over Y and a family of connections $\{A_T : T \gg 1\}$ on the associated $\mathbf{P}U(r)$-bundle, such that each A_T is an irreducible unobstructed G_2-instanton over (Y, ϕ_T).

The asymptotic stability assumption guarantees finite energy of Hermitian bundle metrics on $\mathcal{E}_\pm|_{V_\pm}$ (see [33, Sect. 2.2]), which are equivalent to asymptotically translation-invariant HYM connections $A_\pm \rightsquigarrow A_{\infty,\pm}$, under the Chern correspondence (cf. Theorem 3.15). The maps λ_+ and $\bar{\mathfrak{r}}^* \circ \lambda_-$ can be seen geometrically as linearisations of the natural inclusions of the moduli of asymptotically stable bundles $\mathcal{M}_{Z_\pm}$ into the moduli of ASD instantons $\mathcal{M}_{\Sigma_+}$ over the $K3$ surface 'at infinity', and we think of $H^1_{A_{\infty,+}}$ as a tangent model of $\mathcal{M}_{\Sigma_+}$ near the ASD instanton $A_{\infty,+}$. Then the transversality condition asks that the actual inclusions intersect transversally at $A_{\infty,+} \in \mathcal{M}_{\Sigma_+}$. That the intersection points are isolated reflects that the resulting G_2-instanton is rigid, since it is unobstructed and the deformation problem has index 0.

Remark 3.2 If $H^1(\Sigma_+, \mathcal{E}nd_0(\mathcal{E}_+|_{\Sigma_+})) = \{0\}$, then (3.3) is vacuous. If, moreover, the topological bundles underlying $\mathcal{E}_\pm$ are isomorphic, then the existence of $\bar{\mathfrak{r}}$ is guaranteed by [15, Theorem 6.1.6].

Furthermore, condition (3.2) yields a short exact sequence, which is self-dual under Serre duality:

$$0 \to H^1(Z_\pm, \mathcal{E}nd_0(\mathcal{E}_\pm)) \to H^1(\Sigma_\pm, \mathcal{E}nd_0(\mathcal{E}_\pm|_{\Sigma_\pm})) \to H^2(Z_\pm, \mathcal{E}nd_0(\mathcal{E}_\pm)(-\Sigma_\pm)) \to 0.$$

This implies [36, p. 176 ff.] that each

$$\mathrm{im}\,\lambda_\pm \subset H^1_{A_{\infty,\pm}}$$

is a complex Lagrangian subspace with respect to the complex symplectic structure induced by $\Omega_\pm := \omega_{J,\pm} + i\omega_{K,\pm}$ or, equivalently, Mukai's complex symplectic structure on $H^1(Z_\pm, \mathcal{E}nd_0(\mathcal{E}_\pm))$. Under the assumptions of Theorem 3.1 the moduli space $\mathscr{M}_{\Sigma_+}$ of holomorphic bundles over Σ_+ is smooth near $[\mathcal{E}_+|_{\Sigma_+}]$ and so are the moduli spaces $\mathscr{M}_{Z_\pm}$ of holomorphic bundles over $Z_\pm$ near $[\mathcal{E}_\pm]$. Locally, $\mathscr{M}_{Z_\pm}$ embeds as a complex Lagrangian submanifold into $\mathscr{M}_{\Sigma_\pm}$. Since $\mathfrak{r}^*\omega_{K,-} = -\omega_{K,+}$, both $\mathscr{M}_{Z_+}$ and $\mathscr{M}_{Z_-}$ can be viewed as Lagrangian submanifolds of $\mathscr{M}_{\Sigma_+}$ with respect to the symplectic form induced by $\omega_{K,+}$. Equation (3.3) asks for these Lagrangian submanifolds to intersect transversely at the point $[\mathcal{E}_+|_{\Sigma_+}]$. If one thinks of G_2-manifolds arising via the twisted connected sum construction as analogues of 3-manifolds with a fixed Heegaard splitting, then this is much like the geometric picture behind Atiyah–Floer conjecture in dimension three [2].

In Sect. 4, we will review a constructive method to obtain explicit examples of such instanton gluing in many interesting cases.

3.1 Hermitian Yang-Mills Connections on ACyl CY 3-Folds

Suppose (W, ω, Ω) is Calabi–Yau 3-fold and $(Y := \mathbb{R} \times W, \phi := \mathrm{d}t \wedge \omega + \mathrm{Re}\,\Omega)$ is the corresponding cylindrical G$_2$-manifold. In this section we relate translation-invariant G$_2$-instantons over Y with Hermitian–Yang–Mills connections over W.

Definition 3.3 Let (W, ω) be a Kähler manifold and let E be a $\mathbf{PU}(n)$-bundle over W. A connection $A \in \mathscr{A}(E)$ on E is *Hermitian–Yang–Mills (HYM) connection* if

$$F_A^{0,2} = 0 \quad \text{and} \quad \Lambda F_A = 0. \tag{3.4}$$

Here Λ is the dual of the Lefschetz operator $L := \omega \wedge \cdot$.

Remark 3.4 Instead of working with $\mathbf{PU}(n)$-bundles, one can also work with $\mathrm{U}(n)$-bundles and instead of the second part of (3.4) require that ΛF_A be equal to a constant. These view points are essentially equivalent.

Remark 3.5 By the first part of (3.4) a HYM connection induces a holomorphic structure on E. If W is compact, then there is a one-to-one correspondence between gauge equivalence classes of HYM connections on E and isomorphism classes of polystable holomorphic bundles $\mathcal{E}$ whose underlying topological bundle is E, see Donaldson [9] and Uhlenbeck–Yau [37].

On a Calabi–Yau 3-fold, (3.4) is equivalent to

$$F_A \wedge \mathrm{Im}\,\Omega = 0 \quad \text{and} \quad F_A \wedge \omega \wedge \omega = 0;$$

hence, using $*(\mathrm{d}t \wedge \omega + \mathrm{Re}\,\Omega) = \frac{1}{2}\omega \wedge \omega - \mathrm{d}t \wedge \mathrm{Im}\,\Omega$ one easily derives:

Proposition 3.6 ([33, Proposition 8]) *Denote by $\pi_W \colon Y \to W$ the canonical projection. A is a HYM connection if and only if $\pi_W^* A$ is a* G$_2$*-instanton.*

In general, if A is a G$_2$-instanton on a G-bundle E over a G$_2$-manifold (Y, ϕ), then the moduli space $\mathcal{M}$ of G$_2$-instantons near $[A]$, i.e., the space of gauge equivalence classes of G$_2$-instantons near $[A]$ is the space of small solutions $(\xi, a) \in (\Omega^0 \oplus \Omega^1)(Y, \mathfrak{g}_E)$ of the system of equations

$$\mathrm{d}_A^* a = 0 \quad \text{and} \quad \mathrm{d}_{A+a}\xi - *(F_{A+a} \wedge \psi) = 0$$

modulo the action of $\Gamma_A \subset \mathscr{G}$, the stabiliser of A, assuming either that Y is compact or appropriate control over the growth of ξ and a. The infinitesimal deformation theory of $[A]$ is governed by that equation's linearisation operator

$$L_A := \begin{pmatrix} & \mathrm{d}_A^* \\ \mathrm{d}_A & - * (\psi \wedge \mathrm{d}_A) \end{pmatrix} : \left(\Omega^0 \oplus \Omega^1\right)(Y, \mathfrak{g}_E) \to \left(\Omega^0 \oplus \Omega^1\right)(Y, \mathfrak{g}_E). \tag{3.5}$$

Definition 3.7 A is called *irreducible and unobstructed* if L_A is surjective.

If A is irreducible and unobstructed, then $\mathcal{M}$ is smooth at $[A]$. If Y is compact, then L_A has index zero; hence, is surjective if, and only if, it is invertible; therefore, irreducible and unobstructed G_2-instantons form isolated points in $\mathcal{M}$. If Y is non-compact, the precise meaning of $\mathcal{M}$ and L_A depends on the growth assumptions made on ξ and a and $\mathcal{M}$ may very well be positive-dimensional.

Proposition 3.8 *If A is HYM connection on a bundle E over a* G_2*-manifold $Y := \mathbb{R} \times W$ as in Proposition 3.6, then the operator $L_{\pi_W^* A}$ defined in (3.5) can be written as*

$$L_{\pi_W^* A} = \tilde{I}\partial_t + D_A$$

where

$$\tilde{I} := \begin{pmatrix} & -1 & \\ 1 & & \\ & & I \end{pmatrix}$$

and $D_A\colon \left(\Omega^0 \oplus \Omega^0 \oplus \Omega^1\right)(W, \mathfrak{g}_E) \to \left(\Omega^0 \oplus \Omega^0 \oplus \Omega^1\right)(W, \mathfrak{g}_E)$ is defined by

$$D_A := \begin{pmatrix} & & \mathrm{d}_A^* \\ & & \Lambda \mathrm{d}_A \\ \mathrm{d}_A & -I\mathrm{d}_A & -*\left(\operatorname{Im}\Omega \wedge \mathrm{d}_A\right) \end{pmatrix}. \tag{3.6}$$

Definition 3.9 Let A be a HYM connection on a $\mathbf{PU}(n)$-bundle E over a Kähler manifold (W, ω). Set

$$\mathcal{H}_A^i := \ker\left(\bar{\partial}_A \oplus \bar{\partial}_A^*\colon\ \Omega^{0,i}\left(W, \mathcal{E}nd_0(\mathcal{E})\right) \to \left(\Omega^{0,i+1} \oplus \Omega^{0,i-1}\right)\left(W, \mathcal{E}nd_0(\mathcal{E})\right)\right).$$

$\mathcal{H}_A^0$ is called the space of *infinitesimal automorphisms* of A and $\mathcal{H}_A^1$ is called the space of *infinitesimal deformations* of A.

Remark 3.10 If W is compact, then $\mathcal{H}_A^i \cong H^i(W, \mathcal{E}nd_0(\mathcal{E}))$ where $\mathcal{E}$ is the holomorphic bundle induced by A.

Proposition 3.11 *If (W, ω, Ω) is a compact Calabi–Yau 3-fold and A is a HYM connection on a G-bundle $E \to W$, then*

$$\ker D_A \cong \mathcal{H}_A^0 \oplus \mathcal{H}_A^1$$

where D_A is as in (3.6).

3.2 *Gluing* G_2*-Instantons over ACyl* G_2*-Manifolds*

Definition 3.12 Let (Y, ϕ) be an ACylG_2-manifold and let A be a G_2-instanton on a G-bundle over (Y, ϕ) asymptotic to A_∞. For $\delta \in \mathbb{R}$ we set

$$\mathcal{T}_{A,\delta} := \ker L_{A,\delta} = \left\{ \underline{a} \in \ker L_A : \underline{a} \overset{\delta}{\rightsquigarrow} 0 \right\}.$$

where $\underline{a} = (\xi, a) \in \left(\Omega^0 \oplus \Omega^1\right)(Y, \mathfrak{g}_E)$. Set $\mathcal{T}_A := \mathcal{T}_{A,0}$.

Proposition 3.13 ([32, Propositions 3.22, 3.23]) *Let (Y, ϕ) be an* ACyl G$_2$*-manifold and let A be a* G$_2$*-instanton asymptotic to A_∞. Then there is a constant $\delta_0 > 0$ such that for all $\delta \in [0, \delta_0]$, $\mathcal{T}_{A,\delta} = \mathcal{T}_A$ and there is a linear map $\iota \colon \mathcal{T}_A \to \mathcal{H}^0_{A_\infty} \oplus \mathcal{H}^1_{A_\infty}$ such that*

$$\underline{a} \overset{\delta_0}{\rightsquigarrow} \iota(\underline{a}).$$

In particular, $\ker \iota = \mathcal{T}_{A,-\delta_0}$.

Furthermore,

$$\dim \operatorname{im} \iota = \frac{1}{2} \dim \left(\mathcal{H}^0_{A_\infty} \oplus \mathcal{H}^1_{A_\infty}\right)$$

and, if $\mathcal{H}^0_{A_\infty} = 0$, then $\operatorname{im} \iota \subset \mathcal{H}^1_{A_\infty}$ is Lagrangian with respect to the symplectic structure on $\mathcal{H}^1_{A_\infty}$ induced by ω.

Assume we are in the situation of Proposition 3.13; if moreover $\ker \iota = 0$ and $\mathcal{H}^0_{A_\infty} = 0$, then one can show that the moduli space $\mathcal{M}_Y$ of G$_2$-instantons near $[A]$ which are asymptotic to some HYM connection is smooth. Although the moduli space $\mathcal{M}_W$ of HYM connections near $[A_\infty]$ is not necessarily smooth, formally, it still makes sense to talk about its symplectic structure and view $\mathcal{M}_Y$ as a Lagrangian submanifold. The following theorem shows that transverse intersections of a pair of such Lagrangians give rise to G$_2$-instantons.

Theorem 3.14 ([32, Theorem 3.24]) *Let $(Y_\pm, \phi_\pm)$ be a pair of* ACylG$_2$*-manifolds that match via $f \colon W_+ \to W_-$. Denote by $(Y_T, \phi_T)_{T \geq T_0}$ the resulting family of compact* G$_2$*-manifolds arising from the construction in Sect. 2.3.2. Let $A_\pm$ be a pair of* G$_2$*-instantons on $E_\pm$ over $(Y_\pm, \phi_\pm)$ asymptotic to $A_{\infty,\pm}$. Suppose the following hold:*

- *There is a bundle isomorphism $\bar{f} \colon E_{\infty,+} \to E_{\infty,-}$ covering f such that $\bar{f}^* A_{\infty,-} = A_{\infty,+}$,*
- *The maps $\iota_\pm \colon \mathcal{T}_{A_\pm} \to \ker D_{A_{\infty,\pm}}$ constructed in Proposition 3.13 are injective and their images intersect trivially*

$$\operatorname{im}(\iota_+) \cap \operatorname{im}\left(\bar{f}^* \circ \iota_-\right) = \{0\} \subset \mathcal{H}^0_{A_{\infty,+}} \oplus \mathcal{H}^1_{A_{\infty,+}}. \tag{3.7}$$

Then there exists $T_1 \geq T_0$ and for each $T \geq T_1$ there exists an irreducible and unobstructed G$_2$*-instanton A_T on a G-bundle E_T over (Y_T, ϕ_T).*

Sketch of Proof One proceeds in three steps. We first produce an approximate G$_2$-instanton $\tilde{A}_T$ by an explicit cut-and-paste procedure. This reduces the problem to solving the non-linear partial differential equation

$$\mathrm{d}^*_{\tilde{A}_t} a = 0 \quad \text{and} \quad \mathrm{d}_{\tilde{A}_T + a}\xi + *_T(F_{\tilde{A}_T + a} \wedge \psi_T) = 0. \tag{3.8}$$

for $a \in \Omega^1(Y_T, \mathfrak{g}_{E_T})$ and $\xi \in \Omega^0(Y_T, \mathfrak{g}_{E_T})$ where $\psi_T := *\phi_T$. Under the hypotheses of Theorem 3.14 one can solve the linearisation of (3.8) in a uniform fashion. The existence of a solution of (3.8) then follows from a simple application of Banach's fixed-point theorem. □

3.3 *From Holomorphic Bundles over Building Blocks to* G₂*-Instantons over ACyl* G₂*-Manifolds*

We now briefly explain how one may deduce Theorem 3.1 from Theorem 3.14.

Let (V, ω, Ω) be an ACylCY^3 with asymptotic cross-section $(\Sigma, \omega_I, \omega_J, \omega_K)$. The following theorem can be used to produce examples of HYM connections A on a $\mathbf{PU}(n)$-bundle $E \to V$ asymptotic to an ASD instanton A_∞ on a $\mathbf{PU}(n)$-bundle $E_\infty \to \Sigma$ (here, by a slight additional abuse, we denote by E_∞ and A_∞ their respective pullbacks to $\mathbb{R}_+ \times \mathbb{S}^1 \times \Sigma$). Hence, by taking the product with $\mathbb{S}^1$, it yields examples of G_2-instantons $\pi_V^* A$ asymptotic to $\pi_\Sigma^* A_\infty$ over the ACylG$_2$-manifold $\mathbb{S}^1 \times V$. Denote the canonical projections in this context by

$$\pi_V\colon \mathbb{S}^1 \times V \to V \quad \text{and} \quad \pi_\Sigma\colon \mathbb{T}^2 \times \Sigma \to \Sigma.$$

Theorem 3.15 ([33, Theorem 59] & [19, Theorem 1.1]) *Let Z and Σ be as in Theorem 2.13 and let $(V := Z \setminus \Sigma, \omega, \Omega)$ be the resulting* ACylCY^3. *Let $\mathcal{E}$ be a holomorphic vector bundle over Z and let A_∞ be an ASD instanton on $\mathcal{E}|_\Sigma$ compatible with the holomorphic structure. Then there exists a HYM connection A on $\mathcal{E}|_V$ which is compatible with the holomorphic structure on $\mathcal{E}|_V$ and asymptotic to A_∞.*

Remark 3.16 The last assertion of the exponential decay $A \rightsquigarrow A_\infty$ is claimed in [33, Theorem 59] but its proof in that reference is not satisfactory. That part of the theorem is essentially superseded by [19, Theorem 1.1], which additionally extends this existence result to *singular* G$_2$-instantons, obtained from asymptotically stable reflexive sheaves, following in spirit the argument in the compact case, by [4].

This together with Theorem 3.14 and the following result immediately implies Theorem 3.1.

Proposition 3.17 ([32, Proposition 4.3]) *In the situation of Theorem 3.15, suppose $H^0(\Sigma, \mathcal{E}nd_0(\mathcal{E}|_\Sigma)) = 0$. Then*

$$\mathcal{H}^1_{\pi_\Sigma^* A_\infty} = H^1_{A_\infty} \tag{3.9}$$

and, for some small $\delta > 0$, there exist injective linear maps κ_- and κ such that the following diagram commutes:

$$\begin{array}{ccccc} \mathcal{T}_{\pi_V^* A,-\delta} & \longrightarrow & \mathcal{T}_{\pi_V^* A} & \xrightarrow{\iota} & \mathcal{H}^1_{\pi_\Sigma^* A_\infty} \\ \downarrow \kappa_- & & \downarrow \kappa & & \downarrow \cong \\ H^1(Z, \mathcal{E}nd_0(\mathcal{E})(-\Sigma)) & \longrightarrow & H^1(Z, \mathcal{E}nd_0(\mathcal{E})) & \longrightarrow & H^1(\Sigma, \mathcal{E}nd_0(\mathcal{E}|_\Sigma)). \end{array} \tag{3.10}$$

Sketch of Proof Equation (3.9) is a direct consequence of $\mathcal{H}^0_{A_\infty} = 0$. If A is a HYM connection asymptotic to A_∞ over an ACylCY[3] then there exists a $\delta_0 > 0$ such that, for all $\delta \leq \delta_0$,

$$\mathcal{T}_{\pi_V^* A,\delta} = \left\{ \underline{a} \in \ker D_A : \underline{a} \overset{\delta}{\rightsquigarrow} 0 \right\} \tag{3.11}$$

with D_A as in (3.6). Furthermore, there exists $\delta_1 > 0$ such that, for all $\delta \leq \delta_1$, one has $\mathcal{H}^0_{A,\delta} = 0$ and

$$\mathcal{T}_{\pi_V^* A,\delta} \cong \mathcal{H}^1_{A,\delta}$$

where $\mathcal{H}^i_{A,\delta} := \left\{ \alpha \in \mathcal{H}^i_A : \alpha \overset{\delta}{\rightsquigarrow} 0 \right\}$. □

4 Transversal Examples via the Hartshorne-Serre Correspondence

In [5, 6, 21], building blocks Z are produced by blowing up Fano or semi-Fano 3-folds along the base curve $\mathscr{C}$ of an anticanonical pencil (see Proposition 4.6). By understanding the deformation theory of pairs (X, Σ) of semi-Fanos X and anti-canonical $K3$ divisors $\Sigma \subset X$, one can produce hundreds of thousands of pairs with the required matching (see Sect. 4.3). In order to apply Theorem 3.1 to produce G2-instantons over the resulting twisted connected sums, one first requires some supply of asymptotically stable, inelastic vector bundles $\mathcal{E} \to X$. Moreover, to satisfy the hypotheses of compatibility and transversality, one would in general need some understanding of the deformation theory of triples $(X, \Sigma, \mathcal{E})$. In this Sect. I outline our approach in [24] to the problem of production of ingredients, in the form of gluable pairs of holomorphic bundles over building blocks.

The Hartshorne-Serre construction generalises the correspondence between divisors and line bundles, under certain conditions, in the sense that bundles of higher rank are associated to subschemes of higher codimension. We recall the rank 2 version, as an instance of Arrondo's formulation[3] [1, Theorem 1]:

Theorem 4.1 *Let $\mathscr{S} \subset Z$ be a local complete intersection subscheme of codimension 2 in a smooth algebraic variety. If there exists a line bundle $\mathcal{L} \to Z$ such that*

- $H^2(Z, \mathcal{L}^*) = 0$,

[3]For a thorough justification of this choice of reference for the correspondence, see the Introduction section of Arrondo's notes.

- $\wedge^2 \mathcal{N}_{\mathscr{S}/Z} = \mathcal{L}|_{\mathscr{S}}$, *where* $\mathcal{N}_{\mathscr{S}/Z}$ *denotes the normal bundle of* $\mathscr{S}$ *in Z.*

then there exists a rank 2 holomorphic vector bundle $\mathcal{F} \to Z$ *such that*

1. $\wedge^2 \mathcal{F} = \mathcal{L}$,
2. $\mathcal{F}$ *has one global section whose vanishing locus is* $\mathscr{S}$.

We will refer to such $\mathcal{F}$ *as the* Hartshorne-Serre bundle obtained from $\mathscr{S}$ (and $\mathcal{L}$).

Using the Hartshorne-Serre construction, we can systematically produce families of bundles over the building blocks, which, in favourable cases, are parametrised by the building block's blow-up curve $\mathscr{C}$ itself. This perspective lets us understand the deformation theory of the bundles very explicitly, and it also separates the latter from the deformation theory of the pair (X, Σ). We can therefore *first* find matchings between two semi-Fano families using the techniques from [6], and *then* exploit the high degree of freedom in the choice of the blow-up curve $\mathscr{C}$ (see Lemma 4.7) to satisfy the compatibility and transversality hypotheses.

4.1 A Detailed Example

As a proof of concept, we will henceforth walk through the process of construction of examples, with the particular pair adopted in [24]:

Example 4.2 The product $X_+ = \mathbb{P}^1 \times \mathbb{P}^2$ is a Fano 3-fold. Let $|\Sigma_0, \Sigma_\infty| \subset |-K_{X_+}|$ be a generic pencil with (smooth) base locus $\mathscr{C}_+$ and $\Sigma_+ \in |\Sigma_0, \Sigma_\infty|$ generic. Denote by $r_+ : Z_+ \to X_+$ the blow-up of X_+ in $\mathscr{C}_+$, by $\widetilde{\mathscr{C}_+}$ the exceptional divisor and by ℓ_+ a fibre of $p_1 : \widetilde{\mathscr{C}_+} \to \mathscr{C}_+$. The proper transform of Σ_+ in Z_+ is also denoted by Σ_+, and (Z_+, S_+) is a building block by Proposition 4.6. For future reference, we fix classes

$$H_+ := r_+^*([\mathbb{P}^1 \times \mathbb{P}^1]) \quad \text{and} \quad G_+ = r_+^*([\{x\} \times \mathbb{P}^2]) \in H^2(Z_+).$$

NB.: Clearly $-K_{X_+}$ is very ample, thus also $-K_{X_+|\Sigma_+}$, so X_+ lends itself to application of Lemma 4.7.

Example 4.3 A double cover $\pi : X_- \xrightarrow{2:1} \mathbb{P}^1 \times \mathbb{P}^2$ branched over a smooth $(2, 2)$ divisor D is a Fano 3-fold. Let $|\Sigma_0, \Sigma_\infty| \subset |-K_{X_-}|$ be a generic pencil with (smooth) base locus $\mathscr{C}_-$ and $\Sigma_- \in |\Sigma_0, \Sigma_\infty|$ generic. Denote by $r_- : Z_- \to X_-$ the blow-up of X_- in $\mathscr{C}_-$, and by $\widetilde{\mathscr{C}_-}$ the exceptional divisor. The proper transform of Σ_- in Z_- is also denoted by Σ_-, and (Z_-, S_-) is a building block by Proposition 4.6. For future reference, we fix classes

$$H_- := (r_- \circ \pi)^*([\mathbb{P}^1 \times \mathbb{P}^1]) \quad \text{and} \quad G_- = (r_- \circ \pi)^*([\{x\} \times \mathbb{P}^2]) \in H^2(Z_-)$$

and

$$h_- := \frac{1}{2}(r_- \circ \pi)^*([\{x\} \times \mathbb{P}^1]) \in H^4(Z_-),$$

where x is a point.

In that context, the existence of solutions satisfying the hypotheses of the TCS G_2-instanton gluing theorem takes the following form:

Theorem 4.4 ([24, Theorem 1.3]) *There exists a matching pair of building blocks* $(Z_\pm, \Sigma_\pm)$, *obtained as* $Z_\pm = \mathrm{Bl}_{\mathscr{C}_\pm} X_\pm$ *for* $X_+ = \mathbb{P}^1 \times \mathbb{P}^2$ *and the double cover* $X_- \xrightarrow{2:1} \mathbb{P}^1 \times \mathbb{P}^2$ *branched over a* (2, 2) *divisor, with rank* 2 *holomorphic bundles* $\mathcal{E}_\pm \to Z_\pm$ *satisfying the hypotheses of Theorem 3.1.*

Here's a sketch of the procedure leading to Theorem 4.4:

- We construct holomorphic bundles on building blocks from certain complete intersection subschemes, via the Hartshorne-Serre correspondence (Theorem 4.1), as well as two families of bundles $\{\mathcal{F}_\pm \to X_\pm\}$, over the particular blocks of Theorem 4.4, that are conducive to application of Theorem 3.1.
- Then, in Sect. 4.5, we focus on the moduli space $\mathscr{M}^s_{\Sigma_+,\mathcal{A}_+}(v_{\Sigma_+})$ of stable bundles on Σ_+, where the problems of compatibility and transversality therefore "take place". Here $X_+ = \mathbb{P}^1 \times \mathbb{P}^2$, $\Sigma_+ \subset X_+$ is the anti-canonical $K3$ divisor and, for a smooth curve $\mathscr{C}_+ \in |-K_{X_+|\Sigma_+}|$, the block $Z_+ := \mathrm{Bl}_{\mathscr{C}_+} X_+$ is in the family obtained from Example 4.2.
  It can be shown that $\mathscr{M}^s_{\Sigma_+,\mathcal{A}_+}(v_{\Sigma_+})$ is isomorphic to Σ_+ itself, and that the restrictions of the family of bundles $\mathcal{F}_+$ correspond precisely to the blow-up curve $\mathscr{C}$. Now, given a rank 2 bundle $\mathcal{F}_+ \to Z_+$ such that $\mathcal{G} := \mathcal{F}_{+|\Sigma_+} \in \mathscr{M}^s_{\Sigma_+,\mathcal{A}_+}(v_{\Sigma_+})$, the restriction map

$$\mathrm{res} : H^1(Z_+, \mathscr{E}nd_0(\mathcal{F}_+)) \to H^1(\Sigma_+, \mathscr{E}nd_0(\mathcal{G})) \tag{4.1}$$

  corresponds to the derivative at $\mathcal{F}_+$ of the map between instanton moduli spaces. Combining with Lemma 4.7, which guarantees the freedom to choose $\mathscr{C}_+$ when constructing the block Z_+ from Σ_+, one has:

Theorem 4.5 ([24, Theorem 1.4]) *For every* $\mathcal{G} \in \mathscr{M}^s_{\Sigma_+,\mathcal{A}_+}(v_{\Sigma_+})$ *and every line* $V \subset H^1(\Sigma_+, \mathscr{E}nd_0(\mathcal{G}))$, *there is a smooth base locus curve* $\mathscr{C} \in |-K_{X_+|\Sigma_+}|$ *and an exceptional fibre* $\ell \subset \widetilde{\mathscr{C}}$ *corresponding by Hartshorne-Serre to an inelastic vector bundle* $\mathcal{F}_+ \to Z_+$, *such that* $\mathcal{F}_{+|\Sigma_+} = \mathcal{G}$ *and the restriction map* (4.1) *has image* V.

- Let $\mathfrak{r} : \Sigma_+ \to \Sigma_-$ be a matching between $X_+ = \mathbb{P}^1 \times \mathbb{P}^2$ and $X_- \xrightarrow{2:1} \mathbb{P}^1 \times \mathbb{P}^2$. Then for any $\mathcal{F}_- \to Z_-$ as above we can (up to a twist by holomorphic line bundles $\mathcal{R}_\pm \to Z_\pm$) choose the smooth curve $\mathscr{C}_+ \in |-K_{X_+|\Sigma_+}|$ in the construction of Z_+ so that there is a Hartshorne-Serre bundle $\mathcal{F}_+ \to Z_+$ that matches $\mathcal{F}_-$ transversely. Then the bundles $\mathcal{E}_\pm := \mathcal{F}_\pm \otimes \mathcal{R}_\pm$ satisfy the gluing hypotheses of Theorem 3.1.

4.2 Building Blocks from Semi-Fano 3-Folds and Twisted Connected Sums

For all but 2 of the 105 families of Fano 3-folds, the base locus of a generic anti-canonical pencil is smooth. This also holds for most families in the wider class of 'semi-Fano 3-folds' in the terminology of [5], i.e. smooth projective 3-folds where $-K_X$ defines a morphism that does not contract any divisors. We can then obtain building blocks using [6, Proposition 3.15]:

Proposition 4.6 *Let X be a semi-Fano 3-fold with $H^3(X,\mathbb{Z})$ torsion-free, $|\Sigma_0, \Sigma_\infty| \subset |-K_X|$ a generic pencil with (smooth) base locus $\mathscr{C}$, $\Sigma \in |\Sigma_0, \Sigma_\infty|$ generic, and Z the blow-up of X at $\mathscr{C}$. Then Σ is a smooth $K3$ surface, its proper transform in Z is isomorphic to Σ, and (Z, Σ) is a building block. Furthermore*

1. *the image N of $H^2(Z,\mathbb{Z}) \to H^2(\Sigma,\mathbb{Z})$ equals that of $H^2(X,\mathbb{Z}) \to H^2(\Sigma,\mathbb{Z})$;*
2. *$H^2(X,\mathbb{Z}) \to H^2(\Sigma,\mathbb{Z})$ is injective and the image N is primitive in $H^2(\Sigma,\mathbb{Z})$.*

Let us notice for later use that, whenever $-K_X|_\Sigma$ is very ample, it is possible to 'wiggle' a blow-up curve $\mathscr{C}$ so as to realise any prescribed incidence condition $(x, V) \in T\Sigma$. This fact will play an important role in the transversality argument in Sect. 4.5.

Lemma 4.7 ([24, Lemma 2.5]) *Let X be a semi-Fano, $\Sigma \in |-K_X|$ a smooth $K3$ divisor, and suppose that the restriction of $-K_X$ to Σ is very ample. Then given any point $x \in \Sigma$ and any (complex) line $V \subset T_x\Sigma$, there exists an anticanonical pencil containing Σ whose base locus $\mathscr{C}$ is smooth, contains x, and $T_x\mathscr{C} = V$.*

Finally, note that if $X_\pm$ is a pair of semi-Fanos and $\mathfrak{r} : \Sigma_+ \to \Sigma_-$ is a matching in the sense of Definition 2.7, then $\mathfrak{r}$ also defines a matching of building blocks constructed from $X_\pm$ using Proposition 4.6. Thus given a pair of matching semi-Fanos we can apply Theorem 2.8 to construct closed G_2-manifolds, *but* this still involves choosing the blow-up curves $\mathscr{C}_\pm$.

4.3 The Matching Problem

We now explain in more detail the argument of [6, Sect. 6] for finding matching building blocks $(Z_\pm, \Sigma_\pm)$. The blocks will be obtained by applying Proposition 4.6 to a pair of semi-Fanos $X_\pm$, from some given pair of deformation types $\mathcal{X}_\pm$.

A key deformation invariant of a semi-Fano X is its Picard lattice $\mathrm{Pic}(X) \cong H^2(X;\mathbb{Z})$. For any anticanonical $K3$ divisor $\Sigma \subset X$, the injection $\mathrm{Pic}(X) \hookrightarrow H^2(\Sigma;\mathbb{Z})$ is primitive. The intersection form on $H^2(\Sigma;\mathbb{Z})$ of any $K3$ surface is isometric to $L_{K3} := 3U \oplus 2E_8$, the unique even unimodular lattice of signature $(3, 19)$. We can therefore identify $\mathrm{Pic}(X)$ with a primitive sublattice $N \subset L_{K3}$ of the $K3$ lattice, uniquely up to the action of the isometry group $O(L_{K3})$ (this is usually uniquely determined by the isometry class of N as an abstract lattice).

Given a matching $\mathfrak{r}\colon \Sigma_+ \to \Sigma_-$ between anticanonical divisors in a pair of semi-Fanos, we can choose the isomorphisms $H^2(\Sigma_\pm; \mathbb{Z}) \cong L_{K3}$ compatible with $\mathfrak{r}^*$, hence identify $\mathrm{Pic}(X_+)$ and $\mathrm{Pic}(X_-)$ with a *pair* of primitive sublattices $N_+, N_- \subset L_{K3}$. While the $O(L_{K3})$ class of $N_\pm$ individually depends only on $X_\pm$, the $O(L_{K3})$ class of the pair (N_+, N_-) depends on $\mathfrak{r}$, and we call (N_+, N_-) the *configuration* of $\mathfrak{r}$. Many important properties of the resulting twisted connected sum only depend on the hyper-Kähler rotation in terms of the configuration.

Given a configuration $N_+, N_- \subset L_{K3}$, let

$$N_0 := N_+ \cap N_-, \quad \text{and} \quad R_\pm := N_\pm \cap N_\mp^\perp.$$

We say that the configuration is *orthogonal* if $N_\pm$ are rationally spanned by N_0 and $R_\pm$ (geometrically, this means that the reflections in N_+ and N_- commute). Given a pair $\mathcal{X}_\pm$ of deformation types of semi-Fanos, then there are sufficient conditions for a given orthogonal configuration to be realised by some matching [6, Proposition 6.17],

Proposition 4.8 *i.e., so that there exist $X_\pm \in \mathcal{X}_\pm$, $\Sigma_\pm \in |-K_{X_\pm}|$, and a matching $\mathfrak{r} : \Sigma_+ \to \Sigma_-$ with the given configuration.*

Now consider the problem of finding matching bundles $\mathcal{E}_\pm \to Z_\pm$ in order to construct G$_2$-instantons by application of Theorem 3.1. For the compatibility hypothesis it is obviously necessary that Chern classes match:

$$c_1(\mathcal{E}_+|_{\Sigma_+}) = \mathfrak{r}^* c_1(\mathcal{E}_-|_{\Sigma_-}) \in H^2(\Sigma_+).$$

Identifying $H^2(\Sigma_+; \mathbb{Z}) \cong L_{K3} \cong H^2(\Sigma_-; \mathbb{Z})$ compatibly with $\mathfrak{r}^*$, this means we need

$$c_1(\mathcal{E}_{+|\Sigma_+}) \;=\; c_1(\mathcal{E}_{-|\Sigma_-}) \;\in\; N_+ \cap N_- \;=\; N_0.$$

Hence, if N_0 is trivial, both $c_1(\mathcal{E}_{\pm|\Sigma_\pm})$ must also be trivial, which is a very restrictive condition on our bundles. To allow more possibilities, we want matchings $\mathfrak{r}$ whose configuration $N_+, N_- \subset L_{K3}$ has non-trivial intersection N_0.

Table 4 of [7] lists all 19 possible such matchings with Picard rank 2, among which we can find the pair of building blocks of Examples 4.2 and 4.3, coming from the Fano 3-folds $X_+ = \mathbb{P}^1 \times \mathbb{P}^2$ and the double cover $X_- \xrightarrow{2:1} \mathbb{P}^1 \times \mathbb{P}^2$ branched over a $(2, 2)$ divisor. Several other choices would be possible to produce examples of G$_2$-instantons.

4.4 Hartshorne-Serre Bundles over Building Blocks

4.4.1 The General Construction Algorithm

Let X be a semi-Fano 3-fold and (Z, Σ) be the block constructed as a blow-up of X along the base locus $\mathscr{C}$ of a generic anti-canonical pencil (Proposition 4.6). In [24, Sect. 3.1] a general approach is provided for making the choices of $\mathcal{L}$ and $\mathscr{S}$ in Theorem 4.1, in order to construct a Hartshorne-Serre bundle $\mathcal{F} \to Z$ which, up to a twist, yields the bundle $\mathcal{E}$ meeting the requirements of Theorem 3.1. The approach may be summarised as follows:

Summary 4.9 *Let $(Z_\pm, \Sigma_\pm)$ be the building blocks constructed by blowing-up $N_\pm$-polarised semi-Fano 3-folds $X_\pm$ along the base locus $\mathscr{C}_\pm$ of a generic anti-canonical pencil* (cf. *Proposition 4.6). Let $N_0 \subset N_\pm$ be the sub-lattice of orthogonal matching, as in Sect. 4.3. Let $\mathcal{A}_\pm$ be the restriction of an ample class of $X_\pm$ to $\Sigma_\pm$ which is orthogonal to N_0. We look for the Hartshorne-Serre parameters $\mathscr{S}_\pm$ and $\mathcal{L}_\pm$ of Theorem 4.1, where $\mathscr{S}_+ = \ell$ is an exceptional fibre in Z_+, $\mathscr{S}_-$ is a genus 0 curve in Z_- and $\mathcal{L}_\pm \to Z_\pm$ are line bundles such that:*

1. $c_1(\mathcal{L}_\pm) \in N_0 \mod 2\mathrm{Pic}(\Sigma_\pm)$;
2. $c_1(\mathcal{L}_{\pm|\Sigma_\pm}) \cdot \mathcal{A}_\pm > 0$;
3. $\chi(\mathcal{L}_\pm^*) \leq 0$;
4. $c_1(\mathcal{L}_+) \cdot \mathscr{S}_+ = -1$ *and* $(S_- - c_1(\mathcal{L}_-)) \cdot \mathscr{S}_- = 2$;
5. $c_1(\mathcal{L}_{+|\Sigma_+})^2 = -4$ *and* $\Sigma_- \cdot \mathscr{S}_- - \frac{1}{4}c_1(\mathcal{L}_{-|\Sigma_-})^2 = 2$;

Finally, among candidate data satisfying these constraints, inelasticity must be arranged "by hand'.

The reader who would like to construct other examples might follow this 4-step programme:

Step 1. Find two matching $N_\pm$-polarized semi-Fano 3-folds $X_\pm$ such that:

(i) there exists $x \in N_+$ such $x^2 = -4$ (or more generally $x^2 = 2k - 6$, for a moduli space $\mathscr{M}^s_{\Sigma,\mathcal{A}}(v)$ of dimension $2k$).
(ii) there exists a primitive element $y \in N_0$ such that $y^2 \leq -8$ and 4 divides y^2.

Step 2. Find $\mathcal{L}_\pm$ and $\mathscr{S}_-$ which verify the conditions of Summary 4.9 (perhaps with a computer).

Step 3. The following must be checked by *ad-hoc* methods:

1. $H^2(\mathcal{L}_\pm^*) = 0$, for the Hartshorne-Serre construction (Theorem 4.1);
2. divisors with small slope do not contain $\mathscr{S}$, for asymptotic stability ([16, Proposition 10]);
3. $h^1(\mathcal{L}^*) = h^1(\mathcal{F}) = 0$ and the dimensional constraint (4.2) for your choice of $\dim \mathscr{M}^s_{\Sigma,\mathcal{A}}(v)$, corresponding to inelasticity (Proposition 4.15).

Step 4. Conclude with similar arguments to Sect. 4.5.

4.4.2 Construction of $\mathcal{F}_+$ over $X_+ = \mathbb{P}^1 \times \mathbb{P}^2$ and $\mathcal{F}_-$ over $X_- \xrightarrow{2:1} \mathbb{P}^1 \times \mathbb{P}^2$

In view of the constraints in Summary 4.9, we apply Theorem 4.1 to $Z_+ = \mathrm{Bl}_{\mathscr{C}} X_+$ as above, obtained by blowing up $X_+ = \mathbb{P}^1 \times \mathbb{P}^2$ from Example 4.2, with parameters

$$\mathscr{S} = \ell \quad \text{and} \quad \mathcal{L} = \mathcal{O}_{Z_+}(-\Sigma_+ - G_+ + H_+).$$

Proposition 4.10 ([24, Propositions 3.5, 4.4, 5.7]) *Let (Z_+, Σ_+) be a building block as in Example 4.2, $\mathscr{C}$ a pencil base locus and $\ell \subset Z_+$ an exceptional fibre of $\widetilde{\mathscr{C}} \to \mathscr{C}$. There exists a rank 2 asymptotically stable and inelastic Hartshorne Serre bundle $\mathcal{F}_+ \to Z_+$ obtained from ℓ such that*

1. *$c_1(\mathcal{F}_+) = -\Sigma_+ - G_+ + H_+$, and*
2. *$\mathcal{F}_+$ has a global section whose vanishing locus is a fibre ℓ of $p_1 : \widetilde{\mathscr{C}} \to \mathscr{C}$.*

Similarly, one applies Theorem 4.1 to the building block Z_- obtained by blowing up $X_- \xrightarrow{2:1} \mathbb{P}^1 \times \mathbb{P}^2$, from Example 4.3, with

$$[\mathscr{S}] = h_- \quad \text{and} \quad \mathcal{L} = \mathcal{O}_{Z_-}(G_-).$$

Proposition 4.11 ([24, Propositions 3.9, 4.5, 5.8]) *Let (Z_-, Σ_-) be a building block provided in Example 4.3 and $\mathscr{S}$ a line of class h_-. There exists a rank 2 Hartshorne-Serre bundle $\mathcal{F}_- \to Z_-$ obtained from $\mathscr{S}$ such that:*

1. *$c_1(\mathcal{F}_-) = G_-$, and*
2. *$\mathcal{F}_-$ has a global section whose vanishing locus is $\mathscr{S}$, where $[\mathscr{S}] = h_-$.*

Remark 4.12 In order to check the stability of Hartshorne-Serre bundles over $\Sigma_\pm$, we use a tailor-made instance [16, Proposition 10] of a more general Hoppe-type stability criterion for holomorphic bundles over so-called *polycyclic varieties*, whose Picard group is free Abelian [16, Corollary 4].

In the context above, the moduli spaces of the stable bundles $\mathcal{F}_{\pm|\Sigma_\pm}$ have the 'minimal' positive dimension, for transversal intersection to occur:

Proposition 4.13 *Let $(Z_\pm, \Sigma_\pm)$ be the building block provided in Examples 4.2 and 4.3, and let $\mathcal{F}_\pm \to Z_\pm$ be the asymptotically stable bundles constructed in Propositions 4.10 and 4.11. Let $\mathscr{M}^s_{\Sigma_\pm, \mathcal{A}_\pm}(v_\pm)$ be the moduli space of $\mathcal{A}_\pm$-stable bundles on $\Sigma_\pm$ with Mukai vector $v_\pm = v(\mathcal{F}_{\pm|\Sigma_\pm})$. We have:*

$$\dim \mathscr{M}^s_{\Sigma_\pm, \mathcal{A}_\pm}(v_\pm) = 2.$$

Recall that (see eg. [14]) that the *Mukai vector* of a vector bundle $\mathcal{F} \to \Sigma$ on a $K3$ surface is defined as

$$v(\mathcal{F}) := (\operatorname{rk}\mathcal{F},\ c_1(\mathcal{F}),\ \chi(\mathcal{F}) - \operatorname{rk}\mathcal{F}) \in \left(H^0 \oplus H^2 \oplus H^4\right)(\Sigma, \mathbb{Z}),$$

with $\chi(\mathcal{F}) = \frac{c_1(\mathcal{F})^2}{2} + 2\operatorname{rk}\mathcal{F} - c_2(\mathcal{F})$.

4.4.3 Inelasticity of Asymptotically Stable Hartshorne-Serre Bundles

These results hold for general building blocks and may be of independent interest. Recall that a bundle $\mathcal{F}$ over a building block (Z, Σ) is *inelastic* if

$$H^1(Z, \mathscr{E}nd_0(\mathcal{F})(-\Sigma)) = 0.$$

This condition means that there are no global deformations of the bundle $\mathcal{F}$ which maintain $\mathcal{F}_{|\Sigma}$ fixed at infinity. The following characterisation of inelasticity, in the case of asymptotically stable bundles, relates the freedom to extend $\mathcal{F}$ and the dimension of the moduli space $\mathscr{M}^s_{\Sigma,\mathcal{A}}(v_{\mathcal{F}})$. The proof uses Serre duality and Maruyama's characterisation of the moduli space of stable bundles over a polarised $K3$ surface [23, Proposition 6.9].

Proposition 4.14 *Let (Z, Σ) be a building block and $\mathcal{F}$ an asymptotically stable bundle on Z. Let $\mathscr{M}^s_{\Sigma,\mathcal{A}}(v)$ be the moduli space of $\mathcal{A}$-μ-stable bundles on Σ with Mukai vector $v = v(\mathcal{F}_{|\Sigma})$. Then $\mathcal{F}$ is inelastic if and only if*

$$\dim \operatorname{Ext}^1(\mathcal{F}, \mathcal{F}) = \frac{1}{2}\dim \mathscr{M}^s_{\Sigma,\mathcal{A}}(v).$$

For Hartshorne-Serre bundles of rank 2 satisfying certain topological hypotheses, we may express the half-dimension of the moduli space in terms of the construction data:

Proposition 4.15 *Let (Z, Σ) be a building block, and let $\mathcal{F} \to Z$ be an asymptotically stable Hartshorne–Serre bundle obtained from a genus 0 curve $\mathscr{S} \subset Z$ and a line bundle $\mathcal{L} \to Z$ as in Theorem 4.1. Let $\mathscr{M}^s_{\Sigma,\mathcal{A}}(v)$ be the moduli space of $\mathcal{A}$-μ-stable bundles on Σ with Mukai vector $v = v(\mathcal{F}_{|\Sigma})$. We assume:*

1. $H^1(\mathcal{L}^*) = 0,$
2. $H^1(\mathcal{F}) = 0.$

Then $\mathcal{F}$ is inelastic if and only if

$$\frac{1}{2}\dim \mathscr{M}^s_{\Sigma,\mathcal{A}}(v) = \dim H^0(\mathcal{N}^*_{\mathscr{S}/Z} \otimes \mathcal{L}_{|\mathscr{S}}) - \dim H^0(\mathcal{F}) + 1. \qquad (4.2)$$

4.5 Proof of Theorem 4.5

Let $X_+ = \mathbb{P}^1 \times \mathbb{P}^2$ as in Example 4.2, and $\Sigma_+ \subset X_+$ be a smooth anti-canonical $K3$ divisor. For suitable choices of polarisation $\mathcal{A}_+$ on Σ_+ and Mukai vector v_{Σ_+}, the associated moduli space $\mathscr{M}^s_{\Sigma_+,\mathcal{A}_+}(v_{\Sigma_+})$ of (rank 2) $\mathcal{A}_+$-stable bundles is 2-dimensional. For a smooth curve $\mathscr{C} \in |-K_{X_+|\Sigma_+}|$, let $Z_+ := \mathrm{Bl}_{\mathscr{C}} X_+$ be the building block resulting from Proposition 4.6. Then, for each exceptional fibre $\ell \subset \widetilde{\mathscr{C}}$, the Mukai vector

$$v_{Z_+} := (2, -\Sigma_+ - G_+ + H_+, \ell) \in \left(H^0 \oplus H^2 \oplus H^4\right)(Z_+, \mathbb{Z})$$

has the property that, given a bundle $\mathcal{F}_+ \to Z_+$ as in Proposition 4.10 with $(\mathrm{rk}\,\mathcal{F}_+, c_1(\mathcal{F}_+), c_2(\mathcal{F}_+)) = v_{Z_+}$, the restriction to Σ_+ has Mukai vector v_{Σ_+}, so $\mathcal{G} := \mathcal{F}_{+|\Sigma_+} \in \mathscr{M}^s_{\Sigma_+,\mathcal{A}_+}(v_{\Sigma_+})$. Thus the Hartshorne-Serre construction yields a family of asymptotically stable vector bundles $\{(\mathcal{F}_+)_p \to Z_+ \mid p \in \mathscr{C}\}$ with

$$(\mathrm{rk}\,\mathcal{F}_+, c_1(\mathcal{F}_+), c_2(\mathcal{F}_+)) = v_{Z_+} \tag{4.3}$$

parametrised by $\mathscr{C}$ itself.

One crucial feature of the building block obtained from $X_+ = \mathbb{P}^1 \times \mathbb{P}^2$ is the fact that the moduli space of bundles over the anti-canonical $K3$ divisor Σ_+ is actually isomorphic to Σ_+ itself:

Proposition 4.16 ([24, Lemma 4.7 & Proposition 4.8]) *For each $p \in \Sigma_+$, there exists an $\mathcal{A}_+$-μ-stable and rank 2 Hartshorne-Serre bundle $\mathcal{G}_p \to \Sigma_+$ obtained from p. The induced map*

$$\begin{aligned} g : \Sigma_+ &\longrightarrow \mathscr{M}^s_{\Sigma_+,\mathcal{A}_+}(v_{\Sigma_+}) \\ p &\longmapsto \mathcal{G}_p \end{aligned}$$

is an isomorphism of $K3$ surfaces.

Now let $\mathcal{G} \in \mathscr{M}^s_{\Sigma_+,\mathcal{A}_+}(v_{\Sigma_+})$ and $V \subset H^1(\Sigma_+, \mathscr{E}nd_0(\mathcal{G}))$. From Proposition 4.16, there is $p \in \Sigma_+$ such that $\mathcal{G} = \mathcal{G}_p$ and let $V' = (\mathrm{d}g)_p^{-1}(V)$. Since $-K_{X_+|\Sigma_+}$ is very ample (see Example 4.2), Lemma 4.7 allows the choice of a smooth base locus curve $\mathscr{C} \in |-K_{X_+|\Sigma_+}|$ such that $p \in \mathscr{C}$ and $T_p\mathscr{C} = V'$. By Proposition 4.10, we can find a family $\{(\mathcal{F}_+)_q \to Z \mid q \in \mathscr{C}\}$ of holomorphic bundles parametrised by $\mathscr{C}$, with prescribed topology (4.3) and $(\mathcal{F}_{+|\Sigma})_q = \mathcal{G}_q$. Such a bundle $\mathcal{F}_+$ has therefore all the properties claimed in Theorem 4.5.

Corollary 4.17 ([24, Cor. 6.1]) *In the context of Example 4.2, for every bundle $\mathcal{G} \in \mathscr{M}^s_{\Sigma_+,\mathcal{A}_+}(v_{\Sigma_+})$ and every complex line $V \subset H^1(\Sigma_+, \mathscr{E}nd_0(\mathcal{G}))$, there are a smooth curve $\mathscr{C}_+ \in |-K_{X_+|\Sigma_+}|$ and an asymptotically stable and inelastic vector bundle $\mathcal{E}_+ \to Z_+$ such that $\mathcal{E}_{+|\Sigma_+} = \mathcal{G}$ and* $\mathrm{res} : H^1(Z_+, \mathscr{E}nd_0(\mathcal{E}_+)) \to H^1(\Sigma_+, \mathscr{E}nd_0(\mathcal{G}))$ *has image V.*

Let

$$\mathcal{E}_- := \mathcal{F}_- \otimes \mathcal{O}_{Z_-}(-H_- + 2G_-).$$

Corollary 4.18 ([24, Cor. 6.2]) *In the context of Example 4.3, there exists a family of asymptotically stable and inelastic vector bundles* $\{\mathcal{E}_- \to Z_-\}$, *parametrised by the set of the lines in* X_- *of class* h_-, *such that* $\mathcal{E}_{-|\Sigma_-} \in \mathscr{M}^s_{\Sigma_-,\mathcal{A}_-}(v_{\Sigma_-})$.

We fix a representative $\mathcal{E}_- \to Z_-$ in the family of holomorphic bundles from Corollary 4.18, to be matched by a bundle $\mathcal{E}_+ \to Z_+$ given by Corollary 4.17, so that asymptotic stability and inelasticity hold from the outset.

It remains to address compatibility and transversality. Since the chosen configuration for $\mathfrak{r}$ ensures that $\mathfrak{r}^*$ identifies the Mukai vectors of $\mathcal{E}_{\pm|\Sigma_\pm}$, it induces a map $\bar{\mathfrak{r}}^* : \mathscr{M}^s_{\Sigma_-,\mathcal{A}_-}(v'_{\Sigma_-}) \to \mathscr{M}^s_{\Sigma_+,\mathcal{A}_+}(v'_{\Sigma_+})$. In particular, the target moduli space is 2-dimensional, by Proposition 4.13, and $\mathfrak{r}^*(\text{im res}_-)$ is 1-dimensional, since the bundles $\{\mathcal{E}_-\}$ are parametrised by lines of fixed class h_-. So indeed we apply Corollary 4.17 with $\mathcal{G} = \bar{\mathfrak{r}}^*(\mathcal{E}_{-|\Sigma_-})$ and any choice of a direct complement subspace V such that

$$V \oplus \bar{\mathfrak{r}}^*(\text{im res}_-) = H^1\left(\Sigma_+, \mathscr{E}nd_0\left(\bar{\mathfrak{r}}^*(\mathcal{E}_{-|\Sigma_-})\right)\right).$$

Denoting by $\mathcal{M}_{\Sigma_\pm}(v)$ the moduli space of ASD instantons over $\Sigma_\pm$ with Mukai vector v, the maps $f_\pm$ (cf. (3.1)) in Theorem 3.1 are the linearisations of the Hitchin-Kobayashi isomorphisms

$$\mathscr{M}^s_{\Sigma_\pm,\mathcal{A}_\pm}(v'_{\Sigma\pm}) \simeq \mathcal{M}_{\Sigma_\pm}(v'_{\Sigma\pm}).$$

Therefore, our bundles $\mathcal{E}_\pm$ indeed satisfy $A_{\infty,+} = \bar{\mathfrak{r}}^* A_{\infty,-}$ for the corresponding instanton connections. Moreover, by linearity, $\lambda_+(H^1(Z_+, \mathscr{E}nd_0(\mathcal{E}_+)))$ is transverse in $T_{A_{\infty,+}}\mathcal{M}_{\Sigma_+}(v'_{\Sigma_+})$ to the image of the real 2-dimensional subspace $\lambda_-(H^1(Z_-, \mathscr{E}nd_0(\mathcal{E}_-))) \subset T_{A_{\infty,-}}\mathcal{M}_{\Sigma_-}(v'_{\Sigma_-})$ under the linearisation of $\bar{\mathfrak{r}}^*$.

References

1. Arrondo, E. (2007). A home-made Hartshorne-Serre correspondence. *Revista Matematica Complutense*, *20*, 423–443.
2. Atiyah, M. (1988). New invariants of 3- and 4-dimensional manifolds, *The mathematical heritage of Hermann Weyl* (pp. 285–299). Durham, NC, 1987.
3. Bryant, R. (1985). Metrics with holonomy G_2 or Spin(7). *Lecture Notes in Mathematics*, *1111*, 269–277.
4. Bando, S., & Siu, Y. T. (1994). Stable sheaves and Einstein-Hermitian metrics. *Geometry and Analysis on Complex Manifolds*, *39*, 39–50.
5. Corti, A., Haskins, M., Nordström, J., & Pacini, T. (2013). Asymptotically cylindrical Calabi-Yau 3-folds from weak Fano 3-folds. *Geometry and Topology*, *17*(4), 1955–2059.
6. Corti, A., Haskins, M., Nordström, J., & Pacini, T. (2015). G_2-manifolds and associative submanifolds via semi-Fano 3-folds. *Duke Mathematical Journal*, *164*(10), 1971–2092.
7. Crowley, D., & Nordström, J. (2014). Exotic G_2-manifolds. arXiv:1411.0656 [math.AG].
8. Donaldson, S. K. (2002). Floer homology groups in Yang-Mills theory. *Cambridge Tracts in Mathematics* (Vol. 147). Cambridge: Cambridge University Press. With the assistance of Furuta, M. & Kotschick, D.

9. Donaldson, S. K. (1985). Anti self-dual Yang-Mills connections over complex algebraic surfaces and stable vector bundles. *Proceedings of the London Mathematical Society*, *50*(1), 1–26.
10. Donaldson, S. K. (1990). Polynomial invariants for smooth four-manifolds. *Topology*, *29*(3), 257–315.
11. Donaldson, S. K., & Segal, E. (2011). Gauge theory in higher dimensions, II. *Surveys in Differential Geometry*, *16*, 1–41.
12. Donaldson, S. K., & Thomas, R. P. (1996). *Gauge theory in higher dimensions*, The geometric universe Oxford, 1998, pp. 31–47.
13. Haskins, M., Hein, H.-J., & Nordström, J. (2015). Asymptotically cylindrical Calabi-Yau manifolds. *Journal of Differential Geometry*, *101*(2), 213–265.
14. Huybrechts, D., & Lehn, M. (2010). *The geometry of moduli spaces of sheaves* (2nd ed.). Cambridge: Cambridge Mathematical Library.
15. Huybrechts, D., & Lehn, M. (1997). The geometry of moduli spaces of sheaves. *Aspects of Mathematics*, E31, Friedr. Braunschweig: Vieweg & Sohn.
16. Jardim, M., Menet, G., Prata, D. M., & Sá Earp, H. N. (2017). Holomorphic bundles for higher dimensional gauge theory. *Bulletin of the London Mathematical Society*, *49*(1), 117–132.
17. Joyce, D. D. (2000). Compact manifolds with special holonomy. *Oxford Mathematical Monographs*. Oxford: Oxford University Press.
18. Joyce, D. D. (1996) Compact Riemannian 7-manifolds with holonomy G_2. I, II. *Journal Differential Geometry*, *43*(2), 291–328, 329–375.
19. Jacob, A. & Walpuski, T. (2016). *Hermitian Yang-Mills metrics on reflexive sheaves over asymptotically cylindrical Kähler manifolds*. arXiv:1603.07702v1 [math.DG].
20. Kovalev, A., & Lee, N.-H. (2011). K3 surfaces with non-symplectic involution and compact irreducible G_2-manifolds. *Mathematical Proceedings of the Cambridge Philosophical Society*, *151*(2), 193–218.
21. Kovalev, A. (2003). Twisted connected sums and special Riemannian holonomy. *Journal für die reine und angewandte Mathematik*, *565*, 125–160.
22. Lockhart, R. B., & McOwen, R. C. (1985). Elliptic differential operators on noncompact manifolds. *Scuola Normale Superiore di Pisa, Classe di Scienze (4)*, *12*(3), 409–447.
23. Maruyama, M. (1978). Moduli of stable sheaves ii. *Journal of Mathematics of Kyoto University*, *18*, 557–614.
24. Menet, G., Nordström, J., & Sá Earp, H.N. (2017). Construction of G_2-instantons via twisted connected sums. arXiv:1510.03836 [math.AG]. To appear in Mathematical Research Letters.
25. Maz'ya, V. G., & Plamenevskiĭ, B. A. (1978). Estimates in L_p and in Hölder classes, and the Miranda-Agmon maximum principle for the solutions of elliptic boundary value problems in domains with singular points on the boundary. *Mathematische Nachrichten*, *81*, 25–82.
26. Milnor, J. W., & Stasheff, J. D. (1974). *Characteristic classes*. Princeton: Princeton University Press.
27. Pacini, T. (2012). Desingularizing isolated conical singularities: Uniform estimates via weighted sobolev spaces. arXiv:1005.3511v3. To appear in Communications in Analysis and Geometry.
28. Salur, S. (2006). Asymptotically cylindrical ricci-flat manifolds. *Proceedings of the American Mathematical Society*, *134*(10), 3049–3056.
29. Salamon, S. (1989). Riemannian geometry and holonomy groups. *Pitman Research Notes in Mathematics Series* (Vol. 201).
30. Sá Earp, H. N. (2009). *Instantons on G_2-manifolds*. Ph.D. Thesis.
31. Sá Earp, H. N. (2014). Generalised Chern-Simons theory and G_2–instantons over associative fibrations. *SIGMA - Symmetry, Integrability and Geometry: Methods and Applications*, *10*(083).
32. Sá Earp, H. N. (2015). G_2-instantons over asymptotically cylindrical manifolds. *Geometry and Topology*, *19*(1), 61–111.
33. Sá Earp, H. N., & Walpuski, T. (2015). G_2−instantons over twisted connected sums. *Geometry & Topology*, *19*(3), 1263–1285.
34. Thomas, R. (1997). *Gauge theories on Calabi-Yau manifolds*. D. Phil. thesis, University of Oxford.

35. Tian, G. (2000). Gauge theory and calibrated geometry. I. *Annals of Mathematics (1)*, *151*(1), 193–268.
36. Tyurin, A. (2012) Vector bundles, Universitätsverlag Göttingen. In Fedor Bogomolov, Alexey Gorodentsev, Victor Pidstrigach, Miles Reid and Nikolay Tyurin (Eds.), Collected works (Vol. I, p. 330).
37. Uhlenbeck, K. K., & Yau, S-T. (1986). On the existence of Hermitian-Yang-Mills connections in stable vector bundles. *Communications on Pure and Applied Mathematics*, *39*(S1), S257–S293.
38. Walpuski, T. (2013). G_2–instantons on generalised Kummer constructions. *Geometry and Topology*, *17*(4), 2345–2388.
39. Walpuski, T. (2016). G_2-instantons over twisted connected sums: an example. *Mathematical Research Letters*, *23*(2), 529–544. arXiv:1505.01080.

Complex and Calibrated Geometry

Kim Moore

Abstract This is an expository article based on a talk given by the author at the Fields Institute in August 2017 for the *Workshop on G_2 manifolds and related topics*. The aim of the article is to review some recent results of the author [11] investigating the relationship between calibrated and complex geometry.

1 Introduction

Let M be a four-dimensional Calabi–Yau manifold with Ricci-flat Kähler form ω and holomorphic volume form Ω. Since $\mathrm{Hol}(\omega) \subseteq SU(4) \subseteq Spin(7)$, we can think of M as a $Spin(7)$-manifold. In this case, the $Spin(7)$-form or Cayley calibration is given by

$$\Phi = \frac{1}{2}\omega \wedge \omega + \mathrm{Re}\,\Omega.$$

In particular, we can see from the above expression that (M, Φ) has two special types of Cayley submanifold: two-dimensional complex submanifolds (calibrated by $\frac{1}{2}\omega \wedge \omega$) and special Lagrangian submanifolds (calibrated by $\mathrm{Re}\,\Omega$).

Of course, M may admit Cayley submanifolds that are neither complex nor special Lagrangian. One might ask whether such submanifolds can or must be built out of complex and special Lagrangian submanifolds. In this expository article, we will consider the following problem.

Given a compact complex submanifold N of a Calabi–Yau four-fold M, can one deform N as a Cayley submanifold into a Cayley submanifold N' that is not complex?

Of course, it is straightforward to see that the answer to this question is no by the following result of Harvey and Lawson.

Proposition 1.1 ([3, II.4 Thm 4.2]) *Let X be a Riemannian manifold with calibration α and let Y be a compact α-calibrated submanifold of X. Let Y' be any other compact submanifold of X homologous to Y. Then*

K. Moore (✉)
University College London, London, UK
e-mail: kim.moore@ucl.ac.uk

S. Karigiannis et al. (eds.), *Lectures and Surveys on G_2-Manifolds and Related Topics*, Fields Institute Communications 84, https://doi.org/10.1007/978-1-0716-0577-6_15

$$\mathrm{vol}(Y) \leq \mathrm{vol}(Y'),$$

with equality if, and only if, Y' is also α-calibrated.

Therefore, if N is a two-dimensional compact complex submanifold of a Calabi–Yau four-fold M, given N' a Cayley deformation of N we have that both

$$\mathrm{vol}(N) \leq \mathrm{vol}(N'),$$

applying Proposition 1.1 to N and N' with calibration $\frac{1}{2}\omega \wedge \omega$, and

$$\mathrm{vol}(N') \leq \mathrm{vol}(N),$$

applying Proposition 1.1 to N' and N with the Cayley calibration Φ. But then we must have $\mathrm{vol}(N) = \mathrm{vol}(N')$ and so Proposition 1.1 with calibration $\frac{1}{2}\omega \wedge \omega$ tells us that N and N' must both be complex submanifolds.

In this article, we explore the geometric reasons for this result, and the implications for complex submanifold theory. The material in this article is based on the author's PhD thesis [12] and paper [11].

2 Deformation Theory of Calibrated Submanifolds

Given a manifold with a calibration, one would like to be able to describe its calibrated submanifolds. One way of doing this is to study the *moduli space* of a certain type of calibrated submanifold. The first study of a moduli space of calibrated submanifolds may be attributed to Kodaira [6], who studied the moduli space of compact complex submanifolds of a complex manifold, which we describe in Sect. 2.1 below, although this result predates the definition of calibration by some twenty years! Later, motivated by *Calibrated geometries*, McLean [10] sought to prove analogues of Kodaira's result for calibrated submanifolds inside manifolds with special holonomy. We will review McLean's results on compact Cayley submanifolds in Sect. 2.2.

2.1 Kodaira's Deformation Theory of Compact Complex Submanifolds

Kodaira's approach to the deformation theory of complex submanifolds uses techniques from algebraic geometry. His approach is completely different from the later method of McLean, but it will be interesting to quote and interpret Kodaira's result here in order to compare to our work later.

Let M be a complex manifold with compact complex submanifold N. Denote by $H^k(N, \nu_M^{1,0}(N))$ the kth sheaf cohomology group of the sheaf of holomorphic sections of the holomorphic normal bundle of N in M. Define the moduli space $\mathcal{M}$ of complex deformations of N in M to be the set of complex submanifolds N' of M so that there exists a diffeomorphism $N \to N'$ isotopic to the identity.

Theorem 2.1 ([6, Main Thm]) *Let M be a complex manifold with compact complex submanifold N. If $H^1(N, \nu_M^{1,0}(N)) = 0$, then $\mathcal{M}$ is a smooth manifold of dimension* $\dim_{\mathbb{R}} H^0(N, \nu_M^{1,0}(N))$.

Remark We call $H^0(N, \nu_M^{1,0}(N))$ the *infinitesimal complex deformations* of N, and $H^1(N, \nu_M^{1,0}(N))$ the *obstruction space*. Note that the vanishing of the obstruction space is sufficient, but not necessary.

We can apply Dolbeault's theorem [2, pg 45] and the Hodge decomposition theorem [4, Thm 4.1.13, Cor 4.1.14] to rephrase Kodaira's theorem in terms of a differential operator.

Corollary 2.2 *Let M be a complex manifold with compact complex submanifold N. Then the space of infinitesimal complex deformations of N is isomorphic to the kernel of*

$$\bar{\partial} : C^\infty(\nu_M^{1,0}(N)) \to C^\infty(\Lambda^{0,1}N \otimes \nu_M^{1,0}(N)).$$

Moreover, the obstruction space is isomorphic to the kernel of

$$\bar{\partial} + \bar{\partial}^* : C^\infty(\Lambda^{0,1}N \otimes \nu_M^{1,0}(N)) \to C^\infty(\Lambda^{0,2}N \otimes \nu_M^{1,0}(N) \oplus \nu_M^{1,0}(N)).$$

2.2 McLean's Deformation Theory of Compact Cayley Submanifolds

McLean's goal in his 1998 paper [10] was to prove analogous results to Kodaira's Theorem 2.1 for compact calibrated submanifolds of manifolds with special holonomy. In particular, McLean proved the following result on the moduli space of Cayley deformations of a compact Cayley submanifold that admits a spin structure.

Theorem 2.3 ([10, Thm 6-3]) *Let Y be a compact Cayley submanifold of a Spin(7)-manifold X, and suppose that Y admits a spin structure. Then there exists a rank two complex vector bundle A over Y so that the Zariski tangent space to the moduli space of Cayley deformations of Y in X is given by the kernel of the twisted Dirac operator*

$$D\!\!\!/ : C^\infty(\mathbb{S}_+ \otimes A) \to C^\infty(\mathbb{S}_- \otimes A),$$

where $\mathbb{S}_+$ and $\mathbb{S}_-$ are respectively the bundles of positive and negative spinors on Y.

Here, elements of the kernel of $D\!\!\!/$ are called *infinitesimal deformations*, while the cokernel of $D\!\!\!/$ is called the *obstruction space*. This is because if the obstruction space is trivial, then the moduli space of Cayley deformations is a smooth manifold.

Sketch proof By the work of Harvey and Lawson [3, IV.1.C Cor 1.29] there exists a bundle-valued differential form $\tau \in \Omega^4(\Lambda^2_7)$ on X, where Λ^2_7 is the seven dimensional representation of *Spin*(7) acting on two-forms on X, satisfying for any oriented four-dimensional submanifold W of X,

$$\tau|_W \equiv 0,$$

if, and only if, W is a Cayley submanifold (up to a choice of orientation on W). This bundle-valued four-form can be described succinctly by the following expression. For orthogonal tangent vectors x, y, z, w define

$$\tau(x, y, z, w) = \pi_7(\Phi(\,\cdot\,, y, z, w) \wedge x^\flat), \tag{2.1}$$

where $\pi_7 : \Lambda^2 X \to \Lambda^2_7$ is the projection map given by

$$\pi_7(u^\flat \wedge v^\flat) = \frac{1}{2}(u^\flat \wedge v^\flat + \Phi(u, v, \cdot, \cdot))$$

and $\flat : TX \to T^*X$ denotes the musical isomorphism. Recall that if Y is a Cayley submanifold, then we can view [10, pg 741] $\Lambda^2_+ Y$ as a subbundle of $\Lambda^2_7|_Y$ via the map $\alpha \mapsto \pi_7(\alpha)$. We will denote by E the orthogonal complement to $\Lambda^2_+ Y$ in $\Lambda^2_7|_Y$, so that

$$\Lambda^2_7|_Y \cong \Lambda^2_+ Y \oplus E.$$

The tubular neighbourhood theorem [8, IV Thm 5.1] allows us to identify small normal vector fields on Y with small deformations of Y. If v is a normal vector field on Y, write $\exp_v := \exp \circ v : Y \to X$, with $Y_v := \exp_v(Y)$, the deformation corresponding to v. Then we can identify the moduli space of Cayley deformations of Y in X with the zero set of the following partial differential operator:

$$\begin{aligned} F : C^\infty(\nu_X(Y)) &\to C^\infty(E), \\ v &\mapsto \pi_E(*\exp_v^*(\tau|_{Y_v})), \end{aligned} \tag{2.2}$$

where $*$ denotes the Hodge star of Y and $\pi_E : \Lambda^2_7|_Y \to E$ denotes the projection map. The linear part of the operator at zero is

$$dF|_0(v) = \left.\frac{d}{dt}\right|_{t=0} F(tv) = \pi_E(*\mathcal{L}_v \tau|_Y).$$

This can be computed explicitly as the following operator (see for example [13, Prop 2.3]). Let $\{e_1, e_2, e_3, e_4\}$ be an orthonormal frame for TX, with dual coframe $\{\varrho^1, \varrho^2, \varrho^3, \varrho^4\}$. Define

$$D : C^\infty(\nu_X(Y)) \to C^\infty(E),$$
$$v \mapsto \sum_{i=1}^{4} \pi_7(e^i \wedge \nabla^\perp_{e_i} v), \tag{2.3}$$

where $\nabla^\perp$ is the connection on the normal bundle of Y in X induced by the Levi-Civita connection of X. To deduce McLean's result, first observe that [10, pg 741] there exists a rank two complex vector bundle A so that

$$\nu_X(Y) \otimes \mathbb{C} \cong \mathbb{S}_+ \otimes A$$
$$E \otimes \mathbb{C} \cong \mathbb{S}_- \otimes A.$$

Then McLean's result may be deduced by showing that the following diagram commutes

$$\begin{array}{ccc} C^\infty(\mathbb{S}_+ \otimes A) & \xrightarrow{\not{D}} & C^\infty(\mathbb{S}_- \otimes A) \\ \downarrow & & \downarrow \\ C^\infty(\nu_M(N) \otimes \mathbb{C}) & \xrightarrow{D} & C^\infty(E \otimes \mathbb{C}) \end{array}$$

□

To study the kernel of the operator defined in (2.2), we can extend the map F to some Banach spaces and try to apply the Banach space implicit function theorem [7, Ch 6 Thm 2.1]. To do this, we first need the linear part of F, which is the map D defined in (2.3), to be Fredholm, which it is since D is elliptic and Y is compact. Moreover we need D to surject, which unfortunately is not true in general. The cokernel of D describes the subspace of $C^\infty(E)$ that D does not reach, and hence obstructions to elements of the kernel of D, infinitesimal Cayley deformations of Y, to be extended to true Cayley deformations of Y. However, if the obstruction space vanishes, then we may apply the implicit function theorem and deduce that every infinitesimal Cayley deformation of Y extends to a true Cayley deformation of Y.

3 Cayley Deformations of Compact Complex Submanifolds

We now focus on Cayley deformations of a compact complex surface inside a Calabi–Yau four-fold. We saw in the introduction that it is very easy to see from the work of Harvey and Lawson that there are no Cayley deformations of a compact complex surface in a Calabi–Yau four-fold that are not complex deformations. This method is highly efficient, clean and compact, but doesn't leave us with any geometric intuition for why one cannot deform a compact complex submanifold into a Cayley submanifold that isn't complex. In particular, it is known that without the assumption that the submanifold is complex, we may deform such a submanifold into not only a Cayley submanifold that is not complex, but a special Lagrangian submanifold, as we will see in the following example.

Example ([9, Ex 5.8]) Consider $\mathbb{R}^8$ with the standard $Spin(7)$-structure (Φ_0, g_0). Writing any nonzero point of $\mathbb{R}^8$ as (r, p), where $r \in (0, \infty)$ and $p \in S^7$ we define a G_2-structure (φ, h) on S^7 by

$$\Phi_0|_{(r,p)} = r^3 dr \wedge \varphi|_p + r^4 *_h \varphi|_p,$$

with h the usual round metric. Notice that since Φ_0 is closed, $d\varphi = 4 * \varphi$, so this G_2-structure is not torsion-free. Then it is easy to check that a cone $C = (0, \infty) \times L$ is a Cayley submanifold of $(\mathbb{R}^8, \Phi_0, g_0)$ if, and only if, L is an associative submanifold of (S^7, φ, h). Homogeneous associative submanifolds of S^7 were classified by Lotay [9], including the following family, diffeomorphic to $SU(2)/\mathbb{Z}_3$. The deformation theory of homogeneous associative submanifolds of S^7 was studied by Kawai [5], while a comparative study of deformations of Cayley cones can be found in a paper of the author [13, Sect. 5].

Consider the following action of $SU(2)$ on $\mathbb{C}^4$

$$\begin{pmatrix} z_1 \\ z_2 \\ z_3 \\ z_4 \end{pmatrix} \mapsto \begin{pmatrix} a^3 z_1 + \sqrt{3}a^2 b z_2 + \sqrt{3}ab^2 z_3 + b^3 z_4 \\ -\sqrt{3}a^2\bar{b}z_1 + a(|a|^2 - 2|b|^2)z_2 + b(2|a|^2 - |b|^2)z_3 + \sqrt{3}\bar{a}b^2 z_4 \\ \sqrt{3}a\bar{b}^2 z_1 - \bar{b}(2|a|^2 - |b|^2)z_2 + \bar{a}(|a|^2 - 2|b|^2)z_3 + \sqrt{3}\bar{a}^2 b z_4 \\ -\bar{b}^3 z_1 + \sqrt{3}\bar{a}\bar{b}^2 z_2 - \sqrt{3}\bar{a}^2\bar{b}z_3 + \bar{a}^3 z_4 \end{pmatrix},$$

where $a, b \in \mathbb{C}$ satisfy $|a|^2 + |b|^2 = 1$. We define $L(\theta)$ to be the orbit of the point $(\cos\theta, 0, 0, \sin\theta)^T$ under the above action, that is,

$$L(\theta) := \begin{pmatrix} a^3 \cos\theta + b^3 \sin\theta \\ -\sqrt{3}a^2\bar{b}\cos\theta + \sqrt{3}\bar{a}b^2 \sin\theta \\ \sqrt{3}a\bar{b}^2 \cos\theta + \sqrt{3}\bar{a}^2 b \sin\theta \\ -\bar{b}^3 \cos\theta + \bar{a}^3 \sin\theta \end{pmatrix},$$

where $a, b \in \mathbb{C}$ satisfy $|a|^2 + |b|^2 = 1$. Then for

$$\mathbb{Z}_3 := \left\{ \begin{pmatrix} \zeta & 0 \\ 0 & \bar{\zeta} \end{pmatrix} \in SU(2) \mid \zeta^3 = 1 \right\},$$

$L(\theta)$ is invariant under the action of $\mathbb{Z}_3$ for all θ, therefore $L(\theta) \cong SU(2)/\mathbb{Z}_3$.

We have that $L(\theta)$ is associative for $\theta \in [0, \frac{\pi}{4}]$. It is easy to check that $L(0) = L$ is the real link of a complex cone, whereas $L(\frac{\pi}{4})$ is the link of a special Lagrangian cone. Therefore $C(\theta) := (0, \infty) \times L(\theta)$ defines a family of Cayley cones in $\mathbb{C}^4$. In particular, this example shows that we can deform a complex cone into a special Lagrangian cone through Cayley cones. Notice that Harvey and Lawson's result, Proposition 1.1, doesn't apply in this situation because the cone is not compact.

3.1 The Cayley Operator on a Complex Submanifold

Let (M, J, ω, Ω) be a four-dimensional Calabi–Yau manifold and let N be a two-dimensional compact complex submanifold of M. We want to compare complex and Cayley deformations of N, but we already have some clues to help us. By Kodaira's Theorem 2.1 in combination with Dolbeault's theorem we know that infinitesimal complex deformations of N in M are given by the kernel of

$$\bar{\partial} : C^\infty(\nu_M^{1,0}(N)) \to C^\infty(\Lambda^{0,1}N \otimes \nu_M^{1,0}(N)). \tag{3.1}$$

Meanwhile, the work of McLean tells us that infinitesimal Cayley deformations of (a spin manifold) N in M are given by the kernel of the twisted Dirac operator

$$\not{D} : C^\infty(\mathbb{S}_+ \otimes A) \to C^\infty(\mathbb{S}_- \otimes A). \tag{3.2}$$

At first glance, comparing the kernels of these two operators seems like a fruitless task. However, since N is Kähler, its spin structure and Dirac operator take a special form [1, pg 82]. Given a two-dimensional Kähler manifold with a fixed spin structure, we can identify

$$\mathbb{S}_+ \cong \left(\Lambda^{0,0}N \oplus \Lambda^{0,2}N\right) \otimes S_k,$$
$$\mathbb{S}_- \cong \Lambda^{0,1}N \otimes S_k,$$

where S_k is a holomorphic line bundle satisfying $S_k \otimes S_k = \Lambda^{2,0}N$. Under these identifications, the Dirac operator is

$$\sqrt{2}\left(\bar{\partial} + \bar{\partial}^*\right) : C^\infty(S_k \oplus \Lambda^{0,2}N \otimes S_k) \to C^\infty(\Lambda^{0,1}N \otimes S_k). \tag{3.3}$$

We have already seen that McLean proved that $\nu_M(N) \otimes \mathbb{C} \cong \mathbb{S}_+ \otimes A$ and $E \otimes \mathbb{C} \cong \mathbb{S}_- \otimes A$, for some rank two complex vector bundle A. Comparing the three operators (3.1), (3.2) and (3.3) it is not unreasonable to hope that we can identify

$$\nu_M(N) \otimes \mathbb{C} \cong \nu_M^{1,0}(N) \oplus \Lambda^{0,2}N \otimes \nu_M^{1,0}(N),$$
$$E \otimes \mathbb{C} \cong \Lambda^{0,1}N \otimes \nu_M^{1,0}(N),$$

and for us to be able to show that under these identifications infinitesimal Cayley deformations of N are given by the kernel of the operator

$$\bar{\partial} + \bar{\partial}^* : C^\infty(\nu_M^{1,0}(N) \oplus \Lambda^{0,2}N \otimes \nu_M^{1,0}(N)) \to C^\infty(\Lambda^{0,1}N \otimes \nu_M^{1,0}(N)).$$

This turns out to be true—and despite the heuristic comparison above, in fact N is not required to be spin for the following result, taken from [11, Prop 3.5 and Thm 3.9], to hold.

Theorem 3.1 *Let N be a compact complex surface inside a Calabi–Yau four-fold M. Then infinitesimal Cayley deformations of N in M are given by the kernel of the operator*

$$\bar{\partial} + \bar{\partial}^* : C^\infty(\nu_M^{1,0}(N) \oplus \Lambda^{0,2}N \otimes \nu_M^{1,0}(N)) \to C^\infty(\Lambda^{0,1}N \otimes \nu_M^{1,0}(N)).$$

Moreover, the expected dimension of the moduli space of Cayley deformations of N in M is given by the index of this operator

$$\operatorname{ind}\left(\bar{\partial} + \bar{\partial}^*\right) = \frac{1}{2}\sigma(N) + \frac{1}{2}\chi(N) - [N] \cdot [N],$$

where $\sigma(N)$ is the signature of N, $\chi(N)$ is the Euler characteristic of N and $[N] \cdot [N]$ is the self-intersection number of N.

Sketch proof We have that the complex structure on M induces a natural splitting of the complexified normal bundle of N in M into holomorphic and anti-holomorphic parts

$$\nu_M(N) \otimes \mathbb{C} \cong \nu_M^{1,0}(N) \oplus \nu_M^{0,1}(N).$$

We would like to show that

$$\nu_M^{0,1}(N) \cong \Lambda^{0,2}N \otimes \nu_M^{1,0}(N).$$

To understand why this might be true, we consider the holomorphic volume form Ω of M, which is a nowhere-vanishing, parallel section of the canonical bundle of M, denoted by $K_M := \Lambda^{4,0}M$. Recall that the adjunction formula [4, Prop 2.2.17] says that

$$K_M|_N \cong \Lambda^{2,0}N \otimes \Lambda^2 \nu_M^{*1,0}(N). \tag{3.4}$$

In particular, $\overline{\Omega}|_N$ is a well-defined nowhere-vanishing section of $\Lambda^{0,2}N \otimes \Lambda^2 \nu_M^{*0,1}(N)$. So given any section of $\nu_M^{0,1}(N)$ is is easy to check that

$$\overline{\Omega}(v, \cdot, \cdot, \cdot)|_N, \tag{3.5}$$

is a well-defined section of $\Lambda^{0,2}N \otimes \nu_M^{*0,1}(N)$. Finally, the Riemannian metric on M defines a musical isomorphism $\sharp : \nu_M^{*0,1}(N) \to \nu_M^{1,0}(N)$. It is easy to verify that these objects provide the desired isomorphism.

It is simple to check using local coordinates that $E \cong \Lambda^{0,1}N \otimes \nu_M^{1,0}(N)$, again with the help of the musical isomorphism, and moreover that the following diagram commutes:

$$\begin{array}{ccc} C^\infty(\nu_M^{1,0}(N) \oplus \Lambda^{0,2}N \otimes \nu_M^{1,0}(N)) & \xrightarrow{\bar{\partial}+\bar{\partial}^*} & C^\infty(\Lambda^{0,1}N \otimes \nu_M^{1,0}(N)) \\ \downarrow & & \downarrow \\ C^\infty(\nu_M(N) \otimes \mathbb{C}) & \xrightarrow{\quad D \quad} & C^\infty(E \otimes \mathbb{C}) \end{array}$$

where D was defined in (2.3).

The index formula follows from the Hirzebruch–Riemann–Roch Theorem [4, Cor 5.1.4]. □

Example Let

$$M := \{[z_0 : \cdots : z_5] \in \mathbb{C}P^5 \mid z_0^6 + \cdots + z_5^6 = 0\},$$

and take

$$N = \{z \in M \mid f_1(z) = f_2(z) = 0\},$$

where f_i are irreducible homogeneous polynomials of degree d_i such that the Jacobian of $g = (f_1, f_2)$ has rank two at each point of N. Then we can compute that

$$\begin{aligned} [N] \cdot [N] &= 6d_1^2 d_2^2, \\ \chi(N) &= 90 d_1 d_2 + 6d_1^3 d_2 + 6d_2^2 d_1 + 6d_1^2 d_2^2, \\ \sigma(N) &= -60 d_1 d_2 - 2d_1^3 d_2 - 2d_2^3 d_1, \end{aligned}$$

so that

$$\operatorname{ind}(\bar{\partial} + \bar{\partial}^*) = d_1 d_2 (15 + 2d_1^2 + 2d_2^2 - 3d_1 d_2).$$

Examining this expression, we see that the expected dimension of the moduli space of Cayley deformations of N in M will be strictly positive and even for any $d_1, d_2 \in \mathbb{N}$.

4 Complex Deformations of a Compact Complex Surface

We would like to compare complex and Cayley deformations of a compact complex surface N in a Calabi–Yau four-fold (M, J, ω, Ω). So far we have seen, by Kodaira's Theorem 2.1, that infinitesimal complex deformations of N are given by holomorphic normal vector fields in the kernel of

$$\bar{\partial} : C^\infty(\nu_M^{1,0}(N)) \to C^\infty(\Lambda^{0,1}N \otimes \nu_M^{1,0}(N)), \tag{4.1}$$

with the dimension of the space of infinitesimal complex deformations of N given by the real dimension of the kernel of (4.1).

We saw in Sect. 3 that the infinitesimal Cayley deformations of N are given by forms in the kernel of the operator

$$\bar{\partial} + \bar{\partial}^* : C^\infty(\nu_M^{1,0}(N) \oplus \Lambda^{0,2}N \otimes \nu_M^{1,0}(N)) \to C^\infty(\Lambda^{0,1}N \otimes \nu_M^{1,0}(N)), \tag{4.2}$$

with the dimension of the space of infinitesimal Cayley deformations of N given by the *complex* dimension of the kernel of (4.2) (since we complexified the normal bundle of N in M to find this operator).

At first glance, comparing the above operators this seems to be a mistake, but it turns out that there is an isomorphism between the kernel of (4.1) and the kernel of

$$\bar{\partial}^* : C^\infty(\Lambda^{0,2}N \otimes \nu_M^{1,0}(N)) \to C^\infty(\Lambda^{0,1}N \otimes \nu_M^{1,0}(N)). \tag{4.3}$$

The following result is taken from [11, Lem 4.6].

Lemma 4.1 *Let N be a complex surface in a Calabi–Yau four-fold* (M, J, ω, Ω). *Then the kernels of* (4.1) *and* (4.3) *are isomorphic.*

Proof (*Sketch*) Similar to the proof of Theorem 3.1 where we constructed an isomorphism $\nu_M^{0,1}(N) \to \Lambda^{0,2}N \otimes \nu_M^{1,0}(N)$, we take the map

$$\begin{aligned} \nu_M^{1,0}(N) &\to \Lambda^{0,2}N \otimes \nu_M^{1,0}(N), \\ v &\mapsto \left(\bar{v} \lrcorner \overline{\Omega}\right)^\sharp, \end{aligned}$$

where $\sharp : \nu_M^{*0,1}(N) \to \nu_M^{1,0}(N)$ is the standard musical isomorphism. That this map sends Ker $\bar{\partial}$ to Ker $\bar{\partial}^*$ is essentially a consequence of Ω being parallel. □

What we have seen so far suggests therefore that an infinitesimal Cayley deformation of N that is not an infinitesimal complex deformation of N looks like $v \oplus w \in C^\infty(\nu_M^{1,0}(N) \oplus \Lambda^{0,2}N \otimes \nu_M^{1,0}(N))$ with

$$\bar{\partial}v = -\bar{\partial}^* w \neq 0.$$

We know by Hodge theory that this cannot happen when N is compact—so this would explain why we cannot deform a compact complex surface into a Cayley submanifold that is not complex.

To make these ideas more formal, we will argue in the style of McLean to characterise complex deformations of a compact complex surface. We will first look for a differential form that vanishes exactly when restricted to a complex surface.

Firstly, let us take a Cayley submanifold N' of a Calabi–Yau four-fold (M, J, ω, Ω). We have that

$$\tau|_{N'} \equiv 0,$$

and

$$\Phi|_{N'} = \text{Re } \Omega|_{N'} + \frac{1}{2}\omega \wedge \omega|_{N'} = \text{vol}_{N'}.$$

So to further ensure that N' is complex, we must ask that

$$\text{Re } \Omega|_{N'} \equiv 0.$$

So we see that $v \in C^\infty(\nu_M(N) \otimes \mathbb{C})$ defines a complex deformation of N if, and only if

$$G(v) = (\exp_v^* \tau|_{N_v}, \exp_v^* \operatorname{Re} \Omega|_{N_v}) = (0, 0).$$

We ask how the linearisation of G at zero differs from the linearisation of F defined in (2.2) at zero. Finding the linear part of $\exp_v^* \operatorname{Re} \Omega|_{N_v}$, we see that

$$\left.\frac{d}{dt}\right|_{t=0} \exp_{tv} \operatorname{Re} \Omega|_{N_{tv}} = \mathcal{L}_v \operatorname{Re} \Omega = \frac{1}{2} d(v \lrcorner \Omega + v \lrcorner \overline{\Omega}).$$

Writing $v = v_1 \oplus v_2$, where $v_1 \in C^\infty(\nu_M^{1,0}(N))$ and $v_2 \in C^\infty(\nu_M^{0,1}(N))$, we see that

$$\frac{1}{2} d(v \lrcorner \Omega + v \lrcorner \overline{\Omega}) = \frac{1}{2} d(v_1 \lrcorner \Omega + v_2 \lrcorner \overline{\Omega}) = \frac{1}{2} \bar{\partial}(v_1 \lrcorner \Omega) + \frac{1}{2} \partial(v_2 \lrcorner \overline{\Omega}),$$

since $v_1 \lrcorner \Omega \in C^\infty(\Lambda^{2,0} N \otimes \nu_M^{*1,0}(N))$ and $v_2 \lrcorner \overline{\Omega} \in C^\infty(\Lambda^{0,2} N \otimes \nu_M^{*0,1}(N))$. Therefore a normal vector field $v = v_1 \oplus v_2$ is an infinitesimal complex deformation of N if, and only if, the linearisation of the first component of G vanishes, which by Theorem 3.1 is

$$\bar{\partial} v_1 + \bar{\partial}^*(v_2 \lrcorner \overline{\Omega}) = 0,$$

where we recall the isomorphism of $\nu_M^{0,1}(N)$ and $\Lambda^{0,2} N \otimes \nu_M^{1,0}(N)$ given in Eq. (3.5), and the linearisation of the second component of G vanishes, which as we've just seen is

$$\bar{\partial}(v_1 \lrcorner \Omega) = 0 = \partial(v_2 \lrcorner \overline{\Omega}),$$

since $\bar{\partial}(v_1 \lrcorner \Omega)$ and $\partial(v_2 \lrcorner \overline{\Omega})$ take values in different vector bundles. Similarly to Lemma 4.1, we can show that this is equivalent to

$$\bar{\partial} v_1 = 0 = \bar{\partial}^*(v_2 \lrcorner \overline{\Omega}),$$

which moreover by definition of the isomorphism $\nu_M^{0,1}(N) \to \Lambda^{0,2} N \otimes \nu_M^{1,0}(N)$ in the proof of Theorem 3.1 is equivalent to the set of $v \oplus w \in C^\infty(\nu_M^{1,0}(N) \oplus \Lambda^{0,2} N \otimes \nu_M^{1,0}(N))$ such that

$$\bar{\partial} v = 0 = \bar{\partial}^* w.$$

This informal argument shows that our assertion that infinitesimal Cayley deformations of N expressed in the form $v \oplus w \in C^\infty(\nu_M^{1,0}(N) \oplus \Lambda^{0,2} N \otimes \nu_M^{1,0}(N))$ that are not infinitesimal complex deformations must satisfy

$$\bar{\partial} v = -\bar{\partial}^* w,$$

is correct.

It turns out that we can study the complex deformations of N in M without thinking about Cayley deformations at all.

Let us think about the holomorphic volume form. The adjunction formula tells us that, for N, a complex surface inside a four-dimensional Calabi–Yau manifold M with holomorphic volume form Ω, we have

$$\Omega|_N \in K_M|_N \cong K_N \otimes \Lambda^2 \nu_M^{*1,0}(N).$$

So this tells us that given any three tangent vector fields v_1, v_2 and v_3 on N we must have that

$$\Omega(v_1, v_2, v_3, \cdot) = 0.$$

It is natural to wonder whether conversely, if given any three tangent vector fields v_1, v_2 and v_3 to a real oriented four manifold W in a Calabi–Yau four-fold M we have

$$\Omega(v_1, v_2, v_3, \cdot) = 0,$$

then W must a complex submanifold of M. This is not quite right, but a similar result turns out to be true, as we show in [11, Prop 4.2].

Proposition 4.2 *An oriented four-dimensional real submanifold X of a four-dimensional Calabi–Yau manifold (M, J, ω, Ω) is a complex submanifold if, and only if,*

$$\sigma|_X \equiv 0,$$

where $\sigma \in C^\infty(\Lambda^3 M \otimes T^*M|_X)$ is defined by

$$\sigma(v_1, v_2, v_3) = \text{Re}\,\Omega(v_1, v_2, v_3, \cdot),$$

for any $v_1, v_2, v_3 \in C^\infty(TX)$.

This result is purely an exercise in linear algebra. It suffices to check that the proposition holds for a linear subspace of $\mathbb{C}^4$.

Example Let ω and Ω be the standard Kähler form and holomorphic volume form on $\mathbb{C}^3$. Then we can define a G_2-structure on $\mathbb{R}^7$ by

$$\varphi = dx \wedge \omega + \text{Re}\,\Omega,$$
$$*\varphi = \frac{1}{2}\omega \wedge \omega - dx \wedge \text{Im}\,\Omega,$$

where x is the coordinate on $\mathbb{R}$ in $\mathbb{R}^7 = \mathbb{R} \times \mathbb{C}^3$. Introducing another factor of $\mathbb{R}$ with coordinate t, we can take the following Calabi–Yau structure on $\mathbb{C}^4$

$$\tilde{\omega} = dt \wedge dx + \omega,$$
$$\tilde{\Omega} = (dt + idx) \wedge \Omega.$$

Let N be a four-dimensional real submanifold of $\mathbb{C}^3 \subseteq \mathbb{C}^4$. By Proposition 4.2, N is complex in $\mathbb{C}^4$ if, and only if,

$$\mathrm{Re}\,\tilde{\Omega} = dt \wedge \mathrm{Re}\,\Omega - dx \wedge \mathrm{Im}\,\Omega = 0,$$

as a three-form on N. So in particular, we must have that

$$\mathrm{Re}\,\Omega|_N = 0 = \mathrm{Im}\,\Omega|_N.$$

This in combination with the fact that $N \subseteq \mathbb{C}^3$ shows that

$$\varphi|_N = 0,$$

and so N is a coassociative submanifold of $\mathbb{R}^7$, and moreover

$$\mathrm{vol}_N = *\varphi|_N = \frac{1}{2}\omega \wedge \omega|_N - dx \wedge \mathrm{Im}\,\Omega|_N = \frac{1}{2}\omega \wedge \omega|_N,$$

so if N is a complex submanifold of $\mathbb{C}^4$ and is contained in $\mathbb{C}^3$ then it is also a complex submanifold of $\mathbb{C}^3$.

It turns out that we can generalise this idea to study any complex submanifold of a Calabi–Yau manifold.

Proposition 4.3 *Let (M, J, ω, Ω) be an m-dimensional Calabi–Yau manifold and let $p \in \mathbb{N}$ be such that $p < m - 1$. Then an oriented 2p-dimensional real submanifold X of M is a complex submanifold of M if, and only if,*

$$\sigma|_X = 0,$$

where $\sigma \in C^\infty(\Lambda^{p+1}M \otimes \Lambda^{m-p-1}M)$ is given by

$$\sigma(v_1, \ldots v_{p+1}) = \mathrm{Re}\,\Omega(v_1, \ldots, v_{p+1}, \cdot, \ldots, \cdot),$$

for any $v_1, \ldots v_{p+1} \in C^\infty(TX)$. If $p + 1 = m$, then we must have that

$$\mathrm{Re}\,\Omega|_X = 0 = \mathrm{Im}\,\Omega|_X.$$

Example Applying Proposition 4.3 to the previous example, a complex surface N in $\mathbb{C}^3$ must satisfy

$$\mathrm{Re}\,\Omega|_N = 0 = \mathrm{Im}\,\Omega|_N,$$

so considering N as a submanifold of $\mathbb{C}^4$ this implies that as a three-form

$$\mathrm{Re}\,\tilde{\Omega}|_N = (dx \wedge \mathrm{Re}\,\Omega)|_N - (dx \wedge \mathrm{Im}\,\Omega)|_N = 0,$$

so N is also a complex submanifold of $\mathbb{C}^4$ as one would expect.

Given Proposition 4.3, we can study complex deformations of complex submanifolds of Calabi–Yau manifolds in a similar style to McLean's Theorem 2.3. We focus on the special case of compact complex surfaces inside Calabi–Yau four-folds here, but the result below holds for any compact complex submanifold of a Calabi–Yau manifold, see [11, Prop 4.4, Prop 4.5]. Notice that this recovers a special case of Kodaira's theorem [6, Main Thm] using a completely different method.

Theorem 4.4 *Let N be a two-dimensional compact complex submanifold of a four-dimensional Calabi–Yau manifold M. Then the moduli space of complex deformations of N in M is locally homeomorphic to the zero set of a partial differential operator*

$$G : C^\infty(\nu_M^{1,0}(N) \oplus \nu_M^{0,1}(N)) \to C^\infty(\Lambda^{0,1}N \otimes \nu_M^{1,0}(N) \oplus \Lambda^{1,0}N \otimes \nu_M^{0,1}(N)),$$

with linearisation at zero given by

$$dG|_0(v_1 \oplus v_2) = \bar{\partial} v_1 \oplus \partial v_2. \tag{4.4}$$

Remark Notice that the kernels of $\bar{\partial}$ and ∂ acting on holomorphic and anti-holomorphic vector fields are naturally isomorphic by complex conjugation.

Sketch proof It is clear by Proposition 4.2 that given a complexified normal vector field v, the corresponding deformation N_v is a complex submanifold of M if, and only if,

$$\exp_v^*(\sigma|_{N_v}) = 0.$$

This is a three-form on N that takes values in $T^*M|_N \otimes \mathbb{C}$. A local argument, [11, Prop 4.4] shows that is suffices to check only the parts of this form that take values in the space $\Lambda^{2,1}N \otimes \nu_M^{*1,0}(N) \oplus \Lambda^{1,2}N \otimes \nu_M^{*0,1}(N)$. Denote the projection onto this vector bundle by π. Then N_v is a complex submanifold if, and only if,

$$\pi(\exp_v^*(\sigma|_{N_v})) = 0.$$

Finally, we have that the maps

$$\begin{aligned}
\Lambda^{0,1}N \otimes \nu_M^{1,0}(N) &\to \Lambda^{2,1}N \otimes \nu_M^{*1,0}(N),\\
\Lambda^{1,0}N \otimes \nu_M^{0,1}(N) &\to \Lambda^{1,2}N \otimes \nu_M^{*0,1}(N),
\end{aligned}$$

given respectively by

$$\begin{aligned}
\alpha \otimes v &\mapsto \alpha \wedge (v \lrcorner \Omega)|_N,\\
\tilde{\alpha} \otimes \tilde{v} &\mapsto \tilde{\alpha} \wedge (\tilde{v} \lrcorner \overline{\Omega})|_N,
\end{aligned}$$

define vector bundle isomorphisms [11, Lem 4.3]. Denoting these isomorphisms by Ψ, we finally define the partial differential operator whose kernel can be identified with the moduli space of complex deformations of N in M to be

$$G : C^\infty(\nu_M^{1,0}(N) \oplus \nu_M^{0,1}(N)) \to C^\infty(\Lambda^{0,1}N \otimes \nu_M^{1,0}(N) \oplus \Lambda^{1,0}N \otimes \nu_M^{0,1}(N)),$$
$$v \mapsto \Psi^{-1} \circ \pi(\exp_v^*(\sigma|_{N_v})).$$

A short computation [11, Prop 4.5] shows that the linearisation of G at zero is the operator (4.4) as claimed. □

5 Further and Future Work

5.1 *Can We Describe the Moduli Space of Complex Submanifolds in Any Ambient Complex Manifold Using These Techniques?*

As long as the ambient manifold is Kähler, its complex manifolds are calibrated submanifolds. In the work described here the existence of a parallel, nowhere vanishing $(m, 0)$-form on the ambient (complex m-dimensional) manifold is essential. If the ambient manifold were Kähler with a nowhere vanishing $(m, 0)$-form that was not parallel, one could repeat the above argument, with the linearised operator having additional zero-order terms.

5.2 *Can These Results Be Extended to Noncompact Complex Submanifolds?*

As was mentioned in Sect. 3, we expect a noncompact complex submanifold of a Calabi–Yau four-fold will admit Cayley deformations that are not complex, as evidenced by the given example. The author has studied *conically singular* Cayley and complex submanifolds [13], and has shown that infinitesimal Cayley and complex deformations of a conically singular complex surface in a Calabi–Yau four-fold are still of the same type. Note that Harvey and Lawson's result stated in Proposition 1.1 continues to hold for *currents* with compact support and therefore will apply in the setting of conically singular calibrated submanifolds.

The analytic techniques available for studying deformations of a conically singular submanifold only allow one to consider somewhat rigid classes of deformations. For example, it is not possible to deform a conically singular calibrated submanifold into a non-singular calibrated submanifold using current techniques. An interesting problem would be to produce new techniques to study wider classes of deformations of singular calibrated submanifolds, perhaps inspired by techniques from algebraic geometry.

Acknowledgements The author would like to thank Spiro Karigiannis, Naichung Conan Leung and Jason Lotay for the invitation to the event *Workshop on G_2 manifolds and related topics*, and the Fields Institute for supporting her visit to Toronto.

References

1. Friedrich, T. (2000). *Dirac operators in Riemannian geometry* (Vol. 25). Graduate studies in mathematics. Providence, RI: American Mathematical Society (Translated from the 1997 German original by Andreas Nestke).
2. Griffiths, P., & Harris, J. (1994). *Principles of algebraic geometry*. Wiley Classics Library. New York: Wiley. Reprint of the 1978 original.
3. Harvey, F. R., & Lawson, H. B. (1982). Calibrated geometries. *Acta Mathematica*, *148*, 47–157.
4. Huybrechts, D. (2005). *Complex geometry*. Universitext. Berlin: Springer.
5. Kawai, K. (2017). Deformations of homogeneous associative submanifolds in nearly parallel G_2-manifolds. *Asian Journal of Mathematics*, *21*(3), 429–461.
6. Kodaira, K. (1962). A theorem of completeness of characteristic systems for analytic families of compact submanifolds of complex manifolds. *Annals of Mathematics*, *2*(75), 146–162.
7. Lang, S. (1983). *Real analysis* (2nd ed.). Advanced book program. Reading, MA: Addison-Wesley Publishing Company.
8. Lang, S. (2002). *Introduction to differentiable manifolds* (2nd ed.). Universitext. New York: Springer.
9. Lotay, J. D. (2012). Associative submanifolds of the 7-sphere. *Proceedings of the London Mathematical Society* (3), *105*(6), 1183–1214.
10. McLean, R. C. (1998). Deformations of calibrated submanifolds. *Communications in Analysis and Geometry*, *6*(4), 705–747.
11. Moore, K. (2019). Cayley deformations of compact complex surfaces. *Journal of the London Mathematical Society, 100*(2), 668–691.
12. Moore, K. (2017). *Deformation theory of Cayley submanifolds*. PhD thesis, University of Cambridge.
13. Moore, K. (2019). Deformations of conically singular Cayley submanifolds. *The Journal of Geometric Analysis, 29*, 2147–2216.

Deformations of Calibrated Submanifolds with Boundary

Alexei Kovalev

Abstract We review some results concerning the deformations of calibrated minimal submanifolds which occur in Riemannian manifolds with special holonomy. The calibrated submanifolds are assumed compact with a non-empty boundary which is constrained to move in a particular fixed submanifold. The results extend McLean's deformation theory previously developed for closed compact submanifolds.

1 Preliminaries

Calibrated submanifolds are a particular type of minimal submanifolds introduced by Harvey and Lawson [12] as a generalization of complex submanifolds of Kähler manifolds; for a detailed reference, see *op.cit.* and [11]. Harvey and Lawson [12] also found four new types on calibrations defined on Euclidean spaces and, more generally, on Ricci-flat Riemannian manifolds with reduced holonomy. Further examples of calibrations were subsequently discovered, see [13, Sect. 4.3] and references therein.

McLean [18] studied deformations for the four types of calibrated submanifolds defined in [12] and showed that the deformation problem may be interpreted as a non-linear PDE with Fredholm properties. In some cases, there is a smooth finite-dimensional 'moduli space'. The submanifolds in [18] were assumed compact and without boundary. McLean's deformation theory in [18] was later extended by several authors to more general classes of submanifolds. In this paper, we survey the generalizations of McLean's results to compact submanifolds having non-empty boundary. Proofs will at most be briefly sketched or referred to the original papers.

We begin in this section with some key concepts of calibrated geometry and foundational results concerning deformations of compact submanifolds (possibly with boundary). The remainder of the paper is organized in four sections dealing with the four types of calibrated submanifolds in [12, 18], namely special Lagrangian, coassociative, associative and Cayley submanifolds.

A. Kovalev (✉)
DPMMS, University of Cambridge, Cambridge, England
e-mail: a.g.kovalev@dpmms.cam.ac.uk

S. Karigiannis et al. (eds.), *Lectures and Surveys on G_2-Manifolds and Related Topics*, Fields Institute Communications 84, https://doi.org/10.1007/978-1-0716-0577-6_16

Submanifolds are taken to be embedded and connected, for notational convenience (it can be checked that the results extend to immersed submanifolds). Smooth functions and, more generally, sections of vector bundles on (sub)manifolds with boundary are understood as 'smooth up to the boundary', so at each point of the boundary these have one-sided partial derivatives of any order in the inward-pointing normal direction.

1.1 Calibrations

Definition 1 Let (M, g) be a Riemannian manifold. For any tangent k-plane V, i.e. a k-dimensional subspace V of a tangent space $T_x M$, a choice of orientation on V together with the restriction of g determines a natural *volume form* on V, $\mathrm{vol}_V \in \Lambda^k V^*$.

A differential k-form ϕ on M is called a *calibration* if (i) $d\phi = 0$ and (ii) for each $x \in M$ and every oriented k-dimensional subspace $V \subset T_x M$, $\phi|_V = a \,\mathrm{vol}_V$ for some $a \leq 1$.

An oriented k-dimensional submanifold N of M is said to be *calibrated by* ϕ if the pull-back of ϕ to N coincides with the Riemannian volume form for the metric on N induced by g, i.e. $\phi|_{T_x N} = \mathrm{vol}_{T_x N}$ for each $x \in N$.

The next result shows that calibrated submanifolds are minimal, in fact volume-minimizing (if compact).

Proposition 2 *Let M be a Riemannian manifold and let $\phi \in \Omega^k(M)$ be a calibration on M.*

(a) *If a closed k-dimensional submanifold $X \subset M$ is calibrated by ϕ, then X is volume-minimizing in its homology class. Moreover, if Y is a volume-minimizing closed k-dimensional submanifold of M in the homology class of X, then Y is calibrated by ϕ.*
(b) *Let $W \subset M$ be a submanifold such that $\phi|_W = 0$. If $X \subset M$ is a calibrated compact k-dimensional submanifold with non-empty boundary $\partial X \subset W$, then X is volume-minimizing in the relative homology class $[X] \in H_k(M, W; \mathbb{Z})$. Moreover, if Y is a compact submanifold of M with boundary $\partial Y \subset W$ and Y is volume-minimizing in the relative homology class $[X]$, then Y is calibrated by ϕ.*

The clause (a) is proved in [12, Thm II.4.2] by application of Stokes' theorem. The extension (b) to submanifolds with boundary follows by a similar argument as the hypothesis $\phi|_W = 0$ ensures the vanishing of the additional terms arising from the boundary (cf. [8, pp. 1233–1234]). Suppose that a submanifold Y with boundary in W represents the relative homology class $[X]$. Considering X and Y as chains, we can find a $(k + 1)$-dimensional chain N with boundary in W and a k-dimensional chain P contained in W so that $Y - X = \partial N + P$. We then obtain

$$\mathrm{Vol}(Y) \geq \int_Y \phi = \int_X \phi + \int_{\partial N} \phi + \int_P \phi = \mathrm{Vol}(X),$$

noting in the inequality that ϕ is a calibration, then applying Stokes' theorem and taking account of the vanishing of ϕ on W.

The above argument may be viewed as a generalization of the volume-minimizing property of compact complex submanifolds of Kähler manifolds by application of Wirtinger's inequality. (In a Kähler manifold with Kähler form ω, every k-dimensional complex submanifold is calibrated by $\omega^k/k!$.)

1.2 Normal Deformations of Submanifolds

McLean's deformation theory [18] was originally developed for closed compact submanifolds. The nearby deformations of a given closed submanifold of a Riemannian manifold may be assumed to be *normal deformations*, defined using the Riemannian exponential map on normal vector fields, i.e. C^1 sections of the normal bundle of this submanifold.

When a submanifold has a boundary, we wish to consider the deformation problem as an elliptic boundary value problem. We shall in fact require that the boundary moves in a certain fixed submanifold which, following [4], we call a *scaffold*. In general, we cannot use, as in the case of closed submanifolds, exponential deformations defined using the given metric g on the ambient manifold, since the scaffold may not be preserved under such deformations. We shall define on M a modified metric $\hat{g}$ whose associated exponential map does preserve the scaffold because the scaffold will be totally geodesic with respect to the new metric. (The actual construction of the metric $\hat{g}$ will depend on the considered calibration.)

Proposition 3 (a) *Let P be a closed submanifold of a Riemannian manifold M. There exist an open subset V_P of the normal bundle $N_{P/M}$ of P in M, containing the zero section, and a tubular neighbourhood T_P of P in M, such that the exponential map $\exp_M|_{V_P} : V_P \to T_P$ is a diffeomorphism onto T_P.*

(b) *Let M be a smooth manifold of dimension n and $P \subset M$ a compact submanifold with non-empty boundary ∂P. Let W be a submanifold of M with $\partial P \subset W$ and let $\hat{g}$ be a Riemannian metric on M such that P and W meet orthogonally and W is totally geodesic with respect to $\hat{g}$.*
There exists an open subset V_P of the normal bundle $\hat{N}_{P/M}$ of P in M, containing the zero section, and an n-dimensional submanifold T_P of M with boundary such that $P \subset T_P$ and $\widehat{\exp}_M|_{V_P} : V_P \to T_P$ is a diffeomorphism onto T_P. Furthermore, if a section $\mathbf{v}$ of $\hat{N}_{P/M}$ takes values in V_P, then $\widehat{\exp}_M(\mathbf{v}(x)) \in W$ for all $x \in \partial P$.

The clause (a) is a consequence of the tubular neighbourhood theorem [17, Chap. IV, Thm. 9]. For the extension (b) to submanifolds with boundary and the existence of the adapted metric $\hat{g}$ modifying a given metric on M on an open neighbourhood of ∂P, cf. [4, Prop. 6] or [16, Prop. 4.4]. We stress that the new auxiliary metric $\hat{g}$ in (b) is used *solely* for the purpose of considering the exponential map and applications of the tubular neighbourhood theorem—but *not* for the minimal or volume-minimizing properties of calibrated submanifolds (for which we continue to use the original metric on M).

By *nearby deformations* of a compact submanifold P (with or without boundary) we shall mean submanifolds of the form $P_{\mathbf{v}} = \exp_{\mathbf{v}}(P)$, where $\mathbf{v}$ is a C^1-section of the normal bundle $N_{P/M}$. The section $\mathbf{v}$ is assumed sufficiently small in the C^1 norm, so that $P_{\mathbf{v}}$ is contained in a tubular neighbourhood T_P defined by Proposition 3. We shall call sections of $N_{P/M}$ the *normal vector fields* on P.

We interpret submanifolds as appropriate equivalence classes of the embedding (more generally, immersion) maps and the term 'moduli space of submanifolds' is used below in this sense.

2 Special Lagrangian Submanifolds in Calabi–Yau Manifolds

Let M be a Kähler manifold of complex dimension m, with Kähler form ω and suppose further that the metric on M has holonomy contained in $SU(m)$. Then the canonical bundle of M may be trivialized by a holomorphic $(m, 0)$-form Ω satisfying

$$(-1)^{m(m-1)/2}\,(i/2)^m\,\Omega \wedge \bar{\Omega} = \omega^m/m!\,. \tag{1}$$

It also follows that the Kähler metric is Ricci-flat. Conversely, if a Kähler form ω and a holomorphic $(m, 0)$-form Ω satisfy (1) on M, then this Kähler metric has holonomy in $SU(m)$. We shall call (M, ω, Ω) as above a *Calabi–Yau manifold.*

The real m-form Re Ω is a calibration on M (cf. [12, Thm. III.1.10]) and the submanifolds calibrated by Re Ω are called *special Lagrangian submanifolds*. It will be convenient to use an equivalent definition.

Proposition 4 ([12, Cor. III.1.11]) *Let (M, ω, Ω) be a Calabi–Yau manifold of complex dimension m. A real m-dimensional submanifold $L \subset M$ with some choice of orientation is special Lagrangian if and only if*

$$\omega|_L = 0 \quad \textit{and} \quad \operatorname{Im}\,\Omega|_L = 0. \tag{2}$$

The following result about the deformations of compact special Lagrangian submanifolds without boundary is due to McLean.

Theorem 5 ([18, Thm. 3.6]) *Let M be a Calabi–Yau manifold and L a closed special Lagrangian submanifold in M. Then the moduli space of nearby special Lagrangian deformations of L is a smooth manifold of dimension the first Betti number $b^1(L)$.*

The argument of Theorem 5 uses the equivalent definition (2) of special Lagrangian submanifolds in terms of the vanishing of differential forms.

Applying Proposition 3(a) to L we may write any nearby deformation of L as $L_v = \exp_v L$ for a section $v \in \Gamma(N_{L/M})$ of the normal bundle. On the other hand, there is an isometry of vector bundles

$$j_L : \mathbf{v} \in N_{L/M} \to (\mathbf{v} \lrcorner \omega)|_L \in \Lambda^1 T^* L \tag{3}$$

defined using the Kähler form on M. Thus the nearby deformations of L are equivalently given by 'small' 1-forms on L.

The map

$$F : \alpha \in \Omega^1(L) \to (\exp_{\mathbf{v}}^*(\omega), \exp_{\mathbf{v}}^*(\operatorname{Im}\Omega)) \in \Omega^2(L) \oplus \Omega^m(L), \quad \mathbf{v} = j_L^{-1}(\alpha), \tag{4}$$

is defined for 'small' α, and $F(\alpha) = 0$ precisely if $\exp_{\mathbf{v}}(P)$ is a special Lagrangian deformation.

Proposition 6 *Let L be a special Lagrangian submanifold of a Calabi–Yau manifold and F the 'deformation map' defined in* (4).

(a) *The map F is smooth, with derivative at $\alpha = 0$ given by*

$$dF|_0(\alpha) = (d\alpha, d * \alpha), \qquad \alpha \in \Omega^1(L).$$

(b) *If L is a closed submanifold, then there is a neighbourhood T_L of the zero 1-form such that the image $F(T_L)$ consists of pairs of exact forms.*

Note that in Proposition 6(a) a special Lagrangian L need not be compact.

The nearby special Lagrangian deformations of L correspond to the 1-forms α satisfying a non-linear differential equation $F(\alpha) = 0$ of first order, with $F(0) = 0$. One can show using Hodge theory for a closed manifold L and the implicit function theorem in Banach spaces that the C^1-small solutions of the special Lagrangian deformation problem may be parameterised by the closed and co-closed 1-forms α on L. By Hodge theory again, as L is a closed manifold, the moduli space of special Lagrangian deformations is then locally parameterised by the vector space of harmonic 1-forms on L or, equivalently, by the de Rham cohomology group $H^1(L, \mathbb{R})$ of dimension $b^1(L)$. Theorem 5 follows.

The constraint (1) determines a holomorphic form Ω up to a factor $e^{i\theta}$ for some real constant θ and $\operatorname{Re}(e^{i\theta}\Omega)$ is also a calibration on M. Manifolds calibrated by $\operatorname{Re}(e^{i\theta}\Omega)$ are called special Lagrangian with phase θ. Every submanifold calibrated by $\operatorname{Re}(e^{i\theta}\Omega)$ is Lagrangian (with respect to the symplectic form ω) and minimal (with respect to the metric on M). Conversely, it is known that every connected minimal Lagrangian submanifold in a Calabi-Yau manifold is calibrated by $\operatorname{Re}(e^{i\theta}\Omega)$ for some real constant θ [12, cf. Prop. III.2.17].

If L' is a minimal Lagrangian deformation of L then by the above L' must be calibrated and volume-minimizing. Therefore, L is special Lagrangian by application of Proposition 2(a). It follows that moduli space in Theorem 5 can be equivalently regarded as the space of nearby minimal Lagrangian deformations of L. In particular, there is no loss of generality in restricting attention to submanifolds calibrated by Re Ω, i.e. with $\theta = 0$.

Remark Salur [20] extended the result of Theorem 5 to the situation when the almost complex structure on M is not necessarily integrable. More explicitly, M is a Hermitian symplectic $2m$-manifold with symplectic form ω, the metric on M is Hermitian with respect to an ω-compatible almost-complex structure and there is a (non-vanishing) complex $(m, 0)$-form Ω on M satisfying (1). The main theorem in [20] then asserts that the moduli space of nearby special Lagrangian deformations of L with arbitrary phase is smooth with dimension $b^1(L)$. In the case when d Im $\Omega = 0$, it can be checked that any special Lagrangian deformations of L necessarily have the same phase as L (and can be obtained essentially by McLean's argument in [18]).

Suppose now that a compact special Lagrangian submanifold L has non-empty boundary ∂L. We shall need the following definition from [4, p. 1954].

Definition 7 Let M be a Calabi–Yau manifold and let $L \subset M$ be a submanifold with boundary ∂L. Denote by $\mathbf{n} \in \Gamma(T_{\partial L}L)$ the inward unit normal vector field. A *scaffold* for L is a smooth submanifold W of M with the following properties:

(1) $\partial L \subset W$;

(2) $\mathbf{n} \in \Gamma(T_{\partial L}W)^\omega$ (here, S^ω denotes the symplectic orthogonal complement of a subspace S of a symplectic vector space V, defined by $S^\omega \equiv \{v \in V : \omega(v, s) = 0 \ \forall s \in S\}$);

(3) the bundle $(TW)^\omega$ is trivial.

The deformations of special Lagrangian submanifolds with boundary constrained to be in a fixed scaffold were studied by Butscher who proved the following.

Theorem 8 ([4, Main Theorem]) *Let L be a compact special Lagrangian submanifold of a Calabi-Yau manifold M with non-empty boundary ∂L and let W be a symplectic, codimension two scaffold for L. Then the moduli space of nearby minimal Lagrangian deformations of L with boundary on W is finite dimensional and is locally parameterised by the vector space of closed co-closed 1-forms on L satisfying Neumann boundary conditions*

$$\mathcal{H}^1_{\mathbf{n}}(L) = \{\alpha \in \Omega^1(L) : d\alpha = 0,\ d^*\alpha = 0,\ (\mathbf{n} \lrcorner \alpha)|_{\partial L} = 0\}.$$

Theorem 8 allows special Lagrangian deformations of L with arbitrary phase θ and the proof uses an extended version of the deformation map including θ as an additional variable. The deformation map also requires a construction of an auxiliary metric $\hat{g}$ so that W is totally geodesic for $\hat{g}$ and the appropriate version of the tubular neighbourhood theorem (Proposition 3(b)). (The condition (3) in the definition of the scaffold is used in the construction of $\hat{g}$.) The argument then proceeds by appealing to the implicit function theorem in a similar manner to the deformation problem for closed submanifolds. This requires a version of Hodge theory for compact manifolds with boundary [21] to identify appropriate Banach subspaces of forms so that the linearization of the deformation map be surjective.

The vector space $\mathcal{H}^1_{\mathbf{n}}(L)$ in Theorem 8 is naturally isomorphic to the real cohomology group $H^1(L, \mathbb{R})$ and thus has dimension $b^1(L)$ by Hodge theory for manifolds with boundary, see [5, p. 927].

Observe that the condition (2) in the definition of a scaffold means that $J\mathbf{n}$ is perpendicular to W. This will be automatically satisfied when W is a complex submanifold of positive codimension in M, so the tangent spaces of W are invariant under J. In this case, we also have $\Omega|_W = 0$. Then, applying Proposition 2(b), we obtain that for each special Lagrangian L with boundary in W the minimal Lagrangian deformations of L with boundary confined to W will actually be special Lagrangian. We thus obtain a variant of Butscher's result for the space of special Lagrangian deformations.

Corollary 9 *Let L be a compact special Lagrangian submanifold of a Calabi-Yau manifold M with non-empty boundary ∂L. Let W be a complex codimension one submanifold of M with trivial normal bundle and with $\partial L \subset W$ (in particular, W is a scaffold for L). Then the moduli space of nearby special Lagrangian deformations of L is a smooth manifold of dimension $b^1(L)$.*

3 Coassociative Submanifolds in G_2-Manifolds

The two calibrations considered in this and the next section are defined on 7-dimensional manifolds with a torsion-free G_2-structure. We shall first briefly recall some key definitions. The readers are referred to [12], [13, Chaps. 11, 12] and the article by Karigiannis in this volume for a more detailed account of G_2-structures and the related calibrations.

The group G_2 can be defined, following [2], as the stabilizer, in the standard action of $GL(7, \mathbb{R})$ on $\Lambda^3(\mathbb{R}^7)^*$, of the 3-form

$$\varphi_0 = dx^{123} + dx^{145} + dx^{167} + dx^{246} - dx^{257} - dx^{347} - dx^{356}, \tag{5}$$

where $dx^{123} = dx^1 \wedge dx^2 \wedge dx^3$ and so on, with $x^1, \ldots, x^7$ the usual coordinates on the Euclidean $\mathbb{R}^7$. The 3-form in (5) encodes the cross-product defined by considering $\mathbb{R}^7$ as pure imaginary octonions and setting $\varphi_0(a, b, c) = \langle a \times b, c \rangle$. The group G_2 is a 14-dimensional Lie group and a subgroup of $SO(7)$.

The Hodge dual of φ_0 is a 4-form given by:

$$*\varphi_0 = dx_{4567} + dx_{2367} + dx_{2345} + dx_{1357} - dx_{1346} - dx_{1256} - dx_{1247}.$$

Let M be a 7-dimensional manifold. We say that a differential 3-form φ on M is *positive*, or is a *G_2 3-form*, if for each $p \in M$ there is a linear isomorphism $\iota_p : \mathbb{R}^7 \to T_pM$ with $\iota_p^*(\varphi(p)) = \varphi_0$, where φ_0 is given in (5). Every G_2-structure on M can be induced by a positive 3-form φ and we shall, slightly informally, say that φ *is* a G_2-structure. As $G_2 \subset SO(7)$, every G_2-structure φ induces on M a metric $g(\varphi)$ and orientation and thus also a Hodge star $*_\varphi$.

The intrinsic torsion of a G_2-structure φ on M vanishes precisely when $d\varphi = 0$ and $d{*_\varphi}\varphi = 0$ [6]. In this case, we call (M, φ) a *G_2-manifold.*

The 4-form $*_\varphi\varphi$ defines on each G_2-manifold M a calibration (cf. [12, Sect. IV.1.B]). In fact, the results in this section only require that the G_2 3-form be closed $d\varphi = 0$. We say that an oriented 4-dimensional submanifold $X \subset M$ is a *coassociative submanifold* if the equality $*_\varphi\varphi|_X = \mathrm{vol}_X$ is attained. If in addition $d{*_\varphi}\varphi = 0$ holds, then X is calibrated by $*_\varphi\varphi$ and we call X a *coassociative calibrated submanifold.*

The following equivalent definition of coassociative submanifolds will be useful.

Proposition 10 (cf. [12, Cor. IV.1.20]) *For an orientable 4-dimensional submanifold X of a 7-manifold M with a G_2-structure $\varphi \in \Omega^3_+(M)$, the equality $*_\varphi\varphi|_X = \mathrm{vol}_X$ holds for some orientation of X if and only if $\varphi|_X = 0$.*

Let φ be a closed G_2 3-form on M and let a submanifold $X \subset M$ be coassociative. Then the infinitesimal deformations of X can equivalently be given by self-dual 2-forms on X via an isometry of vector bundles (cf. [18, Proposition 4.2])

$$\jmath_X : \mathbf{v} \in N_{X/M} \to (\mathbf{v} \lrcorner \varphi)|_X \in \Lambda^2_+ T^*X; \tag{6}$$

where $\Lambda^2_+ T^*X$ denotes the bundle of self-dual 2-forms. (The corresponding statements in [18] use anti-self-dual 2-forms because McLean uses a different sign convention for the G_2 3-form.) The map

$$F : \alpha \in \Omega^2_+(X) \to \exp^*_{\mathbf{v}}(\varphi) \in \Omega^3(X), \qquad \mathbf{v} = \jmath_X^{-1}(\alpha), \tag{7}$$

is defined for 'small' α, and $F(\alpha) = 0$ precisely if $\exp_{\mathbf{v}}(X)$ is a coassociative deformation.

The next theorem summarizes the results obtained by McLean about the deformations of closed coassociative submanifolds.

Theorem 11 (cf. [18, Thm. 4.5], [15, Thm. 2.5]) *Let M be a 7-manifold with a closed G_2-structure φ and $X \subset M$ a coassociative submanifold (not necessarily closed).*

(a) *Then for each $\alpha \in \Omega^2_+(X)$, one has $dF|_0(\alpha) = d\alpha$ and the 3-form $F(\alpha)$ (if defined) is exact.*
(b) *If, in addition, X is compact and without boundary then every closed self-dual 2-form α on X arises as $\alpha = \jmath_X(\mathbf{v})$, for some normal vector field $\mathbf{v}$ tangent to a smooth 1 parameter family of coassociative submanifolds containing X. Thus, in this case, the space of nearby coassociative deformations of X is a smooth manifold parameterized by the space $\mathcal{H}^2_+(X)$ of closed self-dual 2-forms on X.*

Remark Self-dual 2-forms on a compact manifold without boundary are closed precisely if they are harmonic. By Hodge theory, the dimension of $\mathcal{H}^2_+(X)$ is therefore equal to the dimension $b^2_+(X)$ of a maximal positive subspace for the intersection form on X. It is thus a topological invariant.

The hypotheses of Theorem 11 do not include the co-closed condition $d*_\varphi\varphi = 0$. In fact, the argument in [18] constructs a smooth moduli space of closed submanifolds X satisfying $\varphi|_X = 0$, for a closed G_2-structure φ.

There is a certain analogy between McLean's deformation theory of closed coassociative submanifolds of G_2-manifolds and closed special Lagrangian submanifolds of Calabi–Yau manifolds. In both cases, the respective submanifolds are calibrated and minimal and have an equivalent definition in terms of the vanishing of appropriate real differential forms on the ambient manifold. The deformation theory is 'unobstructed' and there is a smooth finite-dimensional moduli space, locally parameterised by some finite-dimensional space of harmonic forms on the submanifold with the dimension a topological invariant obtained by Hodge theory.

On the other hand, when the submanifold has a boundary the deformation theories become rather different. As we explain below, following [16], the deformation problem for compact coassociative submanifolds with boundary cannot possibly be set as a boundary value problem of first order with standard Dirichlet or Neumann boundary conditions. Instead the deformation problem will be 'embedded' in a boundary value elliptic problem of second order.

A suitable choice of boundary considerations is again facilitated by the concept of (a coassociative version of) a scaffold which we now define. By way of preparation, we consider an orientable 6-dimensional submanifold S in a 7-manifold M with a closed G_2 3-form φ on M. The normal bundle of S is trivial and there is a 'tubular neighbourhood' T_S of S diffeomorphic to $S \times \{-\varepsilon < s < \varepsilon\}$, such that S corresponds to $\{s = 0\}$ and $\mathbf{n} = \frac{\partial}{\partial s}$ is a unit vector field on T_S with $\mathbf{n}|_S$ orthogonal to S in the metric $g(\varphi)$. More precisely, we consider a coassociative submanifold with (compact) boundary contained in S and then the required T_S exists after shrinking S to some neighbourhood of this boundary. We can write

$$\varphi|_{T_S} = \omega_s \wedge ds + \Upsilon_s,$$

for some 1-parameter families of 2-forms ω_s and 3-forms Υ_s on S.

The forms $\omega_0 = (\mathbf{n}_S \lrcorner \varphi)|_S$ and $\Upsilon_0 = \varphi|_S$ together define an $SU(3)$-structure on S, in general with torsion. This can be seen point-wise, by a consideration similar to G_2-structures earlier in this section, from the property that the simultaneous stabilizer of ω_0 and Υ_0 in the action of $GL(T_pS)$ at each $p \in S$ is isomorphic to $SU(3)$. In particular, ω_0^3 defines an orientation on S.

Definition 12 An orientable 6-dimensional submanifold S is a *symplectic submanifold* of a 7-manifold (M, φ) with a closed G_2-structure if $d_S\omega_0 = 0$, where ω_0 is as defined above and d_S is the exterior derivative on S.

A 3-dimensional submanifold $L \subset S$ of a symplectic submanifold $S \subset M$ is said to be *special Lagrangian* if $\omega_0|_L = 0$ and $\varphi|_L = 0$.

Every S in the above definition has a Hermitian symplectic structure compatible with the $SU(3)$-structure induced from M. In particular, a non-vanishing $(3,0)$-form on S is obtained as $\Omega_0 = *_S\Upsilon - i\Upsilon$, where the 6-dimensional Hodge start $*_S$ is taken with respect to the orientation ω_0^3 and the induced metric from M. As noted in the previous section, McLean's theory remains valid and any closed special Lagrangian $L \subset S$ has a smooth moduli space of dimension $b^1(L)$ of nearby special Lagrangian deformations.

Definition 13 Let (M, φ) be a 7-manifold with a closed G_2-structure and $X \subset M$ a coassociative submanifold with boundary ∂X. We say that an orientable 6-dimensional submanifold S of M is a *scaffold* for X if

(a) X meets S orthogonally, i.e. $\partial X \subset S$ and the normal vectors to S at ∂X are tangent to X, $\mathbf{n} \in N_{S/M}|_{\partial X}$, and
(b) S is a symplectic submanifold of (M, φ).

One notable property of a scaffold S in Definition 13 is that for each coassociative X meeting S orthogonally, the intersection $L = X \cap S$ is special Lagrangian in S.

The infinitesimal deformations of compact coassociative X with boundary in a fixed submanifold S correspond via (6) to a subspace of self-dual 2-forms on X satisfying boundary conditions. We can write the restriction of any 2-form α on X to a collar neighbourhood $C_{\partial X} = T_S \cap X$ of the boundary as $\tilde{\alpha} = \alpha_\tau + \alpha_\nu \wedge ds$. The Dirichlet and Neumann boundary conditions for α are then given by, respectively, $\alpha_\tau = 0$ and $\alpha_\nu = 0$. When α is self-dual, the two conditions are equivalent and force α and the corresponding normal vector field $j_X^{-1}(\alpha)$ to vanish at each point of ∂X. However, if $d\alpha = 0$ and α vanishes on the boundary then $\alpha = 0$ by [5, Lemma 2]. This may be understood as an extension of [12, Thm. IV.4.3], which states that there is a locally unique coassociative submanifold containing any real analytic 3-dimensional submanifold upon which φ vanishes.

It turns out that a suitable choice of the infinitesimal deformations with boundary condition is given by the following subspace of self-dual 2-forms

$$\Omega^2_+(X)_{\mathrm{bc}} = \{\alpha \in \Omega^2_+(X) : \ \mathbf{n}\lrcorner d\alpha = 0 \text{ and } d_{\partial X}(\mathbf{n}\lrcorner\alpha) = 0 \text{ on } \partial X\}. \qquad (8)$$

The boundary condition $\mathbf{n}\lrcorner d\alpha = 0$ ensures that every harmonic form in $\Omega^2_+(X)_{\mathrm{bc}}$ is closed and the boundary condition $d_{\partial X}(\mathbf{n}\lrcorner\alpha) = d_{\partial X}\alpha_\nu = 0$ means that the boundary ∂X will only move in the space of special Lagrangians in S.

The subspace of $\Omega^2_+(X)_{\mathrm{bc}}$ of the coassociative infinitesimal deformations is given by the harmonic (or closed) forms $(\mathcal{H}^2_+)_{\mathrm{bc}}$. It has a finite dimension $\leq b^1(\partial X)$ and our next theorem asserts that elements of $(\mathcal{H}^2_+)_{\mathrm{bc}}$ 'integrate' to actual coassociative deformations of X.

Theorem 14 ([16, Thm. 1.1]) *Suppose that M is a 7-manifold with a G_2-structure given by a closed 3-form. The moduli space of compact coassociative local deformations of X in M with boundary ∂X in a scaffold S is a finite-dimensional smooth manifold parameterized by $(\mathcal{H}^2_+)_{bc}$. The dimension of this moduli space is not greater than $b^1(\partial X)$.*

Here is an example when strict inequality $\dim(\mathcal{H}^2_+)_{\mathrm{bc}} < b^1(\partial X)$ occurs.

Example 15 ([16, p. 72]) A Kähler complex 3-fold (Z, ω) is called *almost Calabi–Yau* if it admits a nowhere vanishing holomorphic $(3, 0)$-form Ω. Then the 7-manifold $M = Z \times S^1$ has a closed G_2-structure $\omega \wedge d\theta + \operatorname{Re}\Omega$, where θ is a coordinate on S^1. Let $X = L \times S^1 \subset M$ be a compact coassociative 4-fold. Then L is special Lagrangian in Z. We can think of X as an embedding of a manifold $L \times [0, 1]$ whose two boundary components, $L \times \{0\}$ and $L \times \{1\}$, are mapped to L in Z. (If Z is Calabi–Yau, i.e. if (1) also holds, then X is a calibrated coassociative.) It is not difficult to see that $Z \times \mathrm{pt}$ is a scaffold for X. Theorem 14 gives us that X has a smooth moduli space of coassociative deformations with dimension $\leq 2b^1(L)$.

Let $\alpha \in \Omega^2_+(X)$. Then $\alpha = \xi_\theta \wedge d\theta + *_L \xi_\theta$, for some path of 1-forms ξ_θ on L. It follows from [21, Thm. 3.4.10] that a harmonic self-dual 2-form on X is uniquely determined by its values ξ_0, ξ_1 on the boundary. The subspace of harmonic $\alpha \in \Omega^2_+(X)$ such that ξ_0 and ξ_1 are harmonic on L has dimension $2b^1(L)$ and corresponds precisely to the paths $\xi_\theta = (1-\theta)\xi_0 + \theta\xi_1$. On the other hand, $\alpha \in (\mathcal{H}^2_+)_{\mathrm{bc}}$ if and only if α is harmonic and $\partial\xi_\theta/\partial\theta = 0$, so $\xi_0 = \xi_1$. Thus $\dim(\mathcal{H}^2_+)_{\mathrm{bc}} = b^1(L) < b^1\big((L \times \{0\}) \sqcup (L \times \{1\})\big)$ in this example.

This can also be seen geometrically. If the deformations of the aforementioned two boundary components coincide in $Z \times \mathrm{pt}$ then, by taking a product with S^1, we obtain a coassociative deformation of $X = L \times S^1$ defining a point in the moduli space in Theorem 14. On the other hand, if a coassociative deformation $\tilde{X}$ of X is such that the deformations $\tilde{L}_0$ and $\tilde{L}_1$ of $L \times \{0\}$ and $L \times \{1\}$ are special Lagrangian but *distinct* then $\tilde{X}$ and $\tilde{L}_0 \times S^1$ are two distinct coassociative 4-folds intersecting in a real analytic 3-fold on which φ vanishes, which contradicts [12, Thm. IV.4.3]. Therefore, the moduli space in this example is identified with special Lagrangian deformations of L in the almost Calabi–Yau manifold Z. As we noted earlier, these deformations have a smooth moduli space of dimension $b^1(L)$.

4 Associative Submanifolds in G_2-Manifolds

Let M be again a 7-dimensional manifold with a G_2-structure given by a positive 3-form φ, as defined in the previous section. If $d\varphi = 0$ then φ is a calibration on M as $\varphi_p|_V \leq \mathrm{vol}_V$ for each oriented 3-plane in T_pM (cf. [12, Sect. IV.1.A]). The equality is attained precisely when V is an *associative subspace* of T_pM, i.e. is closed under the cross-product on T_pM induced by the G_2-structure φ. (The corresponding subalgebra $(V, \times)$ is isomorphic to $\mathbb{R}^3$ with the standard vector product.)

The deformation theory discussed in this section does not always require the G_2 form φ to be closed. Similarly to the discussion of coassociative submanifolds in the previous section, we shall define the term *associative submanifold* Y in (M, φ) for an arbitrary G_2 structure φ, meaning a 3-dimensional submanifold Y satisfying $\varphi|_Y = \mathrm{vol}\, Y$. If also the G_2-structure is closed, $d\varphi = 0$, then we shall call Y an *associative calibrated submanifold*; indeed, in this case Y is calibrated by φ.

McLean [18] studied the deformation theory of closed associative calibrated submanifolds in G_2-manifolds. His results were later extended by Akbulut and Salur [1] to arbitrary G_2-structures on 7-manifolds. We assume a torsion-free G_2-structure to simplify some details.

As in the previous sections, it is useful to first note that associative submanifolds of M can be equivalently defined by the vanishing of an appropriate differential form on M. In the present case, it is a 3-form χ with values in TM,

$$\chi = \sum_{k=1}^{7} (\eta_k \lrcorner *_\varphi \varphi) \otimes \eta_k,$$

for any local orthonormal positively oriented frame field $(\eta_k)_{k=1}^{7}$ on M [8, p. 1217].

Proposition 16 (cf. [18, Sect. 5]) *A 3-dimensional submanifold Y, with some choice of orientation, is associative if and only if $\chi|_Y = 0$.*

Suppose that Y is a closed associative submanifold in a G_2-manifold (M, φ) and let $\mathbf{v} \in \Gamma(N_{Y/M})$ be a normal vector field along Y. The deformation map for Y is defined as

$$F : \mathbf{v} \in \Gamma(N_{Y/M}) \to \exp_{\mathbf{v}}^{*} \tau \in \Omega^3(Y, TM|_Y).$$

The linearization of F at $\mathbf{v} = 0$ is given by

$$D\mathbf{v} = \sum_{i=1}^{3} e_i \times \nabla^{\perp}_{e_i} \mathbf{v}, \tag{9}$$

where e_1, e_2, e_3 is any positively oriented local orthonormal frame field of TY (thus $e_3 = e_1 \times e_2$), the cross-product is induced by φ and a connection $\nabla^{\perp}$ on $N_{Y/M}$ is induced by the Levi–Civita connection of $(M, g(\varphi))$. Since Y is 3-dimensional and associative, both TY and $N_{Y/M}$ are trivial vector bundles [14, Remark 2.14] and the expression (9) is valid globally over Y. There is an invariant interpretation of $N_{Y/M}$ as a vector bundle associated with a principal Spin(4)-bundle over Y via the tensor product of a spin representation and some other representation. Then D becomes the respective Dirac type operator (meaning that the principal symbol of D^2 is $\sigma(D^2)(p, \xi) = \|\xi\|^2$, for all $p \in Y$).

The map F makes sense for an arbitrary G_2-structure φ on M, but when the G_2-structure is not torsion-free the expression (9) for D then has extra terms of order zero. So in this more general case D is still a Dirac type operator with the same principal symbol.

We thus obtain.

Theorem 17 ([1, 18]) *For a closed associative submanifold Y in a G_2-manifold (M, φ), the Zariski tangent space to associative deformations of Y is finite-dimensional, given by the kernel of the Dirac type operator D in* (9), *an elliptic operator of index* 0. *In particular, Y is either rigid or the associative deformations* *of Y are obstructed, i.e. a section $\mathbf{v}$ with $D\mathbf{v} = 0$ need not arise as $\dot{s}_0$ from any 1-parameter family $Y_t = \exp s_t$ of associative submanifolds with $s_0 = 0$.*

The deformations of compact associative submanifolds with boundary contained in a fixed submanifold (scaffold) were investigated by Gayet and Witt [9], see also Gayet [8]. An appropriate choice of scaffold in this situation is given by a 4-dimensional submanifold X such that no tangent space T_pX contains associative 3-planes. In particular, X may be any coassociative submanifold and we shall assume this below for technical convenience. In this case, any associative calibrated submanifold with boundary in X is volume minimizing in its relative homology class as $\varphi|_X = 0$ by Propositions 2(b) and 10.

Let Y be a compact associative submanifold with boundary ∂Y contained in a coassociative submanifold X. Denote by $\mathbf{n} \in \Gamma(TY|_{\partial Y})$ the inward-pointing unit normal along ∂Y. For each point $p \in \partial Y$ the cross-product of the G_2-structure φ

$$J(v) = \mathbf{n}_p \times v$$

defines an (orthogonal) complex structure on the orthogonal complement $(\mathbf{n}_p)^\perp \subset T_pM$. Further, J acts on the fibres of the normal bundle $N_{\partial Y/X}$, making it into a complex line bundle. Note that $N_{\partial Y/X}$ is a subbundle of $N_{Y/M}|_{\partial Y}$; the respective orthogonal complement $\mu_{\partial Y}$ is also invariant under J and can be considered as a complex line bundle. The tangent spaces of ∂Y are also preserved by J and in this way ∂Y is made into a compact Riemann surface. The latter complex line bundles satisfy an adjunction-type relation $\bar{\mu}_{\partial Y} \cong N_{\partial Y/X} \otimes_{\mathbb{C}} T\partial Y$ [9, Lemma 3.2].

The infinitesimal associative deformation problem for Y with boundary confined to X can be expressed as

$$D\mathbf{v} = 0, \quad B(\mathbf{v}|_{\partial Y}) = 0, \qquad \mathbf{v} \in \Gamma(N_{Y/M}), \tag{10}$$

where D is the Dirac type operator in (9) and the zero order operator B is induced by the orthogonal projection $N_{Y/M}|_{\partial Y} \to \mu_{\partial Y}$ with kernel $N_{\partial Y/X}$.

Gayet and Witt prove.

Theorem 18 ([9, Thm. 4.4, Cor. 4.5]) *Let (M, φ) be a 7-manifold endowed with a G_2-structure and let $Y \subset M$ be a compact associative submanifold with boundary contained in a coassociative submanifold $Y \subset M$.*

Then the linear operator $D \oplus B$ in (10) defines an elliptic boundary value problem with finite Fredholm index

$$\operatorname{index}(D \oplus B) = \sum_j \Big(\int_{\Sigma_j} c_1(N_{\Sigma_j/X}) + 1 - g_j\Big). \tag{11}$$

Here Σ_j denote the boundary components of ∂Y with g_j the genus of Σ_j, and $c_1(N_{\Sigma_j/X})$ is the first Chern class of the complex line bundle $N_{\Sigma_j/X}$.

The key point in the proof of Theorem 18 is that the index in question can be computed as the index of the Cauchy–Riemann operator $\bar{\partial}_{\partial Y/X}$ associated with the complex structure J on $N_{\partial Y/X}$.

Example 19 ([9, p.2364]) When the associative submanifold has a boundary, the index in Theorem 18 can be positive. One simple example uses a construction by Bryant and Salamon [3] of torsion-free G_2-structure φ inducing a complete metric with holonomy G_2 on $\mathcal{S} = S^3 \times \mathbb{R}^4$, the total space of the spinor bundle over the standard round 3-sphere. The zero section $S^3 \times \{0\}$ is an associative submanifold, being the fixed locus of the G_2-involution acting as -1 on the fibres (note [13, Prop. 12.3.7]). Take $Y \subset S^3 \times \{0\}$ to be a 3-dimensional ball, so $\partial Y = S^2$. Let a be a nowhere vanishing section of $\mathcal{S}|_{\partial Y} = \partial Y \times \mathbb{R}^4 \to \partial Y$. Then $a,\ Ja$ (with J as defined above) generate a trivial complex line bundle, denote its total space by $\tilde{X}$. It can be checked that there is a local coassociative submanifold $X \subset \mathcal{S}$ containing ∂Y and with $T_pX = T_p\tilde{X}$ at each $p \in \partial Y$. Theorem 18 then applies and the deformation problem has index 1. This example generalizes, under additional assumptions, to a complex line bundle $\tilde{X}$ having positive degree n, then the respective index is $n+1$ [9, p.2364].

Gayet and Witt also gave a generalization of Theorem 18 where X is only required to contain no associative 3-planes in its tangent spaces. The latter property is preserved under small perturbations of the G_2-structure. The deformation problem remains elliptic and the index formula (11) still holds (with appropriate modification of the definition of the boundary operator B). This does not guarantee a smooth moduli space of associative deformations even when the index of $D \oplus B$ is non-negative. However, Gayet [8, Thm. 1.4] proved that a smooth moduli space of dimension index $D \oplus B$ can be obtained by arbitrary small generic perturbation of the scaffold X.

5 Cayley Submanifolds in Spin(7)-manifolds

The calibration considered in this section is defined on 8-dimensional manifolds with a torsion-free Spin(7)-structure. We begin with a short summary of the Spin(7)-structure and the Cayley calibration and refer to [12] and [13, Chap. 11, 12] for further details. There is a certain, though only partial, analogy with the 'geometries' considered in the previous sections.

The group Spin(7) can be defined, following [2], as the stabilizer, in the standard action of $GL(8,\mathbb{R})$ on $\Lambda^4(\mathbb{R}^8)^*$, of the 4-form Φ_0 written in standard coordinates as

$$\begin{aligned}\Phi_0 = {} & dx_{1234} + dx_{1256} + dx_{1278} + dx_{1357} - dx_{1368} - dx_{1458} - dx_{1467} \\ & - dx_{2358} - dx_{2367} - dx_{2457} + dx_{2468} + dx_{3456} + dx_{3478} + dx_{5678}, \end{aligned} \tag{12}$$

where $dx^{1234} = dx^1 \wedge dx^2 \wedge dx^3 \wedge dx^4$ and so on. The form Φ_0 arises by considering $\mathbb{R}^8$ as the (normed) algebra of octonions, or Cayley numbers, and setting

$\Phi_0(x, y, z, w) = \frac{1}{2}\langle x(\bar{y}z) - z(\bar{y}x), w\rangle$. In this way, Spin(7) is also identified as a subgroup of $SO(8)$. This form is also self-dual $*\Phi_0 = \Phi_0$ with respect to the standard Euclidean metric and orientation.

Given an oriented 8-manifold M, define a subbundle of 4-forms $\mathcal{A}M \subset \Lambda^4 T^*M$ with the fibre $\mathcal{A}_pM$ at each $p \in M$ being the set of all 4-forms that can be identified with Φ_0 via an orientation-preserving isomorphism $T_pM \to \mathbb{R}^8$. The fibres of $\mathcal{A}M$ are diffeomorphic to the orbit $GL_+(8, \mathbb{R})/\operatorname{Spin}(7)$ of Φ_0, a 43-dimensional submanifold of the 70-dimensional vector space $\Lambda^4(\mathbb{R}^8)^*$.

A choice of 4-form $\Phi \in \Gamma(\mathcal{A}M)$ is equivalent to a choice of a Spin(7)-structure on M. By the inclusion $\operatorname{Spin}(7) \subset SO(8)$, every such Φ induces on M a metric $g = g(\Phi)$, an orientation and a Hodge star $*_\Phi$, with $*_\Phi\Phi = \Phi$. We shall sometimes refer to Φ as a Spin(7)-structure.

When a form $\Phi \in \mathcal{A}M$ is closed, $d\Phi = 0$, we say that the Spin(7)-structure Φ is torsion-free and that (M, Φ) is a Spin(7)-*manifold*. The condition $d\Phi = 0$ is equivalent to the metric $g(\Phi)$ being Ricci-flat with reduced holonomy contained in Spin(7) [7]. In this case, Φ defines a calibration on M.

Given a Spin(7)-structure on an 8-manifold M, we say that an oriented 4-dimensional submanifold $P \subset M$ is a *Cayley submanifold* if $\Phi|_P = \operatorname{vol}_P$. We say that P is a *Cayley calibrated submanifold* if in addition $d\Phi = 0$, i.e. precisely if P is calibrated by Φ.

Let P be a compact Cayley calibrated submanifold of a Spin(7)-manifold (M, Φ). The infinitesimal Cayley deformations of P are given by the kernel of a first order elliptic operator $D : \Gamma(N_{P/M}) \to \Gamma(E)$, for a rank 4 vector bundle E over P,

$$E = \{\alpha \in \Lambda^2_7 M|_P : \alpha|_{TP} = 0\}.$$

Here $\Lambda^2_7 M \subset \Lambda^2 T^*M$ is a subbundle corresponding to an irreducible representation of Spin(7) on the space of 2-forms $\Lambda^2(\mathbb{R}^8)^*$. The following result was proved by McLean.

Theorem 20 ([18, Thm. 6.3]) *Let P be a Cayley submanifold of a* Spin(7)*-manifold (M, Φ). Then the Zariski tangent space to Cayley deformations of P is finite-dimensional, given by the kernel of the elliptic operator*

$$D : \mathbf{v} \in \Gamma(N_{P/M}) \to \sum_{i=1}^{4} e_i \times \nabla^{\perp}_{e_i}\mathbf{v} \in \Gamma(E), \tag{13}$$

where e_1, e_2, e_3, e_4 is any positively oriented local orthonormal frame field of TP, the cross-product is induced by the Spin(7)*-structure Φ and a connection $\nabla^\perp$ on $N_{P/M}$ is induced by the Levi–Civita connection of $(M, g(\Phi))$.*

Remarks When P is a spin manifold, there is an invariant interpretation of D using a spin structure on P [18, Sect. 6]. Denote by $\mathcal{S}_+$ and $\mathcal{S}_-$ the positive and negative spinor bundles over P. Then

$$N_{P/M} \otimes_{\mathbb{R}} \mathbb{C} \cong \mathcal{S}_+ \otimes_{\mathbb{C}} F \text{ and } E \otimes_{\mathbb{R}} \mathbb{C} \cong \mathcal{S}_- \otimes_{\mathbb{C}} F, \tag{14}$$

for some quaternionic line bundle F over P. The operator D is identified, via (14), with a positive Dirac type operator associated with a connection on F.

Theorem 20 was extended to arbitrary Spin(7)-structures in [10, Sect. 13]. When the Spin(7)-structure is not torsion-free, the expression (13) for D has extra terms of order zero. This does not affect the principal symbol or the index of D.

Deformations of Cayley submanifolds were further investigated by Ohst and included the following.

Theorem 21 ([19, Prop. 3.4 and Thm. 3.10]) *Let (M, Φ) be an 8-manifold with a* Spin(7)*-structure and $P \subset M$ a closed Cayley submanifold. Then*

(a) *the index of the operator* (13) *associated with P is*

$$\text{index}\, D = \tfrac{1}{2}\chi(P) + \tfrac{1}{2}\sigma(P) - [P]\cdot[P], \tag{15}$$

where $\chi(P)$ is the Euler characteristic, $\sigma(P)$ is the signature and $[P]\cdot[P]$ is the self-intersection number of P.

(b) *For every generic* Spin(7)*-structure $\tilde{\Phi}$ on M such that $\|\tilde{\Phi} - \Phi\|$ is sufficiently small and $\tilde{\Phi}$ induces the same metric $g(\tilde{\Phi}) = g(\Phi)$ the following holds. The moduli space of Cayley submanifolds with respect to $\tilde{\Phi}$ which are $C^{1,\alpha}$-close to P $(0 < \alpha < 1)$ is either empty or a smooth manifold of dimension* index D *(if* index $D \geq 0$*).*

Remarks The submanifold P need not be Cayley with respect to $\tilde{\Phi}$, thus P need not be in the respective moduli space. The norm $\|\tilde{\Phi} - \Phi\|$ can be taken to be the $C^{1,\alpha}$-norm on a compact neighbourhood of P.

The variant of Theorem 21 also holds with $\tilde{\Phi}$ generic in the set of all Spin(7)-structures close to Φ, i.e. without the restriction on the metric $g(\tilde{\Phi})$.

We next turn to deformations of compact Cayley submanifolds with boundary in a fixed submanifold (scaffold). The result given below is again due to Ohst and allows a range of dimensions of the scaffold.

Theorem 22 ([19, Thm. 4.18]) *Let (M, Φ) be a Spin(7)-manifold and W a submanifold of M with $3 \leq \dim W \leq 7$. Let P be a compact, connected Cayley submanifold of M with non-empty boundary $\partial P \subseteq W$ such that P and W meet orthogonally.*

Then for every generic torsion-free Spin(7)-structure $\tilde{\Phi}$ which is $C^{2,\alpha}$-close to Φ, the moduli space of all Cayley (calibrated) submanifolds in $(M, \tilde{\Phi})$ which are $C^{2,\alpha}$-close to P and have boundary contained in W and meet W orthogonally in the metric $g(\tilde{\Phi})$ is a finite set (possibly empty). Here $0 < \alpha < 1$.

The proof of Theorem 22 uses a second order elliptic boundary problem implied by the linearization D in (13) of the Cayley deformation map. This is because D admits no suitable elliptic boundary conditions [19, Prop. 4.21].

The condition $\Phi|_W = 0$ required in Proposition 2(b) for the volume-minimizing property in the relative homology class of P can only hold if $\dim W \leq 4$. Suppose that ∂P is a deformation retract of W. If dim $W = 5$, then $v = d(*_W(\Phi|_W))^\flat$ restricts to a vector field on ∂P. If, further, v is parallel with respect to the induced metric on W, then P is volume-minimizing among the nearby deformations in its relative homology class.

When dim $W = 6$ and $d(*_W(\Phi|_W)) = 0$, the scaffold W is a symplectic submanifold. Then P minimizes the volume among all the submanifolds P' in its relative homology class, with boundary $\partial P' \subset W$ a Lagrangian nearby deformation of ∂P. In all of the above situations, the minimal volume in the relative homology class of P is attained precisely by Cayley calibrated submanifolds [19, Sect. 5.1].

References

1. Akbulut, S., & Salur, S. (2008). Deformations in G_2 manifolds. *Advances in Mathematics*, *217*, 2130–2140.
2. Bryant, R.L. (1987). Metrics with exceptional holonomy. *Annals of Mathematics*, (2) *126*, 525–576.
3. Bryant, R. L., & Salamon, S. M. (1989). On the construction of some complete metrics with exceptional holonomy. *Duke Mathematical Journal*, *58*, 829–850.
4. Butscher, A. (2002). Deformations of minimal Lagrangian submanifolds with boundary. *Proceedings of the American Mathematical Society*, *131*, 1953–1964.
5. Cappell, S., DeTurck, D., Gluck, H., & Miller, E. Y. (2006). Cohomology of harmonic forms on Riemannian manifolds with boundary. *Forum Mathematicum*, *18*, 923–931.
6. Fernández, M., & Gray, A. (1982). Riemannian manifolds with structure group G_2. *Annali di Matematica Pura ed Applicata*, (4) *132*, 19–45.
7. Fernández, M. (1986). A classification of Riemannian manifolds with structure group Spin(7). *Annali di Matematica Pura ed Applicata*, (4) *143*, 101–122.
8. Gayet, D. (2014). Smooth moduli spaces of associative submanifolds. *Quarterly Journal of Mathematics*, *65*, 1213–1240.
9. Gayet, D., & Witt, F. (2011). Deformations of associative submanifolds with boundary. *Advances in Mathematics*, *226*, 2351–2370.
10. Gutowski, J., Ivanov, S., & Papadopoulos, G. (2003). Deformations of generalized calibrations and compact non-Kähler manifolds with vanishing first Chern class. *Asian Journal of Mathematics*, *7*, 39–79.
11. Harvey, R. (1990). *Spinors and calibrations*. Boston: Academic.
12. Harvey, R., & Lawson, H. B. (1982). Calibrated geometries. *Acta Mathematica*, *148*, 47–152.
13. Joyce, D. D. (2007). *Riemannian holonomy groups and calibrated geometry*. Oxford: OUP.
14. Joyce, D. D., & Karigiannis, S. A new construction of compact torsion-free G_2-manifolds by gluing families of Eguchi-Hanson spaces. *Journal of Differential Geometry*. arXiv:1707.09325.
15. Joyce, D. D., & Salur, S. (2005). Deformations of asymptotically cylindrical coassociative submanifolds with fixed boundary. *Geometry and Topology*, *9*, 1115–1146.
16. Kovalev, A. G., & Lotay, J. D. (2009). Deformations of compact coassociative 4-folds with boundary. *Journal of Geometry and Physics*, *59*, 63–73.
17. Lang, S. (1972). *Differential manifolds*. Reading, Massachusetts: Addison-Wesley.
18. McLean, R. C. (1998). Deformations of calibrated submanifolds. *Communications in Analysis and Geometry*, *6*, 705–747.
19. Ohst, M. (2015). *Deformations of Cayley submanifolds*. Ph.D. thesis, University of Cambridge. https://doi.org/10.17863/CAM.16246.

20. Salur, S. (2000). Deformations of special Lagrangian submanifolds. *Communications in Contemporary Mathematics*, *2*, 365–372.
21. Schwarz, G. (1995). *Hodge decomposition - a method for solving boundary value problems*. Berlin: Springer-Verlag.

GPSR Compliance
The European Union's (EU) General Product Safety Regulation (GPSR) is a set of rules that requires consumer products to be safe and our obligations to ensure this.

If you have any concerns about our products, you can contact us on

ProductSafety@springernature.com

In case Publisher is established outside the EU, the EU authorized representative is:

Springer Nature Customer Service Center GmbH
Europaplatz 3
69115 Heidelberg, Germany

www.ingramcontent.com/pod-product-compliance
Ingram Content Group UK Ltd.
Pitfield, Milton Keynes, MK11 3LW, UK
UKHW021833270726
14058UKWH00001B/117